T0211570

Lecture Notes in Computer Science 10421

Commenced Publication in 1973
Founding and Former Series Editors:
Gerhard Goos, Juris Hartmanis, and Jan van Leeuwen

Editorial Board

More information about this series at http://www.springer.com/series/7407

Victor Malyshkin (Ed.)

Parallel Computing Technologies

14th International Conference, PaCT 2017
Nizhny Novgorod, Russia, September 4–8, 2017
Proceedings

 Springer

Editor
Victor Malyshkin
Russian Academy of Sciences
Novosibirsk
Russia

ISSN 0302-9743   ISSN 1611-3349   (electronic)
Lecture Notes in Computer Science
ISBN 978-3-319-62931-5   ISBN 978-3-319-62932-2   (eBook)
DOI 10.1007/978-3-319-62932-2

Library of Congress Control Number: 2017946060

LNCS Sublibrary: SL1 – Theoretical Computer Science and General Issues

Printed on acid-free paper

This Springer imprint is published by Springer Nature
The registered company is Springer International Publishing AG
The registered company address is: Gewerbestrasse 11, 6330 Cham, Switzerland

Preface

The 2017 International Conference on Parallel Computing Technologies (PaCT) was a four-day event held in Nizhny Novgorod (Russia). This was the 14th international conference in the PaCT series. The conferences are held in Russia every odd year. The first conference, PaCT 1991, was held in Novosibirsk (Academgorodok), September 7–11, 1991. The next PaCT conferences were held in Obninsk (near Moscow), August 30 to September 4, 1993; in St.-Petersburg, September 12–15, 1995; in Yaroslavl, September, 9–12 1997; in Pushkin (near St. Petersburg), September, 6–10, 1999; in Academgorodok (Novosibirsk), September 3–7, 2001; in Nizhny Novgorod, September, 15–19, 2003; in Krasnoyarsk, September 5–9, 2005; in Pereslavl–Zalessky, September 3–7, 2007; in Novosibirsk, August 31–September 4, 2009; in Kazan, September 19–23, 2011; in St. Petersburg, September 30 to October 4, 2013; in Petrozavodsk, August 31 to September 4, 2015. Since 1995 all the PaCT proceedings have been published by Springer in the LNCS series. PaCT 2017 was jointly organized by the Institute of Computational Mathematics and Mathematical Geophysics (Russian Academy of Sciences), Lobachevsky State University of Nizhny Novgorod, Novosibirsk State University, and Novosibirsk State Technical University.

The aim of PaCT 2017 was to give an overview of new developments, applications, and trends in parallel computing technologies. We sincerely hope that the conference will help our community to deepen the understanding of parallel computing technologies by providing a forum for an exchange of views between scientists and specialists from all over the world. The topics of PaCT conferences are progressively changing, reflecting the modern trends in the area of parallel computing technologies. For example, traditionally, in the area of knowledge accumulation on the methods of parallel implementation of large-scale numerical models, papers describing numerical algorithms and their implementation were accepted for PaCT. Today, most of these papers were rejected because they did not contribute new knowledge to the PaCT community. However, papers describing complex parallel implementation of large-scale numerical models were accepted. Another progressively growing trend is the development of systems of automatic construction of parallel programs on the basis of axiomatic descriptions of an object domain. Papers describing fundamental algorithms of dynamic distributed program construction were accepted, particularly those on algorithms of the dynamic distribution of distributed data for distributed multi-computers (DDD algorithms)

The conference attracted about 100 participants from around the world. Authors from 13 countries submitted 93 papers. Of the papers submitted, 49 were selected for the conference after being reviewed by an international Program Committee. Many thanks to our sponsors: the Russian Academy of Sciences, Federal Agency for

---

In an older version of the front matter of these proceedings, the word "Nizhny" had been written inconsistently. This has now been corrected.

Scientific Organization, Ministry of Education and Science of the Russian Federation, Russian Foundation for Basic Research, Advanced Micro Devices, Inc., RSC Technologies, and Intel Corporation.

September 2017   Victor Malyshkin

Organization

PaCT 2017 was organized by the Supercomputer Software Department, Institute of Computational Mathematics and Mathematical Geophysics Siberian Branch, Russian Academy of Science (ICM&MG), Lobachevsky State University of Nizhny Novgorod (UNN), Novosibirsk State University, and Novosibirsk State Technical University.

Organizing Committee

Conference Co-chairs

Victor Malyshkin	ICM&MG, Novosibirsk
Victor Gergel	UNN, Nizhny Novgorod

Conference Secretary

Maxim Gorodnichev	ICM&MG, Novosibirsk

Organizing Committee Members

Svetlana Achasova	ICM&MG
Sergey Arykov	ICM&MG
Olga Bandman	ICM&MG
Konstantin Barkalov	UNN
Ekaterina Goldinova	UNN
Vladimir Grishagin	UNN
Sergey Kireev	ICM&MG
Valentina Kustikova	UNN
Elena Malkina	UNN
Valentina Markova	ICM&MG
Yur Medvedev	ICM&MG
Joseph Meyerov	UNN
Mikhai Ostapkevich	ICM&MG
Vladislav Perepelkin	ICM&MG
Georgy Schukin	ICM&MG
Dmitry Shaposhnikov	UNN
Alain Travina	UNN

Steering Committee

Viktor Kazantsev	UNN
Alexandr Moskovsky	RSC Technologies
Grigory Osipov	UNN
Vadim Saigin	UNN
Roman Strongin	UNN

Program Committee

V. Malyshkin   M. Raynal   A. Simbuerger
V. Markova   G. Schukin   B. Steinberg
I. Menshov   U. Schwiegelshohn   A. Tkacheva
V.Perepelkin   Y. Sergeyev   V. Toporkov
D. Petcu   V. Shakhov   R. Wyrzykowski

Sponsoring Institutions

Ministry of Education and Science of the Russian Federation
Russian Academy of Sciences
Federal Agency for Scientific Organizations
Advanced Micro Devices, Inc.
RSC Technologies
Intel Corporation

Contents

Cellular Automata and Discrete Event Systems

Organization of Parallel Computation

Parallel Computing Applications

Mainstream Parallel Computing

Experimenting with a Context-Aware Language

Chiara Bodei, Pierpaolo Degano, Gian-Luigi Ferrari, and Letterio Galletta[(⊠)]

Dipartimento di Informatica, Università di Pisa, Pisa, Italy
{chiara,degano,giangi,galletta}@di.unipi.it

Abstract. Contextual information plays an increasingly crucial role in concurrent applications in the times of mobility and pervasiveness of computing. Context-Oriented Programming languages explicitly treat this kind of information. They provide primitive constructs to adapt the behaviour of a program, depending on the evolution of its operational environment, which is affected by other programs hosted therein independently and unpredictably. We discuss these issues and the challenges they pose, reporting on our recent work on ML_{CoDa}, a language specifically designed for adaptation and equipped with a clear formal semantics and analysis tools. We will show how applications and context interactions can be better specified, analysed and controlled, with the help of some experiments done with a preliminary implementation of ML_{CoDa}.

1 Introduction

Today there is a growing trend in having software systems able to operate *every time* and *everywhere*, and applications are working side by side, either in a cooperative or in a competitive way. Ubiquitous and pervasive computing scenarios are typical of the Internet of Things (IoT), a cyber-physical communication infrastructure, made of a wide variety of interconnected and possibly mobile devices. As a consequence, modern software systems have to cope with changing operational environments, i.e. their *context*. At the same time, they must never compromise their intended behaviour and their non-functional requirements, typically security or quality of service. Thus, programming languages need effective mechanisms to become *context-aware*, so as to detect the changes in the context where the application is plugged in, and to properly *adapt* to them, with little or no user involvement. Accordingly these mechanisms must maintain the functional and non-functional properties of applications after the adaptation steps. For example, suppose you want to have just a quick look at your mail and at your social platforms when in a hotel: you would like to connect in a secure way, but without bothering with all the details of the wireless connection, ideally in a fully transparent manner.

The context is crucial for adaptive software and typically it includes different kinds of computationally accessible information coming from both outside (e.g. sensor values, available devices, and code libraries offered by the environment), and from inside the application boundaries (e.g. its private resources, user profiles, etc.). The literature proposes many different programming languages

© Springer International Publishing AG 2017
V. Malyshkin (Ed.): PaCT 2017, LNCS 10421, pp. 3–17, 2017.
DOI: 10.1007/978-3-319-62932-2_1

that support dynamic adjustments and tuning of programs, e.g. [17,18,20,24–26] (a detailed discussion on the great deal of work in this area is in [13,23]). In this field, Context Oriented Programming (COP) [1,9,15,16] offers a neat separation between the working environment and the application. Indeed, the COP linguistic paradigm explicitly deals with contexts, by providing programming adaptation mechanisms to support dynamic changes of behaviour, in reaction to changes in the context (see [2,23] for an overview). In this paradigm, programming adaptation is specified using *behavioural variations*, chunks of code that can be automatically selected depending on the current context hosting the application, dynamically modifying its execution.

To address adaptivity we defined $\mathrm{ML_{CoDa}}$ [3,10,11,13], a core of ML with COP features. It has two tightly integrated components: a declarative constituent for programming the context and a functional one for computing. The bipartition reflects the separation of concerns between the specific abstractions for describing contexts and those used for programming applications [22]. The context in $\mathrm{ML_{CoDa}}$ is a knowledge base implemented as a (stratified, with negation) Datalog program [19,21]. Applications inspect the contents of a context by simply querying it, in spite of the possibly complex deductions required. The behavioural variations of $\mathrm{ML_{CoDa}}$ are a sort of pattern matching with Datalog goals as selectors. They are a first class, higher-order construct that can then be referred to by identifiers, and used as parameters in functions. This fosters dynamic, compositional adaptation patterns, as well as reusable, modular code. The selection of a goal is done by the *dispatching* mechanism that inspects the actual context and makes the right choices. Note that the choice depends on both the application code and the "open" context, unknown at development time. If no alternative is viable then a *functional failure* occurs, as the application cannot adapt to the current context. *Non-functional failures* are also possible, when the application does not meet some requirements, e.g. about quality of service or security.

The execution model of $\mathrm{ML_{CoDa}}$ assumes that the context is the interface between each application it hosts and the system running it. Applications interact with the system using a predefined set of APIs that provide handles to resources and operations on them. Also, they interact with each other via the context. The system and the applications do not trust each other, and may act maliciously, e.g. one application can alter some parts of the context so driving another in an unsafe state. The application designer would like to detect both functional and non-functional failures as early as possible, and for that $\mathrm{ML_{CoDa}}$ has a two-phase static analysis, one at compile and one at load-time [4,11,13,14], briefly summarised below. The static analysis takes care of failures in adaptation to the current context (*functional failures*), dealing with the fact that applications operate in an "open" environment. Indeed, the actual value and even the presence of some elements in the current context are only known when the application is linked with it at run time. The first phase of our static analysis is based on a type and effect system that, at compile time, computes a safe over-approximation of the application behaviour, namely an *effect*. Then the effect is used at load time to verify that the resources required by the application are available in the actual context, and in its future modifications. To do that, the effect of the application is suitably combined with the effect of the APIs provided by the context that are computed by the same type and effect system. If an application passes this analysis, then no functional failure can arise at run time. The results of the static analysis also drive an instrumentation of the original code, so as to monitor its execution and block dangerous activities [4].

In addition to the formal aspects of ML_{CoDa}, a main feature of our approach is that a single and fairly small set of constructs is sufficient enough for becoming a practical programming language, as shown in [7]. ML_{CoDa} can easily be embedded in a real programming eco-system as .NET, so preserving compatibility with future extensions and with legacy code developed within this framework. Being part of a well supported programming environment minimises the learning cost and lowers the complexity of deploying and maintaining applications. In [7] a prototypical implementation of ML_{CoDa} is presented as an extension of the (ML family) functional language F#. Indeed, no modifications at all were needed to the available compiler and to its runtime. The F# metaprogramming facilities are exploited, such as code introspection, quotation and reflection, as well as all the features provided by .NET, including a vast collection of libraries and modules. In particular, we used the Just-In-Time mechanism for compiling to native code. As a consequence, ML_{CoDa} is implemented as a standard .NET library. In the path towards the implementation a crucial role has been played by the formal description of the language and by its formal semantics, which highlight and explain how the two components of ML_{CoDa} interact. Furthermore they helped in identifying and describing the crucial parts of the implementation toolchain, compilation, generated code and runtime structures.

Here, we will survey on some applications we developed in ML_{CoDa} to assess our language, showing how context interactions can be better specified, analysed and controlled. We also discuss some extensions that will make our language more expressive and applicable. The next section introduces ML_{CoDa}, with the help of our first case study. Two more case studies are summarised in Sect. 3. Section 4 shortly illustrates the Just-In-Time compiler of ML_{CoDa}. In Sect. 5 we conclude and discuss the planned extensions, in particular those required to handle many applications running concurrently.

2   A First Example: An e-Healthcare System

Here, we illustrate the main features of ML_{CoDa} by considering an e-healthcare system with a few aspects typical of the Internet of Things. A more detailed description of this case study is in [7], and its full executable definition is in https://github.com/vslab/fscoda.

In our scenario each physician can retrieve a patient's clinical record using a smartphone or a tablet, which also tracks the current location. Got the relevant data, the doctor decides which exams the patient needs and the system helps scheduling them. In addition, the system checks whether the doctor has the competence and the permission to actually perform the required exam, otherwise it suggests another physician who is enabled to, possibly coming from another department. Moving from a ward to another, the operating context changes and allows the doctor to access the complete clinical records of the patients therein. The application must adapt to the new context and it may additionally provide different features, e.g. by disabling rights to use some equipment and by acquiring access to new ones. Indeed, location-awareness of devices is exploited to tune access policies.

The e-healthcare context. We consider below a small part of the context, in particular that for storing and making some data available about the doctors' location, information on their devices, the patients' records and the ward medical equipment. Some basic data are represented by Datalog facts, and one can retrieve further information using the inference machinery of Datalog, which uses logical rules, also stored in the context.

For example, the fact that Dr. Turk is in the cardiology ward is rendered as

```
physician_location("Dr. Turk" "Cardiology").
```

The following inference rule permits to deduce that the clinical data of patients can be accessed by the doctors in the same department where patients are. It states that the predicate on the left hand-side of the implication operator :- holds when the conjunction of the predicates (physician_location and patient_location) in the right hand-side yields true, i.e. when the physician and patient's location coincide.

```
physician_can_view_patient(Physician, Patient) :-
        physician_location(Physician, Location),
        patient_location(Patient, Location).
```

The ML$_{CoDa}$ context is quite expressive and can model fairly complex situations. Typically, some medical exams can only be performed after some others. To compute this list of exams, all the dependencies among them are to be considered. This could be expressed by the following recursive rules:

```
patient_needs_result(Patient, Exam) :-
        patient_has_been_prescribed(Patient, Exam).

patient_needs_result(Patient, Exam) :-
        exam_requirement(TargetExam, Exam),
        patient_needs_result(Patient, TargetExam).
```

The first rule means that the prescription of an exam implies that the involved patient needs the results of the test. The second rule says that whenever a patient needs an exam, so are also needed all the screenings the exam depends on. Datalog can conveniently model recursive relations like the dependency among exams, which may require involved queries with standard relational databases.

The next rule dictates that a patient has to do an exam if the two clauses in the right hand-side are true. The first has been already discussed above, while the second clause says that a patient should *not* do an exam if its results are already known (in the rule below the operator \+ denotes the logical *not*, dealt with in our version of Datalog [8]).

```
patient_should_do(Patient, Exam) :-
        patient_needs_result(Patient, Exam),
        \+ patient_has_result(Patient, Exam).
```

In addition, we can declaratively describe physical objects in quite a similar, homogeneous manner. The following (simplified) rule specifies when a device can display a certain exam, by checking whether it has the needed capabilities:

```
device_can_display_exam(Device, Exam) :-
    device_has_caps(Device, Capability),
    exam_view_caps(Exam, Capability).
```

By listing a set of facts, we can easily assert the capabilities of a device, e.g.

```
device_has_caps('iPhone 5', '3D acceleration').
device_has_caps('iPhone 5', 'Video codec').
device_has_caps('iPhone 5', 'Text display').
device_has_caps('Apple Watch', 'Text display').
```

Adaptation constructs. Now we focus on context-dependent bindings and behavioural variations. These adaptation constructs allow specifying program behaviour, which depends on the context in our e-healthcare system. When entering a ward, the patients' records under treatment can be displayed on the doctor's personal device. Moreover, the e-healthcare system computes the list of the clinical exams a patient should do and that the doctor can perform. The following code (in a F#-like syntax) shows how the adaptation constructs are used to implement these functionalities. The display function, given a doctor phy and a patient pat, prints the information about the patient's exams on the screen.

```
1  let display phy pat =
2    match ctx with
3    | _ when !- physician_can_view_patient(phy, pat) ->
4      match ctx with
5      | _ when !- patient_has_result(pat, ctx?e) ->
6        printfn "%s sees that %s has done:" phy pat
7        for _ in !-- patient_has_result(pat, ctx?exam) do
8          display_exam phy ctx?exam
9      | _ ->
10       printfn "%s sees that %s has done no exam" phy pat
11
12     let next_exam = "no exam" |- True
13     let next_exam = ctx?exam |-
14                       (physician_exam(phy, ctx?exam),
15                        patient_active_exam(pat, ctx?exam))
16       printfn "%s can submit %s to %s" phy pat next_exam
17    | _ ->
18      printfn "%s cannot view details on %s" phy pat
```

Behavioural variations change the program flow according to the current context. They have the form **match** ctx **with** | _ **when** !- Goal -> expression, where the sub-expression **match** ctx **with** explicitly refers to the context; the part | _ **when** !- Goal introduces the goal to solve; and -> expression is the sub-expression to evaluate when the goal is true.

Using the outermost behavioural variation (starting at line 2), we check whether the doctor phy is allowed to access the data of the patient pat, when the goal physician_can_view_patient(phy, pat) at line 3 holds.

With the nested behavioural variation (line 4), we check if the patient has got the results of some exams, using the predicate patient_has_result. If this is the case, the **for** construct extracts the list of exam results from the context (line 7). The statement **for** _ **in** !-- Goal **do** expression iterates the evaluation of expression over all the solutions of the Goal. It works as an iterator on-the-fly, driven by the solvability of the goal in the context. The predicate patient_has_result at line 7 contains the *goal variable* ctx?exam: if the query succeeds, at each iteration ctx?exam is bound to the current value satisfying Goal. A goal variable is introduced in a goal, defining its scope, using the syntax ctx?var_name.

Finally, through **let** x = expression1 |- Goal [**in**] expression2 (the context dependent binding), the function display shows an exam that the physician phy can do on the patient pat. At lines 12–13 we declare by cases the parameter next_exam, referred to in line 16. Only at run time when the actual context is known, we can determine which case applies and which value will be bound to next_exam when the parameter is used. If the goal in lines 14–15 holds, then next_exam assumes the value retrieved from the context, otherwise it gets the default value "no exam".

Note that it may happen that no goal is satisfied in a context while executing a behavioural variation or resolving a parameter. This means that the application in not able to adapt, either because the programmer assumed at design time the presence of functionalities that the current context lacks, or because of design errors. We classify this new kind of runtime errors as *adaptation failures*. For example, the following function assumes that given the identifier of a physician, it is always possible to retrieve the physician's location from the context using the physician_location predicate:

```
let find_physician phy =
  let loc = ctx?location |-
            physician_location(phy, ctx?location) in
  loc
```

The context-dependent binding may find no solution for the goal, e.g. when find_physician is invoked on a physician whose location is not in the context. If this is the case, the current implementation throws a runtime exception.

```
let find_physician phy =
  try
    let loc = ctx?location |-
              physician_location(phy, ctx?location) in   loc
```

```
with e -> printfn "WARNING: cannot locate %s:\n%A" phy e
             "unknown location"
```

As described in [11], we may adopt a more sophisticated approach where for statically determining whether the adaptation might fail and reporting it before running the application.

Finally, the interaction with the Datalog context is not limited to queries: it is possible indeed to program the modifications to the knowledge base on which it performs deduction, by adding or removing facts with the **tell** and **retract** operations, as in:

```
tell <| patient_has_result("Jordan", "CT scan")
```

Some execution examples. We now show how the functions defined above give different results when invoked in different contexts, parts of which are only described intuitively. For instance, in a context where Dr. Turk is not in the same ward as Bob, the result of the invocation `display "Dr. Turk""Bob"` is `Dr. Turk cannot view details on Bob`. This is because physicians are only allowed to see data about the patients in the department where they are. Indeed, the behavioural variation introduced at line 3 on `physician_can_view_patient` finds out that accessing data is not allowed. If instead Dr. Cox is in the same department where Bob is, the call `display "Dr. Cox""Bob"` correctly prints the details about Bob (actually stored in the Datalog knowledge base):

```
Dr. Cox sees that Bob has done no exam
Dr. Cox can submit Bob to Blood test
```

In this case the outermost behavioural variation (starting at line 2) confirms that Dr. Cox can view the data. The nested one (starting at line 4), driven by `patient_has_result`, finds no exam for Bob, hence the function displays the no-exam message (line 10). Furthermore, the program finds out that Dr. Cox could do a blood test on Bob, as he is enabled to; then it additionally finds out that Bob needs no pre-screening and so that exam can be done immediately, because the predicate at line 15 holds.

Suppose now to have a slightly more complex situation, in which the context itself is modified. Patient Jordan has already performed an EEG test, and doctors prescribed her a CT and nothing else. Dr. Kelso is in Jordan's room, is enabled to do only CT tests and carries a device on which he can visualise the results. In this context, the invocation `display "Dr. Kelso""Jordan"` outputs

```
Dr. Kelso sees that Jordan has done:  EEG
Dr. Kelso can submit Jordan to CT scan
```

Differently from the case above, Jordan has already performed an exam, listed by the iteration construct. Once Dr. Kelso have performed a CT scan on Jordan, the context has to be accordingly changed, by asserting the fact

```
tell <| patient_has_result("Jordan", "CT scan")
```

Now the query `display "Dr. Kelso""Jordan"` has a different output in the modified context: besides displaying a longer list of exam results, the application shows Dr. Kelso that Jordan needs him to perform no other exam:

```
Dr. Kelso sees that Jordan has done:  EEG, CT scan
Dr. Kelso can submit Jordan to no exam
```

Suppose now that Dr. Cox moves to Jordan room and checks her medical report, but he has a device that cannot show CT images. The `display_exam` function warns the doctor and possibly presents the results in a more limited form, e.g. a static thumbnail. So, the result of the query `display "Dr. Cox""Jordan"` is

```
Dr. Cox sees that Jordan has done: EEG, CT scan
                 (current device cannot display the exam data)
Dr. Cox can submit Jordan to no exam
```

3   Further Case Studies

The following case studies illustrate how ML_{CoDa} can be used to specify small-sized real context-aware applications. Afterwards, we outline some internals of our preliminary compiler.

Fsc-Rover. We now briefly describe the implementation in ML_{CoDa} of a small rover robot, endowed with two wheels, engine control, foto and video camera and a distance sensor, done by Riccardo Rolla, a master student of our research group (see https://github.com/riccardorolla/rpi-iot-fscoda). The rover moves in a building, and detects the objects therein and some of their features. Also, it interacts with other applications that use the information it collects, by exchanging messages on the Internet. The rover can perform either its actions, called *local*, or actions issued by other applications, called *remote*. Each kind of action has a different set of parameters and the rover has to identify the right values for them, by inspecting the properties of the objects in the context.

Besides getting a formal executable specification of a rover, differently from the other case studies this one makes it evident that the context provides effective support to uniformly handle both local and remote activities. The control loop of the rover, shown below, is really quite standard: it repeats the following until *no-request* is found

- Add to the rover program all the remote actions read from the context;
- Execute asynchronously local and remote actions;
- Collect and process data and store the results in the context;
- Send responses to remote applications.

```
while (not (get_detected "exit")) do
    for _ in !-- request(ctx?idchat,ctx?cmd) do
        array_cmd <- array_cmd |> Array.append [|ctx?cmd|]
      for _ in !-- next(ctx?cmd) do
        array_cmd <- array_cmd |> Array.append [|ctx?cmd|]
```

```
listresult <- Async.Parallel
            [for c in  array_cmd -> execute c]
                    |> Async.RunSynchronously
for r in listresult do
    match r with
      |cmd,res -> for _ in !-- result(cmd,ctx?out) do
                    retract <| Fsc.Facts.result(cmd, ctx?out)
                    tell <| Fsc.Facts.result(cmd, res)
....
    match ctx with
    | _ when !- (request(ctx?idchat,ctx?cmd),result(ctx?cmd,ctx?out))
              -> do
                let result=send_message ctx?idchat ctx?cmd
                retract<|Fsc.Facts.request(ctx?idchat, ctx?cmd)
    | _ ->  printfn "no request"
....
  run()
```

The query request(ctx?idchat,ctx?cmd) extracts information from the context to assemble the list of commands to be executed. This is done by checking for messages arriving at the context from the Internet. The tag idchat identifies a remote application. Note that both rover commands and results are modelled as suitable facts inside the context through the **tell** and **retract** operations. The function run sets up the context, e.g. it turns on/off the video camera and the distance sensor. Its code not displayed here also invokes a configuration function that sets the sequence of local actions.

The behaviour of the rover also depends on the *obstacles* identified by the camera in the current environment. The following function is used to detect the nature of the obstacles by inspecting the context. The idea is that the objects are suitable facts in the context and that object recognition in the current image depends on the parameters of confidence of objects, such as size, rotation, etc.

```
let infoimage = get_out "discovery" |> imagerecognition
      for tag in infoimage.tags do
          discovery tag.name tag.confidence
      for _ in !-- recognition(ctx?obj,ctx?value) do
```

FSEdit editor. Here, we briefly survey the implementation of FSEdit, a context-aware text editor implemented in ML_{CoDa}. This case study was a workbench for testing how our implementation deals and interacts with pure F# code, in particular against the standard GUI library provided by .NET. Besides playing with contexts, this case study also helped finding some little flaws in the way our compiler treated some pieces of code using the object oriented features of F#. Furthermore, it also allowed us to identify some programming patterns that may be considered as idiomatic of ML_{CoDa} programs (see below).

The editor supports three different execution modes: *rich text editor*, *text editor* and *programming editor*. A context switch among the different modes changes the GUI of the editor, by offering e.g. different tool-bars and menus.

In the first mode, the GUI allows the user to set the size and the face of a font; to change the color of text; and to adjust the alignment of the paragraphs. In the second mode, the editor becomes very minimalistic and allows the user to edit pure text files, where no information of the text formatting can change. Finally, in the programming mode, the editor shows file line numbers and provides a simple form of syntax highlighting for C source files.

The context of FSEdit contains the current execution mode and other information that directly depend on it, as shown by the predicates below:

```
tokens(TS) :- tokens_(TS), execution_mode(programming).

file_dialog_filter(F) :- execution_mode(M),
                         file_dialog_filter_(F,M).
```

For example, the predicate `tokens` only holds in the programming mode and returns the keywords of the programming language selected by the user to perform syntax highlighting. For simplicity, the editor currently supports the C programming language only. The second piece of information is about the kind of files supported by the editor in the different modes. For instance, *.rtf files in rich text mode, *.txt files in text mode and *.c in programming mode.

As said before, the execution mode affects the behaviour of the editor. For instance, in the following piece of code we invoke the syntax highlighter procedure when the user changes the text, if the editor is in the right mode.

```
let textChanged (rt : RichTextBox) = // dlet
  let def_behaviour () = ...        // code not shown

  let f_body = def_behaviour () |- True    // Basic behaviour

  let f_body  = (f_body ; syntaxHighlighter rt) |-
                              execution_mode("programming")
f_body
```

As anticipated, this code snippet is interesting because it shows an idiomatic use of the context dependent binding. Indeed, there are two definitions of the identifier `f_body`: the first one represents the basic behaviour of the editor that is independent of the context; the second one extends the basic behaviour with the features that are to be provided when the editor is in the programming mode. Notice in particular the use of `f_body` on the right-hand side in the last-but-one line of the snippet. Although it may seem a recursive definition it is not; it is instead an invocation of `f_body` defined in the previous line, i.e. the one specifying the basic behaviour of the editor.

4   A Glimpse on ML$_{\text{CoDa}}$ Compiler

The ML$_{\text{CoDa}}$ compiler ypc is based on the integration of the functional language F# with a customised version of YieldProlog[1] serving as Datalog engine.

---

[1] Available at https://github.com/vslab/YieldProlog.

Our compiler ahead-of-time compiles each Datalog predicate into a .NET method, whose code enumerates one by one the solutions, i.e. the assignments of values to variables that satisfy the predicate. In this way, the interaction and the data exchange between the application and the context is fully transparent to the programmer because the .NET type system is uniformly used everywhere.

The functional part of ML_{CoDa} that extends F# is implemented through just-in-time compilation. To do that, a programmer annotates these extensions with *custom attributes*, among which the most important is CoDa.Code. When a function annotated by it is to be executed, the ML_{CoDa} runtime is invoked to trigger the compilation step. Since the operations needed to adapt the application to contexts are transparently handled by our runtime support, the compiler fsharpc works as it is. Actually, CoDa.Code is an alias for the standard ReflectedDefinitionAttribute that marks modules and members whose abstract syntax trees are used at runtime through reflection. Of course ML_{CoDa} specific operations are only allowed in methods marked with this attribute; otherwise an exception is raised when they are invoked.

5   Conclusions, Discussion and Open Problems

We have surveyed the COP language ML_{CoDa} and we have reported on the experiments carried on some case studies. These proved ML_{CoDa} expressive enough to support the designer of real applications, although admittedly simplified in some details. The formal description of the dynamic and the static semantics of ML_{CoDa} drove a preliminary implementation of a compiler and of an analysis tool. Especially, we found that the bipartite nature of ML_{CoDa} permits the designer to clearly separate the design of the context from that of the application, yet maintaining their inter-relationships. This is particularly evident in the rover case study of Sect. 3, where the context provides the mechanism to virtualise and abstract from the communication infrastructure, thus making the logic of the control of the rover fully independent from the actual features of the communication infrastructure.

At the same time, the people working with ML_{CoDa} asked for more functionalities to make ML_{CoDa} more effective, and below we discuss some lines of improvement, both pragmatic and theoretical.

Non-functional properties. A crucial aspect that arose when designing the e-healthcare system concerns context-aware security and privacy, which we approached still from a formal linguistic viewpoint. We equipped ML_{CoDa} with security policies and with mechanisms for checking and enforcing them [4]. It turns out that policies are just Datalog clauses and that enforcing them reduces to asking goals. As a matter of fact, the control of safety properties, like access control or other security policies, requires extensions to the knowledge base and to its management that are not too heavy. We also extended the static analysis mentioned above to identify the operations that may violate the security policies in force. Recall that this step can only be done at load time, because the execution context is only known when the application is about to run, and thus our static analysis cannot be completed at compile time. Yet we have been able to instrument the code of an application, so as to incorporate in it an *adaptive reference monitor*, ready to stop executions when a policy to be enforced is about to be violated. When an application enters a new context, the results of the static analysis mentioned above are used to suitably drive the invocation of the monitor that is switched on and off upon need.

Further work will investigate other non-functional properties that however are of interest in real applications. A typical example is quality of service, which requires enriching both our logical knowledge base and our applications with quantitative information, *in primis* time. Such an extension would also provide the basis for evaluating both applications and contexts. For instance, statistical information about performance can help in choosing the application that better fits our needs, as well as statistical information on the usage of contexts or reliability of resources therein can be used for suggesting the user the context that guarantees more performance. A further approach to statically reason about resource usage, typically acquisition and release, is in [5].

Coherency of the context and interference. Other issues concern the context, in particular the operations to handle it and to keep it coherent. When developing and testing the e-healthcare system discussed in Sect. 2, it was necessary to extend the Datalog deduction machinery in order to get the entire list of the solutions to a given query.

Pursuing coherency at any cost can instead hinder adaptation, e.g. when an application *e* can complete its task even in a context that became partially incoherent. This is a pragmatically very relevant issue. Consider for example the case when a resource becomes unavailable, but was usable by *e* in the context when a specific behavioural variation started. At the moment we implemented our language in a strict way that prevents *e* even to start executing. For sure this guarantees that no troubles will arise, but also precludes to run an application that only uses that resource when available, e.g. at the very beginning, and never again, so no adaptation error will show up at run time. While a continuous run time monitoring can handle this problem, but at a high cost, finding a sound and efficient solution to this issue is a hard challenge from a theoretical point of view. Indeed, it involves a careful mix of static analysis and of run time monitoring of the applications that are executing in a context. Also, "living in an incoherent context" is tightly connected with the way one deals with the needed recovery mechanisms that should be activated without involving the users.

Concurrency. The above problem is critical in the inherently concurrent systems we are studying. Indeed, an application does not perform its task in isolation, rather it needs some resources offered by a context where plenty of other applications are running therein (often competing for those resources). For instance, the implementation of the rover described in Sect. 3 posed concurrency issues, because the control activity of the robot is performed in parallel with the collection and analysis of data coming from the context. The current *ad hoc* solution exploits the management of threads offered by the operative system, and it is not yet fully integrated in ML_{CoDa}.

A first extension of ML_{CoDa} with concurrency is in [12], where there is a two-threaded system: the context and the application. The first virtualises the resources and the communication infrastructure, as well as other software components running within it. Consequently, the behaviour of a context, describing in particular how it is updated, abstractly accounts for all the interactions of the entities it hosts. The other thread is the application and the interactions with the other entities therein are rendered as the occurrence of asynchronous events that represent the relevant changes in the context. A more faithful description of concurrency requires to explicitly describing the many applications that execute in a context, that exchange information using it and that asynchronously update it. This is the approach followed in [6]. Nonetheless, the well known problem of interference now arises, because one thread can update the context possibly making unavailable some resources or contradicting assumptions that another thread relies upon. Classical techniques for controlling this form of misbehaviour, like locks, are not satisfying, because they contrast with the basic assumption of having an open world where applications appear and disappear unpredictably, and freely update the context. However, application designers are only aware of the relevant fragments of the context and cannot anticipate the effects a change may have. Therefore, the overall consistency of the context cannot be controlled by applications, and "living in an incoherent context" is unavoidable. The semantics proposed in [6] addresses this problem using a run time verification mechanism. Intuitively, the effects of the running applications are checked to guarantee that the execution of the selected behavioural variation will lead no other application to an inconsistent state, e.g. by disposing a shared resource. Dually, also the other threads are checked to verify that they are harmless with respect to the application entering in a behavioural variation.

Recovery mechanisms. We already mentioned briefly the need of recovery mechanism when run time errors arise, in particular when adaptation failures prevent an application to complete its task. Recovery mechanisms are especially needed to adapt applications that raise security failures, in case of policy violations. Recovery should be carried on with little or no user involvement, and this imposes on the system running the applications to execute parts of their code "atomically." A typical way is to consider those pieces of code as all-or-nothing transactions, and to store auxiliary information for recovering from failures. If the entire transaction is successfully executed, then the auxiliary information can be disposed, otherwise it has to be used to restore the application in a consistent state, e.g. the one holding at the start of the transaction. To this end, we plan to investigate recovery mechanisms appropriate for behavioural variations, to allow the user to undo some actions considered risky or sensible, and force the dispatching mechanism to make different, alternative choices.

However, in our world the context might have been changed in the meanwhile, and such a state might not be consistent any longer. A deep analysis is therefore needed of the interplay between the way applications use contextual information to adapt or to execute, and the highly dynamic way in which contexts change. A possible line of investigation can be giving up with the quest for a coherent *global* context, while keeping coherent portions of it, i.e. *local* contexts where applications run and, so to speak, posses for a while.

References

1. Appeltauer, M., Hirschfeld, R., Haupt, M., Masuhara, H.: ContextJ: context-oriented programming with Java. Comput. Softw. **28**(1), 272–292 (2011)
2. Appeltauer, M., Hirschfeld, R., Haupt, M., Lincke, J., Perscheid, M.: A comparison of context-oriented programming languages. In: International Workshop on Context-Oriented Programming (COP 2009), pp. 6:1–6:6. ACM, New York (2009)
3. Bodei, C., Degano, P., Ferrari, G.-L., Galletta, L.: Last mile's resources. In: Probst, C.W., Hankin, C., Hansen, R.R. (eds.) Semantics, Logics, and Calculi. LNCS, vol. 9560, pp. 33–53. Springer, Cham (2016). doi:10.1007/978-3-319-27810-0_2
4. Bodei, C., Degano, P., Galletta, L., Salvatori, F.: Context-aware security: linguistic mechanisms and static analysis. J. Comput. Secur. **24**(4), 427–477 (2016)
5. Bodei, C., Dinh, V.D., Ferrari, G.L.: Checking global usage of resources handled with local policies. Sci. Comput. Program. **133**, 20–50 (2017)
6. Busi, M., Degano, P., Galletta, L.: A semantics for disciplined concurrency in COP. In: Proceedings of the ICTCS 2016. CEUR Proceedings, vol. 1720, pp. 177–189 (2016)
7. Canciani, A., Degano, P., Ferrari, G.L., Galletta, L.: A context-oriented extension of F#. In: FOCLASA 2015. EPTCS, vol. 201, pp. 18–32 (2015)
8. Ceri, S., Gottlob, G., Tanca, L.: What you always wanted to know about datalog (and never dared to ask). IEEE Trans. Knowl. and Data Eng. **1**(1), 146–166 (1989)
9. Costanza, P.: Language constructs for context-oriented programming. In: Proceedings of the Dynamic Languages Symposium, pp. 1–10. ACM Press (2005)
10. Degano, P., Ferrari, G.L., Galletta, L.: A two-component language for COP. In: Proceedings of the 6th International Workshop on Context-Oriented Programming. ACM Digital Library (2014)
11. Degano, P., Ferrari, G.-L., Galletta, L.: A two-phase static analysis for reliable adaptation. In: Giannakopoulou, D., Salaün, G. (eds.) SEFM 2014. LNCS, vol. 8702, pp. 347–362. Springer, Cham (2014). doi:10.1007/978-3-319-10431-7_28
12. Degano, P., Ferrari, G.L., Galletta, L.: Event-driven adaptation in COP. In: PLACES 2016. EPTCS, vol. 211 (2016)
13. Degano, P., Ferrari, G.L., Galletta, L.: A two-component language for adaptation: design, semantics and program analysis. IEEE Trans. Softw. Eng. (2016). doi:10.1109/TSE.2015.2496941
14. Galletta, L.: Adaptivity: linguistic mechanisms and static analysis techniques. Ph.D. thesis, University of Pisa (2014). http://www.di.unipi.it/~galletta/phdThesis.pdf
15. Hirschfeld, R., Costanza, P., Nierstrasz, O.: Context-oriented programming. J. Object Technol. **7**(3), 125–151 (2008)
16. Kamina, T., Aotani, T., Masuhara, H.: EventCJ: a context-oriented programming language with declarative event-based context transition. In: Proceedings of the 10 International Conference on Aspect-Oriented Software Development (AOSD 2011), pp. 253–264. ACM (2011)

17. Kephart, J.O., Chess, D.M.: The vision of autonomic computing. IEEE Comput. **36**(1), 41–50 (2003)
18. Kiczales, G., Hilsdale, E., Hugunin, J., Kersten, M., Palm, J., Griswold, W.G.: An overview of AspectJ. In: Knudsen, J.L. (ed.) ECOOP 2001. LNCS, vol. 2072, pp. 327–354. Springer, Heidelberg (2001). doi:10.1007/3-540-45337-7_18
19. Loke, S.W.: Representing and reasoning with situations for context-aware pervasive computing: a logic programming perspective. Knowl. Eng. Rev. **19**(3), 213–233 (2004)
20. Magee, J., Kramer, J.: Dynamic structure in software architectures. SIGSOFT Softw. Eng. Notes **21**(6), 3–14 (1996)
21. Orsi, G., Tanca, L.: Context modelling and context-aware querying. In: Moor, O., Gottlob, G., Furche, T., Sellers, A. (eds.) Datalog 2.0 2010. LNCS, vol. 6702, pp. 225–244. Springer, Heidelberg (2011). doi:10.1007/978-3-642-24206-9_13
22. Salehie, M., Tahvildari, L.: Self-adaptive software: landscape and research challenges. ACM Trans. Auton. Adapt. Syst. **4**(2), 14:1–14:42 (2009)
23. Salvaneschi, G., Ghezzi, C., Pradella, M.: Context-oriented programming: a software engineering perspective. J. Syst. Softw. **85**(8), 1801–1817 (2012)
24. Spinczyk, O., Gal, A., Schröder-Preikschat, W.: AspectC++: an aspect-oriented extension to the C++ programming language. In: CRPIT 2002, pp. 53–60. Australian Computer Society, Inc. (2002)
25. Walker, D., Zdancewic, S., Ligatti, J.: A theory of aspects. SIGPLAN Not. **38**(9), 127–139 (2003)
26. Wand, M., Kiczales, G., Dutchyn, C.: A semantics for advice and dynamic join points in aspect-oriented programming. ACM Trans. Program. Lang. Syst. **26**(5), 890–910 (2004)

Generating Maximal Domino Patterns
by Cellular Automata Agents

Rolf Hoffmann[1] and Dominique Désérable[2]([⊠])

[1] Technische Universität Darmstadt, Darmstadt, Germany
hoffmann@informatik.tu-darmstadt.de
[2] Institut National des Sciences Appliquées, Rennes, Rennes, France
domidese@gmail.com

Abstract. Considered is a 2D cellular automaton with moving agents. The objective is to find agents controlled by a Finite State Program (FSP) that can form domino patterns. The quality of a formed pattern is measured by the degree of order computed by counting matching 3×3 patterns (templates). The class of domino patterns is defined by four templates. An agent reacts on its own color, the color in front, and whether it is blocked or not. It can change the color, move or not, and turn into any direction. Four FSP were evolved for multi-agent systems with 1, 2, 4 agents initially placed in the corners of the field. For a 12×12 training field the aimed pattern could be formed with a 100% degree of order. The performance was also high with other field sizes. Livelocks are avoided by using three different variants of the evolved FSP. The degree of order usually fluctuates after reaching a certain threshold, but it can also be stable, and the agents may show the termination by running in a cycle, or by stopping their activity.

Keywords: Cellular automata agents · Multi-agent system · Pattern formation · Evolving FSM behavior · Spatial computing

1 Introduction

Pattern formation is an area of active research in various domains as in physics, chemistry, biology, computer science or natural and artificial life. There exists a lot of examples, namely in polymer composites, laser trapping, spin systems, self-organization, growth processes, morphogenesis, excitable media and so forth [1–10]. Cellular automata (CA) make suitable and powerful tools for catching the influence of the microscopic scale onto the macroscopic behavior of such complex systems [11–13]. At the least, the 1–dimensional Wolfram's "Elementary" CA can be viewed as generating a large diversity of 2–dimensional patterns whenever the time evolution axis is considered as the vertical spatial axis, with patterns depending or not on the random initial configuration [14]. A similar evolution process is observed in the Yamins–Nagpal "$1D$ spatial computer" generating the roughly radial striped pattern of the *Drosophila melanogaster* [15,16]. But the authors emphasize therein how the local-to-global CA paradigm can turn into

V. Malyshkin (Ed.): PaCT 2017, LNCS 10421, pp. 18–31, 2017.
DOI: 10.1007/978-3-319-62932-2_2

the inverse global-to-local question, namely "given a pattern, which agent rules will robustly produce it?"

Based upon our experience from previous works dealing with CA agents and with FSM–agents driven by Finite State Machines and generating spatial patterns [17–19], we focus herein on the problem of generating an optimal configuration of *domino* patterns in an $n \times n$ field, from four 3×3 Moore-neighborhood domino templates. "Optimal" means that the configuration must have neither gap nor overlap. Although the objective in [19] was to form long orthogonal *line* patterns, some similarity will be observed between both configurations as related to alignments in spin systems. A more down-to-earth application for the domino pattern is the problem of packing encountered in different logistics settings, such as the loading of boxes on pallets, the arrangements of pallets in trucks, or cargo stowage [20]. Another application is the construction of a sieve for rectangular particles with a maximum flow rate.

Related Work. (i) *Pattern formation.* A programming language is presented in [15] for pattern-formation of locally-interacting, identical agents – as an example, the layout of a CMOS inverter is formed by agents. Agent-based pattern formations in nature and physics are studied in [21,22]. In [23] a general framework is proposed to discover rules that produce special spatial patterns based on a combination of machine learning strategies including genetic algorithms and artificial neural networks.

(ii) *FSM–controlled agents.* We have designed evolved FSM–controlled CA agents for several tasks, like the *Creature's Exploration Problem* [24,25], the *All-to-All Communication Task* [25–27], the *Target Searching Task* [28], the *Routing Task* [29,30]. The FSM for these tasks were evolved by genetic algorithms mainly. Other related works are a multi-agent system modeled in CA for image processing [31] and modeling the agent's behavior by an FSM with a restricted number of states [32]. An important pioneering work about FSM–controlled agents is [33] and FSM–controlled robots are also well known [34].

This work extends the issues presented in [17–19] with a different class of patterns herein and unlike in [19] only two colors and neither markers nor additional communication signals are used. Furthermore, agents are now able to find patterns with the maximum degree of order. In Sect. 2 the class of target patterns is defined and in Sect. 3 the multi-agent system is presented. In Sect. 4 livelock situations and the termination problem are described. The used genetic algorithm is explained in Sect. 5 and the effectiveness and efficiency of selected FSP are evaluated in Sect. 6 before Conclusion. The CA agents used herein are implemented from the *write* access CA–w concept [35–37].

2   Domino Patterns and Degree of Order

Given a square array of $(n + 2) \times (n + 2)$ cells including border, we focus on the problem of generating an optimal configuration of *domino* patterns in the $n \times n$ enclosed field, from four domino templates (Fig. 1). The role of the border, with a perimeter of $4n + 4$ white cells, is to facilitate the work of the agents, thus

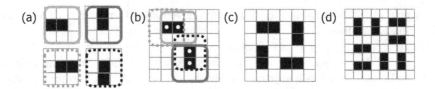

Fig. 1. (a) The four 3×3 domino templates define the domino pattern class. (b) A pattern with 4 hits. It can be tiled (with overlaps) by matching templates. Each matching template produces a hit (dot) in the center. (c) A pattern for a 4×4 field (plus border) with the maximal degree or order 8. (d) A pattern for a 6×6 field with the maximal degree or order 16.

moving within an uniform field. The four possible 3×3 Moore-neighborhood domino templates around a central black cell are displayed in Fig. 1a, showing our so-called spin-like *left* ($\leftarrow$), *up* ($\uparrow$), *right* ($\rightarrow$), *down* ($\downarrow$) dominos. They define the domino pattern *class*.

The templates are tested on each of the n^2 sites (i_x, i_y) of the $n \times n$ field. So each template is applied in parallel on each cell, which can be seen as a classical CA rule application. If a template fits on a site, then a hit (at most one) is stored at this site. Then the sum of all hits is computed which defines the *degree of order h*. A pattern with 4 hits is displayed in Fig. 1b: the top-left horizontal domino is generated by matching the *right* template centered at $(0,0)$ with the *left* template centered at $(1,0)$ then producing two hits. In the same way the bottom-right vertical domino is generated by matching the *down* template with the *up* template, thus giving altogether a pattern with order $h = 4$. Dominos are isolated in the sense that neither contact nor overlap is allowed; in other words, a black domino must be surrounded by ten white cells.

Domino Enumeration. For an even side length n, let h_{max} be the maximum expected order. Hereafter we give an evaluation of this optimal order by induction in a non formal way. In (c) and (d) two optimal patterns are displayed respectively for a 4×4 field and a 6×6 field. They are redisplayed in Fig. 2,

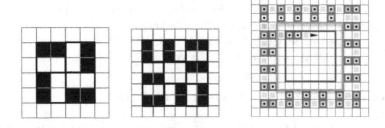

Fig. 2. From left to right: 1. Tiling the 4×4 field with 4 tetraminos. 2. Tiling the 6×6 field with 9 tetraminos. 3. The agent entering the central 6×6 subfield, *with* border, in the 12×12 field: fifth snapshot of Fig. 5a.

showing the patterns now tiled with square 2×2 tetraminos. Such a 4–mino may either contain a domino or be empty. So, a $n \times n$ field (n even) can be tiled by exactly $\xi_n^* = n^2/4$ tetraminos. That gives an upper bound for the maximal order. Note that the central 4–mino in the 6×6 field is empty.

Let us now observe the 12×12 field in Fig. 2 showing one agent generating the pattern. Starting from the top-left corner, the agent generates 4 rows of 5 aligned dominos, moving clockwise, before entering a central 6×6 subfield, *with* border. We are now ready for the induction.

Let us call a "*void*" a cell belonging to an *inner* border and let ν_n be the void index in a $n \times n$ field; we claim that $\nu_0 = \nu_2 = \nu_4 = 0$ and

$$\nu_n = 4(n-5) + \nu_{n-6} \qquad (n > 4) \tag{1}$$

and give an informal proof. The first term of the sum is the perimeter (in number of cells) of the inner border surrounding the central $(n-6) \times (n-6)$ subfield, the second term denotes the number of voids in that subfield. Setting $m = n/2$ and $p = \lfloor m/3 \rfloor$ we get

$$\nu_n = \begin{cases} 4p\,(3p-2) & (m \equiv 0) \\ 4p\,(3p) & (m \equiv 1) \\ 4p\,(3p+2) & (m \equiv 2) \end{cases} \quad (\mathrm{mod}\ 3). \tag{2}$$

The number ξ_n of non-empty 4–minos and bounded by ξ_n^* is then given by

$$\xi_n = \frac{n^2 - \nu_n}{4} \tag{3}$$

Table 1. Domino enumeration for $n \times n$ fields: upper bound ξ_n^*, void index ν_n, domino number ξ_n, optimal degree h_{max}.

n	m	p	ξ_n^*	ν_n	ξ_n	h_{max}
0	0	0	0	0	0	0
2	1	0	1	0	1	2
4	2	0	4	0	4	8
6	3	1	9	4	8	16
8	4	1	16	12	13	26
10	5	1	25	20	20	40
12	6	2	36	32	28	56
14	7	2	49	48	37	74
16	8	2	64	64	48	96
18	9	3	81	84	60	120
20	10	3	100	108	73	146
22	11	3	121	132	88	176
24	12	4	144	160	104	208

namely the domino number, and therefore the maximum expected order is $h_{max} = 2\xi_n$ and the relative order is $h_{rel} = h/h_{max}$.   □

The quantities for the first even values of the field size n are displayed in Table 1.

3   Modeling the Multi-agent-System

Compared to classical CA, moving agents with a certain "intelligence" have to be modeled. Therefore the cell rule becomes more complex. Different situations have to be taken into account, such as an agent is situated on a certain cell and is actively performing actions, or an agent is blocked by another agent or by a border cell in front. The *cell state* is modeled as a record of several data items:

$CellState = (Color, Agent)$
   $Color\ L \in \{0, 1\}$
   $Agent = (Activity, Identifier, Direction, ControlState)$
      $Activity \in \{\texttt{true}, \texttt{false}\}$
      $Identifier\ ID \in \{0, 1, ..., k - 1\}$
      $Direction\ D \in \{0, 1, 2, 3\} \equiv \{\texttt{toN}, \texttt{toE}, \texttt{toS}, \texttt{toW}\}$
      $ControlState\ S \in \{0, 1, ..., N_{states} - 1\}$.

This means that each cell contains a potential agent, which is either active and visible or passive and not visible. When an agent is moving from A to B, its whole state is copied from A to B and the *Activity* bit of A is set to false. The agent's structure is depicted in Fig. 3. The finite state machine (FSM) realizes the "brain" or control unit of the agent. Embedded in the FSM is a state table which defines the actual behavior. The state table can also be seen as a program or algorithm. Therefore the abbreviations FSP (*finite state program*) or AA (*agent's algorithm*) are preferred herein. Outputs are the actions and the next control state. Inputs are the control state s and defined input situations x.

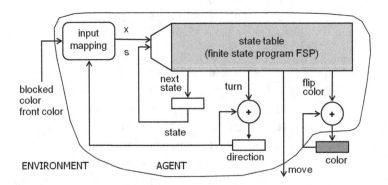

Fig. 3. An agent is controlled by a finite state machine (FSM). The state table defines the agent's next control state, its next direction, and whether to move or not. The table also defines whether the color shall be toggled $(0 \rightarrow 1)$ or $(1 \rightarrow 0)$.

An input mapping function is used in order to limit the size of the state table. The *input mapping* reduces all possible input combinations to an index $x \in X = \{0, 1, \ldots, N_x - 1\}$ used in combination with the control state to select the actual line of the state table.

The capabilities of the agents have to be defined before designing or searching for an AA. The main capabilities are: the perceivable inputs from the environment, the outputs and actions an agent can perform, the capacity of its memory (number of possible control and data states) and its "intelligence" (useful proactive and reactive activity). Here the intelligence is limited and carried out by a mapping of its state and inputs to the next state, actions and outputs.

An agent can react on the following inputs:

- **control state:** agent's control state s,
- **direction:** agent's direction D,
- **color:** color L of the cell the agent is situated on,
- **front color:** color L_F of the cell in front,
- **blocked:** the blocking condition caused either by a border, another agent in front, or in case of a conflict when another agent gets priority to move to the front cell. The inverse condition is called *free*.

An agent can perform the following actions:

- **next state:** $state \leftarrow nextstate \in \{0, \ldots, N_{states} - 1\}$.
- **move:** $move \in \{0, 1\} \equiv \{wait, go\}$.
- **turn:** $turn \in \{0, 1, 2, 3\}$.
  The new direction is $D(t + 1) \leftarrow (D(t) + turn) \bmod 4$.
- **flip color:** $flipcolor \in \{0, 1\}$.
  The new color is $L(t + 1) \leftarrow (L(t) + flipcolor) \bmod 2$.

An agent has a moving direction D that also selects the cell in front as the actual neighbor. What can an agent observe from a neighboring cell? In our model it can only detect the blocking condition and the color in front. So the agents' sensing capabilities are weak.

All actions can be performed in parallel. There is only one constraint: when the agent's action is *go* and the situation is *blocked*, then an agent cannot move and has to wait, but still it can turn and change the cell's color. In case of a moving conflict, the agent with the lowest identifier (ID = $0...k-1$) gets priority. Instead of using the identifier for prioritization, it would be possible to use other schemes, e.g. random priority, or a cyclic priority with a fixed or space-dependent base. The following input mapping was used, $x \in \{0, 1, \ldots, 7\}$:

$$x = 0 + 4b, \text{ if } color = 0 \text{ and } frontcolor = 0$$
$$x = 1 + 4b, \text{ if } color = 1 \text{ and } frontcolor = 1$$
$$x = 2 + 4b, \text{ if } color = 0 \text{ and } frontcolor = 1$$
$$x = 3 + 4b, \text{ if } color = 1 \text{ and } frontcolor = 0$$

where $b = 0$ if *free*, otherwise $b = 1$ if *blocked*. This mapping was designed by experience from former work. Of course, other input mappings are possible, with more or less x codes, or other assignments, e.g. more neighbors could be taken into account, the blocking conditions could be distinguished (by border, by agent, by conflict), or a part of the agent's private control state could be presented to the neighbors. Note that the sensing capabilities are quite limited, and that makes the given task difficult to solve.

4   Livelock and Termination

The Livelock Problem. During the work of evolving FSP, it turned out that livelocks may appear for systems with more than one agent. In a livelock the agents act in a way that there is no more progress in the system's global state towards the aimed pattern. An analogy is when two people meet head-on and each tries to step around the other, but they end up swaying from side to side, getting in each other way as they try to get out of the way. Here livelocks appeared when 2 or 4 agents were placed symmetrically in space. Then the state/actions sequence was the same and the agents got stuck in cyclic paths. Fortunately we found a simple way to avoid them. Three variants of an FSP are used. Agents start in three different control states, depending on the agent's identifier: *initial state = ID mod* 3. By this technique we were able to find FSP without livelocks, agents can now show three different behaviors. As we cannot influence the structure of the evolved FSP, the FSP state's graph may have different prefix state sequences, or the FSP may even fall into three separate graphs (co-evolution of up to three FSP). This means that the genetic algorithm automatically finds the best choice of more equal or more distinct FSP under the restriction of a given maximal number of states N_{states}.

The Termination Problem. How can the multi-agent system be stopped in a decentralized way after having reached the required degree of order? One idea is to communicate the hits all-to-all. Thereby the difficulty is that pattern and degree of order are usually changing over time, and the transportation of the hit information is delayed in space. So it would be more elegant, if the system state (pattern or hit-count) reaches automatically a fixed point. We define for our multi-agent system that has reached a certain degree of order

(1) *Soft-termination:* The pattern is stable, and there exists one agent that is active (moves and/or changes direction).
(2) *Hard-termination:* The pattern is stable, and all agents are passive (not moving and/or not changing direction).

The termination problem has been studied for distributed systems, and now it is under research also for multi-agent systems [38].

5   Evolving FSP by a Genetic Algorithm

An ultimate aim could be to find an FSP that is optimal for all possible initial configurations on average. This aim is very difficult to reach because it needs a huge amount of computation time. Furthermore, it depends on the question whether all-rounders or specialists are favored. Therefore, in this work we searched only for *specialists* optimized for (i) a fixed field size of $N = n \times n$, $n = 12$, (ii) 4 special initial configurations with 1, 2, 4 agents where the agents are placed in the corners of the field. The number of different FSP which can be coded by a state table is $Z = (|s||y|)^{(|s||x|)}$ where $|s|$ is the number of control states, $|x|$ is the number of inputs and $|y|$ is the number of outputs. As the search space increases exponentially, we restricted the number of inputs to $|x| = 8$ and the number of states to $|s| = N_{states} = 18$. Experiments with lower numbers of states did not yield the aimed quality of solutions.

A relatively simple genetic algorithm similar to the one in [17] was used in order to find (sub)optimal FSP with reasonable computational cost. A possible FSP solution corresponds to the contents of the FSM's state table. For each input combination (x, state) = j, a list of actions is assigned:

```
actions(j) = (nextstate(j), move(j), turn(j), flipcolor(j))
```

as displayed on the FSP genome in Fig. 4.

The fitness is defined as the number t of time steps which is necessary to emerge successfully a target pattern with a given degree h_{target} of order, averaged over all given initial random configurations. "Successfully" means that a target pattern with $h \geq h_{target}$ was found. The fitness function F is evaluated by simulating the system with a tentative FSP_i on a given initial configuration. Then the mean fitness $\overline{F}(FSP_i)$ is computed by averaging over all initial configurations of the training set. $\overline{F}$ is then used to rank and sort the FSP.

In general it turned out that it was very time consuming to find good solutions with a high degree of order, due to the difficulty of the agents' task in relation to their capabilities. Furthermore the search space is very large and difficult to explore. The total computation time on a Intel Xeon QuadCore 2 GHz was around 4 weeks to find all needed FSP.

Evolved Finite State Programs. The used fields are of size $N = n \times n$. The cell index (i_x, i_y) starts from the top left corner $(0, 0)$ to the bottom right corner $(n - 1, n - 1)$. The top right corner is $(n - 1, 0)$. The index K defines a set of initial configurations. Here only 4 initial configurations are used:

$K = 1$: 1 agent with direction $\rightarrow$, placed at $(0, 0)$

$K = 2$: 2 agents, one placed like in configuration $K = 1$, and another with direction $\leftarrow$ placed at $(n - 1, n - 1)$

$K = 4$: 4 agents, two of them placed like in configuration $K = 2$, and another with direction $\downarrow$ placed at $(n - 1, 0)$, and another with direction $\uparrow$ placed at $(0, n - 1)$

$K = 124$: This index specifies a set of configurations, the union of $K = 1$, $K = 2$, and $K = 4$

```
state      0  1  2  3  4  5  6  7  8  9 10 11 12 13 14 15 16 17    0  1  2  3  4  5  6  7  8  9 10 11 12 13 14 15 16 17
          /x=0                                                 \  /x=1                                                 \
nextstate 10  4 15  1  2 10 10  1  1 12  7  6 12  9  5 11  3  2    5  1  3 11  2  5  7  6 10 13  7  8  6  3  1 17 11 10
flipcolor  0  1  0  0  1  0  1  1  0  0  0  0  0  0  1  0  0  0    0  0  0  0  0  0  0  1  1  1  1  0  0  1  1  0  1  0
move       0  1  0  0  0  0  0  1  0  0  1  0  1  0  0  1  0  1    0  0  0  0  0  1  1  0  1  1  0  1  1  0  0  0  0  0
turn       1  1  2  1  3  1  2  1  3  3  2  3  3  2  2  2  2  0    0  1  2  3  1  3  0  1  1  2  3  0  2  1  0  0  1  2
          /x=2                                                 \  /x=3                                                 \
nextstate  3  6 10 16  2 13  9  7  9  5  4  0 17 15 17  4 15 17    1  5  5  6  5  4  1 13 17  8  0 11 17  2 12  5  4  7
flipcolor  1  0  0  0  1  0  1  1  1  1  0  0  1  0  0  1  0  1    0  0  1  1  0  1  0  0  1  1  1  1  0  0  0  0  0  1
move       1  1  1  0  1  1  1  0  1  1  1  0  1  0  0  0  1  1    1  0  1  1  1  1  1  0  1  1  1  0  1  1  0  0  1  0
turn       2  3  2  0  2  1  2  0  1  1  1  3  3  3  0  1  1  3    0  1  0  2  0  3  0  1  3  2  3  1  3  1  3  1  0  2
          /x=4                                                 \  /x=5                                                 \
nextstate  4 16 12 16 15 10  8  1  1  5  7  1  6  0 15 17 17  6   10  1 12 16 17 16  7 10  5  7  8  6 14  3 12  3  0  6
flipcolor  1  0  1  1  0  1  1  0  1  1  1  0  1  1  1  1  1  0    1  1  0  1  0  0  0  1  1  0  0  1  1  0  0  0  0  0
move       1  0  1  0  0  0  1  1  1  1  0  0  0  0  1  1  0  0    1  0  0  0  0  1  1  0  1  1  0  1  1  0  0  1  0  0
turn       0  1  1  1  2  0  3  0  1  2  1  2  1  2  1  3  0  0    1  0  1  3  0  2  0  2  1  1  1  1  2  2  3  2  2  0
          /x=6                                                 \  /x=7                                                 \
nextstate  0 12  7  2  7 11  0 15  0  3  5  2  0  2 10  7  5 16   17 17  9  2 11  5  6  2  0 10  2  8 10  4  6  5  4  8
flipcolor  1  0  0  1  1  0  0  0  0  1  1  0  0  1  0  0  1    0  0  1  1  1  1  1  1  0  0  1  0  0  1  1  0  1  0  0
move       1  0  0  1  1  1  1  0  0  1  1  1  0  1  0  0  1    1  0  0  0  1  0  0  1  0  0  1  1  1  0  0  1  1  0
turn       1  2  2  0  2  2  3  1  2  1  1  0  0  3  3  2  2  1    0  2  0  1  1  2  0  3  2  2  2  1  2  1  3  1  1  0
```

Fig. 4. FSP genome with $N_{states} = 18$ states and $N_x = 8$ inputs.

The best found FSP is denoted by

$\boldsymbol{FSP}^{n,K,h}$: for field size n, configuration K, and a reached order h_{max}. The reached order can also be given relatively as h_{rel} with percent suffix

The following four FSP were evolved by the genetic algorithm:

$$FSP^{12,1,100\%} = FSP^{12,1,56}$$
$$FSP^{12,2,100\%} = FSP^{12,2,56}$$
$$FSP^{12,4,100\%} = FSP^{12,4,56}$$

and the more general mixed one $FSP^{12,124,100\%}$ that is 100% successful on each of the 3 initial fields for $K = 1, 2, 4$. Its genome is displayed in Fig. 4. Note that $h_{max} = 56$ for $n = 12$ according to Table 1 whence $h_{rel} = 100\%$.

6   Simulation and Performance Evaluation

Simulation. Firstly, the agent-system was simulated and observed for the first three evolved programs ($FSP^{12,1,56}$, $FSP^{12,2,56}$, $FSP^{12,4,56}$). Figure 5 shows the time evolution of the domino pattern for the system with 1, 2, and 4 agents. The strategy of 1 agent is to move along the border clockwise (Fig. 5a) and then after one cycle moving inwards. Roughly the path is close to a spiral. Looking to the path in detail, the agent moves more or less back and forth in order to build the optimal pattern. Thereby already built dominoes can be destructed and rebuilt in a different way. An optimal pattern with $h_{max} = 56$ is built at $t = 215$.

The systems with two agents (Fig. 5b) and four agents (Fig. 5c) follow a similar strategy, but the work is shared and each agent cooperates in building the optimal pattern. The cooperation is achieved by detecting dominoes already in place and then rearrange them in a better way or move just inwards to the empty area in order to create new dominoes. The optimal pattern is built at $t = 154$ for the 2-agent-system, and at $t = 51$ for the 4-agent-system.

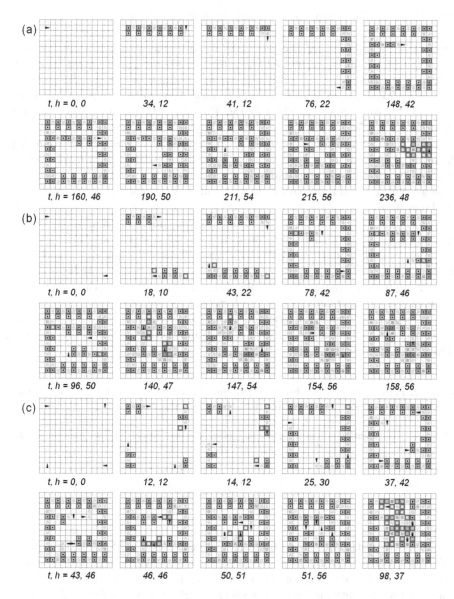

Fig. 5. Dots are marking the hits. Inner squares in light grey are marking visited cells, the darker the more often visited. (a) 1-agent-system. The agent starts in the left corner and moves mainly clockwise, and from the border to the centre. At $t = 215$ an optimal pattern with $h = 56$ is formed. For $t \geq 236$ the pattern remains stable with $h = 48$. (b) 2-agent-system. The agents are building the pattern together. Agent 0 and 1 use a slightly different algorithm, see configuration at $t = 18$. For $t \geq 158$ the agents run in a cycle without changing the optimal pattern. (c) 4-agent-system. The agents 0, 1, 2 use slightly different algorithms, see configuration at $t = 12$. At $(t = 98, h = 37)$ all agents have stopped their activities.

Termination. What happens after having built the optimal pattern?

1-agent-system: During $t = 215 \ldots 235$ the agent continues its walk in the direction of the right border, thereby changing the pattern's order in the sequence $h = 56, 52, 56, 54, 56, 54, 52, 50, 48$. Then, for $t \geq 236$ the pattern remains stable with $h = 48$, and then for $t \geq 240$ the agent is running a 4-step-cycle within a block of 2×2 cells. So we have a *non-optimal soft-termination* with $h_{rel} = 48/56$.

2-agent-system: For $t = 154, 155, 156, 157, 158^+$ the agents change slightly the order to $h = 56, 52, 52, 52, 56$. Then for $t \geq 158$ each of them runs in a cycle of period 8 following a square path within a block of 3×3 cells without changing the optimal pattern. This means an *optimal soft-termination* with $h_{rel} = 100\%$. This result was not expected and was not explicitly forced by the genetic. But it shows that optimal terminations in such multi-agent-systems are possible and can be evolved.

4-agent-system: For $t = 51, 52, \ldots 98^+$ the agents are reducing the order to $h = 56, 50, \ldots 37$ with fluctuations. At $(t = 60, h = 49)$ agent 1 stops its activities, and then at $(t = 63, h = 47)$ agent 0 stops its activities, and then at $(t = 75, h = 44)$ agent 3 stops its activities, and then at $(t = 98, h = 37)$ agent 2 stops its activities. For $t \geq 98$ all agents have stopped their activities, this means a *non-optimal hard-termination* (see last snapshot in Fig. 5c). Agent with $ID = 0, 1, 2, 3$ started at position $(0,0), (n-1, 0), (n-1, n-1), (0, n-1)$ respectively. Agents 0 and 3 are using the same variant of the FSP whereas agents 1 and 2 use other variants.

Table 2. Performance of the k-agent systems, especially evolved for each k. The 4-agent system is almost 4 times faster than the 1-agent system.

Program	Agents k	Time t_k	Time per cell t_k/N	Speedup $S = t_1/t_k$	Efficiency S/k
$FSP^{12,1,100\%}$	1	215	1.49	1.00	1.00
$FSP^{12,2,100\%}$	2	154	1.07	1.40	0.70
$FSP^{12,4,100\%}$	4	54	0.38	3.98	0.99

Comparison with the More General, Mixed $FSP^{12,124,100\%}$. The mixed FSP was evolved to work with 1, 2, or 4 agents, therefore it is more general. Now the time to reach 100% success is longer, $t_{124} = 255, 180, 105$ for $k = 1, 2, 4$. Compared to the optimal time of the special FSP presented before and given in Table 2 the ratio is $t_{124}/t(k) = 1.19, 1.17, 1.94$. That means that special evolved algorithms may save significant computation time, in our example up to 94%.

Performance of the Mixed FSP for Other Field Sizes. Now it was tested how sensitive the mixed FSP is against a change of the field size. It was required that all k-agent systems ($k = 1, 2, 4$) are successful to the up most reachable degree h_{rel}^{max}. It was found by incrementing h_{rel} to the point where at least one agent-system was not successful. Table 3 shows the times t_k to order the systems up to h_{rel}^{max} (refer to Eq. 3 and Table 1 for h_{max}). In order to compare

Table 3. Used was the mixed FSP evolved for field size 12×12. The times t_k for different field sizes was recorded for the maximal reachable degree of order.

Field size	4×4	6×6	8×8	10×10	12×12	14×14	16×16
Reached h_{rel}^{max}	$4/8 = 50\%$	$14/16 = 88\%$	$22/26 = 85\%$	$34/40 = 85\%$	$56/56 = 100\%$	$72/74 = 97\%$	$88/96 = 92\%$
t_1	10	52	82	125	255	303	351
t_2	13	30	42	62	180	216	272
t_4	4	19	21	32	105	123	161

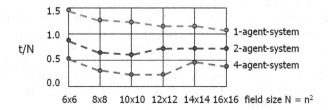

Fig. 6. Time steps per cell needed to order the fields with a degree $h_{rel} \geq 85\%$.

the performance for different field sizes the metric t/N (time steps per cell) was used. Furthermore a fixed bound for $h_{rel} = 85\%$ was used. Then the time was measured for this bound. The outcome is depicted in Fig. 6. The normalized time t/N is minimal at $n = 16, 10, 10$ for $k = 1, 2, 4$ and the 4-agent system is around 3 times faster than the 1-agent-system.

7 Conclusion

The class of the aimed domino patterns was defined by four templates (3×3 local patterns). Four FSP were evolved for multi-agent systems with 1, 2, 4 agents initially placed in the corners of the field. The reached degree of order was 100% for the 12×12 training field, and greater than 85% for field sizes between 6×6 and 16×16. Livelocks were avoided by using up to three different variants of the FSP depending on the agent's identifier. These variants use different initial control states and may show a totally different individual behavior. This can be interpreted as a co-evolution of three cooperating behaviors. It was observed that the achieved pattern can reach a stable fixed point, and then the agents run in small cycles or even stop their activities totally. Further work is directed to the termination problem, the co-evolution, and the problem of finding robust multi-agent systems that can order fields of any size perfectly.

References

1. Shi, D., He, P., Lian, J., Chaud, X., Bud'ko, S.L., Beaugnon, E., Wang, L.M., Ewing, R.C., Tournier, R.: Magnetic alignment of carbon nanofibers in polymer composites and anisotropy of mechanical properties. J. App. Phys. **97**, 064312 (2005)
2. Itoh, M., Takahira, M., Yatagai, T.: Spatial arrangement of small particles by imaging laser trapping system. Opt. Rev. **5**(1), 55–58 (1998)
3. Jiang, Y., Narushima, T., Okamoto, H.: Nonlinear optical effects in trapping nanoparticles with femtosecond pulses. Nat. Phys. **6**, 1005–1009 (2010)
4. Niss, M.: History of the Lenz-Ising model, 1920–1950: from ferromagnetic to cooperative phenomena. Arch. Hist. Exact Sci. **59**(3), 267–318 (2005)
5. Press, D., Ladd, T.D., Zhang, B., Yamamoto, Y.: Complete quantum control of a single quantum dot spin using ultrafast optical pulses. Nature **456**, 218–221 (2008)
6. Bagnold, R.E.: The Physics of Blown Sand and Desert Dunes. Chapmann and Hall, Methuen, London (1941)
7. Turing, A.M.: The chemical basis of morphogenesis. Philos. Trans. R. Soc. Lond. **B237**, 37–72 (1952)
8. Tyson, J.J.: The Belousov-Zhabotinskii Reaction. Lecture Notes in Biomathematics. Springer, Heidelberg (1976). doi:10.1007/978-3-642-93046-1
9. Greenberg, J.M., Hastings, S.P.: Spatial patterns for discrete models of diffusion in excitable media. SIAM J. Appl. Math. **34**(3), 515–523 (1978)
10. Progogine, I., Stengers, I.: Order out of Chaos. Heinemann, London (1983)
11. Chopard, B., Droz, M.: Cellular Automata Modeling of Physical Systems. Cambridge University Press, Cambridge (1998)
12. Deutsch, A., Dormann, S.: Cellular Automaton Modeling of Biological Pattern Formation. Birkäuser, Boston (2005)
13. Désérable, D., Dupont, P., Hellou, M., Kamali-Bernard, S.: Cellular automata in complex matter. Complex Syst. **20**(1), 67–91 (2011)
14. Wolfram, S.: Statistical mechanics of cellular automata. Rev. Mod. Phys. **55**(3), 601–644 (1983)
15. Nagpal, R.: Programmable pattern-formation and scale-independence. In: Minai, A.A., Bar-Yam, Y. (eds.) Unifying Themes in Complex Sytems IV, pp. 275–282. Springer, Heidelberg (2008). doi:10.1007/978-3-540-73849-7_31
16. Yamins, D., Nagpal, R., Automated global-to-local programming in 1-D spatial multi-agent systems. In: Proceedings of the 7th International Conference on AAMAS, pp. 615–622 (2008)
17. Hoffmann, R.: How agents can form a specific pattern. In: Wąs, J., Sirakoulis, G.C., Bandini, S. (eds.) ACRI 2014. LNCS, vol. 8751, pp. 660–669. Springer, Cham (2014). doi:10.1007/978-3-319-11520-7_70
18. Hoffmann, R.: Cellular automata agents form path patterns effectively. Acta Phys. Pol. B Proc. Suppl. **9**(1), 63–75 (2016)
19. Hoffmann, R., Désérable, D.: Line patterns formed by cellular automata agents. In: El Yacoubi, S., Wąs, J., Bandini, S. (eds.) ACRI 2016. LNCS, vol. 9863, pp. 424–434. Springer, Cham (2016). doi:10.1007/978-3-319-44365-2_42
20. Birgin, E.G., Lobato, R.D., Morabito, R.: An effective recursive partitioning approach for the packing of identical rectangles in a rectangle. J. Oper. Res. Soc. **61**, 303–320 (2010)
21. Bonabeau, E.: From classical models of morphogenesis to agent-based models of pattern formation. Artif. Life **3**(3), 191–211 (1997)

22. Hamann, H., Schmickl, T., Crailsheim, K.: Self-organized pattern formation in a swarm system as a transient phenomenon of non-linear dynamics. Math. Comput. Mod. Dyn. Syst. **18**(1), 39–50 (2012)

23. Bandini, S., Vanneschi, L., Wuensche, A., Shehata, A.B.: A neuro-genetic framework for pattern recognition in complex systems. Fundam. Inf. **87**(2), 207–226 (2008)

24. Halbach, M., Hoffmann, R., Both, L.: Optimal 6-state algorithms for the behavior of several moving creatures. In: Yacoubi, S., Chopard, B., Bandini, S. (eds.) ACRI 2006. LNCS, vol. 4173, pp. 571–581. Springer, Heidelberg (2006). doi:10.1007/11861201_66

25. Ediger, P., Hoffmann, R.: Optimizing the creature's rule for all-to-all communication. In: Adamatzky, A., et al. (eds.) Automata 2008, pp. 398–412 (2008)

26. Ediger, P., Hoffmann, R.: Solving all-to-all communication with CA agents more effectively with flags. In: Malyshkin, V. (ed.) PaCT 2009. LNCS, vol. 5698, pp. 182–193. Springer, Heidelberg (2009). doi:10.1007/978-3-642-03275-2_19

27. Hoffmann, R., Désérable, D.: All-to-all communication with cellular automata agents in $2D$ grids. J. Supercomput. **69**(1), 70–80 (2014)

28. Ediger, P., Hoffmann, R.: CA models for target searching agents. Elec. Notes Theor. Comput. Sci. **252**, 41–54 (2009)

29. Ediger, P., Hoffmann, R., Désérable, D.: Routing in the triangular grid with evolved agents. J. Cell. Autom. **7**(1), 47–65 (2012)

30. Ediger, P., Hoffmann, R., Désérable, D.: Rectangular vs triangular routing with evolved agents. J. Cell. Autom. **8**(1–2), 73–89 (2013)

31. Komann, M., Mainka, A., Fey, D.: Comparison of evolving uniform, non-uniform cellular automaton, and genetic programming for centroid detection with hardware agents. In: Malyshkin, V. (ed.) PaCT 2007. LNCS, vol. 4671, pp. 432–441. Springer, Heidelberg (2007). doi:10.1007/978-3-540-73940-1_43

32. Mesot, B., Sanchez, E., Peña, C.-A., Perez-Uribe, A.: SOS++: finding smart behaviors using learning and evolution. In: Artificial Life VIII, pp. 264–273. MIT Press (2002)

33. Blum, M., Sakoda, W.J.: On the capability of finite automata in 2 and 3 dimensional space. In: SFCS 1977, pp. 147–161 (1977)

34. Rosenberg, A.L.: Algorithmic insights into finite-state robots. In: Sirakoulis, G., Adamatzky, A. (eds.) Robots and Lattice Automata. Emergence, Complexity and Computation, vol. 13, pp. 1–31. Springer, Cham (2015). doi:10.1007/978-3-319-10924-4_1

35. Hoffmann, R.: The GCA-w massively parallel model. In: Malyshkin, V. (ed.) PaCT 2009. LNCS, vol. 5698, pp. 194–206. Springer, Heidelberg (2009). doi:10.1007/978-3-642-03275-2_20

36. Hoffmann, R.: Rotor-routing algorithms described by CA-w. Acta Phys. Pol. B Proc. Suppl. **5**(1), 53–67 (2012)

37. Hoffmann, R., Désérable, D.: Routing by cellular automata agents in the triangular lattice. In: Sirakoulis, G., Adamatzky, A. (eds.) Robots and Lattice Automata. Emergence, Complexity and Computation, vol. 13, pp. 117–147. Springer, Cham (2015). doi:10.1007/978-3-319-10924-4_6

38. Lahlouhi, A.: MAS-td: an approach to termination detection of multi-agent systems. In: Gelbukh, A., Espinoza, F.C., Galicia-Haro, S.N. (eds.) MICAI 2014. LNCS, vol. 8856, pp. 472–482. Springer, Cham (2014). doi:10.1007/978-3-319-13647-9_42

Automated Parallelization of a Simulation Method of Elastic Wave Propagation in Media with Complex 3D Geometry Surface on High-Performance Heterogeneous Clusters

Nikita Kataev[1]([⊠]), Alexander Kolganov[1,2], and Pavel Titov[3]

[1] Keldysh Institute of Applied Mathematics RAS, Moscow, Russia
kaniandr@gmail.com
[2] Moscow State University, Moscow, Russia
alex-w900i@yandex.ru
[3] Institute of Computational Mathematics and Mathematical Geophysics SB RAS,
Novosibirsk, Russia
tapawel@gmail.com
http://dvm-system.org/en/, http://www.msu.ru/en/, http://icmmg.nsc.ru/en

Abstract. The paper considers application of DVM and SAPFOR in order to automate mapping of 3D elastic waves simulation method on high-performance heterogeneous clusters. A distinctive feature of the proposed method is the use of a curved three-dimensional grid, which is consistent with the geometry of free surface. Usage of curved grids considerably complicates both manual and automated parallelization. Technique to map curved grid on a structured grid has been presented to solve this problem. The sequential program based on the finite difference method on a structured grid, has been parallelized using Fortran-DVMH language. Application of SAPFOR analysis tools simplified this parallelization process. Features of automated parallelization are described. Authors estimate efficiency and acceleration of the parallel program and compare performance of the DVMH based program with a program obtained after manual parallelization using MPI programming technology.

Keywords: Automation of parallelization · Heterogeneous computational cluster · 3D modeling · Curvilinear grid · GPU · Xeon Phi

1 Introduction

3D modeling of elastic waves in media of various structures is an important aspect of a geophysical 3D models creation and studying the characteristics of wave fields. Often, to solve the inverse problem of geophysics is difficult, and

---

The reported study was funded by RFBR according to the research projects 17-01-00820, 16-07-01067, 16-07-01014, 17-41-543003, 16-01-00455, 16-07-00434.

V. Malyshkin (Ed.): PaCT 2017, LNCS 10421, pp. 32–41, 2017.
DOI: 10.1007/978-3-319-62932-2_3

one of the methods is to solve a set of direct problems with varying parameter values and the geometry of the medium by comparing the actual data with the simulation results.

The most used method for solving the direct problem is finite difference method [1,2]. It should be noted that the investigated area can have a complex 3D surface geometry, therefore important feature point is to construct a curved 3D mesh. For example, the object of study can be magmatic volcano. The studying of the medium structure and the monitoring of this object are an important practical task that requires high performance computer power to produce results quick enough. The theory of construction and application of curvilinear grids to actual problems solving are well described in [3,4]. Such an approach to numerical modeling of elastic waves involves working with a lot of 3D data. Given the scale of the field in solving real-world problems numerical modeling task becomes feasible on a personal workstation even with GPU.

There is an increased interest in the use of multi-core and heterogeneous computational systems to achieve maximum performance in calculating the computationally intensive tasks of given class. At the same time the ability to write parallel programs that can effectively tap full potential of these systems requires programmers to thoroughly understand underlying hardware architecture as well as programming models. The situation is drastically complicated with the widespread acceptance of new architectures, such as NVIDIA GPU or Intel Xeon Phi coprocessors. Multiple technologies of parallel programming should be used simultaneously in order to exploit all levels of parallelism.

The matter becomes even more complicated if users map legacy sequential code to a parallel architecture. Significant program transformations are often necessary to avoid drawbacks which prevent program parallelization. Especially important is simultaneous support of two versions of a program: sequential and parallel. Development of tools that facilitate parallel code development and furthermore development of tools that can identify parallelism in an automated way can make an invaluable contribution to the evolution of supercomputing industry and can dramatically reduce efforts required to implement parallel programs.

Automated parallelization relies on the ability to identify regions of code, which can be run asynchronously with other parts of the program, and requires an accurate static and dynamic dependence analysis. The use of interactive tools is essential for automation of parallel programming [8–11]. Some of them rely on automatic parallelizing compilers [10]. The other tools are based on parallelism identification techniques (static and dynamic dependence analysis, source code profiling) and assume that parallelism will be exploited by the user manually [8]. The SAPFOR system was primarily designed to simplify the mapping of sequential programs to a parallel architectures with distributed memory. The system implements static and dynamic analysis techniques and focused on DVMH model of the parallel programming for heterogeneous computational clusters [12,13].

The rest of the paper is organized as follows. Section 2 discusses the mathematical problem. Section 3 describes parallel algorithm features and its software implementation based on MPI technology. Section 4 focuses on usage of DVMH model to exploit parallelism and discusses features of obtained program. Section 5 presents performance of parallel programs execution on heterogeneous computational clusters. The section illustrates exploration of strong and weak scaling, as well as comparison of the results of DVMH and MPI parallel program execution. Section 6 concludes and summarizes the paper.

2   Problem Statement

Solution of the direct problems of geophysics is related to the solution of equations of elasticity theory in the case of 3D medium. In this paper we consider the elastic waves. Medium model is defined by three parameters: Lame coefficients λ, μ and density ρ. For example, in [1] the equations are represented in terms of displacement velocities. In this formulation, we must operate with 9 equations and hence with 9 parameters at the stage of numerical implementation. Using this approach will be sufficient resource intensive. It was therefore decided to use the system, which is described by the three components of the displacement vector $(u, v, w)^T$. Such a method is more optimal in terms of saving memory and computation time.

It is necessary to build a 3D model of the medium to solve the direct problem namely to determine the size and shape of the field, and set the parameters λ, μ, ρ for each of the components of the medium (the medium can be inhomogeneous).

In this paper the authors use curved 3D mesh construction. The most important point is the orthogonal edges of cells near the free surface: all intersecting edges of each cell near the curved free surface are locally orthogonal. This means that in every point of the surface the vertical ribs of every cell are perpendicular to the plane tangent to the surface at this point. In these same points the edges corresponding to horizontal direction are also orthogonal. Such an approach to solve this dynamic geophysical problem is used by one of the authors of this paper for the first time. It allows to increase the accuracy of the approximation of the boundary conditions for the numerical solution of the problem (for the case of curved free surface area – from the first order when one uses a regular hexahedral grid to the second order when one uses a curvilinear cells). The most preferred method is proved to be a transfinite interpolation [3]. After getting mesh built, it is necessary to convert to a new coordinate system, where the grid is a regular hexahedral.

Numerical solution of the problem is based on the finite difference method. This method has proved itself well to create on its basis 3D effective parallel algorithm [6, 7]. The basis of the formulas were taken from [5], adapted to the 3D-case. The scheme has a second-order approximation in space and time.

3   Parallel Algorithm and It's Software Implementation Using MPI

To perform parallel calculations for each unit, it is necessary for nine 3D arrays to be placed and stored in a memory. Those arrays have values for the component displacements u, v, w, environment parameters λ, μ, ρ, as well as the coordinates of the curved grid X, Y, Z. For this work the sequential meshing program and the sequential program that implements the calculations for the difference scheme have been developed in Fortran 95. Those two were combined into one program and obtained program was parallelized using MPI parallel programming technology. Thus, you don't have to record the coordinates of grid points on a hard drive at the stage of the mesh-builder work, and then read them on the stage of the main program work. The proposed approach significantly accelerates the program, depending on the mesh size and hard drive access speed.

For arrays with u, v, w need to organize shadow data exchange between adjacent blocks. Each block has shadow edges, in which it receives remote data from its neighbors. It sends its data in similar shadow edges of its neighbors. Communications made through the created 3D-topology. Each computing process is assigned a triple room: Cartesian coordinates in this topology. It is important to note that data exchange are not only between processes according to one of the adjacent coordinate directions. There are exchanges between the processes, neighboring in all diagonal directions. Total number of exchanges is 26.

Using the finite difference method allows communication between processes using non-blocking MPI exchanges. It uses a buffer array where the values of u, v, w are copied. That allows for two neighboring blocks to exchange all data in one message instead of three. Next, the calculation of the components u, v, w for each block is divided into two independent parts: inside the block and on the shadow edges of the block. Such partitioning enables to start asynchronous communications earlier.

4   Parallelization of Given Program Using SAPFOR and DVM Systems

4.1   Using Static Analyzer of SAPFOR System

The system for automated parallelization SAPFOR [10] is a software development suit that is focused on cost reduction of manual program parallelization. SAPFOR can be used to produce a parallel version of a program in a semi-automatic way according to DVMH model of the parallel programming for heterogeneous computational clusters. Fortran-DVMH language is currently supported. The system was extremely helpful to explore information structure of the proposed algorithm. Data dependencies were discovered and classified according to the relevant approach to a related issue. The visual assistance tool was used to examine analysis results. Types of data dependencies which are supported by SAPFOR are considered further.

Write after read and write after write dependencies produce writing conflicts between different iterations of a loop which modify the same variable. Variable privatization technique removes such conflicts and is natively supported by DVMH languages with the use of PRIVATE specification followed by the list of variables that should be privatized. This specification is not required if a program is executed in the compute systems with distributed memory but in case of intra-node parallelism or GPU computing their absence will lead to data races.

The source of private variables essentially are computations localized within a single parallel loop iteration. In this case such variables are used to store temporary intermediate results of calculations. With a large volume of calculations in a loop the amount of temporary data can be significant thus may hinder the insertion of PRIVATE specification manually. In DVMH version of proposed algorithm the number of private variables within a single loop amounts to 53. Automatic insertion of specifications and correctness verification of a user guidelines can reduce the cost of developing and of debugging of parallel programs [14].

A source of another type of loop carried dependencies which can be eliminated in order to enable parallel program execution is reduction operations, such as accumulation of the sum of all elements in a set. Reduction produces read after write dependency through value accumulation in some variable. The parallel code will accumulate partial values and combine them after loop exit since reduction operations are associative. SAPFOR analysis tools proved to be useful to recognize reduction operations which find a minimum element from a given list of expressions. To specify their in DVMH languages REDUCTION construct can be used.

The other important feature of our system is investigation of regular loop carried dependencies, the read after write dependencies with distances limited constant. For each loop iteration this constant determines a number of previous loop iterations required to execute it. Given the ACROSS specification which is natively supported by DVMH languages that kind of loops can be executed in parallel, the DVMH compiler will split a loop iteration space into stages to implement pipeline parallelism. Static analysis implemented in SAPFOR system revealed no regular dependencies in the developed program.

Static analysis of a code is essential pass for automatic parallelization. Even so, it has a significant drawback, it is necessary to take conservative decisions. This means that if you can not reliably test the absence of a data dependence, in order to ensure correct execution of the program static analyzer concludes on its presence. The use of static analysis to investigate loop carried dependencies in conjunction with SAPFOR GUI and subsequent manual parallelization in Fortran-DVMH language allowed to overcome this difficulty. It is worth noting that the use of co-design in the development of the proposed algorithm provides a minimum number of false positives (in case of 6% of loops). The emergence of false positive dependencies was mainly caused by the presence of branches, for which the analysis was unable to determine jump conditions. In more complex situations SAPFOR system allows the use of dynamic analysis, which determines the absence of dependencies at runtime for certain data sets.

4.2   Development of a Parallel Program in DVMH

To exploit parallelism with DVMH the following main specifications should be placed in the program. To distribute data between compute nodes DISTRIBUTE directive is used. The ALIGN directive implements alignment of an array with some other distributed array, so the first one will be also distributed. The PARALLEL directive should be placed on each loop which accesses distributed data. The PARALLEL directive matches each iteration of the parallel loop with some array element. It means that the loop iteration will be executed on the processor where the corresponding array element is located.

In order to identify suitable arrays alignment, it is necessary to explore all loops in a program. DVMH model requires the execution of the rule of own computations for parallel loops. It means that a processor must modify only its own data which are located in its local memory. Therefore it is enough to examine a relation of loop parameter with array subscripts for all arrays modified in this loop. The information obtained from analysis of all loops determine the best array alignment to minimize the interprocessor communication.

In total, more than 100 arrays are used in the program. However not all of them can be distributed. Mapping of a curvilinear grid on a structured grid requires replication of some computations and data. The reason is that the algorithm uses indirect array access, which is not supported in DVMH model. The number of distributed arrays in the code amounts to 51. Since the configuration of arrays used in the program is known before program execution, it is possible to distinctly specify arrays alignment. For example, the following directives

!DVM$ DISTRIBUTE DQ1_DX (BLOCK, BLOCK, BLOCK)

!DVM$ ALIGN (i, j, k) WITH DQ1_DX(2 * i, 2 * j, 2 * k) :: alambda

describes distribution of the three-dimensional array DQ1_DX. Each array dimension is distributed mainly by equal blocks. An element (i, j, k) of alambda array should be placed on the processor that stores the element (2 * i, j * 2, 2 * k) of DQ1_DX array. This allocation allows us to align the array alambda, which in each dimension is twice as little than the array DQ1_DX. Data that are not specified by these directives are automatically distributed on each processor (replicated data) and will have the same value on each processor.

If rules for the distribution and alignment have been successfully determined, it is essential to place PARALLEL directive on each loop to distribute its iterations between available processors. With the information received from SAPFOR, it will be no trouble to add the necessary clauses PRIVATE, REDUCTION ACROSS to this directive. To specify shadow data exchange (SHADOW_RENEW clause) all reading operations from distributed arrays should be analyzed to determine presence of nonlocal elements usage. The method for determining shadow edges is described in the example on the website of DVM system [15]. The total number of parallelized loops is 63 from 91.

To achieve correctness of functional execution of the program it may be useful to perform functional debugging after program parallelization. For this purpose the program must be compiled as a parallel DVMH program (DVMH directives are Fortran language comments for standard compilers therefore DVMH program can be processed as usual serial program) in a special mode with debug information. Then the compiled program should be run in debug mode to obtain diagnostics in case of incorrect placement of any PARALLEL directives.

And finally, each parallel loop or a group of contiguous parallel loops should be enclosed in the region, a part of the program (with one entrance and one exit) that may be executed on one or several computational devices (for example, GPU and CPU). For the GPU it is also necessary to specify the actualization directive (ACTUAL, GET_ACTUAL), which controls data movement between a random access memory of CPU and memories of accelerators.

5   Results

The effectiveness of the resulting FDVMH-program and parallel program using MPI technology (MPI program) was evaluated on the K100 supercomputer [16] with Intel Xeon X5670 CPU and NVIDIA Tesla C2050 GPUs (Fermi architecture), and on a local server with Intel Xeon E5 1660 v2 CPU and NVIDIA GeForce GTX Titan GPU (Kepler architecture). We also explored weak and strong scaling. Usage of the different generations of the GPUs made it possible to show the impact of restrictions of Fermi architecture on program performance.

Each node of K100 has two 6-cores processors and three GPUs. Two processors are linked by a shared memory (NUMA architecture). As a result of the experiments, it was found out that the optimal configuration to run FDVMH version of proposed algorithm consists of two MPI processes per node (one process to one physical CPU) and 6 cores and 1 GPU for each MPI process. The parallel program with MPI uses 12 processes on each node. FDVMH program was also launched with 12 MPI processes on each node to ensure correct comparison of FDVMH program with MPI program.

Both programs use the same number of CPU cores. DVMH program uses the following configurations: only MPI processes, MPI + OpenMP and MPI + OpenMP + CUDA. A tools for automatic performance balancing between CPU cores and GPU on each node is a part of DVM system. It was applied to the last configuration. The result was a variation of proportions of computations performed by the CPU and GPU from node to node, that was cause by unstable working of K100 compute nodes. Number of iterations to be performed to complete the calculation depends on the size of the input data. Therefore, for a correct comparison of total time of programs execution we choose the number of iterations per second as a metric. The total execution time of the MPI program is 3000 s on a single compute node, approximately.

We used 2 GB data per core and 24 GB data per node for weak scaling measurement. Weak scaling results are given in Table 1. The table shows the number of iterations per second for each configuration (the bigger, the better). In total, we were able to use 480 CPU cores and 80 GPUs (40 nodes of K100), the size of the problem in this case was approximately of 1000 GB. The Table 1 shows that the DVMH program is not concede on execution speed to the MPI program.

The difference between the MPI program and DVMH program using only the CPU cores does not exceed 3% in favor of DVMH despite the fact that the manual parallelization uses MPI asynchronous transfers, but DVMH program uses only the synchronous one. The exchanges between neigh-boring processes

Table 1. The week scaling (the number of iterations per second)

Nodes	1	2	3	4	5	6	7	8	10	20	40
DVMH (MPI)	0.86	0.85	0.85	0.85	0.84	0.84	0.83	0.82	0.82	0.82	0.82
DVMH (MPI/OpenMP)	0.86	0.85	0.88	0.89	0.84	0.88	0.84	0.88	0.85	0.85	0.84
DVMH (MPI/ OpenMP/CUDA)	1.56	1.53	1.53	1.50	1.51	1.49	1.51	1.51	1.50	1.48	1.49
MPI program	0.85	0.85	0.85	0.85	0.85	0.85	0.83	0.83	0.82	0.82	0.82

are lasting a very small time compared to the time of the main calculations so this prevents advantages of asynchronous transfers usage. Also the manual parallelization program has the following problems: the barrier synchronizations, the shadow edges exchange, an initialization of asynchronous communications and a factor of communication environment architecture of K100.

The use of GPUs on each process makes the DVMH program execution two times faster in comparison with the same DVMH program which is run without using GPUs. The DVM system allows to get detailed statistics of the resources usage for each loop in a source code. This statistic shows the compiler tries to use all available resources on GPU. The Kepler architecture provides a significant acceleration of DVMH program execution and allows to use more number of registers per thread (256 registers on Kepler compared to 64 registers on Fermi).

Table 2. Comparison of the Fermi and Kepler architecture (the number of iterations per second)

Variant of running of DVMH program	Performance
DVMH (K100 OpenMP (6 cores))	0.43
DVMH (K100 GPU Fermi)	0.34
DVMH (Xeon E5 OpenMP (6 cores))	0.52
DVMH (Xeon E5 GPU Kepler)	3.4

The results of hybrid launches with different GPU generations are presented in Table 2. The Table 2 shows that the performance of different generations of CPUs practically does not differ (20% in favor of a more modern CPU). But the involvement of the more modern GPU architecture makes it possible to obtain 6.5x times acceleration over all CPU cores and 10x times acceleration compared to the Fermi architecture. Also, we can see that the performance of two GPUs and the 12 cores of CPU of one K100 node is twice lower than the performance of a single GPU Kepler architecture (ref. Table 1).

The DVMH program was launched on 1, 2, 4, 6 and 12 threads of 6-cores Intel Xeon E5 for the strong scaling evaluation. These results are presented in Table 3. We can achieve the linear strong scaling on the DVMH program relative to parallel DVMH program running on the single thread and the almost linear strong scalability relative to the serial program.

Table 3. The strong scaling (the number of iterations per second)

Numbers of threads	1	2	4	6	12
DVMH (Xeon E5 OpenMP)	0.08	0.15	0.31	0.45	0.52
Sequential program	0.09	N	N	N	N

6   Conclusion

In this paper the authors examine the possibility of using SAPFOR [10] and DVM [12] systems to automate the development of parallel programs on the example of the numerical modeling of 3D elastic waves simulation with complex free surface geometry. The use of curvilinear grids greatly complicates both manual and automated parallelization. Mapping of a curvilinear grid on a structured grid has been proposed to solve this problem. The co-design approach enables us to apply automated analysis and parallelization techniques to enable concurrent execution of this program.

SAPFOR was used to analyze the program and to get suggestions about what directives of Fortran-DVMH language should be placed in the source code. We use DVM system profiling capabilities to tune performance of the parallel program. The DVM system also gives us a dynamic tool for a functional debugging which is suited to verify correctness of all the DVMH directives in the program. The manual check of DVMH directives correctness is very difficult due to the fact that the program consists of more than 4000 lines. To ensure the correctness of program execution on accelerators DVMH comparative debugging was used. This is a special mode of a DVMH program execution. In this mode the output data obtained in region during execution on GPU are compared with the data obtained in the region during execution on CPU with a given degree of accuracy.

The results of program launches demonstrate good strong and weak scaling (almost linear) both for DVMH program and MPI program. DVMH program proved to be effective as the MPI program when used under the same conditions. The DVMH program allows us to use in more efficient way all available resources on any cluster by automatically load balancing methods implemented in DVMH runtime system.

The considered tools included in SAPFOR and DVM systems can significantly reduce the effort required to produce effective parallel programs that can be mapped to different architectures. They can also help in the development and optimization of scalable algorithms for supercomputers.

The source code of the developed programs are available here [17].

References

1. Glinskiy, B.M., Karavaev, D.A., Kovalevskiy, V.V., Martynov, V.N.: Numerical modeling and experimental research of the "Karabetov Mountain" mud volcano by vibroseismic methods (in Russian). Numer. Methods Program. **11**, 95–104 (2010)
2. Graves, R.W.: Simulating seismic wave propagation in 3D elastic media using staggered grid finite differences. Bull. Seismol. Soc. Am. **86**(4), 1091–1106 (1996)
3. Liseykin, V.D.: Difference Grid, Theory and Applications (in Russian), p. 3254. FUE Publishing House SB RAS, Novosibirsk (2014)
4. Khakimzyanov, G.S., Shokin, Y.I.: Difference schemes on adaptive grids (in Russian). Publishing Center NGU, Novosibirsk (2005)
5. Appelo, D., Petersson, N.A.: A stable finite difference method for the elastic wave equation on complex geometries with free surfaces. Commun. Comput. Phys. **5**(1), 84–107 (2009)
6. Komatitsch, D., Erlebacher, G., Goddeke, D., Michea, D.: High-order finite-element seismic wave propagation modeling with MPI on a large GPU cluster. J. Comput. Phys. **229**(20), 7692–7714 (2010)
7. Karavaev, D.A., Glinsky, B.M., Kovalevsky, V.V.: A technology of 3D elastic wave propagation simulation using hybrid supercomputers. In: CEUR Workshop Proceedings 1st Russian Conference on Supercomputing Days 2015, vol. 1482, pp. 26–33 (2015)
8. Intel Parallel Studio: http://software.intel.com/en-us/intel-parallel-studio-home
9. Sah, S., Vaidya, V.G.: Review of parallelization tools and introduction to easypar. Int. J. Comput. Appl. **56**(12), 17–29 (2012)
10. Bakhtin, V.A., Borodich, I.G., Kataev, N.A., Klinov, M.S., Kovaleva, N.V., Krukov, V.A., Podderugina, N.V.: Interaction with the programmer in the system for automation parallelization SAPFOR. Vestnik of Lobachevsky State University of Nizhni Novgorod **5**(2), 242–245 (2012). Nizhni Novgorod State University Press, Nizhni Novgorod (in Russian)
11. ParaWise Widening Accessibility to Efficient and Scalable Parallel Code. Parallel Software Products White Paper WP-2004-01 (2004)
12. Konovalov, N.A., Krukov, V.A., Mikhajlov, S.N., Pogrebtsov, A.A.: Fortan DVM: a language for portable parallel program development. Program. Comput. Softw. **21**(1), 35–38 (1995)
13. Bakhtin, V.A., Klinov, M.S., Krukov, V.A., Podderugina, N.V., Pritula, M.N., Sazanov, Y.: Extension of the DVM-model of parallel programming for clusters with heterogeneous nodes. Bulletin of South Ural State University. Series: Mathematical Modeling, Programming and Computer Software, vol. 18 (277), no. 12, pp. 82–92. Publishing of the South Ural State University, Chelyabinsk (2012). (in Russian)
14. Kataev, N.A.: Static analysis of sequential programs in the automatic parallelization environment SAPFOR. Vestnik of Lobachevsky University of Nizhni Novgorod, vol. 5(2), pp. 359–366. Nizhni Novgorod State University Press, Nizhni Novgorod (2012)
15. Exmaple of program parallelization using DVMH-model. http://dvm-system.org/en/examples/
16. Heterogeneous cluster K100. http://www.kiam.ru/MVS/resources/k100.html
17. Source code. https://bitbucket.org/dvm-system/elastic-wave-3d

Parallel Algorithm with Modulus Structure for Simulation of Seismic Wave Propagation in 3D Multiscale Multiphysics Media

Victor Kostin[1], Vadim Lisitsa[1], Galina Reshetova[2],
and Vladimir Tcheverda[1]([⊠])

[1] Institute of Petroleum Geology and Geophysics SB RAS,
3, prosp. Koptyug, 630090 Novosibirsk, Russia
vova_chev@mail.ru
[2] Institute of Computational Mathematics and Mathematical Geophysics SB RAS,
6, prosp. Lavrentiev, 630090 Novosibirsk, Russia

Abstract. This paper presents a problem-oriented approach, designed for the numerical simulation of seismic wave propagation in models containing geological formations with complex properties such as anisotropy, attenuation, and small-scale heterogeneities. Each of the named property requires a special treatment that increases the computational complexity of an algorithm in comparison with ideally elastic isotropic media. At the same time, such formations are typically relatively small, filling about 25% of the model, thus the local use of computationally expensive approaches can speed-up the simulation essentially. In this paper we discuss both mathematical and numerical aspects of the hybrid algorithm paying most attention to its parallel implementation. At the same time essential efforts are spent to couple different equations and, hence, different finite-difference stencils to describe properly the different nature of seismic wave propagation in different areas. The main issue in the coupling is to suppress numerical artifacts down to the acceptable level, usually a few tenth of the percent.

Keywords: Finite-difference schemes · Local grid refinement · Domain decomposition · MPI · Group of processor units · Master processor unit · Coupling of finite-difference stencils

1 Introduction

Numerical simulation of seismic wave propagation in realistic 3D isotropic ideal-elastic media has become a common tool in seismic prospecting. In particular, the reverse time migration is based on the solution of multiple forward problems for the scalar wave equation; the full-wave form inversion assumes massive simulations as a part of the misfit minimization algorithms. Typically, these simulations are performed by means of finite differences, as this method combines a high efficiency with a suitable accuracy [14], and, in particular, a standard

ⓒ Springer International Publishing AG 2017
V. Malyshkin (Ed.): PaCT 2017, LNCS 10421, pp. 42–57, 2017.
DOI: 10.1007/978-3-319-62932-2_4

staggered grid scheme (SSGS) is used [12] with parameters modification when simulation is done for heterogeneous media [5,13]. However, if a model is complicated by anisotropy, viscoelasticity or small-scale heterogeneities, then more complex and computationally expensive approaches should be used which narrows the applicability of the numerical methods. Specifically, in order to take into account seismic attenuation, a generalized standard linear solid (GSLS) model is commonly used [1], which needs additional memory variables. As a result, the computational intensity of the algorithm doubles in comparison with an ideal-elastic medium. If anisotropy is present in a model, then advanced finite difference schemes such as the Lebedev scheme (LS) [4] or the rotated staggered grid scheme [11] are needed. For the latter the number of variables per grid cell and the number of floating point operations (flop) are four times as large as those for the SSGS used for isotropic elastic media. The presence of small-scale heterogeneities brings about the necessity of using sufficiently small grid steps to match the scale, which may drastically increase the size of the problem up to several orders [3,6].

At the same time, the formations with the above-mentioned properties are typically small, up to 25% of the model due to the geological conditions under which they were formed. As a result it is reasonable to use the computationally intense approaches locally and to apply the efficient SSGS elsewhere in the model. The problem of a proper coupling of different numerical techniques is considered and studied in this paper. Note that the main numerical method used for simulation of seismic wave propagation in realistic models is time domain explicit finite differences. Thus, we deal with stencil computations which can be efficiently parallelized by the domain decomposition techniques [3]. On the other hand, peculiarities of the coupling and imbalance in computational workloads lead to a strong inhomogeneity thus requiring a detailed study of balancing, scaling and efficiency of a parallel algorithm.

Note that the algorithm presented is aimed at the simulation of wave propagation as applied to the exploration seismology. Thus the physical size of a typical model is about 500 by 500 by 200 wavelengths, which results in the discretization of about 5000 by 5000 by 2000 points in each spatial direction or 5×10^{10} grid points. Approximately the same number of degrees of freedom is needed for the finite element approximation. For anisotropic viscoelastic models the total number of variables to be stored per grid cell should be about 150, which leads to the evaluation of memory required in 30 Tb. Assume that a standard cluster has one Gb of RAM per core and a typical simulation time (wall-clock time) for one shot is about 6 h with the full use of the available RAM. A typical acquisition system has about 10000 source positions. Thus, the total machine time needed for the full experiment is 1.8×10^9 core-hours. Note that the provided estimations are valid up to $\pm 20\%$ for any numerical technique used for the full waveform simulation. On the other hand, if the major part of the model is isotropic ideal-elastic, then the number of parameters to be stored for the finite difference approximation drops down to 12, which reduces significantly the RAM requirements, the number of flops, and, consequently, the machine time needed for simulation.

Below we consider separately each case (attenuation, anisotropy, small-scale inhomogeneity), provide a mathematical description of the hybrid algorithm and study its computational aspects in a considerable detail.

2   Elastic Media with Attenuation

2.1   Mathematical Formulation

The generalized standard linear solid (GSLS) model governing seismic wave propagation in viscoelastic media is described by the equations:

$$\rho \frac{\partial u}{\partial t} = \nabla \sigma$$

$$\frac{\partial \varepsilon}{\partial t} = \left(\nabla u + \nabla u^T\right)$$

$$\frac{\partial \sigma}{\partial t} = C_1 \varepsilon + \sum_{l=1}^{L} r^l \tag{1}$$

$$\tau_{\sigma,l} \frac{\partial r^l}{\partial t} = C_2 \varepsilon r^l$$

where ρ is the mass density; C_1 and C_2 are fourth order tensors, defining the model properties; u is the velocity vector; σ and ε are the stress and strain tensors, respectively; r^l are the tensors of memory variables. Note that the number of memory variable tensors is L, which is typically two or three. Proper initial and boundary conditions are assumed.

For the ideal-elastic models, the tensor C_2 is zero which means that the solution of the last equation is trivial if zero initial conditions are imposed. Thus, memory variable tensors can be excluded from the equations and the system turns into that for an ideal-elastic wave equation. This means that there is no need to allocate random access memory (RAM) for the memory variables in the ideal-elastic parts of the model.

Assume now a subdomain $\Omega \subseteq R^3$, where the full viscoelastic wave equation is used, while the ideal-elastic wave equation is valid over the rest of the space. It is easy to prove that the conditions at the interface $\Gamma = \partial \Omega$ are

$$[\sigma \cdot \boldsymbol{n}]|_{\Gamma=0}, \quad [u]|_{\Gamma=0}, \tag{2}$$

where $\boldsymbol{n}$ is the vector of outward normal and $[f]$ denotes a jump of the function f at the interface Γ. These conditions are the same as those for the elastic wave equation at the interface. Moreover, if a standard staggered grid scheme (SSGS) [12] is used, these conditions are automatically satisfied [5,9]. Thus, the coupling of the models does not require any special treatment of conditions at the interface and can be implemented by allocating RAM for memory variables and solving equations for them in the viscoelastic part of the model.

2.2   Parallel Implementation

The parallel implementation of the algorithm has been carried out using static domain decomposition. It has the following features:

- the amount of RAM and flops per grid cell varies for elastic and viscoelastic parts of the model; thus individual domain decomposition is needed;
- operators used to update the solution at the interfaces are local in all spatial directions, thus there is a one-to-one correspondence between two adjacent subdomains of different types (elastic and viscoelastic);
- computing one time step includes two different types of synchronization points.

The first statement is self-evident. The second means that it is natural to change the size of subdomains only in one direction, that is, normal to the interface. Typically, this is the vertical direction due to the structure of geological models, that is variation of parameters in the vertical direction is much stronger than in the horizontal one. After that, no special treatment at the interface is needed, one just has to allocate RAM for memory variables in the subdomains where the viscoelastic wave equation is solved. The last statement affects the efficiency of the algorithm the most. The first type of synchronization point occurs immediately after the velocity components have been updated. This stage requires the same amount of flops per grid cell both for elastic and viscoelastic parts of the algorithm. Another synchronization should be applied when the stresses are computed. This part is strongly different for the two models. This means that regardless of the ratio of the elementary subdomains associated with a single core (node) for elastic and viscoelastic subdomains some of the cores will have a latency period.

The construction of the optimal domain decomposition is based on minimization of the overall computational time (core-hours) of the algorithm. Assume a computational domain of the volume V, where the viscoelastic and the elastic wave equations are solved in the subdomains of volumes $V^v = \alpha V$ and $V^e = (1 - \alpha)V$, respectively with $\alpha \in [0, 1]$. Denote the elementary volumes assigned to a single core for the viscoelastic and the elastic parts of the model are $\tilde{V}^v$ and $\tilde{V}^e = \beta \tilde{V}^v$, respectively. The total number of elementary subdomains (cores) equal to:

$$N = V^v / \tilde{V}^v + V^e / \tilde{V}^e = \left(\alpha + \frac{1 - \alpha}{\beta}\right) V / \tilde{V}^v.$$

The computational time is equal to the sum of the times T_u and T_σ needed to update the velocity and the stress components respectively multiplied by the number of time steps. Denote by t_σ^v and t_σ^e the time needed for updating the stress tensor in the viscoelastic and elastic parts of the model per one grid point and assume $t_\sigma^e = \gamma t_\sigma^v$. Similarly, introduce $t_u^v = t_u^e = \delta t_\sigma^v$, which are the times to update the velocity components. Having assumed a uniform mesh, a uniform cores of the cluster, and taking into account the introduced notations, obtains the estimation of the computational time of the algorithm:

$$T(\alpha, \beta) = C \left[\delta \max(1, \beta) + \max(1, \beta\gamma) \right] \times \left(\alpha + \frac{1-\alpha}{\beta} \right)$$

where C is a parameter depending on the total volume, the number of time steps etc., but it does not depend on the domain decomposition. The parameters $\gamma = 0.33$ and $\delta = 0.32$ are difficult to estimate analytically, therefore they are measured experimentally. A series of simulations are carried out using four different clusters on the base of the Intel processors. They are: NKS-30T of the Siberian Supercomputer center, MVS-100 K of the Joint Supercomputer Center of the Russian Academy of Sciences, the clusters SKIF "Tchebyshev" and "Lomonosov" of Moscow State University, and supercomputer HERMIT at Stuttgart University. The absolute values of the time needed to update the velocity and stresses in the elastic and viscoelastic parts of the model vary for different machines, but their ratios γ and δ are close to constants. Now a minimization problem can be formulated as finding β delivering a minimum of $T(\alpha, \beta)$ for a given α. The plots of $T(\beta)$ for three different α are provided in Fig. 1, scaled so that $T = 1$.

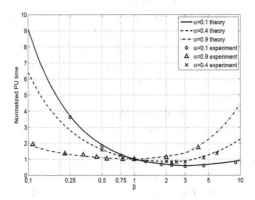

Fig. 1. Normalized core-hours with respect to β.

To confirm the correctness of the derived formula, a series of experiments were conducted for fixed α and β. The results are plotted in Fig. 1 by the markers, confirming the correctness of the theoretical estimation. According to this study if a relative volume of the viscoelastic part is less than 25%, the optimal ratio of elementary subdomains β should be $1/\gamma = 3$. In this case, the computations on the base of the hybrid algorithm are two times faster with respect to the pure viscoelastic simulation.

3   Anisotropy

3.1   Finite Difference Approximation

In this section we consider the simulation of seismic wave propagation in ideal-elastic anisotropic media. Assume system (1) with a zero attenuation term, that is $C_2 \equiv 0$. If a medium is isotropic, the stiffness tensor C_1 has a specific form:

$$C_1 = \begin{pmatrix} c_{11} & c_{12} & c_{13} & 0 & 0 & 0 \\ c_{12} & c_{22} & c_{23} & 0 & 0 & 0 \\ c_{13} & c_{23} & c_{33} & 0 & 0 & 0 \\ 0 & 0 & 0 & c_{44} & 0 & 0 \\ 0 & 0 & 0 & 0 & c_{55} & 0 \\ 0 & 0 & 0 & 0 & 0 & c_{66} \end{pmatrix}$$

This means that the stress-strain relation (the third equation in (1) decouples into four independent equations which can be solved within different spatial points. As a result, one may define different components of the wavefield, that is the velocity vector and the stress tensor components, at different grid points. After that, standard central differences can be used to approximate system 1. Consequently one needs to store one copy of each variable per grid cell, 9 values in total. This scheme is known as a standard staggered grid scheme (SSGS) [11].

For anisotropic media, the stiffness tensor C_1 has no special structure except the symmetry $c_{ij} = c_{ji}$. As a result, no separation of variables can be applied, and all the components of the stress tensor should be defined at the same grid points. Similarly, all the components of the velocity vector should be stored at each grid point, different from those for stresses. There are two schemes for waves' simulation in anisotropic media: the rotated staggered grid scheme [11] (Saenger et al., 2010) and the Lebedev scheme (LS) [4]. The detailed comparison of these approaches one can see in ([7], where the authors have proved that implementation of the LS is preferred due to a lower demand for computational resources. However, as compared to the SSGS, the Lebedev scheme requires four times RAM as much as to store the wavefield variables - four copies of each component per grid cell (see Figs. 2 and 3 for a greater detail).

3.2   Coupling

Assume the elastic anisotropic wave equation holds in a subdomain $\Omega \subseteq R^3$, while the isotropic wave equation holds over the rest of the space. If differential equations are considered, the conditions at the interface $\Gamma = \partial\Omega$ are standard and provided by formula 2. However, if a finite difference approximation is applied, the number of variables for the LS is four times greater than for

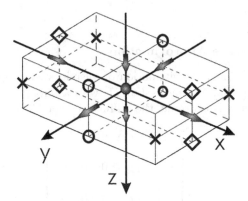

Fig. 2. A grid cell for the SSGS. Filled circles correspond to the points where σ_{xx}, σ_{yy} and σ_{zz} are defined, σ_{yz} is stored at empty circles, σ_{xz} is at rhombi, σ_{xy} is at crosses. Velocity components are defined at arrows pointing to corresponding directions.

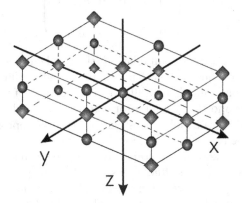

Fig. 3. A grid cell for the LS. All components of the stress tensor are defined at circles, all velocity components are stored at rhombi.

the SSGS. This gives additional degrees of freedom and leads to the existence of spurious modes for the LS. As was shown in [8], the LS based approximation gives four times more unknown and possesses 12 plane-wave solutions instead of three. As a result, the wavefield, computed with the LS, can be decoupled into four parts: u^{++}, u^{+-}, u^{-+}, and u. Here u^{++} is the velocity vector corresponding to the true solution, while the others are numerical artifacts. A similar representation is valid for the stress tensor also. Note that this decomposition holds at each grid point. Thus, the correct conjugation conditions at the interface Γ should be:

for the true components

$$u^V = u^{++}|_\Gamma, \quad \sigma^V \boldsymbol{n} = \sigma^{++}\boldsymbol{n}|_\Gamma$$

for each artificial mode all components should vanish

$$u^{\pm\pm}|_\Gamma = 0, \quad \sigma^{\pm\pm} n|_\Gamma = 0.$$

Here n is the vector of external normal to Ω at the interface Γ. From the physical standpoint these conditions permit the true solution to pass the interface with no reflections but to keep all the artifacts inside the domain Ω without transformation to any other type of waves.

The finite difference approximation of the presented conditions leads to the necessity of wavefield interpolation with respect to tangential directions. This interpolation has to be done at each time step. Our suggestion is to use the fast Fourier transform (FFT)-based interpolation, as it possesses exponential accuracy. Moreover, a typical discretization for the finite difference simulation of wave propagation is about 10 points per wavelength, which means that about 30% in 1D (90% in 2D) of the FFT spectra of the solution corresponding to high frequencies are trivial. This property will be used in the parallel implementation of the algorithm.

Thus, the formulae to update the solution at the interface Γ, where the SSGS and the LS are coupled, are local in normal direction, but non-local in tangential directions, requiring the use of the full wavefield in the vicinity of the interface. On the other hand, if the FFT-based interpolation is applied, only 10% of the spectrum may be used to reduce the amount of data to be exchanged between processor units.

3.3 Parallel Implementation

The main features of the algorithm are:

- the amount of RAM and flops for the anisotropic simulation is four times as great as that for solving the isotropic problem either for updating velocities or stresses;
- formulae to update the solution at the interface are local in normal direction and non-local in tangential ones;
- the FFT-based interpolation is used, thus the transferred data can be compressed.

The anisotropic formations are typically sub-horizontal layers elongated throughout the model. Thus, it is reasonable to define the domain Ω, where the anisotropic elastic wave equation is solved, as a horizontal layer embedding in the formation. As a result, one obtains two interfaces (for each anisotropic domain), they are: $z = z_+$ and $z = z_-$. Within each subdomain, independent 3D domain decomposition is applied to ensure a high balancing level. Taking into account the first two peculiarities of the algorithm, the optimal ratio of the isotropic to anisotropic elementary subdomain is four. To organize the data transfer from the isotropic to the anisotropic domain, the master processors are allocated in each group at both sides of the interfaces. These processors receive the wavefield to be interpolated from the corresponding side of the interface. After that,

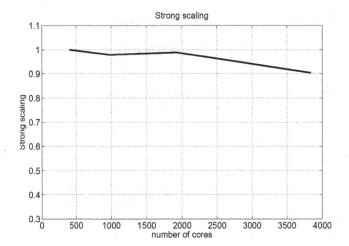

Fig. 4. Strong scaling of the hybrid isotropic-anisotropic algorithm. Along the vertical axis there is the value $\dfrac{nt(n)}{n_0 t(n_0)}$, where n is the current number of processor units, n_0 is some initial value. In our experiments $n_0 = 400$.

the 2D FFT is applied by the master processors, and a significant part of the spectra is sent to the master processor from the other side of the interface, where the interpolation and inverse FFT are performed. Finally, the interpolated data are sent to the processors which need these data. This part is the bottle-neck of the algorithm; however this is a reasonable price for the non-local data transfer. Moreover, the use of the non-blocking procedures Isend/Irecv allows overlapping communications and the update of the solution in the interior of the subdomains. The strong scaling of this implementation is close to 93% for the use of up to 4000 cores, see Fig. 4.

3.4 Numerical Experiment

For testing the isotropic/anisotropic coupling we used anisotropic Gullfaks 2.5D model. The model: P-wave velocity and Thomsen's parameter ε, is presented in Figs. 5 and 6. Note that if $\varepsilon = 0$, the medium is isotropic, thus there is one anisotropic layer in this model. The wave propagation in the anisotropic Gullfaks model was simulated by the hybrid algorithm (Fig. 7) that was based on the LS, after that the difference was computed (Fig. 8). One may note that the error caused by the use of the coupled scheme is about 0.2%, which is an acceptable level for seismic simulations. The speed-up of the hybrid algorithm is 2.5 in comparison with the purely anisotropic simulation.

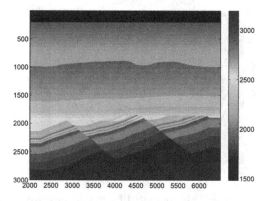

Fig. 5. P-wave velocity for Gullfaks model

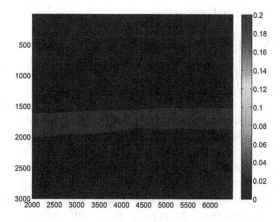

Fig. 6. Thomsem's parameter ε for the Gullfaks model.

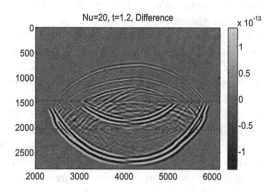

Fig. 7. Wavefield computed by the hybrid algorithm.

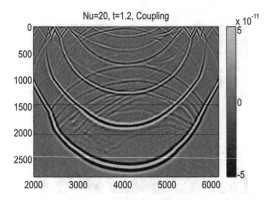

Fig. 8. Difference in the wavefields computed by hybrid algorithm and that based on the LS.

4   Multi-scale Simulation

4.1   Mathematical Formulation

The algorithm is developed to study scattering and diffraction of seismic waves on clusters of small-scale heterogeneities such as fractures, cavities and karstic formations. In such a situation, a detailed description of each fracture is not reasonable but their distribution within a relatively small volume is of interest. We assume that the reservoir model is given on a sufficiently fine grid, which possesses grid steps of the size of several centimeters. A typical seismic wave has a wavelength of about several dozen meters, with the grid steps of the background model being about several meters. Thus, a local mesh refinement is used to perform the full waveform simulation of long wave propagation through a reservoir with a fine structure. As explicit finite differences are used, the size of the time step strongly depends on the spatial discretization and so the time stepping should be local. As a result, the problem of simulation of seismic wave propagation in models containing small-scale structures becomes a mathematical problem of the local time-space mesh refinement.

Consider how coarse and fine grids are coupled. The necessary properties of the finite difference method, based on a local grid refinement are stability and an acceptable level of artificial/numerical reflections. Scattered waves have an amplitude about 1% of the incident wave, thus the amplitude of artifacts should be at most 0.1% of the incident wave. If we refine the grid simultaneously with respect to time and space, then the stability of finite difference schemes can be provided via coupling of a coarse and a fine grids on the basis of energy conservation, which leads to an unacceptable level (more than 1%) of artificial reflections [2]. We modify this approach so that the grid is refined with respect to time and space in turn on two different surfaces surrounding the target area with microstructure. This allows decoupling of the temporal and spatial grid refinements independently and on this way to provide a desired level of artifacts.

Refinement with Respect to Time. A refinement in time with a fixed 1D spatial discretization is shown in Fig. 9. Its modification for 2D and 3D is straightforward.

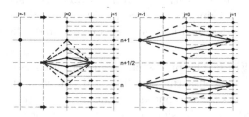

Fig. 9. Embedded stencils for local time step mesh.

Refinement with Respect to Space. To fit different spatial grids, a FFT-based interpolation is used. The procedure is explained for a 2D problem. The mutual disposition of a coarse and a fine spatial grids is illustrated in Fig. 10, which corresponds to updating the stresses. As can be seen, to update the solution on a fine grid on the uppermost line it is necessary to know the wavefield at the points marked with black, these points do not exist on a given coarse grid. Using the fact that all of the points are on the same line (a plane in 3D), we seek for the values of missing nodes by the FFT-based interpolation. This procedure allows us to provide a required low level of artifacts (about 0.001 with respect to the incident wave) generated at the interface of these two grids.

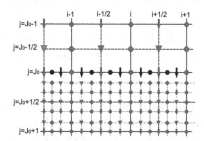

Fig. 10. Spatial steps refinement. Black marks correspond to the points where the coarse-grid solution is interpolated.

4.2 Implementation of Parallel Computations

The use of the local space-time grid stepping makes difficult to ensure a uniform work load for processors in the domain decomposition. In addition, the user should be allowed to locate the reservoir anywhere in the background.

For implementation of parallel computations we introduce two groups of processors: one for 3D heterogeneous environment (a coarse grid), another for a fine mesh describing the reservoir (Fig. 11). There is a need for two level of interactions between processor units: within each group and between the groups as well. The data exchange within a group is done via faces of the adjacent subdomains by non-blocking iSend/iReceive MPI procedures. The interaction between the groups is designed for coupling a coarse and a fine grids.

From Coarse to Fine. The processors aligned to a coarse grid form a group along each of the faces contacting the fine grid. At each of the faces, Master Processor (MP) gathers the computed current values of stresses/displacements, applies the FFT and sends the significant part of the spectrum to the relevant MP on a fine grid (see Fig. 11). All the subsequent data processing, including interpolation and inverse FFT, are performed by the relevant MP in the fine grid group. Subsequently, this MP sends interpolated data to each processor in its subgroup.

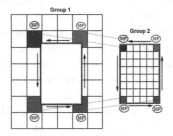

Fig. 11. Processor units for a coarse (left) and a fine (right) grids.

Exchange of the part of the FFT spectrum and interpolation performed by the MP of the second group essentially decreases the amount of sent/received data.

From Fine to Coarse. Processors from the second group compute the solution on the fine grid. For each face of the fine grid block we also design a MP. This MP collects data from a relevant face, performs the FFT, and sends a part of the spectrum to the corresponding MP of the first group (a coarse grid). Formally, the FFT can be excluded and the data to be exchanged can be obtained as a projection of the fine grid solution onto the coarse grid, however the use of truncated spectra decreases the amount of data to be exchanged and ensures stability as it acts as a high-frequency filter.

4.3   Numerical Experiment

The designed algorithm was implemented to study the wave propagation in a fractured reservoir embedded in a complex model with a buried viscoelastic

Fig. 12. The buried channel model, top view.

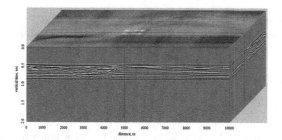

Fig. 13. 3D zero off-set cubes.

channel, see Fig. 12. Zero off-set seismic data are provided in Fig. 13, illustrating the presence of scattered waves associated with a fractured reservoir. Next, these scattered waves are used in some special imaging techniques in order to reveal subseismic heterogeneities [10].

5   Conclusion

We have presented the algorithm for the numerical simulation of seismic wave propagation in complex media. The algorithm is intended for the exploration seismology and is based on finite differences as they combine a high computational efficiency with a desired accuracy [3,14]. The main stream-flow in the modern exploration seismology is the local complication of the models. Thus, the majority of the models are, in general, isotropic elastic having anisotropic and viscoelastic intrusions or clusters of small-scale heterogeneities. Implementation of a brutal-force universal approach brings unrealistic demands on computational resources up to Petabytes of RAM needed and up to 10^9 core-hours required for the simulation. The proposed algorithm is based on the multi-physics background: the complex techniques are used only locally, while the solution in the major part of the model is computed by an efficient standard staggered grid finite-difference scheme. Therefore acceleration of the family of the presented hybrid algorithms strongly depends on the relative volume of the subdomains filled with the media with complicated geometrical and physical properties (anisotropy, viscoelasticity, subseismic heterogeneities): the larger it is the less acceleration is. But of course the way of parallel organization of these computations is also very important to provide solid scalability of the software. We use two main tricks. We use two independent group of processor units - one for the regular background medium (say, isotropic elastic without of small-scale intrusions), another anisotropic/viscoelastic/with microheterogeneities. The special way of data exchange applied to organize cooperation of two groups provides acceptable level of scalability. As one could see it is about 0.8 for a large number of units involved. It is worth mentioning that to optimize parallelization within groups we use non-blocking exchange procedures iSend/iReceive and hide waiting time under computations.

Acknowledgements. This research is supported by the RSCF grant 17-17-01128. The simulations were done on the Siberian Supercomputer Center, Joint Supercomputer Center of RAS and on the supercomputer "Lomonosov" of Moscow State University.

References

1. Blanch, J., Robertson, A., Symes, W.: Modeling of a constant Q: methodology and algorithm for an efficient and optimally inexpensive viscoelastic technique. Geophysiscs **60**, 176–184 (1995)
2. Collino, F., Fouquet, T., Joly, P.: A conservative space-time mesh refinement method for the 1-D wave equation. Part I: construction. Numer. Math. **95**, 197–221 (2003)
3. Kostin, V., Lisitsa, V., Reshetova, G., Tcheverda, V.: Local time-space mesh refinement for simulation of elastic wave propagation in multi-scale media. J. Comput. Phys. **281**, 669–689 (2015)
4. Lebedev, V.I.: Difference analogies of orthogonal decompositions of basic differential operators and some boundary value problems. Sov. Comput. Math. Math. Phys. **4**, 449–465 (1964)
5. Lisitsa, V., Podgornova, O., Tcheverda, V.: On the interface error analysis for finite difference wave simulation. Comput. Geosci. **14**, 769–778 (2010)
6. Lisitsa, V., Reshetova, G., Tcheverda, V.: Finite-difference algorithm with local time-space grid refinement for simulation of waves. Comput. Geosci. **16**, 39–54 (2011)
7. Lisitsa, V., Vishnevskiy, D.: Lebedev scheme for the numerical simulation of wave propagation in 3D anisotropic elasticity. Geophys. Prospect. **58**, 619–635 (2010)
8. Lisitsa, V., Vishnevsky, D.: On specific features of the Lebedev scheme in simulating elastic wave propagation in anisotropic media. Numer. Anal. Appl. **4**, 125–135 (2011)
9. Moczo, P., Kristek, J., Vavrycuk, V., Archuleta, R.J., Halada, L.: 3D heterogeneous staggered-grid finite-differece modeling of seismic motion with volume harmonic and arithmetic averaginng of elastic moduli and densities. Bull. Seismol. Soc. Am. **92**, 3042–3066 (2002)
10. Protasov, M., Reshetova, G., Tcheverda, V.: Fracture detection by Gaussian beam imaging of seismic data and image spectrum analysis. Geophys. prospect. **64**, 68–82 (2016)

11. Saenger, E.H., Gold, N., Shapiro, S.A.: Modeling the propagation of the elastic waves using a modified finite-difference grid. Wave Motion **31**, 77–92 (2000)
12. Virieux, J.: P-SV wave propagation in heterogeneous media: velocity-stress finite-difference method. Geophysics **51**, 889–901 (1986)
13. Vishnevsky, D., Lisitsa, V., Tcheverda, V., Reshetova, G.: Numerical study of the interface error of finite difference simulation of seismic waves. Geophysics **79**, T219–T232 (2014)
14. Virieux, J., Calandra, H., Plessix, R.-E.: A review of the spectral, pseudo-spectral, finite-difference and finite-element modelling techniques for geophysical imaging. Geophys. Prospect. **59**, 794–813 (2011)

Performance Evaluation of Two Load Balancing Algorithms on a Hybrid Parallel Architecture

Tiago M. do Nascimento, Rodrigo W. dos Santos, and Marcelo Lobosco$^{(\boxtimes)}$

Graduate Program on Computational Modeling,
Federal University of Juiz de Fora, Juiz de Fora, Brazil
tiago.nascimento@uab.ufjf.br, {rodrigo.weber,marcelo.lobosco}@ufjf.edu.br

Abstract. Accelerated Processing Units (APUs) are an emerging architecture that integrates, in a single silicon chip, the traditional CPU and the GPU. Due to its heterogeneous architecture, APUs impose new challenges to data parallel applications that want to take advantage of all the processing units available on the hardware to minimize its execution time. Some standards help in the task of writing parallel code for heterogeneous devices, but it is not easy to find the data division between CPU and GPU that will minimize the execution time. In this context, this work further extends and details load balancing algorithms designed to be used in a data parallel problem. Also, a sensitivity analysis of the parameters used in our models was performed. The results have shown that the algorithms are effective in their purpose of improving the performance of an application on an heterogeneous environment.

Keywords: Load balancing · Hybrid parallel architectures · APU · HPC

1 Introduction

Computers are becoming heterogeneous, parallel devices. Multicore CPUs have now the company of accelerators, such as GPUs. While in some computers accelerators are attached as off-board devices, in other computers they appear integrated to the processor. APU [2] is an example of new processor that merge, in a single chip, the functionality of GPUs with the traditional multicore CPUs.

It is not a trivial task to take advantage of all parallelism available on APUs, not only due to the heterogeneity of the architecture, but also due to the distinct types of parallelism their devices were designed to. Multicore CPUs were designed to deal with Instruction Level Parallelism (ILP) and Thread Level Parallelism (TLP), while GPUs were designed to deal with Data Level Parallelism (DLP)[3]. ILP is automatically explored by the hardware or, in some architectures, by the hardware with the help of the compiler. However, programmers must structure their code in order to explore TLP and DLP. Some tools, such as those based on the OpenCL [6] and the OpenACC [1] standards, can help

---

The authors would like to thank UFJF, FAPEMIG, CAPES, and CNPq.

V. Malyshkin (Ed.): PaCT 2017, LNCS 10421, pp. 58–69, 2017.
DOI: 10.1007/978-3-319-62932-2_5

programmers to write code to execute in heterogeneous architectures. OpenCL's application programming interface (API) is used to control the platform and execute programs on the compute devices using TLP and DLP. Some issues, however, remain open, such as the optimal load balancing (LB) between GPUs and CPUs in order to explore TLP and DLP simultaneously. An automatic LB scheme, based on the master-worker parallel pattern [5,6], can be implemented in OpenCL. However, this parallel pattern is particularly suited for problems based on TLP [5].

In previous works [7,8] we proposed two distinct solutions based on an in-order execution for problems based on DLP. In the first algorithm [7], some simulation steps are used to measure the relative performance of GPUs and CPUs. This relative performance ratio is then used to divide data between the GPU and the CPU for the remaining steps. The main idea of the second algorithm [8] is that the measurement process and the data division can be repeated during the execution until the changes in the relative performance ratio are below a given threshold. The first algorithm was named static because once the data is divided between GPU and CPU, this division is kept until the computation finishes. The second algorithm was named dynamic because data division can change along the execution to adjust to the load. This paper has three contributions. The first one is the extension done on both algorithms in order to reduce the costs of their probe phases. The second one is sensitivity analysis of their key parameters. The last contribution is to show that applications that seems to be regular parallel applications, such as the HIS, may suffer from irregular execution time phases during their execution due to the values used on their computations, which may impact the LB algorithm. Although HIS executes exactly the same set of instructions on different data sets, its execution time changed a lot depending on the values used in the computation. We suspect that this fluctuation in the execution time is due to hardware optimizations, and consequently they can affect other applications with similar characteristics.

The remaining of this work is organized as follows. In Sect. 2 we review and extend the two LB algorithms. Section 3 presents the performance evaluation. Finally, Sect. 4 presents our conclusions and plans for future works.

2   Load Balancing Algorithms

From a software perspective, there are two main parallel programming models: Single Program, Multiple Data (SPMD) and Multiple Program, Multiple Data (MPMD). In the SPMD model, all the processing units execute the same code, on multiple data sets. Each processing unit has it own ID, which is used to define the subset of data it must compute. In the MPMD model, different processing units execute different codes, on different data. While the SPMD model maps directly in a hardware that implements the DLP, such as a GPU, the MPMD maps in a hardware that implements the TLP, such as a multicore CPU.

In this scenario, heterogeneous computers represent a big challenge to the development of parallel applications, since it can be hard to use all distinct devices simultaneously to execute a single program due to their distinct characteristics. In fact, heterogeneous computing on CPUs and GPUs using architectures like CUDA [4] and OpenCL [6] has fixed the roles for each device: GPUs have been used to handle data parallel work while CPUs handle all the rest. Although this fixed role for each device appears to be fine, CPUs are idle while GPUs are handling the data parallel work. Since the CPUs do not block after starting a computation on the GPUs, they could also handle part of the work submitted to the GPU. However, it is not common to observe this alternative, specially if data dependency exists in the code executed on GPUs and CPUs. In summary, the use of this fixed role model underutilizes the system, since CPUs are idle while GPUs are computing.

In previous works we have presented two distinct LB algorithms [7,8] to be used by applications that use DLP. The algorithms were designed to extract more performance from an heterogeneous device, such as an APU. The key idea behind the two algorithms is similar: data is split into two parts, one of which will be computed by the CPUs, while the other one will be computed by the GPUs. The amount of data that will be assigned to the CPU and GPU depends on their relative computing capabilities, which is measured in both LB algorithms during the execution of the application.

The algorithms are generic in the sense they can be used in a wide variety of data parallel applications. Usually these applications have at least two aligned loops, in which the inner loop performs the same operations on distinct data items, as Algorithm 1 shows. Each step of the inner loop (or a collection of loops, in the case the data structure has more than one dimension) could be executed in any order, since no data dependency occurs between two distinct loop iterations. The number of steps the outer loop iterates is determined by the nature of the problem, but usually a dependency exists between two consecutive steps: the results of a previous step must be available before the start of a new one, since these results will be used during computation. In many applications the outer loop is related to the progress of a simulation over time, and for this reason will be referred in this work as time-steps. The algorithms are tailored to decide the amount of data each processing element will receive to compute in the inner loop. During the computation of each data item, some algorithms require also access to its neighbors data, which can be located at distinct memory spaces due to data splitting between CPUs and GPUs. These data, called boundaries, must be updated between two consecutive iteration of the outer loop. This update requires the introduction of synchronization operations and the explicit copy of data. Both data copy and synchronization operations are expensive, deteriorating performance, and for this reason should be avoided.

2.1   Static Load Balancing Algorithm

The first LB algorithm [7] works as follows. For the first n percent of the time-steps, both GPU and CPU receive a given percentage of data (P_{GPU} and P_{CPU}) and the time required to compute them, including the time spent in communication, is recorded. This information is then used to compute the relative

```
1 for all time-steps do
2 |   for each data item do
3 |   |   call cpus/gpus devices to compute a piece of data;
4 |   end
5 |   send/receive boundaries;
6 |   synchronize devices;
7 end
```

Algorithm 1. Data parallel algorithm

computing power between CPUs and GPUs and consequently determine the
amount of data each device will receive for the remaining time-steps. Equation 1
is used for this purpose.

$$P_G = \frac{T_c}{(T_g + T_c)},$$ (1)

where T_c is given by:

$$T_c = T_{CPU} \times P_{CPU},$$ (2)

and T_g is given by:

$$T_g = T_{GPU} \times P_{GPU}.$$ (3)

P_G is the percentage of data that the GPU will receive to compute in the
remaining time-steps, so $1 - P_G$ is the percentage that the CPU will receive.
T_{CPU} and T_{GPU} are respectively the time CPU and GPU spent to compute
the first n percent of the time-steps and P_{CPU} and P_{GPU} are respectively the
percentage of data that CPU and GPU received to compute in the first n percent
of the time-steps. This algorithm was called static LB because once the data is
divided between GPU and CPU (P_G and $1 - P_G$, respectively), the division
remains the same until the computation finishes.

In the original algorithm [7], the percentage of time-steps used to measure
T_{CPU} and T_{GPU} was fixed in 1%. This work generalizes the original algorithm
since the value of n can be chosen (Algorithm 2, line 12). This modification led
us to do another slight modification in the original algorithm [7]: the initial
percentage of data P_{GPU} and P_{CPU} that were defined by the programmer, are
now computed using a probe time-step (Algorithm 2, lines 5–8). The main reason
to this modification is to speed up the execution of the probe phase. If the value
of n is high and data is not well balanced between CPUs and GPUs, the time
spent in the probe phase can be high. The modification tries to reduce this
imbalance and is implemented as follows. A single time-step is used to probe
the relative computing power between CPUs and GPUs (Algorithm 2, lines 5–8),
distributing for this purpose 50% of the data set to the CPU and 50% to the GPU
(Algorithm 2, lines 2–3). After this single time-step, the remaining n percent of
the time-steps (Algorithm 2, lines 12–17) will use the value computed during the
probe step and the algorithm proceeds its execution as explained above.

```
 1 initialize opencl;
 2 allocate memory in the CPU memory space (50% of the dataset size);
 3 allocate memory in the GPU memory space (the remaining 50%);
 4 start clock;
 5 for a single time-step do
 6 |   call cpu/gpu to compute their data;
 7 |   synchronize queue;
 8 end
 9 finish clock;
10 compute P_GPU and P_CPU and reallocate memory accordingly;
11 start clock;
12 for n% of the time-steps do
13 |   call cpu/gpu to compute their data;
14 |   synchronize queue;
15 |   send/receive boundaries;
16 |   synchronize queue;
17 end
18 finish clock;
19 recompute P_GPU and P_CPU and reallocate memory accordingly;
20 for all the remaining time-steps do
21 |   call cpu/gpu to compute their data;
22 |   synchronize queue;
23 |   send/receive boundaries;
24 |   synchronize queue;
25 end
26 finalize opencl;
```

Algorithm 2. The static LB algorithm

2.2   Dynamic Load Balancing Algorithm

The dynamic LB algorithm [8] is similar to the static one in the sense that the same equations are used to determine the amount of data that GPUs and CPUs will receive. The difference is that this process will be repeated for each p percent of the time-steps or until P_G does not change its value by more than a given threshold t. In the original work [8], p and t assumed a constant value, both equal to 1%. This work generalizes the original algorithm, since distinct values can be chosen to p (Algorithm 3, line 13) and t (Algorithm 3, line 22). Also, the values of P_{GPU} and P_{CPU} are determined by the algorithm in the same way just described for the static LB version. The Algorithm 3 presents the new dynamic version.

Observe that the dynamic algorithm requires the reallocation of memory (Algorithm 3, line 20) while the difference between old and new P_{GPU}, in percentage terms, is greater than the threshold t. The memory reallocation is required since data is redistributed between GPU and CPU due to the new value of P_{GPU} and P_{CPU}. This is the main overhead of this algorithm, specially if the amount of memory that must be reallocated is huge.

3   Performance Evaluation

This section evaluates the performance of the two LB algorithms presented in this work using for this purpose a simulator of the Human Immune System [9, 10].

```
1   initialize opencl;
2   allocate memory in the CPU memory space (50% of the dataset size);
3   allocate memory in the GPU memory space (the remaining 50%);
4   start clock;
5   for a single time-step do
6   │   call cpu/gpu to compute their data;
7   │   synchronize queue;
8   end
9   finish clock;
10  compute P_GPU and P_CPU and reallocate memory accordingly;
11  repeat
12  │   start clock;
13  │   for p% of the time-steps do
14  │   │   call cpu/gpu to compute their data;
15  │   │   synchronize queue;
16  │   │   send/receive boundaries;
17  │   │   synchronize queue;
18  │   end
19  │   finish clock;
20  │   recompute P_GPU and P_CPU and reallocate memory accordingly;
21  │   compute d, the difference between old and new P_GPU, in percentage terms;
22  until d>t;
23  for all the remaining time-steps do
24  │   call cpu/gpu to compute their data;
25  │   synchronize queue;
26  │   send/receive boundaries;
27  │   synchronize queue;
28  end
29  finalize opencl;
```

Algorithm 3. The dynamic LB algorithm

HIS was chosen because it is a representative of DLP algorithm: the same set of operations must be executed in a large amount of data.

In order to evaluate the performance, the following parameters of the load-balancing algorithms has been varied: (a) the percentage of time-steps n and p used to measure T_{CPU} and T_{GPU} and (b) the threshold t. We also evaluated the impact in performance of the total number of time-steps to be simulated. This section starts describing the benchmark used to compare the load-balancing algorithms. Next, the computational environment is described. Finally, the results obtained are presented.

3.1 Benchmark

A three dimensional HIS simulator [9,10] was used to evaluate the performance of the two load-balancing algorithms. The simulator implements a mathematical model that uses a set of eight Partial Differential Equations (PDEs) to describe how some cells and molecules involved in the innate immune response, such as neutrophils, macrophages, protein granules, pro- and anti-inflammatory cytokines, reacts to an antigen. The mathematical model simulates the temporal and spatial behavior of the antigen, as well as the immune cells and molecules. In the model, an antigen is represented by the lipopolysaccharide. The diffusion of some cells and molecules are described by the mathematical model, as well as the process of chemotaxis. Chemotaxis is the movement of immune cells in response to chemical stimuli by pro-inflammatory cytokine. Neutrophils and macrophages move towards the gradient of pro-inflammatory cytokine concentration. A detailed discussion about the mathematical model can be found in [9], and details about its implementation can be found in [10]. This previous work used C and CUDA in the implementation, using just GPUs in the computation, while this work uses C and OpenCL, using both CPUs and GPUs available in the APU in the computation.

The numerical methods used in this work are regular but requires that, at each time-step, the kernels that execute on CPUs have access to the boundary points computed by the GPUs on the previous time-step, and vice-versa.

3.2   Computational Platform

Computational experiments were performed on an A10-5800K Radeon APU. A10-5800K is composed by one CPU and one GPU. The CPU has four 3.8 GHz cores, with 16 KB of L1 data cache per core, and 2×2 MB of L2 cache, so two cores share a single L2 cache. The GPU has 384 cores running at 800 MHz. The system has 16 GB of main memory, 2 GB of which are assigned to the exclusive use of the GPU. Unfortunately this APU model does not allow true memory sharing between CPU and GPU, in the sense that memory operations such as loads and stores cannot be used to establish direct communication between processes running on the CPU and GPU. Instead, explicit memory copy operations implemented in OpenCL API must be used to exchange data between processes on CPU and GPU. The machine runs Linux 3.11.0-15. OpenCL version 1.2 AMD and gcc version 4.6.3 were used to compile the codes.

3.3   Results

An additional feature that was implemented in this work is the way data division is implemented. In our previous work [8], the amount of data that were assigned to CPUs and GPUs considered the number of planes in $X \times Y \times Z$, not the total number of elements to be computed. In order to minimize data transfer due to the exchange of boundaries, the percentage of data (P_{GPU}, P_{CPU}, and P_G) were always applied to the plane with smaller size. Since a plane is composed by thousands of elements, a unique plane could be responsible for the imbalance between CPUs and GPUs [8]. This work implements another way to divide data, using individual elements, instead of planes. Figure 1 depicts these two ways to divide data.

In previous works [7,8], the percentage of time-steps used to measure T_{CPU} and T_{GPU} in both algorithms was fixed in 1%. In this work, in order to execute the sensitivity analysis, this value was varied from 1% to 10%, in both algorithms. Also, for the dynamic load-balancing algorithm, the value of the threshold, that was also fixed at 1% in [8], was varied from 0.0025% to 10%. Two distinct simulation time-steps values were used in the simulation: 10, 000 and 1, 000, 000. In order to evaluate the impact of the data size in performance, two mesh sizes were used in the evaluation: $50 \times 50 \times 50$ and $200 \times 200 \times 200$. We also evaluate

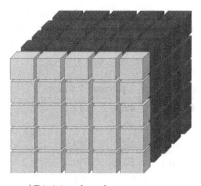

a)Division by planes.   b)Division by individual elements.

Fig. 1. Two ways to divide data: division by planes (a) and division by individual elements (b). The division by individual elements allows the algorithm to use of a fine-grain data partition in the LB.

the impacts of the way data is partitioned between CPUs and GPUs in the dynamic LB algorithm. The results for the version that uses the division of data in planes will be referred as coarse grain version, while the version that divides data by elements will be referred as fine grain version. Since the mesh sizes are equal in all dimensions, we choose the Z plane to divide data in the coarse grain version. The choice of other dimension would not impact performance, since the same amount of data transfers due to boundaries exchange would occur. The HIS application was executed at least 3 times for each LB algorithm, and all standard deviations of the execution time were below 1.7%.

The values used for initial conditions and to set all parameters used in the simulations are the same used in our previous work [7]. The parallel version of the code that uses only CPUs, without using our load-balancing algorithms, executes in 324 s and 13, 320 s for meshes of size $50 \times 50 \times 50$ and $200 \times 200 \times 200$, respectively. The parallel version of the code that uses only GPUs, also without using our load-balancing algorithms, executes in 59 s and 3, 576 s for meshes of size $50 \times 50 \times 50$ and $200 \times 200 \times 200$, respectively.

Tables 1 and 2 present the parallel results obtained using a total of 10, 000 simulation time-steps for meshes of size $50 \times 50 \times 50$ and $200 \times 200 \times 200$, respectively. Each line of the tables presents the results for distinct values of n and p used for the percentage of time-steps in the static and dynamic algorithms, respectively. Each column represents the distinct versions: static algorithm using fine-grain data partition (SF), static algorithm using coarse-grain data partition (SC), dynamic algorithm using fine-grain data partition (DF) and dynamic algorithm using coarse-grain data partition (DC). In the case of the dynamic algorithm, each column presents the execution time obtained for a distinct threshold value. In the fine-grain versions, the threshold values were chosen in the sensitivity analysis to represent a single mesh line (composed by 50 elements for a mesh of size $50 \times 50 \times 50$ or 200 elements for a mesh of size $200 \times 200 \times 200$), a quarter

Table 1. Execution time using a total of 10, 000 simulation time-steps for a mesh of size 50 × 50 × 50. All results are in seconds. The values in boldface are the best results obtained.

Time-steps	SF	SC	DF			DC		
			Line	Quarter	Half	1 plane	3 planes	5 planes
1%	50.9	51.5	**48.4**	**48.4**	**48.4**	50.1	49.3	49.3
2%	50.4	52.0	48.5	**48.4**	48.5	50.0	49.5	49.5
3%	50.5	52.2	48.7	48.6	48.7	50.5	49.8	49.8
5%	50.5	52.7	48.8	49.0	49.1	50.1	50.2	50.3
10%	51.1	52.0	50.0	50.2	50.4	51.2	51.4	51.3

Table 2. Execution time using a total of 10, 000 simulation time-steps for a mesh of size 200 × 200 × 200. All results are in seconds. The value in boldface is the best result obtained.

Time-steps	SF	SC	DF			DC		
			Line	Quarter	Half	1 plane	3 planes	5 planes
1%	2,995.7	2,934.0	**2,729.8**	2,802.4	2,827.2	2,825.1	2,799.8	2,823.0
2%	3,027.3	2,986.2	2,754.6	2,786.5	2,796.4	2,820.6	2,794.5	2,803.4
3%	2,833.3	2,841.5	2,764.0	2,797.6	2,793.4	2,821.1	2,809.8	2,800.2
5%	2,832.3	2,856.6	2,776.7	2,815.9	2,815.0	2,818.5	2,820.6	2,819,9
10%	2.896.1	2,893.1	2,837.5	2,855.7	2,855.7	2,880.6	2,881.1	2,876.9

of the plane (625 or 10, 000 elements) or half a plane (1, 250 or 20, 000 elements). In the coarse-grain versions, the threshold values were chosen in the sensitivity analysis to represent one plane (composed by 2, 500 elements for a mesh of size 50 × 50 × 50 or 40, 000 elements for a mesh of size 200 × 200 × 200), three planes (7, 500 or 120, 000 elements) or five planes (12, 500 or 200, 000 elements).

As one can observe, for all mesh sizes, the best values are obtained using the dynamic algorithm using fine-grain data partition (DF): 1% of the time-steps and a single mesh line seems to be an adequate configuration to obtain a good result for this benchmark. If the best values obtained by the dynamic algorithm (48.4 and 2, 729.8) are compared with the best values obtained by the static one (50.4 and 2, 832.3), we observe performance gains vary from 3.7% to 4.2%. Compared to the parallel versions that do not use the LB, the dynamic algorithm executes 6.7 times faster than the parallel CPU version and 1.2 times faster than the parallel GPU version for the small mesh. For the larger mesh, the dynamic algorithm executes 4.9 times faster than the parallel CPU version and 1.3 times faster than the parallel GPU version.

The best performance obtained by the dynamic algorithm surprised us because of its memory copies and reallocation costs. At first, we would expected that the static algorithm would outperform the dynamic one. Neither memory access times nor cache misses explain the dynamic LB algorithm performance

Table 3. Execution time using a total of 10,000 simulation time-steps for a mesh of size $50 \times 50 \times 50$, initialized with non-zero values. All results are in seconds.

Time-steps	SF	SC	DF			DC		
			Line	Quarter	Half	1 plane	3 planes	5 planes
1%	62.1	61.7	**53.8**	59.9	67.7	67.9	67.9	66.5
2%	62.5	62.3	54.8	60.0	60.2	59.6	64.9	64.9
3%	62.5	61.6	55.6	55.7	60.9	60.4	63.0	63.0
5%	61.8	61.6	57.6	62.8	62.4	62.0	61.5	61.5
10%	62.7	62.2	60.0	60.0	60.1	63.1	63.0	63.0

Table 4. Execution time using a total of 10,000 simulation time-steps for a mesh of size $200 \times 200 \times 200$, initialized with non-zero values. All results are in seconds.

Time-steps	SF	SC	DF			DC		
			Line	Quarter	Half	1 plane	3 planes	5 planes
1%	6736,4	6169,7	**4692,3**	4811,3	4753,3	4698,8	5164,0	6812,3
2%	6129,6	5884,4	4696,5	4766,2	4750,0	4791,9	4898,2	6710,8
3%	5879,7	5912,3	4794,0	4737,5	4784,3	4838,3	5203,7	5202,7
5%	5723,7	5794,9	5021,1	4932,6	5060,3	5015,3	4924,6	6724,3
10%	5607,6	5602,1	5347,1	5339,9	5345,8	5358,8	5379,6	6473,8

since the values are closer to the ones obtained by the static algorithm. We further investigate the cause of this problem and discovered that this effect occurs due to an imbalance in the execution time in the initial and final time-steps of the application. The cause of the imbalance is a hardware optimization done by the CPU: in the beginning of the execution, due to the values used as initial conditions, there are a lot of float-point multiplications by zero, which is detected by the CPU hardware that take an early exit since the execution of the entire processing pipelining is not required to obtain the final result. However, for the non-zero values, the multiplication demands the execution of the entire processing pipelining, which takes longer. So, some OpenCL processing units can be idle while others are very busy. If a slice of the mesh, composed by zero values, is allocated in the CPU, for example, its computing time will be recorded with a lower value, and the CPU will receive more data than it can in fact handle. We observed that the number of non zero values increase as time goes by, so this hardware optimization is not taken again and the computation time becomes regular, but just the dynamic load-balancing algorithm can adjust to this new information. In the final of the computation, the number of zero values increase again, and both algorithms suffer its effects.

To confirm the impact of the zero values to the imbalance, two final experiments were conducted. First, we executed again the same set of experiments described previously, but replacing the zero in the initial conditions to a tiny

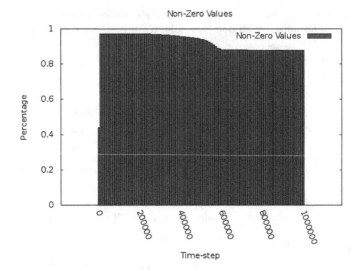

Fig. 2. Non-zero values in a mesh of size $50 \times 50 \times 50$. The results were collected at each $10,000$ time-steps.

value different from zero (1.0×10^{-10}). As a result, the execution time increased, as Tables 3 and 4 show. For the $50 \times 50 \times 50$ mesh, the increase in the execution time ranges from 11% to 39%, while for the $200 \times 200 \times 200$ mesh, the increase ranges from 66% to 141%. This point explains why the dynamic algorithm, despite of its memory costs, performed better than the static algorithm, and confirms that computing with zero values is faster than computing with non-zero values. Finally, we collected the execution time for a mesh of size $50 \times 50 \times 50$, but now running $1,000,000$ time-steps. We also collected the number of non-zero positions in the mesh, as time goes by. The results were collected at each $10,000$ simulation steps. Figure 2 presents the results. One can observe that in the first $10,000$ time-steps, the number of non-zero values is equal to 44%; this number increases during the execution to 97% and then drops to 88%.

4 Conclusion and Future Works

The heterogeneous APU architecture imposes new challenges to data parallel applications that want to take advantage of all the processing units available on the hardware to minimize its execution time. LB algorithms can help in this task. This work has presented an extension done on two LB algorithms, a static and a dynamic ones, in order to reduce the costs of their probe phases. We have also presented a sensitivity analysis of the parameters used in our models. The results have shown that the dynamic algorithm using fine-grain data partition (one line) and 1% of the time-steps to probe the best data division, obtained the best performance. This configuration outperformed in 4.2% the best values obtained by the static algorithm. Compared to the parallel versions of the benchmark that do not use the LB techniques, the dynamic algorithm executes up to 6.7 times faster. We have also shown that applications that seems to be regular parallel applications, such as the one used in the performance evaluation, suffer from irregular execution time phases during their execution due to hardware optimizations done in the CPU, which impacts the static LB algorithm. We plan, as future works, to evaluate the proposed LB algorithms using other benchmarks as well as in other architectures, such a heterogeneous cluster composed by distinct CPUs and GPUs.

References

1. The OpenACC application programming interface - version 2.5. Technical report (2015). OpenAcc.org
2. Branover, A., Foley, D., Steinman, M.: AMD fusion APU: Llano. IEEE Micro **32**(2), 28–37 (2012)
3. Hennessy, J.L., Patterson, D.A.: Computer Architecture: A Quantitative Approach, 5th edn. Morgan Kaufmann Publishers Inc., San Francisco (2011)
4. Kirk, D.B., Hwu, W.W.: Programming Massively Parallel Processors: A Hands-on Approach, 2nd edn. Morgan Kaufmann Publishers Inc., San Francisco (2013)
5. Mattson, T., Sanders, B., Massingill, B.: Patterns for Parallel Programming, 1st edn. Addison-Wesley Professional, Reading (2004)
6. Munshi, A., Gaster, B., Mattson, T.G., Fung, J., Ginsburg, D.: OpenCL Programming Guide, 1st edn. Addison-Wesley Professional, Reading (2011)
7. do Nascimento, T.M., de Oliveira, J.M., Xavier, M.P., Pigozzo, A.B., dos Santos, R.W., Lobosco, M.: On the use of multiple heterogeneous devices to speedup the execution of a computational model of the human immune system. Appl. Math. Comput. **267**, 304–313 (2015)
8. do Nascimento, T.M., dos Santos, R.W., Lobosco, M.: On a dynamic scheduling approach to execute opencl jobs on apus. In: Osthoff, C., Navaux, P.O.A., Barrios Hernandez, C.J., Silva Dias, P.L. (eds.) CARLA 2015. CCIS, vol. 565, pp. 118–128. Springer, Cham (2015). doi:10.1007/978-3-319-26928-3_9
9. Pigozzo, A.B., Macedo, G.C., Santos, R.W., Lobosco, M.: On the computational modeling of the innate immune system. BMC Bioinform. **14**(Suppl. 6), S7 (2013)
10. Rocha, P.A.F., Xavier, M.P., Pigozzo, A.B., M. Quintela, B., Macedo, G.C., Santos, R.W., Lobosco, M.: A three-dimensional computational model of the innate immune system. In: Murgante, B., Gervasi, O., Misra, S., Nedjah, N., Rocha, A.M.A.C., Taniar, D., Apduhan, B.O. (eds.) ICCSA 2012. LNCS, vol. 7333, pp. 691–706. Springer, Heidelberg (2012). doi:10.1007/978-3-642-31125-3_52

Accelerated Analysis of Biological Parameters Space Using GPUs

Marco S. Nobile[1,2(✉)] and Giancarlo Mauri[1,2]

[1] Department of Informatics, Systems and Communication,
University of Milano-Bicocca, Viale Sarca 336, 20126 Milano, Italy
nobile@disco.unimib.it
[2] SYSBIO.IT Centre of Systems Biology, Milano, Italy

Abstract. Mathematical modeling and computer simulation represent a valuable mean to integrate experimental research for the study of biological systems. However, many computational methods—e.g., sensitivity analysis—require the execution of a massive number of simulations to investigate the model behavior in physiological or perturbed conditions, which can be a computationally challenging task. This huge amount of simulations is necessary to collect data in the vast space of kinetic parameters. This paper provides the state-of-the-art of biochemical simulators relying on Graphics Processing Units (GPUs) in the context of Systems Biology. Moreover, we discuss two examples of integration of such simulators into computational methods for parameter sweep and sensitivity analysis, both implemented using the Python language.

1 Introduction

Computational investigation of biological systems is nowadays a valuable and integrative approach to classic laboratory experiments. Specifically, mechanism-based modeling and simulation represents the most likely candidate to achieve a detailed comprehension of cellular processes [1]. Unfortunately, sophisticated techniques created on top of this approach—e.g., sensitivity analysis [2] or parameter estimation [3,4]—are often characterized by huge computational costs, since they require large amounts of simulations, hampering their practical applicability. This huge amount of simulations is necessary, because both methods require the exploration of the kinetic parameters space, in order to determine how the system reacts or which parameterizations are closer to biological reality. In principle, the problem can be mitigated by the combined use of parallel computing architectures and advanced simulation algorithms.

There exist multiple options for parallel computation: compute clusters, grid computing, cloud computing, many-cores architectures like Intel MICs and massively multi-cores architectures like modern Graphics Processing Units (GPUs). Whereas the other architectures can be exploited with very limited modifications to existing softwares, GPUs are characterized by different programming models, execution hierarchy and memory organization. In particular, GPUs are based on a "relaxed" SIMD (i.e., *Same Instruction, Multiple Data*) paradigm [5],

© Springer International Publishing AG 2017
V. Malyshkin (Ed.): PaCT 2017, LNCS 10421, pp. 70–81, 2017.
DOI: 10.1007/978-3-319-62932-2_6

which radically differs from the traditional multi-threaded execution, so that algorithms generally require a complete redesign to fully leverage their peculiar architecture. Nevertheless, the reasons why GPUs represent an attractive opportunity are manifold:

- GPUs are massively multi-core, integrating thousands computation units (e.g., Nvidia's GeForce Titan Z contains 2×2880 cores);
- thanks to the high number of cores, GPUs provide access to tera-scale performances on common workstations (e.g., the GPU Nvidia GeForce Titan X has a theoretical peak power of 10.9 TFLOPS for single precision calculations);
- despite the relevant computational power, GPUs are relatively cheap and more power-efficient with respect to other architectures. For instance, the GPU Nvidia GeForce 1080 costs $ 599 at the time of writing, has a computational power of 9 TFLOPS and a power consumption of 180 W;
- GPUs allow to keep the data "local", without the need for data transfers and job scheduling;
- it is possible to create compute clusters based on GPUs, like TITAN at the Oak Ridge National Laboratory [6], credited as the third fastest supercomputer in the planet at the time of writing.

All these advantages counter-balance the difficulty of GPUs programming, motivating the intense ongoing research in the fields of Bioinformatics, Computational Biology and Systems Biology [7]. In the latter years, many GPU-powered methods have been developed, so that life scientists without computational expertise can leverage such implementations for their investigations. This paper presents a survey of the published tools for GPU-powered simulation in Systems Biology. Moreover, the manuscript provides two practical examples of killer-applications implemented in Python: parameter sweep analysis and sensitivity analysis. The goal is to prove the feasibility, simplicity and the computational advantages of the investigation of biochemical systems by means of GPUs.

The paper is structured as follows. Section 2 introduces the basics of biochemical modeling and simulation. Section 3 presents the state-of-the-art of GPU-powered biochemical simulation and related issues. Section 4 shows two examples of how to integrate a GPU-powered simulator into computational methods for the investigation of biological systems. Finally, Sect. 5 concludes the paper with a discussion of the results and potential future developments.

2   Methods

We can formalize a reaction-based model (RBM) of a biochemical system by specifying the set $\mathcal{S} = \{S_1, \ldots, S_N\}$ of chemical species occurring in the system and the set $\mathcal{R} = \{R_1, \ldots, R_M\}$ of chemical reactions [8]. We define reactions as:

$$R_j : \sum_{i=1}^{N} \alpha_{ji} S_i \xrightarrow{k_j} \sum_{i=1}^{N} \beta_{ji} S_i, \tag{1}$$

where $\alpha_{ji}, \beta_{ji} \in \mathbb{N}$ are the stoichiometric coefficients associated, respectively, with the i-th reactant and the i-th product of the j-th reaction, with $i = 1, \ldots, N$, $j = 1, \ldots, M$. The value $k_j \in \mathbb{R}^+$ is the kinetic (or stochastic) constant associated to reaction R_j.

By assuming constant volume and temperature, and relying on mass-action kinetics [8], a RBM can be automatically converted to a system of coupled Ordinary Differential Equations (ODEs). As a matter of fact, for each chemical species $s \in S$, it is possible to derive an ODE describing the variation in time of its concentrations, according to the structure of the reactions in which it is involved [9]. This procedure creates a set of N coupled ODEs which can be simulated using ODE numerical solvers [10]. Among the existing integration algorithms, LSODA [11] is one of the most popular, thanks to its capability of dealing with stiff systems by automatically switching between explicit (i.e., the Adams' method) and implicit integration methods (i.e., backward differentiation formulae).

When only a few molecules of chemical species are present, however, ODEs generally fail to capture the emergent effects of biochemical stochastic processes [12], like state-switching or multi-stability. In such a case, stochastic simulation algorithms (e.g., exact methods like Gillespie's Stochastic Simulation Algorithm (SSA) [8] and its optimized variants [13], or approximate methods like tau-leaping [14,15]) can be employed to produce an accurate simulated trajectory of the RBMs. Since two independent stochastic simulations of the same model yield different outcomes, the single output dynamics is rarely informative of a system's overall behavior. Hence, multiple repetitions are usually necessary to assess the empirical distribution of the underlying Chemical Master Equation.

Regardless the simulation approach, the computational investigation of biological systems is a computationally challenging task, because of the huge space of possible model parameterizations which give rise to possibly different emergent behaviors. This circumstance motivates the need for advanced computing architectures, most notably GPUs, described in the following section.

3   GPU-Powered Simulators

The simulation of mathematical models allows to determine the quantitative variations of molecular species in time and in space. Simulation can be performed with deterministic, stochastic or hybrid algorithms [12]; the last typology is usually based on a (possibly dynamic) partitioning of reactions that allows to limit the use of pure stochastic simulation, which is more computationally expensive than plain ODEs integration [16–18]. In any case, the simulation method should be chosen according to the scale of the modeled system (i.e., number of molecules) and the possible role of biological noise, not according to performance motivations[1].

---

[1] Simulations can also be extended to keep track of the spatial positioning of species, which is mandatory in the case of biological system whose components are not uniformly distributed in the reaction volume or its compartments. Spatial simulation will not be discussed in this paper.

There are two possible ways to leverage GPUs in the case of biological simulation: *fine*-grained and *coarse*-grained simulation. In the first case, the calculations of a single simulation are distributed over the available resources, to accelerate the simulation of a large model; in the second case, a large number of parallel simulations (using the same or different parameterizations) are executed as individual threads and distributed over the available GPU cores. In the following section, the state-of-the-art of deterministic and stochastic GPU-powered simulation is summarized.

3.1 Deterministic Simulation

When the concentrations of molecular species are high, and the effect of noise can be neglected, ODEs represent the typical modeling approach for biological systems. Given a model parameterization (i.e., the initial state of the system and the set of kinetic parameters), the output dynamics of a system is calculated using some numerical integrator [10].

Ackermann *et al.* [19] proposed a GPU-powered coarse-grained simulator, designed to execute massively parallel simulations of biological molecular networks. Their method automatically converts a model, formalized using the Systems Biology Markup Language (SBML), into a CUDA implementation of the Euler numerical integrator able to simulate that specific system. The execution on a Nvidia GeForce 9800 GX2 showed a speed-up between 28× and 63×, compared to the execution on a CPU Intel Xeon 2.66 GHz (not leveraging any multi-threading nor vector instructions). In a similar vein, a CUDA implementation of the LSODA algorithm, named cuda-sim, was presented by Zhou *et al.* [20]. The cuda-sim simulator performs the so-called "just in time" (JIT) compilation (that is, the creation, compilation and linking at *run-time* of automatically created source code) by converting a SBML model into CUDA code. Authors claim that cuda-sim achieves a 47× speed-up, with respect to a serial execution performed with the LSODA implementation contained in the numpy library [21]. Nobile *et al.* [22] created an alternative coarse-grained simulator based on the LSODA algorithm, named cupSODA, designed to execute a large number of simultaneous deterministic simulations and accelerate typical tasks like parameter estimation and reverse engineering. Given a RBM model, cupSODA automatically determines the corresponding system of ODEs and the related Jacobian matrix, assuming a strictly mass-action kinetics. Differently from cuda-sim, though, cupSODA prevents JIT compilation, relying on a GPU-side parser. Using this strategy, cupSODA achieves an acceleration up to 86× with respect to COPASI [21], here considered as reference CPU-based LSODA biochemical simulator.

Considering the fine-grained simulation, the only simulator is LASSIE, proposed by Tangherloni *et al.* [23], LASSIE, is able to leverage implicit and explicit integrations methods, and distributes the calculations that are required to simulate a large-scale model characterized by thousands reactions and chemical species over the available GPU cores. Using a GPU Nvidia GeForce GTX Titan Z, LASSIE outperforms a sequential simulation running on a CPU Intel i7-4790K 4.00 GHz, achieving up to 90× speed-up.

3.2 Stochastic Simulation

When the effect of biological noise cannot be neglected, randomness can be described either by means of Stochastic Differential Equations (SDEs), which extend ODEs with an additional term representing a stochastic process. Alternatively, the biochemical process can be defined by using explicit mechanistic models, such as RBMs [8]. In this second case, the temporal evolution of the species can be simulated by means of Monte Carlo procedures like SSA [8].

A CUDA version of SSA was developed by Li and Petzold [24]: it achieved a $50\times$ speed-up with respect to a common single-threaded CPU implementation. Sumiyoshi et al. [25] extended this methodology by performing a combined coarse- and fine-grained parallelization. This version achieved a $130\times$ speed-up with respect to the sequential simulation on the host computer. Finally, Komarov and D'Souza [26] designed GPU-ODM, a fine-grained simulator of large-scale models based on SSA, which exploits special data structures and CUDA functionalities. Thanks to these optimizations, GPU-ODM outperformed the most advanced (even multi-threaded) CPU-based implementations of SSA.

The τ-leaping algorithm allows a faster generation of the dynamics of stochastic models with respect to SSA, by calculating multiple reactions over longer simulation steps [15,27]. Komarov et al. [28] proposed a fine-grained τ-leaping implementation, which becomes efficient in the case of extremely large biochemical networks (i.e., characterized by more than 10^5 reactions). Nobile et al. [29] proposed cuTauLeaping, a coarse-grained implementation of the optimized version of τ-leaping proposed by Cao et al. [15]. Thanks to the optimization of data structures in low-latency memories, the use of advanced functionalities and the splitting of the algorithm into multiple phases, cuTauLeaping was up to three orders of magnitude faster on a GeForce GTX 590 GPU than the CPU-based implementation of τ-leaping contained in COPASI [21], executed on a CPU Intel Core i7-2600 3.4 GHz.

4 Applications of GPU-Powered Simulators

This section describes two examples of applications of GPU-powered simulators: parameter sweep and sensitivity analysis. All tests were performed on a GPU Nvidia GeForce Titan Z, 2880 cores, 876 MHz. The CPU was an Intel i7-3537U, 2.5 GHz. The OS was Microsoft Windows 8.1 Enterprise 64 bit, CUDA v7.5 with driver 353.90, COPASI v4.15 build 95. In all tests, we exploit the cuTauLeaping simulator [29], v1.0.0, which can be downloaded from GitHub (https://github.com/aresio/cuTauLeaping). In order to replicate the experiments, it is necessary to also download the following files from the aforementioned GitHub repository:

- the python script for the conversion of SBML files to cuTauLeaping's input format `SBML2BSW.py`;
- the example SBML file `schloegl.xml`.

We want to stress the fact that the experiments can be performed on different models, provided that the SBML is based on mass-action kinetics. For the sake of simplicity, in all tests that follow, it is assumed that the cuTauLeaping executable binary file is copied in the input folders created by SBML2BSW.py script.

4.1 Parameter Sweep Analysis

Parameter Sweep Analysis (PSA), also known as parameter scan analysis, is the systematic modification of a model's parameterization in order to investigate the impact of one or more synergistic perturbations. PSA represents the most straightforward application of coarse-grained simulation: the GPU loads the multiple parameterizations of the model and performs the simulations in parallel.

In order to present a simple example of GPU-powered PSA on a stochastic model, we exploit cuTauLeaping [29]. This tool does not natively import SBML files and relies on its proprietary format; however, a simple python script can be used to convert an arbitrary SBML into the proprietary format. The conversion is performed with the following statement:

```
> python SBML2BSW.py <sbmlfile.xml> <output_directory>
```

where sbmlfile.xml is the input SBML file to be converted and output_directory is the directory that will contain the input files for cuTauLeaping. This command will create the base of the model that will be used for the parallel simulations. However, in order to perform a PSA, we need to add four additional files to the input folder:

- c_matrix: this file must have $32 \times B$ lines, $B \in \mathbb{N}_{>0}$, one for each parameterization to be tested. Each line has M values, one for each reaction, defining the values of the stochastic constants in that specific parameterization. All values must be tab-separated;
- MX_0: this file must have $32 \times B$ lines, $B \in \mathbb{N}_{>0}$, one for each parameterization to be tested. Each line has N values, one for each chemical species, defining the values of the initial amounts in that specific parameterization. All values must be tab-separated;
- M_feed (optional): this file must have $32 \times B$ lines, $B \in \mathbb{N}_{>0}$, one for each parameterization to be tested. Each line has N values, one for each chemical species, defining the initial amounts of the species whose amount must be fixed (i.e., constant) throughout the simulation, in that specific parameterization. All values must be tab-separated;
- t_vector: this file has $T \in \mathbb{N}_{>0}$ rows and specifies the number T of times instants in which the chemical species must be sampled.

These four files can be programmatically generated using any programming language. In the case of stochastic models, a subset of the parameterizations can be identical, in order to assess the frequency distributions of some chemical species, given a specific parameterization.

When the input files are completed, we can finally launch cuTauLeaping and execute the parallel simulations. cuTauLeaping is a command-line tool which can be run using the following arguments:

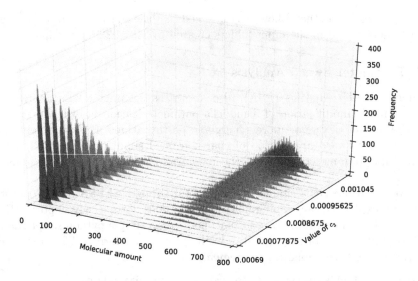

Fig. 1. Example of GPU-powered stochastic PSA on the Schlögl model.

```
> ./cuTauLeaping input_dir tpb bpg gpu 0 output_dir pref
```

where input_folder is the directory which contains the input files; cuTauLeaping will perform tpb × bpg simulations, organized in bpg blocks, each block
composed of tpb threads; output_dir is the directory in which output files will
be written; the files will be named pref_z, with $z = 0, \dots, \text{tpb} \times \text{bpg} - 1$. One
example is reported in Fig. 1, which shows a PSA on the Schlögl model [30], a
known system characterized by bi-stability. In this experiment, a single parame-
ter (i.e., the kinetic constant c_3) is varied in the interval $[6.9 \cdot 10^{-4}, 1.4 \cdot 10^{-3}]$.
Specifically, 20 different values in this interval are selected and used to perform
2^{13} simulations, for a total of 163 840 simulations. Figure 1 shows the frequency
distributions of the chemical species "X" at time $t = 10$ [a.u.], obtained with each
parameterization, highlighting the change of the bi-modal distribution using different values for c_3. The total running time to execute this PSA-1D was 7.34 s
with cuTauLeaping, and 911.74 s with the tau-leaping implementation contained
in the software COPASI [21], thus achieving a speed-up of more than two orders
of magnitude on GPUs[2].

4.2 Sensitivity Analysis

Sensitivity Analysis (SA) is a computational method which investigates how the
uncertainty in the output of a given mathematical model can be apportioned
to different sources of uncertainty in its inputs [31]. The result of a SA is generally a ranking of the sensitivity coefficients. A variety of SA methods can be

---

[2] Notably, the creation of the figure required approximatively 3 min.

employed for the investigation of biochemical systems (e.g., elementary-effects [32,33], variance-based sensitivity [34], derivative-based sensitivity [35]).

Even though SA represents a powerful means to understand a system's behavior, a proper investigation of the multi-dimensional parameters space implies a combinatorial explosion of kinetic configurations to be simulated. Fortunately, since all simulations are mutually independent, the huge computational effort can be mitigated by means of a parallel architecture and, notably, by adopting coarse-grained GPU-powered simulators.

In Listing 1.1, it is shown how the stochastic simulator cuTauLeaping can be integrated with the python library SAlib [36] to perform a SA of the Schlögl model using Sobol's method (line 41), generating $2048 \cdot (4 \cdot 2 + 2) = 20480$ samples of the search space using Saltelli's approach (line 16), considering boundaries of the parameters space equal to $\pm 10\%$ the nominal value (rows 9–10). The SA will determine the ranking of the sensitivity of stochastic parameters, with respect to the amount of chemical species "X" (lines 27), here considered as the output of the SA. We exploit the input directory created in the previous section, creating *ex novo* new coarse-grained input files (rows 21–27). Finally we determine and print the sensitivity coefficients (rows 41–42).

Listing 1.1. Example of SA performed with SAlib and cuTauLeaping.

```
1  from SALib.sample import saltelli
2  from SALib.analyze import sobol, morris
3  import numpy as np
4  from subprocess import check_output
5
6  problem = {
7      'num_vars' : 4 ,
8      'names'    : [ 'c1' , 'c2' , 'c3' , 'c4' ],
9      'bounds'   : [[2.7e-7, 3.3e-7], [9e-5, 1.1e-4],
10                    [9e-4, 1.1e-3], [3.15, 3.85]]
11 }
12
13 N = 2048
14
15 # Generate N*(num_vars*2+2) samples
16 param_values = saltelli.sample(problem, N, calc_second_order=True)
17 np.savetxt ("c_matrix" , param_values , delimiter="\t")
18
19 # Generate cuTauLeaping's input files
20 MX_0 = [[100000, 250, 200000]]*len(param_values)
21 np.savetxt ("MX_0", MX_0, delimiter="\t", fmt="%d\t%d\t%d")
22 MX_feed = [[100000, 0, 200000]]*len(param_values)
23 np.savetxt ("MX_feed", MX_feed, delimiter="\t", fmt="%d\t%d\t%d")
24 t_vector = [5.0]
25 np.savetxt ("t_vector", t_vector)
26 np.savetxt ("cs_vector", [1])
27
28 # Run model with cuTauLeaping
29 BPG = len(param_values) / 32
30 ret = check_output(["../cuTauLeaping.exe", ".", "32", str(BPG),
31       "0", "0", "output", "sch_sa", "0", "0", "2"])
32 ret = ret.split("\r\n") # split rows
33 ret = map(lambda x: x.split("\t"), ret) # split columns
34 ret = filter(lambda x: len(x)>1, ret) # remove empty lines
35
36 # Read back the results
37 Y = np.empty([param_values.shape [0]])
38 for i, X in enumerate( param_values ):
39     Y[i]=ret[i][1] # take amount and discard time
40
41 # Perform analysis with Sobol's method and
42 # print first-order sensitivity indices
43 Si = sobol.analyze(problem, Y, print_to_console=True)
44 print Si ['S1']
```

The sensitivity indices of the four stochastic constants, calculated by SAlib, are equal to: 0.389, 0.115, 0.025 and 0.319, corresponding to the ranking

c_1, c_4, c_2, c_3, which is consistent with published literature [37]. The integration of GPU-powered simulation in SAlib is pretty straightforward, and allows a relevant reduction of the overall running time (less than 10 s, including input files generation).

5   Discussion and Conclusions

General-purpose GPU computing represents a powerful alternative to classic parallel or distributed computing (e.g., multi-threading, compute clusters, grid computing) for scientific computation. GPUs are massively power-efficient multicore devices, cheaper than classic parallel architectures, providing tera-scale performances on common workstations. Moreover, since GPUs keep the computation "local", they prevent many issues affecting the alternative field of cloud computing: security, privacy, data "lock-in", and sensible data transfer.

Although GPUs are difficult to program—due to their peculiar memory and execution hierarchies—many GPU-powered tools are already available for Systems Biology, Computational Biology, and Bioinformatics [7]. This paper provides a survey of GPU-powered biochemical simulators, summarizing the deterministic and stochastic simulators presented in literature. The article also shows two applications of coarse-grained simulation—i.e., parameter sweep analysis and sensitivity analysis—explaining how to seamlessly integrate a GPU-powered simulator (i.e., cuTauLeaping [29]) into such tools, implemented with the Python language. Thanks to this approach, the execution time of such activities was strongly reduced. It is worth noting that coarse-grained simulation can accelerate further tasks like parameter estimation [3,4] or the reverse engineering of kinetic biochemical models [38,39], in which the structure of the RBM is completely or partially unknown.

The example code provided in this paper shows how the simulators can be easily integrated and exploited to perform useful tasks; however, these activities required the implementation of specific portions of code to create the necessary input files. In order to further simplify the adoption of GPUs, there are some projects aimed to further reduce any residual complexity. For instance, the Computational Systems Biology platform COSYS [40] was designed to provide scientists with a user-friendly visual interface to GPU-powered tools. Morever, the applications of GPUs in Systems Biology are not limited to biochemical simulation: GPUs can indeed be exploited to perform different tasks in Systems Biology [7], notably large-scale network analytics using the nvGRAPH library [41] which can be applied to genome-wide data, like the Cancer Genome Atlas.

It is worth noting that there exists a gap in the state-of-the-art of GPU-powered biochemical simulation, represented by the lack of hybrid simulation methods. These simulators partition the reactions set into "slow" and "fast" regimes (including meso-scale regimes [16]), which are then simulated using deterministic and stochastic approaches [17,18]. Hybrid simulation is more adequate with such models than pure "exact" stochastic simulation for two reasons: *(i)* stochastic simulation proceeds by calculating one reaction at a time, with a time-step that is inversely proportional to the number of molecules. When some chemical species have high concentrations, stochastic simulations becomes unfeasible; *(ii)* hybrid simulation is also more accurate than strict deterministic simulation, as the latter completely loses the emergent phenomena due to stochasticity. A GPU-powered hybrid simulator—able to dynamically re-partition the reactions set to adapt to system's behavior—would easily take the place of all existing simulators, introducing a methodology to efficiently simulate any kind of RBM without the need for any modeling and simulation expertise.

The cuTauLeaping simulator considered in our tests is composed of three different kernels: one kernel implements the tau-leaping simulations; one kernel implements SSA (performed when the estimation of τ value is too small); a third kernel verifies the termination criterion for the simulation and terminates the run. Thus, the CPU is basically unused during kernels' execution and its only responsibility is to transfer data between the host and the device at the beginning and at the end of the simulations, respectively. As a future improvement of the simulator, we will consider the use of multi-threading to leverage the additional cores on the CPU to further increase the level of parallelism.

References

1. Cazzaniga, P., Damiani, C., Besozzi, D., Colombo, R., Nobile, M.S., Gaglio, D., Pescini, D., Molinari, S., Mauri, G., Alberghina, L., Vanoni, M.: Computational strategies for a system-level understanding of metabolism. Metabolites **4**, 1034–1087 (2014)
2. Saltelli, A., Ratto, M., Andres, T., Campolongo, F., Cariboni, J., Gatelli, D., Saisana, M., Tarantola, S.: Analysis, Global Sensitivity Analysis: The Primer. Wiley-Interscience, Hoboken (2008)
3. Moles, C.G., Mendes, P., Banga, J.R.: Parameter estimation in biochemical pathways: a comparison of global optimization methods. Genome Res. **13**(11), 2467–2474 (2003)
4. Nobile, M.S., Besozzi, D., Cazzaniga, P., Mauri, G., Pescini, D.: A GPU-based multi-swarm PSO method for parameter estimation in stochastic biological systems exploiting discrete-time target series. In: Giacobini, M., Vanneschi, L., Bush, W.S. (eds.) EvoBIO 2012. LNCS, vol. 7246, pp. 74–85. Springer, Heidelberg (2012). doi:10.1007/978-3-642-29066-4_7
5. Nvidia: Nvidia CUDA C Programming Guide 8.0 (2016)
6. Bland, A.S., Wells, J.C., Messer, O.E., et al.: Titan: early experience with the Cray XK6 at oak ridge national laboratory. In: Proceedings of Cray User Group Conference (CUG 2012) (2012)
7. Nobile, M.S., Cazzaniga, P., Tangherloni, A., Besozzi, D.: Graphics processing units in bioinformatics, computational biology and systems biology. Brief. Bioinform. (2016)
8. Gillespie, D.T.: Exact stochastic simulation of coupled chemical reactions. J. Comput. Phys. **81**, 2340–2361 (1977)
9. Wolkenhauer, O., Ullah, M., Kolch, W., Kwang-Hyun, C.: Modeling and simulation of intracellular dynamics: choosing an appropriate framework. IEEE Trans. Nanobiosci. **3**(3), 200–207 (2004)

10. Butcher, J.C.: Numerical Methods for Ordinary Differential Equations. Wiley, Chichester (2003)
11. Petzold, L.: Automatic selection of methods for solving stiff and nonstiff systems of ordinary differential equations. SIAM J. Sci. Stat. Comput. **4**, 136–148 (1983)
12. Wilkinson, D.: Stochastic modelling for quantitative description of heterogeneous biological systems. Nat. Rev. Genet. **10**(2), 122–133 (2009)
13. Gibson, M.A., Bruck, J.: Efficient exact stochastic simulation of chemical systems with many species and many channels. J. Phys. Chem. A **104**(9), 1876–1889 (2000)
14. Rathinam, M., Petzold, L.R., Cao, Y., Gillespie, D.T.: Stiffness in stochastic chemically reacting systems: the implicit tau-leaping method. J. Chem. Phys. **119**, 12784–12794 (2003)
15. Cao, Y., Gillespie, D.T., Petzold, L.R.: Efficient step size selection for the tau-leaping simulation method. J. Chem. Phys. **124**(4), 044109 (2006)
16. Re, A., Caravagna, G., Pescini, D., Nobile, M.S., Cazzaniga, P.: Approximate simulation of chemical reaction systems with micro, meso and macro-scales. In: Proceedings of the 13th International Conference on Computational Intelligence Methods for Bioinformatics and Biostatistics (CIBB2016) (2016)
17. Harris, L.A., Clancy, P.: A "partitioned leaping" approach for multiscale modeling of chemical reaction dynamics. J. Chem. Phys. **125**(14), 144107 (2006)
18. Salis, H., Kaznessis, Y.: Accurate hybrid stochastic simulation of a system of coupled chemical or biochemical reactions. J. Chem. Phys. **122**(5), 054103 (2005)
19. Ackermann, J., Baecher, P., Franzel, T., Goesele, M., Hamacher, K.: Massively-parallel simulation of biochemical systems. In: Proceedings of Massively Parallel Computational Biology on GPUs, Jahrestagung der Gesellschaft für Informatik e.V, pp. 739–750 (2009)
20. Zhou, Y., Liepe, J., Sheng, X., Stumpf, M.P.H., Barnes, C.: GPU accelerated biochemical network simulation. Bioinformatics **27**(6), 874–876 (2011)
21. Hoops, S., Sahle, S., Gauges, R., et al.: COPASI - a COmplex PAthway SImulator. Bioinformatics **22**, 3067–3074 (2006)
22. Nobile, M.S., Besozzi, D., Cazzaniga, P., Mauri, G.: GPU-accelerated simulations of mass-action kinetics models with cupSODA. J. Supercomputing **69**(1), 17–24 (2014)
23. Tangherloni, A., Nobile, M.S., Besozzi, D., Mauri, G., Cazzaniga, P.: LASSIE: simulating large-scale models of biochemical systems on GPUs. BMC Bioinform. **18**(1), 246 (2017)
24. Li, H., Petzold, L.R.: Efficient parallelization of the stochastic simulation algorithm for chemically reacting systems on the graphics processing unit. Int. J. High Perform. Comput. Appl. **24**(2), 107–116 (2010)
25. Sumiyoshi, K., Hirata, K., Hiroi, N., et al.: Acceleration of discrete stochastic biochemical simulation using GPGPU. Front. Physiol. **6**(42) (2015)
26. Komarov, I., D'Souza, R.M.: Accelerating the gillespie exact stochastic simulation algorithm using hybrid parallel execution on graphics processing units. PLoS ONE **7**(11), e46693 (2012)
27. Gillespie, D.T., Petzold, L.R.: Improved leap-size selection for accelerated stochastic simulation. J. Chem. Phys. **119**, 8229–8234 (2003)
28. Komarov, I., D'Souza, R.M., Tapia, J.: Accelerating the gillespie τ-leaping method using graphics processing units. PLoS ONE **7**(6), e37370 (2012)
29. Nobile, M.S., Cazzaniga, P., Besozzi, D., et al.: cuTauLeaping: a GPU-powered tau-leaping stochastic simulator for massive parallel analyses of biological systems. PLoS ONE **9**(3), e91963 (2014)

30. Wilhelm, T.: The smallest chemical reaction system with bistability. BMC Syst. Biol. **3**(1), 90 (2009)
31. Saltelli, A., Ratto, M., Tarantola, S., Campolongo, F.: Sensitivity analysis for chemical models. Chem. Rev. **105**, 2811–2827 (2005)
32. Morris, M.D.: Factorial sampling plans for preliminary computational experiments. Technometrics **33**(2), 161–174 (1991)
33. Campolongo, F., Cariboni, J., Saltelli, A.: An effective screening design for sensitivity analysis of large models. Environ. Model. Softw. **22**(10), 1509–1518 (2007). Modelling, computer-assisted simulations, and mapping of dangerous phenomena for hazard assessment
34. Saltelli, A., Annoni, P., Azzini, I., Campolongo, F., Ratto, M., Tarantola, S.: Variance based sensitivity analysis of model output. Design and estimator for the total sensitivity index. Comput. Phys. Commun. **181**(2), 259–270 (2010)
35. Sobol, I.M., Kucherenko, S.: Derivative based global sensitivity measures and their link with global sensitivity indices. Math. Comput. Simul. **79**(10), 3009–3017 (2009)
36. Usher, W., Herman, J., Whealton, C., Hadka, D.: Salib/salib: Launch!, October 2016
37. Degasperi, A., Gilmore, S.: Sensitivity analysis of stochastic models of bistable biochemical reactions. In: Bernardo, M., Degano, P., Zavattaro, G. (eds.) SFM 2008. LNCS, vol. 5016, pp. 1–20. Berlin, Heidelberg (2008). doi:10.1007/ 978-3-540-68894-5_1
38. Nobile, M.S., Besozzi, D., Cazzaniga, P., Pescini, D., Mauri, G.: Reverse engineering of kinetic reaction networks by means of cartesian genetic programming and particle swarm optimization. In: 2013 IEEE Congress on Evolutionary Computation, vol. 1, pp. 1594–1601. IEEE (2013)
39. Koza, J.R., Mydlowec, W., Lanza, G., Yu, J., Keane, M.A.: Automatic computational discovery of chemical reaction networks using genetic programming. In: Džeroski, S., Todorovski, L. (eds.) Computational Discovery of Scientific Knowledge. LNCS, vol. 4660, pp. 205–227. Springer, Heidelberg (2007). doi:10.1007/ 978-3-540-73920-3_10
40. Cumbo, F., Nobile, M.S., Damiani, C., Colombo, R., Mauri, G., Cazzaniga, P.: COSYS: computational systems biology infrastructure. In: Proceedings of the 13th International Conference on Computational Intelligence Methods for Bioinformatics and Biostatistics (CIBB2016) (2016)
41. Nvidia: nvGRAPH v8.0 (2016)

Parallel Models and Algorithms
in Numerical Computation

Fragmentation of IADE Method
Using LuNA System

Norma Alias[1] and Sergey Kireev[2,3(✉)]

[1] Faculty of Science, Ibnu Sina Institute,
Universiti Teknologi Malaysia, Johor Bahru, Malaysia
`norma@ibnusina.utm.my`
[2] ICMMG SB RAS, Novosibirsk, Russia
`kireev@ssd.sscc.ru`
[3] Novosibirsk National Research University, Novosibirsk, Russia

Abstract. The fragmented programming system LuNA is based on the Fragmented Programming Technology. LuNA is a platform for building automatically tunable portable libraries of parallel numerical subroutines. This paper focuses on the parallel implementation of the IADE method for solving 1D partial differential equation (PDE) of parabolic type using LuNA programming system. A fragmented numerical algorithm of IADE method is designed in terms of the data-flow graph. A performance comparison of different algorithm's implementations including LuNA and Message Passing Interface are given.

Keywords: Fragmented Programming Technology · LuNA system · Algorithm fragmentation · IADE method

1 Introduction

Today's rapid development of parallel computing systems makes accumulation of a portable numerical library of parallel subroutines an important and difficult issue. A variety of parallel computer architectures including clusters, multicore and many-core CPUs, accelerators, grids and hybrid systems makes it difficult to develop a portable parallel algorithm and implement it efficiently for given hardware. The Fragmented Programming Technology (FPT) [1] is an approach to parallel programming that is aimed to solve the problem of parallel numerical library accumulation [2,3]. It suggests to write a numerical algorithm in an architecture-independent form of a data-flow graph and to tune it to a given computer system in an automated way.

The purpose of the paper is to demonstrate the current implementation of FPT on a solution of a certain problem and to evaluate the obtained performance. The paper considers a model problem of 1D parabolic equation solution

The work has been supported by research grant FRGS/1/2015/TK10/UTM/02/7 and RAS Presidium Programs II.2Π/I.3-1 and II.2Π/I.4-1.

© Springer International Publishing AG 2017
V. Malyshkin (Ed.): PaCT 2017, LNCS 10421, pp. 85–93, 2017.
DOI: 10.1007/978-3-319-62932-2_7

using modification of IADE method [4] as an example of FPT application. A fragmented algorithm for the considered problem was developed and implemented using fragmented programming system LuNA [5]. A performance comparison of different algorithm implementations including LuNA and MPI was made.

2   Fragmented Programming Technology

The FPT defines a representation of an algorithm and a process of its optimization to a certain parallel architecture. In FPT the representation of an algorithm (called "fragmented algorithm") have the following peculiarities:

- *portability* – the fragmented algorithm does not depend on a certain parallel architecture of a multicomputer,
- *automated tunability* – the fragmented algorithm is able to be semi-automatically tuned to a supercomputer of a certain class.

These properties allow to accumulate a library of parallel numerical subroutines with high portability among present and future parallel architectures. In order to satisfy these properties in FTP the following decisions were made:

- The fragmented algorithm is defined as a bipartite data-flow graph with nodes being single assignment variables (called "data fragments") and single execution operations (called "fragments of computation"). Each fragment of computation is a designation of execution of some pure function called "code fragment". Arcs in the graph correspond to data dependencies originated from numerical algorithm. So, the fragmented algorithm does not have any implementation specific elements, and is thus portable. Declarative concurrency of the algorithm representation allows to execute fragments of execution in any order that does not contradict to data dependencies.
- Data fragments of a fragmented algorithm are actually aggregates of atomic variables. The sizes of the data fragments are parameters of the algorithm. Consequently, the fragments of computation are aggregates of atomic variables and operations. On execution of a given fragmented algorithm the sizes of fragments should be tuned to characteristics of a specific parallel computer, for example, to fit a cache memory size.
- The problem of automated tuning of a fragmented algorithm to a certain parallel computer is supposed to be solved in FPT by a special execution system. In order to transfer all existing algorithms to a new supercomputer architecture a new execution system should be implemented.

LuNA programming system [5–7] is a realization of FPT for a class of multicomputers with multicore computing nodes. It comprises LuNA language and LuNA execution system. LuNA language is based on the structure of fragmented algorithm with addition of means for working with enumerated sets of fragments, for organization of structured code fragments that are similar to subroutines in common programming languages, and provides an interface with code fragments written in C++. LuNA execution system consists of a compiler, a generator and a runtime system [5]. LuNA compiler receives a fragmented algorithm written in LuNA language as input and makes common static optimizations. LuNA generator in turn makes architecture specific static optimizations. Then, LuNA runtime system executes the tuned algorithm on a certain multicomputer in semi-interpreted mode providing necessary dynamic properties such as workload balancing.

3 IADE-RB-CG Method

Iterative Alternating Decomposition Explicit (IADE) method has been used for solution of multidimensional parabolic type problems since 90s [4,8,9]. Being the second-order accurate in time and fourth-order accurate in space, the IADE method is proved to be more accurate, more efficient, and has better rate of convergence than the classical fourth-order iterative methods [8]. The method is fully explicit, and this feature can be fully utilized for parallelization.

To approximate the solution of diffusion equation, the IADE scheme employs the fractional splitting resulting in a two-stage iterative process. On the first half-step of the i-th iteration the approximation solution $x^{i+\frac{1}{2}}$ is computed using values x^i, and on the second half-step the new values x^{i+1} are computed using $x^{i+\frac{1}{2}}$. In basic sequential algorithm for 1D problem solution the value $x_j^{i+\frac{1}{2}}$ depends on $x_{j-1}^{i+\frac{1}{2}}$, whereas the value x_j^{i+1} depends on x_{j+1}^{i+1} (j is a spatial index) [4]. To avoid dependency situation, several parallelization strategies are developed to construct non-overlapping subdomains [9]. In this paper, a Red-Black ordering approach was used for 1D problem, resulting in two half-sweeps for Red (even) and Black (odd) elements. In addition, a Conjugate Gradient (CG) acceleration was employed to improve convergence [9].

4 Fragmentation of IADE-RB-CG Method

A solution of 1D parabolic problem [4] is considered as an example of IADE-RB-CG method application. The process of solution is a sequence of time steps; on each step a system of linear equations with the same matrix and a new right-hand side is solved by an iterative process.

The problem of creating a fragmented algorithm consists in decomposing the algorithm into data fragments and fragments of computations with their sizes being parameters of the algorithm and in defining all the necessary dependencies between them. The sizes of fragments should be approximately equal to ease the load balancing. Therefore, the global domain is divided into a number of subdomains of equal sizes. Each subdomain contains Red and Black elements, grouped in separate data fragments to reduce the number of data dependencies. For example, the solution vector x of size M divided into m subdomains is represented in fragmented algorithm as a set of data fragments $xR_0, xR_1, \ldots, xR_{m-1}$ (for Red), and $xB_0, xB_1, \ldots, xB_{m-1}$ (for Black). The size of each data fragment (except the last ones) is $S \approx M/(2m)$.

The scheme of the fragmented algorithm of IADE-RB-CG method is shown in Fig. 1a as a sequence of time steps $n = 1, \ldots, N$. Circles denote sets of data fragments, rectangles are code fragments. Black circles represent input data fragments, white ones are output and gray ones are intermediate data fragments for the time step. Code fragment "init_f" calculates right-hand side vector values in data fragments fR_n, fB_n. Code fragment "solve" calculates the solution for the next time step. Data fragments on the left side of Fig. 1a hold coefficients used for calculation. Data fragments yR, yB hold the half-step solution $x^{i+\frac{1}{2}}$. Hereinafter, the initialization phase in the algorithm representation is omitted. In Fig. 1b an implementation of a "solve" code fragment is shown. It contains an iterative process continuing until convergence. Data fragments rR, rB and zR, zB hold vectors used for CG acceleration.

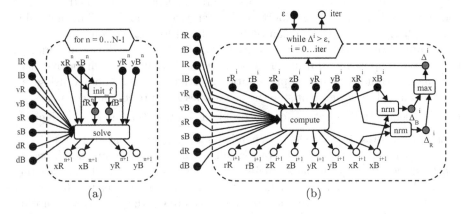

Fig. 1. Fragmented algorithm of IADE-RB-CG method (a) and fragmented algorithm of "solve" code fragment (b)

Code fragment "compute" performs one iteration of IADE-RB-CG method (Fig. 2). Notice that circles here are sets of m data fragments corresponding to m subdomains. Rectangles consequently correspond to sets of fragments of computation. Thus rectangle labeled "iadeR" denotes a set of m fragments of computation, implemented by the same code fragment called "iadeR". The j-th fragment of computation gets j-th data fragments from the sets of input data fragments and produce j-th data fragments from the sets of output data fragments without interaction between subdomains. The same goes for the "iadeB", "cgR", and "cgB". This property allows parallel execution of these fragments of computation in the case of necessary resources availability.

The only interaction between subdomains occurs in the code fragments "left" and "right" where boundaries exchange is performed (Fig. 3). Since all subdomains (except the last one) has an even number of elements, only the Red elements on exchange go to the right subdomain (Fig. 3a), and only the Black elements go to the left subdomain (Fig. 3b).

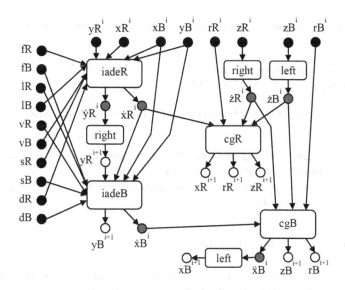

Fig. 2. Fragmented algorithm for i-th iteration of IADE-RB-CG method

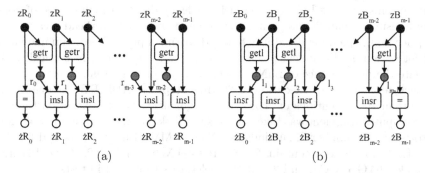

Fig. 3. Fragmented algorithms of boundaries exchange: right (a) and left (b)

5   Fragmented Algorithm Implementation Using LuNA Language

The fragmented algorithm was implemented with LuNA fragmented programming system using LuNA language. LuNA stands for Language for Numerical Algorithms. LuNA program is just a textual representation of a bipartite dataflow graph of an algorithm. Here, an example of LuNA subroutine corresponding to the algorithms on Fig. 3a is presented.

```
sub send_right(#in int m, name sizeR, name zR_in, #out name zR_out)
{ df r;
  for j=0..m-2
    get_R_right_border(#in sizeR[j],zR_in[j], #out r[j]);
  copy(#in zR_in[0], #out zR_out[0]);
  for j=1..m-1
    set_R_left_border(#in sizeR[j],zR_in[j],r[j-1], #out zR_out[j]);
}
```

get_R_right_border, copy and set_R_left_border denote fragments of computation with input and output data fragments shown in parenthesis. The keyword for defines a set (unordered) of fragments of computation for a given range of values of index variable (j).

6   Performance Evaluation

Having such a fragmented algorithm, as presented in Sect. 4, resource allocation and execution order control can be done automatically during execution and dynamically adjusted to available resources. Dataflow-based parallel programming systems, such as LuNA, often lack efficiency due to a high degree of non-determinism of a parallel program execution and execution overhead it causes. It is a price that is paid for the ability to avoid writing a parallel program manually and for obtaining dynamical properties of parallel execution automatically. To evaluate the performance of LuNA system a series of tests was made.

Each test run is an execution of 20 time steps of IADE-RB-CG algorithm (more than 100 executions of "compute" fragment). Average execution time of one "compute" fragment was taken as a result. Task parameters are: M - solution vector size, m - number of subdomains. Cluster MVS-10P [10] was used for the tests. Each cluster node contains 2×8-core Intel Xeon E5-2690 2.9 GHz (16 cores per node), 64 GB RAM, and $2 \times$ MIC accelerators (not used in tests).

The first test evaluates the performance characteristics of LuNA runtime system depending on problem size and number of processor cores used. Currently, the LuNA runtime system is implemented as a set of MPI processes each running one or more working threads. Two variants of execution were compared:

– "Processes" - number of MPI processes is equal to the total number of processor cores used, each process running only one working thread,
– "Threads" - number of MPI processes is equal to the number of cluster nodes, each process running the same number of working threads as the processor cores used per node.

The number of subdomains here is equal to the total number of cores used. The results (Fig. 4) show that the variant "Threads" runs faster, so it will be used in the following experiments. One can also see that with the current implementation of LuNA runtime system the execution time grows rapidly with increasing number of cluster nodes due to the execution overhead. The overhead is expected to be reduced in future LuNA system releases.

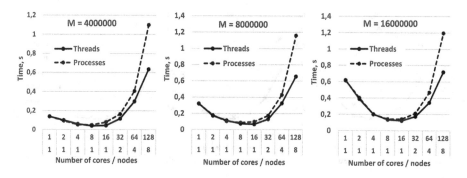

Fig. 4. LuNA runtime performance characteristics

The performance of a parallel program depends largely on the availability of work for the available computational resources, which in turn depends on the degree of algorithm fragmentation. The second test demonstrates how the execution time depends on the number of subdomains. It is known that the resulting graph should be a U-shaped with an optimal fragmentation degree giving minimal execution time [2]. Results in Fig. 5 correspond to that and show optimal fragmentation degrees for different problem sizes and different computing resources.

The last test compares three different implementation of the IADE-RB-CG fragmented algorithm with different degrees of the automation of parallel execution.

- "LuNA" is implementation of the algorithm in LuNA system using execution parameters from previous tests that give the best execution time for given resources. It is the most automated implementation since the LuNA runtime system makes most decisions on fragmented algorithm execution dynamically.
- "LuNA-fw" is a semi-automated implementation of the fragmented algorithm using a simple event-driven MPI-based runtime system with manually written control program. By means of the control program the programmer specifies resource allocation and data fragments' management.
- "MPI" is a manual MPI implementation of the fragmented algorithm with the least execution automation.

Results on Fig. 6 show that the "LuNA" implementation has a considerable execution overhead when using several cluster nodes, and shows performance close to "LuNA-fw" implementation within one node. "LuNA" and "LuNA-fw" implementations both have a noticeable overhead compared to the "MPI" implementation due to necessity to maintain single-assignment data fragments, which leads to redundant memory usage and hence worse cache usage.

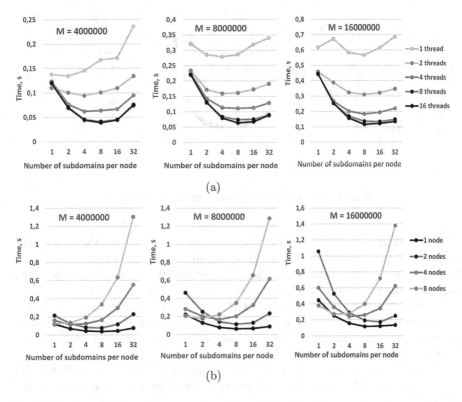

(a)

Fig. 5. The dependence of the execution time of the algorithm on the degree of fragmentation: using one cluster node and different number of threads (a), using different number of nodes with 16 threads each (b)

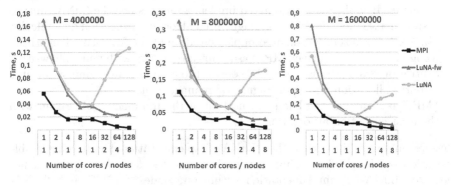

Fig. 6. Comparison of different implementations of the IADE-RB-CG fragmented algorithm

7   Conclusion

Fragmented Programming Technology (FPT) is a promising approach leading to accumulation of numerical algorithms in a portable and tunable form. A solution of a 1D problem by IADE-RB-CG method using LuNA fragmented programming system was presented as an example of FPT application. A fragmented algorithm for the considered problem is proposed.

The performance comparison of different fragmented algorithm implementations shows that LuNA system has a considerable execution overhead when using several cluster nodes, while the LuNA-fw implementation based on event-driven control program has much lower overhead. Thus, a promising direction for the future development and optimization of the LuNA system is an automated generation of a LuNA-fw-like control program.

References

1. Kraeva, M.A., Malyshkin, V.E.: Assembly technology for parallel realization of numerical models on MIMD-multicomputers. Int. J. Future Gener. Comput. Syst. **17**(6), 755–765 (2001). Elsevier Science, NH

2. Kireev, S., Malyshkin, V.: Fragmentation of numerical algorithms for parallel subroutines library. J. Supercomput. **57**(2), 161–171 (2011)

3. Kireev, S., Malyshkin, V., Fujita, H.: The LuNA library of parallel numerical fragmented subroutines. In: Malyshkin, V. (ed.) PaCT 2011. LNCS, vol. 6873, pp. 290–301. Springer, Heidelberg (2011). doi:10.1007/978-3-642-23178-0_26

4. Sahimi, M.S., Ahmad, A., Bakar, A.A.: The Iterative Alternating Decomposition Explicit (IADE) method to solve the heat conduction equation. Int. J. Comput. Math. **47**, 219–229 (1993)

5. Malyshkin, V.E., Perepelkin, V.A.: LuNA fragmented programming system, main functions and peculiarities of run-time subsystem. In: Malyshkin, V. (ed.) PaCT 2011. LNCS, vol. 6873, pp. 53–61. Springer, Heidelberg (2011). doi:10.1007/978-3-642-23178-0_5

6. Malyshkin, V., Perepelkin, V.: Optimization methods of parallel execution of numerical programs in the LuNA fragmented programming system. J. Supercomput. **61**(1), 235–248 (2012)

7. Malyshkin, V., Perepelkin, V.: The PIC implementation in LuNA system of fragmented programming. J. Supercomput. **69**(1), 89–97 (2014)

8. Mansor, N.A., Zulkifle, A.K., Alias, N., Hasan, M.K., Boyce, M.J.N.: The higher accuracy fourth-order IADE algorithm. J. Appl. Math. **2013**, 1–13 (2013)

9. Alias, N., Sahimi, M.S., Abdullah, A.R.: Parallel strategies for the iterative alternating decomposition explicit interpolation-conjugate gradient method in solving heat conductor equation on a distributed parallel computer systems. In: Proceedings of the 3rd International Conference on Numerical Analysis in Engineering, vol. 3, pp. 31–38 (2003)

10. Joint Supercomputer Center of the Russian Academy of Sciences. http://www.jscc.ru/eng/index.shtml. Accessed 12 May 2017

Performance Aspects of Collocated and Staggered Grids for Particle-in-Cell Plasma Simulation

Sergey Bastrakov[1], Igor Surmin[1], Evgeny Efimenko[1,2], Arkady Gonoskov[1,2,3], and Iosif Meyerov[1(✉)]

[1] Lobachevsky State University of Nizhni Novgorod, Nizhni Novgorod, Russia
meerov@vmk.unn.ru
[2] Institute of Applied Physics of the Russian Academy of Sciences,
Nizhni Novgorod, Russia
[3] Chalmers University of Technology, Gothenburg, Sweden

Abstract. We present a computational comparison of collocated and staggered uniform grids for particle-in-cell plasma simulation. Both types of grids are widely used, and numerical properties of the corresponding solvers are well-studied. However, for large-scale simulations performance is also an important factor, which is the focus of this paper. We start with a baseline implementation, apply widely-used techniques for performance optimization and measure their efficacy for both grids on a high-end Xeon CPU and a second-generation Xeon Phi processor. For the optimized version the collocated grid outperforms the staggered one by about 1.5 x on both Xeon and Xeon Phi. The speedup on the Xeon Phi processor compared to Xeon is about 1.9 x.

Keywords: Performance optimization · Xeon Phi · SIMD · Plasma simulation · Particle-in-cell

1 Introduction

Various studies in theoretical and applied physics employ numerical simulation of plasmas based on the particle-in-cell method. Large-scale 3D simulations are typically performed on clusters and supercomputers using specialized parallel codes. There are well-established approaches to implementation of the particle-in-cell method on shared and distributed memory parallel systems with excellent scaling up to at least tens of thousands of cores, including issues of domain decomposition and load balancing [1–5].

The rise of manycore architectures, most notably GPUs and Intel Xeon Phi, has created new challenges for efficient implementation of computational kernels. Due to the increased number of cores, greater parallelism is required to take full advantage of the manycore architectures. Additionally, such systems are typically more fastidious about memory access pattern because of lesser memory

© Springer International Publishing AG 2017
V. Malyshkin (Ed.): PaCT 2017, LNCS 10421, pp. 94–100, 2017.
DOI: 10.1007/978-3-319-62932-2_8

throughput per core and, sometimes, other limitations. Finally, achieving optimum performance on modern architectures requires SIMD-level parallelism.

There is a fairly widely used approach to efficient implementation of the particle-in-cell method on GPUs [6,7]. As for Xeon Phi, several implementations, including our previous work, has demonstrated a moderate speedup over a multicore CPU [8–10]. As pointed out in several studies [1,9,11], the main issue for implementations of the particle-in-cell method on both CPUs and Xeon Phi currently seems to be efficient utilization of SIMD-level parallelism.

Particle-in-cell simulations are performed using various combinations of grids, Maxwell's solvers and numerical schemes. Numerical properties of widely used solvers are rather well-studied [12,13], which allows to choose a scheme suitable for the problem at hand. However, for large-scale simulations performance is also an important factor, particularly on manycore architectures. A typical combination for simulations of laser-driven particle acceleration seems to be either a finite-difference Maxwell's solver on a uniform staggered grid [14,15] or a pseudo-spectral solver on a uniform collocated grid [16,17], both providing a good balance between accuracy and speed.

This paper presents a computational comparison of uniform collocated and staggered grids in the particle-in-cell method in terms of performance. We do not consider numerical properties and instead focus solely on performance aspects of particle–grid operations, which are usually the most time-consuming part of the method. The implementation is done using the particle-in-cell code PICADOR [9]. We evaluate results on a multicore Intel Xeon CPU and an Intel Xeon Phi processor of the Knight's Landing generation. This study could allow getting a better trade-off between accuracy and speed when choosing a configuration for other simulations. Additionally, this paper presents new performance data on a second-generation Xeon Phi processor, which, together with our previous work [10], is among the first published performance results for particle-in-cell simulations on this kind of hardware.

2   Overview of the Particle-in-Cell Method

The particle-in-cell method [18] is widely used for numerical simulation of plasmas. The method operates on two major sets of data: a set of charged particles and the electromagnetic field defined on a spatial grid. Each particle in a simulation, often called a macro-particle, represents a cloud of real particles, such as electrons or ions, of the same kind and has their combined charge and mass. Charge distribution inside a cloud is defined by a particle form factor; there are several widely used options [19]. In this paper we use the cloud-in-cell form factor, which corresponds to the constant charge density inside a cloud. For the rest of the paper we write particles instead of macro-particles for brevity.

Particles in the electromagnetic field are affected by the Lorenz force. Computation of the Lorenz force for each particle involves interpolation of the electromagnetic field based on several nearest grid values, depending on the form factor being used. The dynamic of particles is governed by the relativistic equations of motion. A feature of the particle-in-cell method is that the particles do not interact with one another directly, there are only particle–grid interactions. Grid values of the current density created by particles' movement are computed and, later, used to update the electromagnetic field. The computational loop of the method consists of four basic stages: interpolation of grid values of the field to particles' positions, integrating equations of motion, computing grid values of the current density, updating grid values of the electromagnetic field by integrating the Maxwell's equations. A more detailed description of the method is given, for example, in [19].

There are two widely used approaches to solving the Maxwell's equations in the context of the particle-in-cell method using uniform Cartesian grids. The finite-difference methods [15] employ finite-difference approximations of time and spatial derivatives in the Maxwell's equations. The most popular method of this group is the Yee solver [14,15]. It employs staggering of the components of the field in space and time to achieve the second-order approximation. Using pseudo-spectral methods [16,17] for a particle-in-cell simulation typically involves two grids: one in the coordinate (original) space and another in the Fourier space. In this case field interpolation, particle movement and current deposition are performed in the coordinate space, while the Maxwell's equations are solved in the Fourier space and the updated values of the field transformed back. In this case there is no need for staggering in space and a collocated spatial grid is typically employed. Numerical properties of both finite-difference and pseudo-spectral methods are well-studied [12,13] and are out of the scope of this paper. Instead, we focus on the issues of implementation and performance.

3   Computational Evaluation

3.1   Test Problem

All computational experiments presented in this paper were performed for the following test problem: $40 \times 40 \times 40$ grid, 100 particles per cell, cloud-in-cell particle form factor, direct current deposition [19], finite-difference Maxwell's solver, double precision floating-point arithmetic. We used a node of the Intel Endeavor cluster with an 18-core Intel Xeon E5-2697v4 CPU (codename Broadwell) and a 68-core Intel Xeon Phi 7250 (codename Knight's Landing). On Xeon we ran 1 MPI process with 36 OpenMP threads. Xeon Phi was used in Quadrant cluster mode, flat MCDRAM mode, we ran 8 MPI processes with 34 OpenMP threads per process, which had been previously found to be empirically best for our code [10]. The code was compiled with the Intel C++ Compiler 17.0.

3.2   Baseline Implementation

This study was done using the PICADOR code for particle-in-cell plasma simulation on cluster systems with CPUs and Xeon Phi. In this subsection we briefly describe the organization of the code on shared memory. A more detailed description of PICADOR is given in our previous work [9,10].

Particles are stored and processed separately for each cell. During the field interpolation stage the grid values used for the particles of the current cell are preloaded to a small local array. In a similar fashion, the computed values of current density are first accumulated in local arrays and later written to the global grid. OpenMP threads process particles in different cells in parallel. To avoid explicit synchronization between threads, the processing is split into several substages organized in the checkerboard order, so that during each substage there is no intersection of the sets of grid values used by particles in different cells.

There is an important distinction between collocated and staggered grids in terms of implementation of the particle–grid operations. For a collocated grid field interpolation and current deposition operate on the same set of grid values for every field component. Therefore, indexes and coefficients are computed once for all field or current density components. On the contrary, for a staggered grid each component of the field or current density is processed separately. In addition to performing more arithmetic operations, this involves a more complicated memory access pattern, which could be detrimental to vectorization efficiency.

Performance results of the baseline version for the collocated and staggered grids on Xeon and Xeon Phi are presented at Table 1. The 'Particle push' stage corresponds to field interpolation and integration of particles' equations of motion; solving Maxwell's equations and applying boundary conditions is denoted as 'Other'. As pointed out earlier, the collocated grid is a little faster compared to the staggered grid. Speedup on Xeon Phi compared to Xeon is 1.83 x on the collocated grid and 1.44 x on the staggered grid.

Table 1. Run time of the baseline version. Time is given in seconds.

Stage	Collocated grid		Staggered grid	
	Xeon	Xeon Phi	Xeon	Xeon Phi
Particle push	30.13	15.90	26.18	15.93
Current deposition	11.43	6.73	16.72	13.63
Other	0.27	0.27	0.19	0.27
Overall	41.83	22.90	43.09	29.83

3.3 Supercells

A widely used approach to improve efficiency of computational kernels is data blocking to fit caches. In terms of the particle-in-cell method, cells can be grouped into supercells. This idea was originally introduced for GPU-based implementations [6], but recently was demonstrated to be potentially beneficial for CPUs as well [11]. Table 2 presents results of the version with supercells. For each configuration we picked the empirically best supercell size. Using supercells yields a 1.09 x to 1.17 x overall speedup on the staggered grid. For the collocated grid the effect of supercells is a bit more clear, with the speedups between 1.23 x and 1.28x.

Table 2. Run time of the version with supercells. For each configuration the empirically best supercell size is used. Time is given in seconds.

Stage	Collocated grid		Staggered grid	
	Xeon	Xeon Phi	Xeon	Xeon Phi
Particle push	20.92	11.42	25.27	14.17
Current deposition	11.46	6.89	13.91	11.03
Other	0.26	0.27	0.18	0.26
Overall	32.64	18.58	39.36	25.46

3.4   Vectorization

In the previous versions the integration of particles' equation of motion and the finite-difference Maxwell's solver are auto-vectorized by the compiler. However, the field interpolation and current deposition stages are not vectorized, since it has been inefficient in our earlier experience [9]. A probable cause is that the data which needed to be operated in a vector register is not located sequentially in the memory. The resulting complicated memory pattern requires scatter and gather operations, less efficient compared to the coalesced operations.

However, on the more modern system used in this study the #pragma simd directive of the Intel C++ compiler yields some benefit for the field interpolation stage. The current deposition stage cannot be vectorized as easily because of data dependencies. Instead, we employ a scheme inspired by [11]. The results of this version are presented at Table 3. On the collocated grid the speedups on Xeon and Xeon Phi are 1.26 x and 1.34 x, respectively. On the staggered grid the speedup on Xeon Phi is 1.24 x, there is no speedup on Xeon. The larger speedups on the collocated grid are due to the more regular memory access pattern.

Table 3. Run time of the version with vectorization. Time is given in seconds.

Stage	Collocated grid		Staggered grid	
	Xeon	Xeon Phi	Xeon	Xeon Phi
Particle push	17.40	9.98	24.98	9.17
Current deposition	8.19	3.62	13.91	11.04
Other	0.24	0.26	0.18	0.25
Overall	25.83	13.86	39.07	20.46

4   Summary

This paper presents a computational comparison of performance of collocated and staggered grids in the particle-in-cell method on CPU and Xeon Phi. We start with the baseline implementation, for which the collocated grid slightly outperforms the staggered one, mainly due to lesser number of arithmetic operations required. Then we apply the supercell modification to improve cache locality, which turns out to be a bit more beneficial for the collocated grid. Finally, we consider a vectorized version, which again yields more speedup on the collocated grid due to a more regular memory access pattern. For the final version the speedup on the collocated grid compared to the staggered one is about 1.5 x on both Xeon and Xeon Phi. The speedup on Xeon Phi compared to Xeon is about 1.9 x. We have also tried other approaches to vectorization, such as to process components of the fields in SIMD fashion, or even to vectorize computations inside a component, but these did not yield any speedup.

Our results show that using supercells is beneficial for both collocated and staggered grids on Xeon and Xeon Phi. The collocated grid makes vectorization easier and more efficient compared to the staggered one. This could be relevant in the context of increasing the width of SIMD registers in modern hardware, given the growing popularity of spectral solvers for particle-in-cell simulation. Our future work includes applying the presented approach to optimize performance of the ELMIS spectral particle-in-cell code [20].

Acknowledgements. The authors (E.E., A.G.) acknowledge the support from the Russian Science Foundation project No. 16-12-10486. The authors are grateful to Intel Corporation for access to the system used for performing computational experiments presented in this paper. We are also grateful to A. Bobyr, S. Egorov, I. Lopatin, and Z. Matveev from Intel Corporation for technical consultations.

References

1. Fonseca, R.A., Vieira, J., Fiuza, F., Davidson, A., Tsung, F.S., Mori, W.B., Silva, L.O.: Exploiting multi-scale parallelism for large scale numerical modelling of laser wakefield accelerators. Plasma Phys. Control. Fusion. **55**(12), 124011 (2013)
2. Bowers, K.J., Albright, B.J., Yin, L., Bergen, B., Kwan, T.J.T.: Ultrahigh performance three-dimensional electromagnetic relativistic kinetic plasma simulation. Phys. Plasmas **15**(5), 055703 (2008)
3. Vay, J.-L., Bruhwiler, D.L., Geddes, C.G.R., Fawley, W.M., Martins, S.F., Cary, J.R., Cormier-Michel, E., Cowan, B., Fonseca, R.A., Furman, M.A., Lu, W., Mori, W.B., Silva, L.O.: Simulating relativistic beam and plasma systems using an optimal boosted frame. J. Phys. Conf. Ser. **180**(1), 012006 (2009)
4. Kraeva, M.A., Malyshkin, V.E.: Assembly technology for parallel realization of numerical models on MIMD-multicomputers. Future Gener. Comp. Syst. **17**, 755–765 (2001)
5. Bastrakov, S., Donchenko, R., Gonoskov, A., Efimenko, E., Malyshev, A., Meyerov, I., Surmin, I.: Particle-in-cell plasma simulation on heterogeneous cluster systems. J. Comput. Sci. **3**, 474–479 (2012)
6. Burau, H., Widera, R., Honig, W., Juckeland, G., Debus, A., Kluge, T., Schramm, U., Cowan, T.E., Sauerbrey, R., Bussmann, M.: PIConGPU: a fully relativistic particle-in-cell code for a GPU cluster. IEEE Trans. Plasma Sci. **38**(10), 2831–2839 (2010)
7. Decyk, V.K., Singh, T.V.: Particle-in-cell algorithms for emerging computer architectures. Comput. Phys. Commun. **185**(3), 708–719 (2014)

8. Nakashima, H.: Manycore challenge in particle-in-cell simulation: how to exploit 1 TFlops peak performance for simulation codes with irregular computation. Comput. Electr. Eng. **46**, 81–94 (2015)

9. Surmin, I.A., Bastrakov, S.I., Efimenko, E.S., Gonoskov, A.A., Korzhimanov, A.V., Meyerov, I.B.: Particle-in-Cell laser-plasma simulation on Xeon Phi coprocessors. Comput. Phys. Commun. **202**, 204–210 (2016)

10. Surmin, I., Bastrakov, S., Matveev, Z., Efimenko, E., Gonoskov, A., Meyerov, I.: Co-design of a Particle-in-Cell plasma simulation code for Intel Xeon Phi: a first look at knights landing. In: Carretero, J., et al. (eds.) ICA3PP 2016. LNCS, vol. 10049, pp. 319–329. Springer, Cham (2016). doi:10.1007/978-3-319-49956-7_25

11. Vincenti, H., Lehe, R., Sasanka, R., Vay, J.-L.: An efficient and portable SIMD algorithm for charge/current deposition in Particle-In-Cell codes. Comput. Phys. Commun. **210**, 145–154 (2017)

12. Godfrey, B.B., Vay, J.-L., Haber, I.: Numerical stability analysis of the pseudo-spectral analytical time-domain PIC algorithm. J. Comput. Phys. **258**, 689–704 (2014)

13. Vincenti, H., Vay, J.-L.: Detailed analysis of the effects of stencil spatial variations with arbitrary high-order finite-difference Maxwell solver. Comput. Phys. Commun. **200**, 147–167 (2016)

14. Yee, K.: Numerical solution of initial boundary value problems involving Maxwell's equations in isotropic media. IEEE Trans. Antennas Propag. **14**(3), 302–307 (1966)

15. Taflove, A.: Computational Electrodynamics: The Finite-Difference Time-Domain Method. Artech House, London (1995)

16. Haber, I., Lee, R., Klein, H., Boris, J.: Advances in electromagnetic simulation techniques. In: Proceedings of the Sixth Conference on Numerical Simulation of Plasmas, pp. 46–48 (1973)

17. Liu, Q.: The Pstd algorithm: a time-domain method requiring only two cells per wavelength. Microw. Opt. Technol. Lett. **15**(3), 158–165 (1997)

18. Hockney, R.W., Eastwood, J.W.: Computer Simulation Using Particles. McGraw-Hill, New York (1981)

19. Birdsal, C., Langdon, A.: Plasma Physics via Computer Simulation. Taylor & Francis Group, New York (2005)

20. Gonoskov, A., Bastrakov, S., Efimenko, E., Ilderton, A., Marklund, M., Meyerov, I., Muraviev, A., Sergeev, A., Surmin, I., Wallin, E.: Extended Particle-in-Cell schemes for physics in ultrastrong laser fields: review and developments. Phys. Rev. E **92**, 023305 (2015)

Technological Aspects of the Hybrid Parallelization with OpenMP and MPI

Oleg Bessonov[✉]

Institute for Problems in Mechanics of the Russian Academy of Sciences,
101, Vernadsky ave., 119526 Moscow, Russia
`bess@ipmnet.ru`

Abstract. In this paper we present practical parallelization techniques for different explicit and implicit numerical algorithms. These algorithms are considered on the base of the analysis of characteristics of modern computer systems and the nature of modeled physical processes. Limits of applicability of methods and parallelization techniques are determined in terms of practical implementation. Finally, the unified parallelization approach for OpenMP and MPI for solving a CFD problem in a regular domain is presented and discussed.

1 Introduction

Previously, only distributed memory computer systems (clusters) were available for parallel computations, and the only practical way of parallelization was the MPI distributed-memory approach (provided a user was privileged enough to have an access to such a system). Currently multicore processors have become widely available, and parallelization is no more an option as before. Users have to parallelize their codes because there is no other way to fully utilize the computational potential of a processor. Besides, cluster nodes are now built on multicore processors too. Thus, the shared-memory OpenMP approach as well as the hybrid OpenMP/MPI model become important and popular.

The progress in the computer development can be illustrated by comparing solution times of the CFD problem [1] with 10^6 grid points and 10^6 time-steps:

- 2005, 1-core processor – 280 h;
- 2009, 4-core processor – 30 h;
- 2011, cluster node with two 6-core processors – 11 h;
- 2013, cluster node with two 10-core processors – 4.5 h;
- 2015, 4 cluster nodes with two 10-core processors – 1.5 h.

We can see the acceleration by two orders of magnitude. This has become possible both due to the development of computers and implementation of new parallel methods and approaches.

In previous papers [1–3] we have analyzed parallelization methods from the mathematical, convergence and efficiency points of view. In this work we will consider practical parallelization techniques taking into account characteristics and limitations of modern computer systems as well as essential properties of modeled physical processes.

V. Malyshkin (Ed.): PaCT 2017, LNCS 10421, pp. 101–113, 2017.
DOI: 10.1007/978-3-319-62932-2_9

2   Parallel Performance of Modern Multicore Processors

Modern multicore microprocessors belong to the class of throughput-oriented processors. Their performance is achieved in cooperative work of processor cores and depends both on the computational speed of cores and on the throughput of the memory subsystem. The latter is determined by the configuration of integrated memory controllers, memory access speed, characteristics of the cache memory hierarchy and capacity of intercore or interprocessor communications.

For example, a typical high-performance processor used for scientific or technological computations has the following characteristics:

- 6 to 12 computational cores, with frequencies between 2.5 and 3.5 GHz;
- peak floating point arithmetic performance 300–500 GFLOPS (64-bit);
- 4 channels of DDR4-2133/2400 memory with peak access rate 68–77 GB/s;
- hierarchy of cache memories (common L3-cache, separate L2 and L1);
- ability to execute two threads in each core (hyperthreading).

Many scientific or technological application programs belong to the memory-bound class, i.e. their computational speed is limited by the performance of the memory subsystem. Thus, with increasing the number of cores, it is necessary to make the memory faster (frequency) and/or wider (number of chsannels).

Importance of the memory subsystem can be illustrated by running several applications on computers with different configurations (Fig. 1): 8-core (3.0 GHz) and 6-core (3.5 GHz) processors with 4 memory channels (68 GB/s), and 6-core processor with 2 channels (34 GB/s). All processors belong to the same family Intel Core i7-5900 (Haswell-E). The first computer system has faster memory than two others. Application programs used in the comparison are the following:

- Cylflow: Navier-Stokes CFD code [1] (regular grid, 1.5 M grid points);
- CG AMG: Conjugate gradient solver with the multigrid preconditioner [2] (Cartesian grid in the arbitrary domain, sparse matrix, 2 M grid points);
- CG Jacobi: Conjugate gradient solver with the explicit Jacobi preconditioner [2] (the same grid).

It is seen from Fig. 1 that there is the saturation of the memory subsystem in all cases. For the CFD code, it is less visible: 4-channel computers look similarly, while the 2-channel system is 1.3 times slower. For the multigrid solver, some difference between first two systems appears, and the third one is about two times slower. For the explicit solver, effect is more significant: the 8-core computer additionally gains owing to its faster memory and larger cache, while the 2-channel system additionally loses. The maximum of performance is achieved in this case if only part of processor cores are active (5, 4 and 3 cores, respectively).

On two-processor configurations, the number of memory channels and their integral capacity is doubled. Due to this, performance of most memory-bound programs can be almost doubled (see example of 1.95-times increase in [3]).

On the other hand, hyperthreading usually doesn't help to such programs. In fact, for the above applications, running twice the number of threads with the active hyperthreading reduces performance by about 10%.

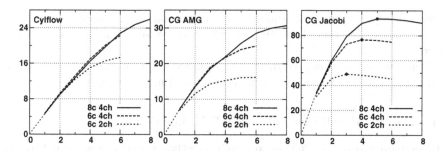

Fig. 1. Parallel performance of application programs (iterations per second) as a function of the number of threads on computers with different memory configurations

3   Properties of Explicit and Implicit Algorithms

There are two main classes of computational algorithms: explicit, with computations like $x = Ay$ (multiplication of a matrix by a vector), and implicit, that look as $Ax = y$ or $Ax \approx y$ (exact or approximate solution of a linear system). Here vectors x and y represent some physical quantities in the discretized domain, and matrix A corresponds to the discretization stencils applied to them.

In the analysis of explicit and implicit algorithms we will consider the lowest level of the computational method. For example, the explicit time integration scheme is an explicit algorithm per se, while the implicit integration scheme may be either resolved by an immediate implicit method, or solved by means of some iterative method employing any sort of the iterative scheme at the lower level.

Explicit algorithms act locally by stencils of the limited size and propagate information with low speed (one grid distance per iteration). Thus, they require $O(n)$ iterations for full convergence where n is the diameter of the domain (in terms of grid distances). On the other hand, implicit algorithms operate globally and propagate information much faster (approximate methods) or even instantly (direct solvers). For example, the Conjugate gradient method with the Modified Incomplete LU decomposition as a preconditioner needs $O(\sqrt{n})$ iterations [4].

Applicability of the methods depends on the nature of underlying physical processes. For example, incompressible viscous fluid flows are driven by three principal mechanisms with different information propagation speeds:

- convection: slow propagation, Courant condition can be applied CFL $= O(1)$ (one or few grid distances per time-step); using an explicit time integration scheme or an iterative solver with few iterations;
- diffusion: faster propagation (tens grid distances per time-step), well-conditioned linear system; using an iterative solver with explicit iterations or an Alternating direction implicit (ADI) solver;
- pressure: instant propagation, ill-conditioned linear system; using an iterative solver with implicit iterations or multigrid or a direct solver.

Parallel properties of computational methods strongly depend on how information is propagated. Iterations of explicit algorithms can be computed independently, in any order, thus giving the freedom in parallelization. In contrast, implicit iterations have the recursive nature and can't be easily parallelized.

Below we will consider parallelization approaches for several variants of computational algorithms of the explicit, implicit and mixed type.

3.1   Natural Parallelization of Explicit Algorithms

The simplest way of parallelizing an explicit method is to divide a computational domain (geometric splitting) or a matrix A (algebraic splitting) into several parts for execution in different threads. It can be easily implemented in Fortran with the OpenMP extension. For example, one-dimensional geometric splitting by the last spatial dimension can be programmed with the use of !$OMP DO statement.

This sort of parallelization is very convenient and is almost automatic. However, it has a natural limitation: with larger number of threads, subdomains become narrow. This can increase parallelization overheads due to load disbalances and increased costs of accesses to remote caches across subdomain boundaries.

For this reason, two-dimensional splitting may become attractive. OpenMP has no natural mechanism for such splitting. Nevertheless, the nested loops can be easily reorganized by the appropriate remap of control variables of two outer loops (Fig. 2). Here, all changes of the original code are shown by capital letters.

The difference between one- and two-dimensional splittings is the placement of data belonging to subdomains. In the first case, each part of data is a single continuous 3-dimensional array. In the second case, data look as a set of 2D arrays, decoupled from each other. The size of each array is $N \times M/P$, where N and M are dimensions, and P is the splitting factor in the second dimension. For $N = M = 100$ and $P = 4$ this corresponds to 2500 data elements, or 20 KBytes.

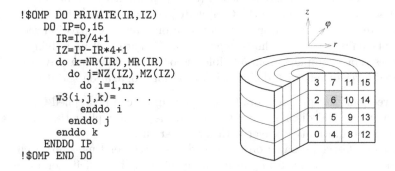

```
!$OMP DO PRIVATE(IR,IZ)
    DO IP=0,15
      IR=IP/4+1
      IZ=IP-IR*4+1
      do k=NR(IR),MR(IR)
        do j=NZ(IZ),MZ(IZ)
          do i=1,nx
            w3(i,j,k)= . . .
          enddo i
        enddo j
      enddo k
    ENDDO IP
!$OMP END DO
```

Fig. 2. Parallelization with the two-dimensional splitting

The main problem of split data is associated with the streamlined prefetch: this mechanism is efficient if arrays are long and continuous, and in the case of piecewise-continuous arrays it stops after the end of each piece of data thus taking additional time for restart and reducing the overall efficiency of prefetch.

For the same reasons, it is not possible to efficiently implement 3-dimensional splitting by the similar way: the size of each piece of data would be N/P, or only 200 Bytes (for the above parameters).

Additional problem appears in the case of a Non-uniform memory computer (NUMA), consisting of two or more processor interconnected by the special links. Each processor controls its own part of memory. Logically, each thread can transparently access any memory location in a system, but remote accesses are much slower than local ones. For this reason, all data should be divided between processors as accurately as possible.

However, data are allocated in a particular processor's memory by pages of the typical size 4 KBytes. Therefore, some data on a boundary between subdomains always fall into the remote memory area (on average, half the page size). For the 1D splitting, this is much less than the size of a boundary array ($N \times M$, or 80 KBytes for the above parameters). In case of the 2D splitting, most boundary arrays are of the size N (only 800 Bytes). Thus, a significant part of each array would fall into the remote memory.

Therefore, for NUMA, splitting should be arranges such that interprocessor communications occur only across boundaries in the last spatial dimension. For the same reason it is not reasonable to use large pages (2 MBytes).

3.2   Parallelization Properties of Implicit Algorithms

Most often implicit algorithms are used as preconditioners for the Conjugate gradient (CG) method [3]. Typically, such preconditioners are built upon variants of the Incomplete LU decomposition (ILU) applied to sparse matrices. This decomposition looks as a simplified form of the Gauss elimination when fill-in of zero elements is restricted or prohibited.

Within the CG algorithm, this preconditioner is applied at each iteration in the form of the solution of a linear system $LU\boldsymbol{x} = \boldsymbol{y}$, that falls into two steps: $L\boldsymbol{z} = \boldsymbol{y}$ and $U\boldsymbol{x} = \boldsymbol{z}$.

These steps are recursive by their nature. There is no universal and efficient method for parallelizing ILU. One well-known approach is the class of domain decomposition methods [5], where the solution of the original global linear system is replaced with the independent solutions of smaller systems within subdomains, with further iterative coupling of partial results. However, this approach is not enough efficient because it makes the convergence slower or impossible at all.

For ill-conditioned linear systems associated with the action of pressure in the incompressible fluid it is important to retain convergence properties of the preconditioning procedure. This can be achieved by finding some sort of parallelization potential, either geometric or algebraic.

For domains of regular shape it is natural to discover a sort of the geometric parallelization. For example, Cartesian discretization in a parallelepiped produces a 7-diagonal matrix, that looks as a specific composition of three 3-diagonal matrices corresponding to three directions. It can be seen that the procedure of twisted factorization of a 3-diagonal system can be naturally generalized to two or three dimensions [4]. Figure 3(a, b) shows factors of the LU-decomposition of a 5-diagonal matrix that corresponds to a 2D domain. Illustration of the computational scheme for three dimensions is shown on Fig. 3(c): the domain is split into 8 octants, and in each octant elimination of non-zero elements for the first step $Lz = y$ is performed from the corner in the direction inwards. For the second step $Ux = z$ data within octants are processed in the reverse order.

Thus, LU-decomposition in a 3D parallelepipedic domain can by parallelized by 8 threads with small amount of inter-thread communications and without sacrifying convergence properties of the iterative procedure.

Additional parallelization within octants can be achieved by applying the staircase (pipelined) method [6,7]. Here, each octant is split into two halves in the direction j (Fig. 3, d), and data in each half are processed simultaneously for different values of the index in the direction k (this looks like a step on stairs).

This method needs more synchronizations between threads, and its application is limited by the factor of 2 or (at most) 4. Thus the resulting parallelization potential for a parallelepipedic domain is limited by 16 or 32 threads.

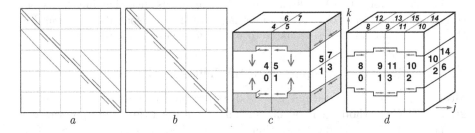

Fig. 3. Nested twisted factorization: factors of a 2D LU-decomposition (a, b); example of the parallel elimination of non-zero elements in a 3D domain (c); illustration of the staircase method (d)

For general domains and non-structured grids there is no more geometric symmetry. Thus the only way of parallelization is to find an algebraic potential. The idea is again to use twisted factorization. Figure 4 (left, center) shows factors of the LU-decomposition of a banded sparse matrix. Each factor consists of two parts, and most calculations in each part can be performed in parallel.

Unfortunately, this way allows to parallelize the solution for only two threads. Additional parallelization can be achieved by applying a variant of the pipelined approach, namely the block-pipelined method [8]. The idea of the approach is to split each part of a factor into pairs of adjacent trapezoidal blocks that have no mutual data dependences and can be processed in parallel (Fig. 4, right). As a result, parallelization of the Gauss elimination will be extended to 4 threads.

Performance of the block-pipelined method depends on the sparsity pattern of a matrix. If the matrix contains too few non-zero elements, the overall effect of the splitting may happen to be low because of synchronization overheads.

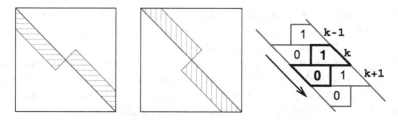

Fig. 4. Twisted factorization of a sparse matrix (left, center); splitting of a subdiagonal part of a factored matrix into pairs of blocks in the block-pipelined method (right)

The above examples demonstrate complexity of parallelization of implicit algorithms that strongly depend on the property of fast propagation of information in underlying physical processes.

3.3   Peculiarities of the Multigrid

There exists a separate class of implicit methods, multigrid, which possess very good convergence and parallelization properties. Multigrid solves differential equations using a hierarchy of discretizations. At each level, it uses a simple smoothing procedure to reduce corresponding error components.

In a single multigrid cycle (V-cycle, Fig. 5), both short-range and long-range components are smoothed, thus information is instantly transmitted throughout the domain. As a result, this method becomes very efficient for elliptic problems that propagate physical information infinitely fast.

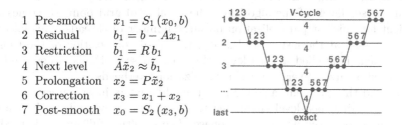

1	Pre-smooth	$x_1 = S_1(x_0, b)$
2	Residual	$b_1 = b - A x_1$
3	Restriction	$\tilde{b}_1 = R\, b_1$
4	Next level	$\tilde{A}\tilde{x}_2 \approx \tilde{b}_1$
5	Prolongation	$x_2 = P\tilde{x}_2$
6	Correction	$x_3 = x_1 + x_2$
7	Post-smooth	$x_0 = S_2(x_3, b)$

Fig. 5. Multigrid algorithm (left) and illustration of V-cycle (right) (Color figure online)

At the same time, multigrid can be efficiently and massively parallelized because processing at each grid level is performed in the explicit manner, and data exchanges between subdomains are needed only at the end of a cycle.

Here we consider the Algebraic multigrid (AMG) approach [2,9] which is based on matrix coefficients rather than on geometric parameters of a domain. This approach is applied in frame of Cartesian discretization in the arbitrary domain. The resulting sparse matrices are stored in the Compressed Diagonal Storage (CDS) format [2] that is more efficient for processing on modern through-put processors than the traditional Compressed Row Storage (CRS).

The main computational operations in an AMG cycle are smoothing (itera-tion of the Gauss-Seidel or SOR method), restriction (fine-to-coarse grid conver-sion by averaging) and prolongation (coarse-to-fine conversion by interpolation).

Formally, iteration of the Gauss-Seidel method looks as an implicit procedure: $(D + L)x_{k+1} = b - Ux_k$ (here D, L and U are diagonal, subdiagonal and superdiagonal parts of the matrix A in the equation $Ax = b$). In order to avoid recursive dependences, the multicolor grid partitioning can be applied. For discretizations with 7-point stencils, two-color scheme is sufficient (red-black partitioning). With this scheme, the original procedure falls into two explicit steps: $D^{(1)}x_{k+1}^{(1)} = b^{(1)} - Ux_k^{(2)}$ and $D^{(2)}x_{k+1}^{(2)} = b^{(2)} - Lx_{k+1}^{(1)}$ (superscripts [1] and [2] refer to red-colored and black-colored grid points, respectively).

To ensure consecutive access to data elements, it is necessary to reorganize all arrays, i.e. to split them into "red" and "black" parts. After that, any appropriate parallelization can be applied, either geometric or algebraic. In particular, the algebraic splitting by more than 200 threads was implemented for Intel Xeon Phi manycore processor [2].

Similarly, the restriction (averaging) procedure can be parallelized. Imple-mentation of the prolongation procedure is more difficult because different inter-polation operators should be applied to different points of the fine grid depending on their location relative to the coarse grid points.

The above considerations are applied to the first (finest) level of the multigrid algorithm. Starting from the second levels, all discretization stencils have 27 points, and 8-color scheme becomes necessary. As a result, computations become less straightforward, with a proportion of indirect accesses.

On coarser levels of the algorithm, the number of grid points becomes not sufficient for efficient parallelization on large number of threads. This effect is most expressed at the last level. Usually, the LU-decomposition or the Conju-gate gradient is applied at this level. However, these methods either can't be parallelized or involve very large synchronization overheads.

To avoid this problem, a solver based on the matrix inversion can be used. Here, the original last-level sparse matrix is explicitly inverted by the Gauss-Jordan method at the initialization phase. The resulting full matrix is used at the execution phase in the simple algorithm of matrix-vector multiplication. This algorithm is perfectly parallelized and doesn't require synchronizations.

Efficiency and robustness of the multigrid algorithm is higher if it is used as a preconditioner in the Conjugate gradient method. In this case, it becomes possible to use the single-precision arithmetic for the multigrid part of the algo-rithm without loosing the overall accuracy. Due to this, the computational cost of the algorithm can be additionally decreased because of reduced sizes of arrays with floating point data and corresponding reduction of the memory traffic.

Thereby, the multigrid method is very efficient and convenient for paralleliza-tion. However, it is very complicated and difficult for implementation, especially for non-structured grids. Its convergence is not satisfactory in case of regular anisotropic grids (though it can be overcome by so-called semi-coarsening [9] when the grid becomes non-structured). In some cases, the behaviour of the method becomes uncertain. Finally, there is no reliable theory and procedure for systems of equations. Thus it is not a universal solution, and applicability of traditional methods remains wide.

3.4 Methods of Separation of Variables and ADI

For solving well-conditioned linear systems, the Alternating direction implicit (ADI) method can be used. If the original matrix is presented as $A = I + L$, where I is the unit matrix and $||L|| \ll 1$, then it can be approximately decomposed as $(I + L) \approx (I + L_x)(I + L_y)(I + L_z)$. The final procedure looks as the solution of several 3-diagonal systems. Also, 3-diagonal systems appear in the direct method of separation of variables for solving the pressure Poisson equation [1].

Parallelization of the solution of a 3-diagonal linear system can be done by applying the twisted factorization for 2 threads, or two-way parallel partition [1] for 4 threads (Fig. 6, left). In the latter method, twisted factorization is applied separately to the first and second halves of a matrix. After two passes (forward and backward) the matrix has only the main diagonal and the column formed due to fill-in. To resolve this system, additional substitution pass is needed.

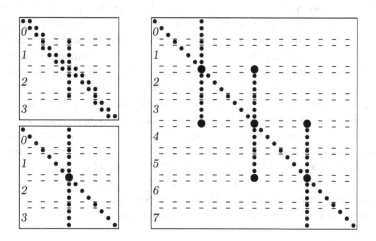

Fig. 6. Illustration of the two-way parallel partition method: matrix view for 4 threads (after the first and the second passes) and for 8 threads (after the second pass)

The two-way parallel partition method can be naturally extended for 8 threads (Fig. 6, right). After two passes of the twisted factorization, the matrix has in this case more complicated structure with 3 partially filled columns. Three equations of the matrix form the reduced linear system (its elements are shown by bold points) that should be resolved before the substitution pass.

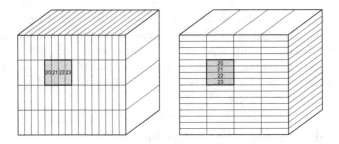

Fig. 7. Two variants of the splitting of a computational domain: 4×16 and 16×4

Computational expenses of this method increase rapidly with the number of threads. Also, there is the sharp increase of the number of exchanges in case of the distributed memory (MPI) parallelization. Therefore, the reasonable level of parallelization for 3-diagonal matrices is limited by 4 or (at most) by 8 threads.

Such limitations can restrict the total level of parallelization. If it is necessary to solve 3-diagonal linear systems in two outer directions with the limit of 4 threads (in each direction), then the total limit will be $4 \times 4 = 16$. To override this, we can use different splittings for different parts of the algorithm. For example, if the splitting of a computational domain is 4×16, all parts of the algorithm can be parallelized except the 3-diagonal systems in the last direction (Fig. 7, left). For solving them, we will use another splitting 16×4 (Fig. 7, right).

For both variants of splitting in this example, there are groups of 4 threads that share the same data (as indicated by shaded rectangles on Fig. 7). Switching from the first splitting to the second one looks like a transposition of data. For the shared memory environment, no real transposition occurs and data are simply accessed in another order. However, for the distributed memory, costly data transfers would take place. For this reason, such groups of subdomains should never be split between cluster nodes. As a consequence, the level of the distributed memory parallelization is limited by the level of parallelization of a 3-diagonal linear system in the last direction, i.e. by 4 or 8 cluster nodes.

The maximal reasonable level of parallelization of the above approach for two-dimensional splitting is between $32 \times 4 = 128$ and $32 \times 8 = 256$, depending on the parallelization scheme for 3-diagonal systems and taking into account the reasonable limitation of 32 threads for each direction.

4 Unified Parallelization Approach for OpenMP and MPI

In the comparison of parallelization environments, it is important to pay attention on the basic characteristics of a distributed memory computer system:

- internode communication speed: $O(1)$ GB/s;
- memory access rate: $O(10)$ GB/s;
- computational speed: $O(10^2)$ GFLOPS, or $O(10^3)$ GB/s.

In fact, not all computations require memory accesses, thus the computational speed expressed in memory units is close to $O(10^2)$ GB/s. Nevertheless, it is clear that the memory subsystem is one order of magnitude slower than the processor, and communications are two orders of magnitude slower.

Therefore exchanges in the MPI model should be kept to a minimum and allowed only on boundaries between subdomains. This applies also to the use of MPI in a shared memory (or multicore) computer though to the less extent. It means that transmission of full (3D) data arrays by MPI should be avoided.

For clusters, it is optimal to use the hybrid parallelization with OpenMP and MPI. Programming with MPI requires serious reorganization of the code: it is necessary to change the natural allocation of data, replacing each monolithic data array with several subarrays in accordance with the splitting and re-adjusting the addressing scheme. These subarrays should be logically overlapped, i.e. contain additional layers (e.g. ghost elements for calculating derivatives). Such complications, together with the need to organize explicit data exchanges between cluster nodes, make development and debug of a code much more difficult. At last, it may become necessary to develop and support two (or more) versions of a code.

These complications can be partly avoided if the splitting between cluster nodes is done only by the last spatial direction. Then, only the numeration in this direction has to be changed. For example, if a 3D domain of the size $L \times M \times N$ is split into 4 subdomains by the last dimension such as $K = N/4$, each MPI process will have to allocate data arrays of the dimensions (L,M,0:K+1). Here, bounds of the last index are expanded to support overlap in such a way, that the slice K of the array in a process corresponds to the slice 0 in the next process, and the same applies to the slices K+1 and 1 in these processes (Fig. 8, left).

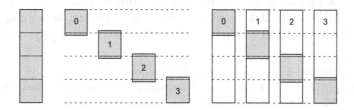

Fig. 8. Illustration of the allocation of a data array in MPI processes: independent subarrays with overlapped areas (left); subsets in the unified address space (right)

As a next step, we can organize the unified address space for the MPI-parallelized program. The simple way is to logically allocate full data array in each MPI process rather than its selected part as in the above example (Fig. 8). This will not lead to the unnecessary occupation of memory because in modern processors only those memory pages are physically allocated which are accessed in a program. In this scheme, addressing of array elements remains unchanged.

In case of dynamic memory allocation, we can avoid logical allocation of full arrays: in Fortran it is possible to allocate an array in a particular process indicating exact bounds of the last index, e.g. ALLOCATE (X(L,M,2*K:3*K+1)). This example corresponds to the process 2 on Fig. 8 (right).

With unified address space, character of the two-dimensional splitting of the computational domain and structure of DO-loops in the hybrid OpenMP/MPI parallelization become very similar to those of the pure OpenMP approach: it is enough to remap bounds of control variables of two outer loops (as shown on Fig. 2) and, after completion of loops, perform exchanges with neighbour processes for sending (receiving) boundary elements of data arrays. Thus, the only difference is that some subdomain boundaries by the last spacial direction become the internode boundaries that require MPI communications.

In order to simplify and unify the computational code, it is worth to organize special "high level" routines for doing exchanges rather than to calls MPI functions directly. Depending on a process number, these routines can determine addresses of data to be transmitted and directions of transmissions. Depending on the total number of MPI processes, they can decide whether the particular data transmission should take place at all. Thus, the code becomes invariant with respect to the number of processes and to the fact of the use of MPI itself.

By avoiding explicit calls of MPI functions, the principal parts of a program become MPI-independent and it becomes possible to compile these parts by any compiler which is not necessarily integrated into an MPI environment. Only a collection the above "high level" routines (grouped into a separate file) works with MPI and should be compiled in the appropriate environment. For convenience, several variants of a file with these routines can be made (for a single-node run without MPI, for 2 nodes, for 4 nodes etc.). By this way, the unified approach for pure OpenMP and for hybrid OpenMP/MPI model is established.

This approach was used for the hybrid parallelization of the CFD code for modeling incompressible viscous flows in cylindrical domains (see [1] for the OpenMP-only version). Below are parameters of a problem for solving on four cluster nodes with two 10-core processors each, 80 cores (threads) total:

- problem size $192 \times 160 \times 160$ (φ, z, r);
- general splitting 4×20 (z, r), size of a subdomain $192 \times 40 \times 8$;
- specific splitting 20×4 (for solving 3-diagonal systems in the last direction), size of a subdomain $192 \times 8 \times 40$.

These splitting parameters correspond to the requirements set out in Subsects. 3.1 and 3.4.

In order to run the above problem correctly, all necessary MPI and OpenMP parameters should be set, such as OMP_NUM_THREADS environment variable for parallelization within a cluster node, number of nodes and their list in the "mpirun" directive, appropriate binding of threads to processor cores in the "taskset" utility etc.

5   Conclusion

In this paper we have considered different practical and technological questions of the parallelization for shared and distributed memory environments. Most examples were done for the geometric approach of parallelization but in many cases they can be extended to the algebraic and other sorts of decomposition.

Special attention was paid on methods with limited parallelization potential that is associated with the essential properties of underlying processes, namely fast propagation of physical information relative to the temporal scale of a numerical method. It is clear that explicit methods while possessing good parallelization properties are not enough efficient for solving such problems. Thus implicit methods remain very important despite they are not always convenient for optimization in general and for parallelization in particular. For this reason, approaches for low and medium scale parallelization are still in demand.

Acknowledgements. This work was supported by the Russian Foundation for Basic Research (projects 15-01-06363, 15-01-02012). The work was granted access to the HPC resources of Aix-Marseille Université financed by the project Equip@Meso (ANR-10-EQPX-29-01) of the program Investissements d'Avenir supervised by the Agence Nationale pour la Recherche (France).

References

1. Bessonov, O.: OpenMP parallelization of a CFD code for multicore computers: analysis and comparison. In: Malyshkin, V. (ed.) PaCT 2011. LNCS, vol. 6873, pp. 13–22. Springer, Heidelberg (2011). doi:10.1007/978-3-642-23178-0_2
2. Bessonov, O.: Highly parallel multigrid solvers for multicore and manycore processors. In: Malyshkin, V. (ed.) PaCT 2015. LNCS, vol. 9251, pp. 10–20. Springer, Cham (2015). doi:10.1007/978-3-319-21909-7_2
3. Bessonov, O.: Parallelization properties of preconditioners for the conjugate gradient methods. In: Malyshkin, V. (ed.) PaCT 2013. LNCS, vol. 7979, pp. 26–36. Springer, Heidelberg (2013). doi:10.1007/978-3-642-39958-9_3
4. Accary, G., Bessonov, O., Fougère, D., Gavrilov, K., Meradji, S., Morvan, D.: Efficient parallelization of the preconditioned conjugate gradient method. In: Malyshkin, V. (ed.) PaCT 2009. LNCS, vol. 5698, pp. 60–72. Springer, Heidelberg (2009). doi:10.1007/978-3-642-03275-2_7
5. Saad, Y.: Iterative Methods for Sparse Linear Systems. PWS Publishing, Boston (2000)
6. Bastian, P., Horton, G.: Parallelization of robust multi-grid methods: ILU-factorization and frequency decomposition method. SIAM J. Stat. Comput. **12**, 1457–1470 (1991)
7. Elizarova, T., Chetverushkin, B.: Implementation of multiprocessor transputer system for computer simulation of computational physics problems (in Russian). Math. Model. **4**(11), 75–100 (1992)
8. Bessonov, O., Fedoseyev, A.: Parallelization of the preconditioned IDR solver for modern multicore computer systems. In: Application of Mathematics in Technical and Natural Sciences: 4th International Conference. AIP Conference Proceedings, vol. 1487, pp. 314–321 (2012)
9. Stüben, K.: A review of algebraic multigrid. J. Comput. Appl. Math. **128**, 281–309 (2001)

Application of Graph Models to the Parallel Algorithms Design for the Motion Simulation of Tethered Satellite Systems

A.N. Kovartsev and V.V. Zhidchenko[✉]

Samara National Research University, Samara, Russia
kovr_ssau@mail.ru, vzhidchenko@yandex.ru

Abstract. Tethered satellite systems (TSS) are characterized by ununiform distribution of mass characteristics of the system and the environment parameters in space, which necessitates the use of mathematical models with distributed parameters. Simulation of such systems is performed with the use of partial differential equations with complex boundary conditions. The complexity of the boundary conditions is caused by the presence of the end-bodies that perform spatial fluctuations, and by the variable length of the tether. As a result computer simulation of TSS motion takes a long time. This paper presents a parallel algorithm for motion simulation of the TSS and representation of this algorithm in the form of a graph model in graph-symbolic programming technology. The main characteristics of the proposed algorithm and the advantages of using graph models of algorithms for modeling the motion of the TSS are discussed.

Keywords: Tethered satellite systems · Parallel computing · Visual programming · Graph models

1 Introduction

Application of tethered satellite systems (TSS) opens up new possibilities in the use of outer space: the creation of artificial gravity, transport operations in space, returning payloads from orbit, the launch of small satellites from the main spacecraft, the use of the Earth's geomagnetic field for orbital maneuvers, creating orbiting power stations, atmospheric probing, study of geomagnetic and gravitational fields, removal of space debris, etc. [1].

Despite of the presence of a large number of works, which deal with various aspects of the space tether systems design, at the present time there is a certain lack of research on the development of methods of analysis and synthesis of controlled and free movement of space tethers of great length. Long tethers are characterized by ununiform distribution of mass characteristics of the system and the parameters of the environment in space. It determines the use of mathematical models with distributed parameters. The apparatus of partial differential equations is used with complex boundary conditions. The complexity of the boundary conditions is caused by the presence of the end-bodies that perform spatial fluctuations, and by the variable length of the tether. All this leads

© Springer International Publishing AG 2017
V. Malyshkin (Ed.): PaCT 2017, LNCS 10421, pp. 114–123, 2017.
DOI: 10.1007/978-3-319-62932-2_10

to considerable time needed for the mathematical modeling of tethered system on the computer. Large systems of ordinary differential equations (ODE) are used for modeling. The number of equations is measured in tens of thousands. Direct numerical solution of such systems is difficult, even for the up to date computing resources.

Existing approaches to the solution of large systems of equations can be divided into two main classes. The first class focuses on parallelization of the known numerical methods (often without changing the methods themselves). However, the results of solving ODE systems using this method on cluster systems are not impressive, because they do not take into account the features of the problem to be solved.

The second class comprises the methods that reduce the computational costs due to special heuristic techniques, which usually use the physical features of the problem. The accuracy of the solution is given less attention. The improvement of the calculation speed is gained by decrease in the accuracy of numerical methods without lowering the quality of the results.

We will consider the application of these approaches to the problem of motion simulation of space tether systems.

2 Mathematical Model of Tether Dynamics in Space

The widely used mathematical model of tethered system motion is a model in which tether is described by a system of partial differential equations. In this case, the mathematical models of continuum mechanics are used to describe the motion of tether in which the tether is considered as extensible (or inextensible) slim body, most often of great length [1]. Derivation of equations of motion of such a system is quite simple. It involves the consideration of the stretched differential element of the tether with the length ΔS and the application of Newton's second law to it:

$$\rho(S)\frac{\partial^2 \vec{r}}{\partial t^2} = \frac{\partial \vec{T}}{\partial S} + \vec{q} \tag{1}$$

where $\rho(S)$ is the linear mass density of the tether, $\vec{r}$ is the position of the stretched differential element, t is the time, $\vec{T}$ is the tension force, $\vec{q}$ is the resultant force acting on the differential element divided to the length of the element.

For a flexible tether, which does not accept transverse loads, tension force is directed tangentially to the tether line, so

$$\vec{T} = T\vec{\tau}, \ \vec{\tau} = \frac{1}{\gamma}\frac{\partial \vec{r}}{\partial S}, \tag{2}$$

where $\vec{\tau}$ is tangent unit vector, $\gamma = \left|\frac{\partial \vec{r}}{\partial S}\right|$.

The tether tension in the simplest case obeys Hooke's law

$$T(\gamma) = EA(\gamma - 1), \tag{3}$$

where E is the modulus of elasticity, A is the cross-sectional area of tether, $\gamma - 1$ is the elongation.

The equation of motion of the flexible tether (1), taking into account the Eqs. (2), (3) is a partial differential equation, which is solved with given boundary and initial conditions of motion.

To integrate the equation of motion (1) conventional numerical methods for solving partial differential equations of the wave type can be used: finite difference method, the method of separation of variables, etc. The solution of partial differential Eq. (1) can be reduced to the solution of a large number of ordinary differential equations. In this case, discretization of (1) is performed along the length of the tether. Then the system of ordinary differential equations is numerically solved to calculate the movement in time of the points for which the tether is broken for.

Integration over time of the initial system of equations arises a computationally intensive task. In order to solve it adequately one should carefully choose the parameters of the numerical methods.

A discrete analogue of the Eq. (1) is a mechanical system consisting of a set of N material points connected by elastic weightless connections (Fig. 1).

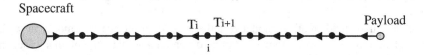

Fig. 1. Discrete model of the tethered satellite system

Various modifications to this model were used by many authors [2–4]. Within the framework of this model the main effects of the TSS motion are taken into account: tether elasticity, spread of longitudinal and transverse vibrations, power dissipation, etc. The boundary and initial conditions of TSS motion are rather simply defined through the definition of corresponding conditions for the end points. As the end-bodies the material points can be considered as well as the bodies of finite size. In the latter case, if necessary, the common equations of rotational motion are considered (dynamic and kinematic Euler's equations).

The equations of motion of M material points with elastic constraints are written as a system of ordinary differential equations

$$\frac{d\vec{r}_k}{dt} = \vec{V}_k, \quad m_k \frac{d\vec{V}_k}{dt} = \vec{F}_k, \tag{4}$$

where $\vec{r}_k$, $\vec{V}_k$, m_k are the position, velocity and mass of the point k respectively. $\vec{F}_k$ is the resultant of the forces acting on the material points $k = 1, 2, \ldots, M$.

The tension force of a tether is determined in the simplest case by the Hooke's law, assuming a linear dependence of the force from the tension. If the tether length becomes smaller than its length in the undeformed state, the tension force is assumed to be zero, because a thin tether does not support compression stress.

3 Numerical Simulation of the Tethered System Motion

Consider some features of the numerical simulation of the TSS as a system with distributed parameters. To do this, we shall use the multi-point model of TSS (Fig. 1). Representation of the tether as a set of material points reduces the problem of integration of the partial differential Eq. (1) to the integration of the system of ordinary differential equations of high order. This problem is a classic Cauchy problem

$$\frac{dX}{dt} = F(X, t),\tag{5}$$

where $X = (x_{ij}) \in R^{M \times n}$ is the phase coordinates matrix for each point of the TSS, n is the number of variables that represent the state of the system, M is the number of points in discrete model of the tether, including the end-bodies (spacecraft and payload), $F = (f_{ij}(X, t)) \in R^{M \times n}$. Initial conditions are described by the matrix $X(t_0) = X_0$ and the simulation is considered on the time interval $t \in [t_0, t_K]$.

Numerical solution of the Eq. (5) is usually computed using the Runge-Kutta fourth-order method. In this case the phase coordinates are calculated as follows:

$$X^{(r+1)} = X^{(r)} + \frac{h}{6}(K_1 + 2K_2 + 2K_3 + K_4),\tag{6}$$

where K_1, K_2, K_3, K_4 are matrices of Runge-Kutta coefficients.

4 Parallel Algorithm for the Motion Simulation of Tethered Satellite Systems

It is obvious that more accurate solution of the system (5) is possible through the increase of the number of points that describe the tether. When $M \to \infty$ the discrete analogue of the tether is converted into a continuous model. However, a large number of points significantly increases the complexity of the problem of motion simulation. Therefore, it is necessary to parallelize the algorithm and use the multiprocessor computing system. This raises a number of difficulties. Firstly, the impact of the links of each point with its right and left neighbors has to be taken into account to consider the cable tension forces (from the point of view of the influence of gravitational and inertial forces the points are independent of each other). Second, matrices K_1, K_2, K_3, K_4 can not be calculated independently of each other, but only sequentially in order of increasing index.

Let $X = X^{(r)}$ be the matrix of initial coordinates. In the first phase for each point of the cable with the number j the matrix K_1 is calculated according to the formula

$$k_1^j = f_j(x_{j-1}, x_j, x_{j+1}),\tag{7}$$

where k_1^j, $f_j(\ )$, x_{j-1}, x_j, x_{j+1} are the rows of the corresponding matrices K_1, F, X. The new value of the phase coordinates which is necessary for the calculation of matrix K_2 can be represented in matrix form

$$\tilde{X} = X + hK_1/2. \tag{8}$$

Similarly to (7) and (8) the Runge-Kutta formulas for phases 2, 3 and 4 may be represented by the following formulas:

$$k_2^j = f_j(\tilde{x}_{j-1}, \tilde{x}_j, \tilde{x}_{j+1}), \quad \tilde{\tilde{X}} = X + hK_2/2, \tag{9}$$

$$k_3^j = f_j(\tilde{\tilde{x}}_{j-1}, \tilde{\tilde{x}}_j, \tilde{\tilde{x}}_{j+1}), \quad \tilde{\tilde{\tilde{X}}} = X + hK_3, \tag{10}$$

$$k_4^j = f_j(\tilde{\tilde{\tilde{x}}}_{j-1}, \tilde{\tilde{\tilde{x}}}_j, \tilde{\tilde{\tilde{x}}}_{j+1}), \tag{11}$$

together with (6), for the fourth phase.

Let's split the tether into p segments. Suppose that the number of points M is divisible by p, so that $m = M/p$. Each tether segment includes m points (Fig. 2).

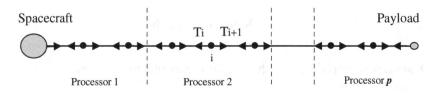

Fig. 2. Splitting the set of tether points into segments

Next we will split the matrix X by rows into p submatrices $X^k = (x_{ij}^k) \in R^{m \times n}$, where $k = 1, \ldots, p$ and $X = \bigcup X^k$. The matrix F^k can be defined the similar way. Now the tether coordinates can be calculated on each of the p processors relatively independently.

To make the parallel algorithm correct it is necessary to supply each processor (tether segment) with the information about the last point of the preceding segment and the first point of the next segment. For this purpose the processor number k must have access to the matrix D^k that stores the phase coordinates of the corresponding points:

$$D^k = \left\| \begin{matrix} x_{m,1}^{k-1} & x_{m,2}^{k-1} \ldots x_{m,n}^{k-1} \\ x_{1,1}^{k+1} & x_{1,2}^{k+1} \ldots x_{1,n}^{k+1} \end{matrix} \right\| \tag{12}$$

In the case where the parallel program is created for shared memory systems, the data for the matrix D^k can be obtained directly from the matrix X. It is only necessary to provide synchronization between the processors so that each processor could receive the coordinates of points from the neighbor segment, calculated by another processor, at the right time. For the problem under consideration it is convenient to choose the barrier synchronization setting the barrier after calculating each matrix K_i. Since the matrices K_1, K_2, K_3, K_4 must be calculated sequentially, the parallel algorithm can be represented as a sequence of four phases of parallel calculations. At each phase K_i is computed in parallel. The processor number k calculates the rows of the matrix K_i that correspond to the points of its tether segment. To do this it uses the matrices X^k and F^k. After the end of the calculation of phase K_i the processes arrive to the barrier to get new phase coordinates $\tilde{X}, \tilde{\tilde{X}}, \tilde{\tilde{\tilde{X}}}$ from the matrices D^k required for the calculation of K_{i+1}.

If the parallel program is created for distributed memory systems, the processes must form the matrices D^k from the messages received from other processes.

5 Graph Model of the Parallel Algorithm

We represent the parallel algorithm graphically. To do this we will use the notation of graph-symbolic programming technology (GSP) [5, 6]. The GSP technology allows to describe the algorithm as a set of control flow diagrams. The program that implements the algorithm is compiled and run automatically. Control flow diagram is a directed graph, where the nodes represent the actions performed on the data, and the arcs represent the sequence of execution of these actions. If some node has several outgoing arcs, then depending on the type of the arc either one or multiple adjacent nodes simultaneously may execute. The arcs of different types have different graphical representations. The advantage of the control flow diagrams is a visual representation of the sequence of computations in the program. The disadvantage is the lack of visibility of data dependencies between the nodes. Figure 3 displays the parallel algorithm for calculating the coefficients of the Runge-Kutta method described above.

On Fig. 3 the arcs, which are marked with the circle in the beginning, represent a transfer of control to the new process. The arcs, the beginning of which is marked by a slash, depict the return of control from another process. At each phase of calculations four processes are used to calculate the different tether segments. The graph nodes in GSP technology can be marked with text or images. For clarity, in Fig. 3 some nodes are labeled with the icons depicting simplified image of a tethered satellite system and calculation phase number. These nodes represent the subprograms for calculation of the individual tether segments. Thus, the graph of the algorithm in Fig. 3 is hierarchical. The nodes marked with images actually represent other graphs that consist of three nodes. The content of one of these graphs is depicted in the right part of the figure. Hierarchical construction of graphs in GSP technology allows you to hide unimportant details of the implementation of various parts of the algorithm. On each level of hierarchy the nodes on the graph describe important elements that help to understand the structure of the algorithm.

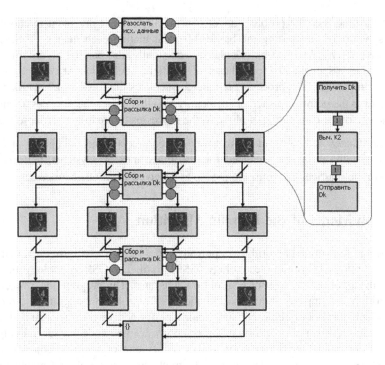

Fig. 3. Graph model of the parallel algorithm for calculating the coefficients of the Runge-Kutta method

6   Evaluation of Speedup of the Parallel Program for Numerical Simulation of the Tethered System Motion

We estimate the speedup of the parallel program that implements the algorithm described above. The calculation of K_i for one point of tether in accordance with formula (7) reduces to computing the value of the function f. This function in the problem under consideration includes about 160 mathematical operations. Let τ be the duration of execution of a single mathematical operation. The calculation of K_i takes $\alpha = 160 \cdot \tau$. The time consumed by the sequential program, which calculates the phase coordinates for M tether points in Nt time steps, in accordance with the formula (6) can be estimated by the following expression:

$$T_1^{sum} \approx 4 \cdot \alpha \cdot M \cdot Nt \qquad (13)$$

Let's estimate the duration of the parallel program execution. In the parallel algorithm discussed above in each phase of the calculation K_i is computed independently on p processors. The duration of each phase is p times smaller compared with the sequential program. The total time needed for the parallel program execution for calculating the phase coordinates of M points in one time step is the sum of the duration of four calculation phases and the duration of data exchange between the segments:

$$T_{\parallel}^1 \approx \frac{4 \cdot \alpha \cdot M}{p} + T_{send} \qquad (14)$$

In programs for the shared memory systems the data exchange time can be neglected. Calculation of the phase coordinates for different time steps is performed sequentially, so the total time consumed by the parallel program is:

$$T_{\parallel shared}^{sum} \approx \frac{4 \cdot \alpha \cdot M}{p} \cdot Nt \qquad (15)$$

Parallel program speedup for the shared memory systems will be the following:

$$S_{shared} = \frac{T_1^{sum}}{T_{\parallel shared}^{sum}} = p \qquad (16)$$

On the distributed memory systems, at the beginning of the first phase of calculations each process must receive the initial data (matrices X^k and F^k). Between the phases the processes must exchange the messages that contain the matrices D^k. As an example, consider a parallel program that uses the Message Passing Interface (MPI) for exchanging data between the processes. The main process of the program that performs input-output of matrices X and F, must send the bands of these matrices to other processes. Let β is the time required to transfer 1 byte of data. Then, to transfer X and F the main process will spend

$$T_{send}^0 = \frac{(20+n) \cdot sizeof(x_i) \cdot M \cdot \beta}{p} \cdot p = (20+n) \cdot sizeof(x_i) \cdot M \cdot \beta \qquad (17)$$

Expression (17) assumes that every point is described by n phase coordinates, and each element of the matrix F^k contains 20 parameters. The $sizeof(x_i)$ function returns the size of the data type, which presents the phase coordinate.

Consider the common case where n = 4 and sizeof(x_i) = 8. As on distributed memory systems, such as clusters, β is much greater than τ, the transfer of initial data takes comparable or larger time than the calculation of all tether points in one time step (see (13)). Therefore, on cluster we should send initial data once, before the calculations.

Data exchange between the phases is as follows. At the beginning of each phase the main process sends the messages containing matrices D^k to other processes. At the end of each phase the processes send the messages containing the new values of the coordinates to the main process. Since the matrix D^k comprises $2 \cdot n$ elements the total time needed to exchange matrices sequentially in four phases is:

$$T_{send}^{Dk} = (2 \cdot n \cdot sizeof(x_i) \cdot \beta \cdot p) \cdot 4 \cdot 2 = 16 \cdot n \cdot sizeof(x_i) \cdot \beta \cdot p \qquad (18)$$

Taking into account expressions (13), (14), (17), (18), we get the following estimation for the parallel program speedup in case of distributed memory system:

$$S_{distr} = \frac{T_1^{sum}}{T_{||distr}^{sum}} = \frac{p}{1 + \frac{(20+n)\cdot sizeof(x_i)\cdot \beta}{4\cdot \alpha \cdot Nt} + \frac{4\cdot n\cdot sizeof(x_i)\cdot \beta \cdot p^2}{\alpha \cdot M}} \tag{19}$$

Table 1 shows the results of experiments made on computing cluster of Samara University which consists of 180 nodes with 360 Intel Xeon 2.8 GHz processors.

Table 1. Experimental results

Number of processes	Duration of calculations, sec	Actual speedup	Estimation of speedup
1	60.000	–	–
5	15.726	3.815	4.314
10	12.986	4.620	7.488
30	6.122	9.800	13.514
100	4.935	12.159	13.235
180	8.062	7.443	10.045
360	13.579	4.419	6.220
450	17.664	3.397	5.202

The calculations were made for 90000 tether points and 400 time steps. The results show the maximum speedup in case of about 100 processes. Further increase of number of processes increases the amount of data being transferred between the calculation phases and decreases the overall speedup. The estimation of speedup calculated in accordance to the expression (19) is close to the experimental data. Expression (19) provides the way to estimate the optimum number of processors for execution of parallel program for motion simulation of the TSS.

7 Conclusion

The paper considers the problem of numerical simulation of the motion of tethered satellite systems using high-performance computing systems. The features of the problem are indicated that affect the possibility of parallelization. A parallel algorithm is proposed for the implementation of the Runge-Kutta method according to the features of the problem. Estimation of speedup of the parallel program that implements the proposed algorithm is given, for the cases of shared and distributed memory systems. The proposed algorithm is described in the form of graphic model with the help of the graph-symbolic programming technology, which provides automatic synthesis of a parallel program based on the model of the algorithm. Graphic model simplifies the analysis of the algorithm and its modification. It contributes to the development of new algorithms and research development in the field of modeling the motion of tethered satellite systems.

Acknowledgements. The work was partially funded by the Russian Federation Ministry of Education and Science and Russian Foundation of Basic Research. Grant #16-41-630637.

References

1. Beletsky, V.V., Levin, E.M.: Dynamics of Space Tether Systems. Univelt, San Diego (1993)
2. Dignat, F., Shilen, V.: Variations control of orbital tethered system (in Russian). J. Appl. Math. Mech. T.64 **5**, 747–754 (2000)
3. Zabolotnov, Y.M., Fefelov, D.I.: Motion of light capsule with the tether in the extra-atmospheric section of deorbit (in Russian). In: Proceedings of the Samara Scientific Center of the Russian Academy of Sciences, T.8, #3, pp. 841–848 (2006)
4. Zabolotnov, Y.M., Elenev, D.V.: Stability of motion of two rigid bodies connected by a cable in the atmosphere. Mech. Solids **48**(2), 156–164 (2013)
5. Kovartsev, A.N., Zhidchenko, V.V., Popova-Kovartseva, D.A., Abolmasov, P.V.: The basic principles of graph-symbolic programming technology (in Russian). In: Open Semantic Technologies for Intelligent Systems, pp. 195–204 (2013)
6. Egorova, D., Zhidchenko, V.: Visual Parallel Programming as PaaS cloud service with Graph-Symbolic Programming Technology. Proc. Inst. Syst. Program. **27**(3), 47–56 (2015). doi:10.15514/ISPRAS-2015-27(3)-3

The DiamondTetris Algorithm for Maximum Performance Vectorized Stencil Computation

Vadim Levchenko and Anastasia Perepelkina$^{(\boxtimes)}$

Keldysh Institute of Applied Mathematics RAS,
Miusskaya sq., 4, Moscow, Russia
lev@keldysh.ru, mogmi@narod.ru

Abstract. An algorithm from the LRnLA family, DiamondTetris, for stencil computation is constructed. It is aimed for Many-Integrated-Core processors of the Xeon Phi family. The algorithm and its implementation is described for the wave equation based simulation. Its strong points are locality, efficient use of memory hierarchy, and, most importantly, seamless vectorization. Specifically, only 1 vector rearrange operation is necessary per cell value update. The performance is estimated with the roofline model. The algorithm is implemented in code and tested on Xeon and Xeon Phi machines.

1 Introduction

The main purpose of supercomputer advancement is scientific computing application, among which stencil schemes are widely used.

The efficient use of all levels of parallelism of a hybrid hardware is essential, yet insufficient to achieve the maximum performance. Since the stencil computation is a memory bound problem, one of the algorithmic challenges lies in mitigating the bottleneck posed by the memory bandwidth to floating point performance ratio.

The most common strategy of the solution of a stencil problem is the stepwise advancement, when the next layer in time is computed only after the previous time iteration has been computed completely. This convention arises from the natural concept of time, but sets an artificial limit on locality of data reference, and data reuse.

The techniques of time-space approach have been introduced decades ago [12]. Some of these are known as temporal tiling, split-tiling [5] (trapezoidal [3], hexagonal [4], diamond [1]), temporal blocking [10], time-skewing [8], wave-front tiling [13]. With the advancement of the new computer hardware, the interest in such techniques increases even more.

The loop optimization in 4 dimensions leads to higher performance, but introduces several complications.

Data dependencies should be traced to ensure the correctness of computations. With the wavefront approach the correctness is apparent, but it limits the optimization freedom.

© Springer International Publishing AG 2017
V. Malyshkin (Ed.): PaCT 2017, LNCS 10421, pp. 124–135, 2017.
DOI: 10.1007/978-3-319-62932-2_11

Diamond blocking offers more locality and parallelization possibilities, but often discarded as complex. The limit here is not only the difficulty of programmers perception of 4D space, but also the overhead in the instruction count.

Another complication is in the programming the algorithm. While the stepwise algorithm may be implemented by several nested "for" loops that are offered by every relevant programming language, temporal blocking with a stencil scheme require much more lines of code. Code generation techniques [9] are developed to negate the undesirable complication and make the visible code human-friendly.

Temporal blocking algorithms rely on the notion of the data locality and reuse, but under the given restrictions there exist a vast number of ways to block the time-space domain.

The search of four-dimensional traversal rule that is optimal for each specific system is an open challenge.

Locally Recursive non-Locally Asynchronous (LRnLA) [7] approach shares similarities with temporal blocking, but evolved separately from the mentioned research. It addressed the following objectives in algorithm optimization:

- Systematize the description of high-dimensional blocking.
- Make a quantititaive measure of algorithm efficiency, which can be used for estimating the resulting performance ona given hardware.
- Maximize the spacial and temporal data locality.

DiamondTetris, the most recent LRnLA algorithm, is introduced in this paper. It is built as an upgrade of a DiamondTorre algorithm for solution of the finite difference cross stencil problems on multicore processors. Specifically, Intel Xeon Phi Knights Landing architecture is chosen for consideration as it becomes the base processing power in the emerging supercomputers.

DiamondTetris features higher locality than DiamondTorre due to the full 4D decomposition in time and space, and higher locality than ConeFold, the previous LRnLA algorithm for CPU, since it takes into account not only the locality, but also the cross shape of the stencil.

The algorithm is built with account for all available levels of parallelism, and uses all levels of memory hierarchy. Among these, when programming for Xeon Phi architecture, it is gravely important to make full use of the vector processing. DiamondTetris solves the problem of vectorizing the computation in stencil scheme by using the minimal vector reshuffle operations.

2   Wave Equation

The wave equation is chosen for a simple illustration of the method

$$\frac{\partial^2 f}{\partial t^2} = c^2 \triangle f \ . \tag{1}$$

Partial differential with respect to time is approximated by the second order finite difference:

$$\frac{f^+ - 2f^0 + f^-}{\triangle t^2} = c^2 \triangle f^0 \ . \tag{2}$$

Here Δt is the time step. To calculate a new value of f the values on the two previous time steps are necessary:

$$f^+ = 2f^0 - f^- + c^2\Delta t^2 \triangle f^0. \tag{3}$$

The computation is conducted as the update of the mesh values. To ensure that for each step we have necessary data, two data arrays are used. f is defined on the even time steps; g is defined on the odd ones. They are updated in turns:

$$f = 2g - f + c^2\Delta t^2 \triangle g,$$
$$g = 2f - g + c^2\Delta t^2 \triangle f.$$

Though the size of the stencil is an adjustable parameter, the discussion here is provided for the specific case. 13-point 4-th order 3D stencil is used for the approximation of the laplacian. The choice is justified by the previous experience in applied computing. For wave simulation 7 points per wavelength is enough for desired accuracy, with little more operations compared to the simplest 2nd order stencil.

We define ShS (Stencil half Size) parameter, which is equals 2 for the chosen stencil. The stencil is cross-shaped. By applying the stencil to a point, then to all the points in the stencil dependencies, and so on, an octahedron is generated.

3 Algorithm

The construction of the algorithm is made by the usual LRnLA principles. We have previously described DiamondTorre construction in x–y–t in [6]. The new algorithm DiamondTetris is based on a time-space decomposition in 4D.

On the $t = 0$ time layer the base form is chosen as a prism (Fig. 1). The faces, that are perpendicular to z-axis are diamond-shaped. The left point of the top face projects on the right point of the bottom face. The incline angle of the prism is $\pi/4$. This shape is chosen so as to adhere to the following criteria: 1. Closely circumscribes the octahedron (in fact, it consists of an octahedron, two half-octahedra (pyramids), 4 tetrahedra to fill the space between them). 2. The simulation domain may be tiled by this shape with no overlaps or gaps This 3D prism in x–y–z is called DiamondTile. It is associated with an algorithm in which all mesh points that are inside the shape are updated once.

Its size is a parameter. For uniformity of description, it is a factor of ShS. We call the scaling factor DTS — Diamond Tile Size. The base of DiamondTile is a diamond that has $2ShS \cdot DTS$ points along x diagonal, $(2ShS \cdot DTS - 1)$ points along y diagonal. DiamondTile is $2ShS \cdot DTS$ points high in z direction.

The second base is a similar shape on the time layer $t = NT \cdot 2 \cdot DTS \cdot \Delta t$, and $2 \cdot NT \cdot DTS \cdot ShS$ mesh steps lower along z axis, where NT is an integer parameter.

The intersection of the dependency cone of the former and the influence cone of the latter is a 4D shape, which can be described by its x–y–z slices. On the i_t time step its slice is the similar DiamondTile $i_t \cdot ShS\Delta t$ mesh steps lower in

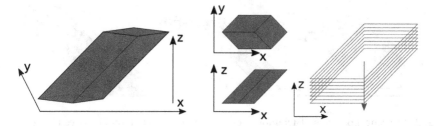

Fig. 1. DiamondTile (left). Its 2D projections (center). DiamondTetris (right). Since the image of the overlapping 3D shapes may be confusing, only the central slice of the falling DiamondTile is depicted. Green (f values computations) and blue (g values) are interchanging. Refer to the electronic version of the paper for colors (Color figure online)

the z axis. It may be associated with the DiamondTile 3D shape 'falling' down along the z axis. So the described 4D shape and the algorithm corresponding to it are named DiamondTetris.

As an algorithm, DiamondTetris for the given equation is described as follows (see (2)):

1. perform stencil calculation for all f values 3D points inside the DiamondTile shape;
2. perform calculation for all g values inside the DiamondTile, shifted by ShS mesh points down in z axis;
3. repeat steps 1–2 ($NT \cdot DTS$) times, each time the DiamondTile is shifted down by ShS.

The construction of the algorithm as an intersection of the influence cone and the dependency cone guarantees that these shapes, tiled in x–y–z–t, have either unilateral data dependencies, or none at all.

DiamondTiles, that are adjacent in x, z or t direction have unilateral dependencies (Fig. 2). A row of DiamondTetrises with the same x, z and t coordinate are asynchronous. They have no data dependencies between each other and may be processed in parallel.

The row along the y axis may be computed concurrently (red in Fig. 2). The computation will be valid in terms of data dependencies, if all DiamondTetrises below and to the right of it are processed. Some closest of them are pictured yellow on Fig. 2. Alternatively, this (red) row may be the first one in the computation, if it intersects the boundary. In this case, computation may differ if the boundary condition applies.

After it is computed, the row shifted by $z = 2ShS \cdot DTS$ and $x = DTS \cdot ShS$ (purple), or by $x = -ShS \cdot DTS$ and $y = ShS \cdot DTS$ (blue) may be processed. Purple and blue row are asynchronous and may be computed independently.

The DiamondTetris 4D shape is a natural generalization of a 3D Diamond-Torre shape in [6,14]. In DiamondTorre a 2D diamond progresses to the right in x direction; in DiamondTetris a 3D DiamondTile progresses down in z direction.

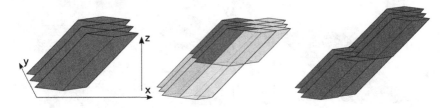

Fig. 2. Starting positions of DiamondTetrises. A row of DiamondTetrises along y axis is asynchronous (red). It may be started when all DiamondTetrises below (yellow, only some are shown) are processed. After it, the next row to the left (blue) and the next row above (purple) may be started. Refer to the electronic version of the paper for colors (Color figure online)

4 Implementation

4.1 Data Structure

The data structure for field storage is optimized with account for uniformity and locality of data access. We define (Fig. 3)

an element as $4ShS^3$ mesh points, situated in a small DiamondTile shape;
an element block as DTS^3 elements;
a cell as 2 blocks with f field data and 2 blocks with g field data.

These data structures are used for storage. Cells are stored in a 3D array. An element size is 256B (double precision).

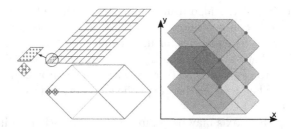

Fig. 3. Element. Element block. Pairs of blocks in a rectangular array.

For computation of DiamondTetris we define (Fig. 4):

A vector as $2DTS \cdot ShS$ points, combined in one data structure. These points are situated diagonally in the x–z slice. Each vector consists of $2ShS$ SIMD vectors.
A vector element as $2ShS^2$ vectors of $2DTS \cdot ShS$ points. It is constructed from DTS elements, standing on top of each other.

A vector block as $(DTS+1)^2$ vector elements comprising a shape that is close to a DiamonTile, but wider.

These data structures are used for computation. For 1 DiamondTetris we need 2 vector blocks at a time: one for the f field, and one for the g field.

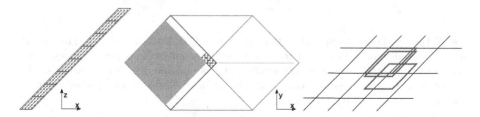

Fig. 4. Vector element in x–z. Vector block in x–y. Vector block data 'falls' in the cell block array

4.2 Computation Flow

Processing flow of 1 DiamondTetris is as follows (Fig. 5):

1. Load cell data to the two (f and g) vectors blocks.
2. With the loaded data, perform all possible calculation for the DTS^2 vector blocks with f values (green color area on Fig. 5). Some stencil points cannot be accessible now due to the deficiency of data in the g vector block.
3. Vector shift: vector elements are updated. The top $2 \cdot ShS$ points of each vector are saved to memory and discarded from the vector block strusture. $2 \cdot ShS$ points below are loaded from memory.
4. The remaining computation for the f values is conducted.
5. Data from the leftmost g field vector elements (a 'hood') are saved to memory and discarded from the vector block. The data from right of the vector block are loaded from memory to take its place.
6. At this point the vector block positions are analogous to the initial one, with f and g swapped. The 2–5 steps are repeated for g computation and f load/save.
7. 2–6 are repeated $NT \cdot DTS$ times.
8. The computed data from the vector blocks that weren't saved before are saved to the memory.

At this point we choose the optimal DTS. At each calculation 2 vectors blocks (f and g) are needed. With $DTS = 8$ the data of two vector blocks (double precision) fits L2 cache.

The choice of $DTS = 8$ is especially convenient now, since 8 double precision values may fit into a 512-bit vector of Xeon Phi. 8 single precision values fit into a 256-bit vector of AVX2 instruction set. of a target platform.

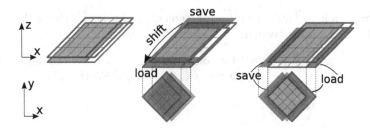

Fig. 5. DiamondTetris computation flow, $DTS = 4$. Left: Data is loaded, f (green filled area) is partially computed. Refer to the electronic version of the paper for colors. Center: vector shift, the remaining computation of f is possible. Right: load/save 'hood'. Return to the initial position with f and g switched. Notice that f vector position stays the same. (Color figure online)

4.3 Vector Computation

The computation with SIMD vectors is implemented so as to minimize vector rearranging operations. Specifically, vector rearrange operations exist only on step 3.

Figure 6 represents the two states of the computation of the each f vector: before and after the shift.

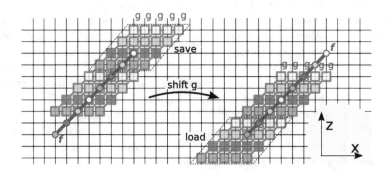

Fig. 6. f (blue circles) and g (green squares) vector positions before (left) and after (right) shift. Refer to the electronic version of the paper for colors. In one vector the shapes with the same filling belong to one SIMD vector. For simplicity of depiction, the vectors are shorter ($DTS = 2$ instead of 8). Only those vectors that are relevant for the computation of 1 f vector are shown, x–z projection. (Color figure online)

One f (green lines) or g (blue lines) vector consists of 4 SIMD vectors. Data belonging to the different SIMD vectors inside one vector are shown by different fill color.

The g value data in the each point of the stencil should be summed with a specific coefficient to calculate the correct update for the f value.

Table 1. Reference for vector summation. Refer to the electronic version of the paper for colors

	$+2\Delta z$	$+\Delta z$	0	$-\Delta z$	$-2\Delta z$
f	g	g	g	g	g (shifted)
f	g	g	g	g (shifted)	g (shifted)
f	g	g	g (shifted)	g (shifted)	g (shifted)
f	g	g (shifted)	g (shifted)	g (shifted)	g (shifted)

At $+2\Delta z$ the cross stencil has one point. For each point in the f vector this data is accessible in the vector block before the shift. At $+\Delta z$ the stencil also has just one point. For SIMD vectors shown by magenta, yellow, white color fill this point is accessible before the shift, for cyan colored SIMD vector all data may be reached after the shift. Each time the summation is a direct vector summation.

At $+0\Delta z$ layer the stencil has 9 points that comprise the finite differences in the x and y directions. These can be computed for SIMD vectors shown by cyan and magenta before the shift, and for the rest after the shift.

The Table 1 shows which vectors at what point are ready for the summation. Not shifted values are summed before the shift (step 2 above), shifted SIMD vectors are added after the shift (step 4 from above).

4.4   Boundary Conditions and Initial Values

The boundary condition that is the most simple to implement is periodicity. It is implemented in the y-axis direction.

However this condition cannot be valid with the DiamondTetris algorithm in x and z direction. In case when the data on different sides of the domain exists on different time steps, Bloch boundary condition is an obvious generalization of the periodicity, and works perfectly in this framework.

For example, to compute the data on the left side in the x direction, 2 left points of the stencil lie outside of the domain. Instead the points from the right side of the domain can be taken. Let $f(x,t)$ be the value of unknown point, and let $f(x + L_x, t + \tau)$ be the value on the right side of the domain. It is shifted by the domain x size L_x in space, and, in the moment when the leftmost data is calculated at the time t, the rightmost data has already progressed to time $t + \tau$, $\tau = 2 \cdot DTS \cdot NT \cdot dt$.

The simulated waves should satisfy:

$$k_x L_x = \omega\tau + 2\pi n, \quad n = \ldots -2, -1, 0, 1, 2 \ldots \tag{4}$$

The Bloch boundaries in the current implementation are ridged, since they follow the blocks of the data structure.

Same applies to the boundary in the z direction.

Due to the inclined shape of DiamondTetris, the starting row of DiamondTetrises should 'sit' on the rows below and to the right of it. Their data are initialized to satisfy this assumption. So, the initial values are set on the range of time steps from 0 to $2 \cdot DTS$. The example configuration of the initial state is shown on Fig. 7.

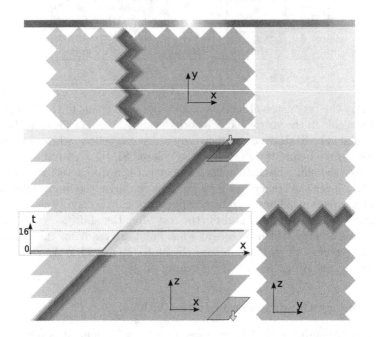

Fig. 7. The example of the initial state for the computation. In 3 2D slices of the 3D domain color shows the time step on which the f and g fields are initialized. White: no data, green to light blue: $t = 0$ to $t = 16$. $DTS = 8$, $NT = 1$, $8 \times 4 \times 9$ cells in data array. The 2 bases of DiamondTetris that may be initially computed are depicted on the x–z slice. (Color figure online)

The snapshot of wave propagation is shown on Fig. 8.

5   Results and Conclusions

The constructed algorithm has several advantages over the state of art effort in the field of stencil implementation for the Intel Xeon Phi architecture [13].

Firstly, it requires less vector rearranging operations in-between SIMD arithmetics, which leads to the more efficient use of SIMD parallelism. 1 vector shift per value update is performed.

The algorithm is tuned to fit L2 cache. One DiamondTetris data (two vector blocks) always fit L2 cache. The data that is to be loaded from memory satisfies the locality requirements since the data structure conforms with the algorithm shape. Furthermore, the computation of 1 vector element (with all 4 vector elements it depends on) fits the L1 cache.

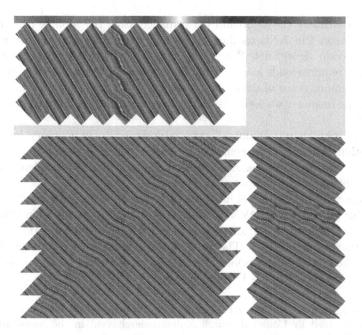

Fig. 8. Computation snapshot of the plain wave propagation. Time steps differ by 16 across the domain

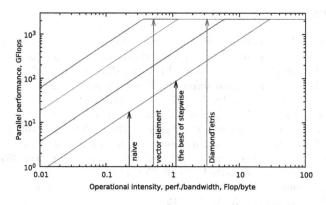

Fig. 9. Roofline model for KNL with stencil algorithm performance estimations

4D temporal blocking provides more asynchronous blocks than wavefront tiling, allowing the use of many parallel levels. The clear distinction between asynchronous and sequential DiamondTetrises allows effortless parallelization between cores. For inter-node parallelism the "Computation window" approach may be used like in [14]. Data transfers between nodes may be concealed by adjusting the NT parameter, so efficient scaling on many-node cluster may be anticipated.

We can estimate the algorithm performance using the roofline model [2,11]. On Fig. 9 Xeon Phi KNL roofline is plotted, with the estimated positions of the stencil code algorithms. "Naive" refers to the algorithm where each stencil application requires loads and stores of all its data. "The best of stepwise" refers to the maximum reuse of data due to stencil overlap. Since these data cannot fit the cache memory, we estimate that the performance is limited by the DDR bandwidth.

On the other hand, DiamondTetris is designed to fit memory levels. The computation of 1 vector element fits L1 cache, and the computation of whole DiamondTetris fits L2 cache. According to the model, it is possible to promote stencil problems to the compute-bound class.

The described algorithm is implemented in code. Two versions of code may run: double precision with AVX-512 instruction set (for use on Xeon Phi Many-Integrated-Core processors); single precision with AVX2 vectorization (for tests on the other modern CPU). The preliminary performance results reach $\sim 1/7$ from the theoretical peak performance and do not depend on the problem size (as long as the data size fits DDR memory).

Acknowledgments. The access to the computing resources with Intel Xeon Phi KNL has been provided by Colfax Research (colfaxresearch.com) in the course of "Deep Dive" HOW series.

References

1. Bertolacci, I.J., Olschanowsky, C., Harshbarger, B., Chamberlain, B.L., Wonnacott, D.G., Strout, M.M.: Parameterized diamond tiling for stencil computations with chapel parallel iterators. In: Proceedings of the 29th ACM on International Conference on Supercomputing, ICS 2015, pp. 197–206. ACM, New York (2015). http://doi.acm.org/10.1145/2751205.2751226
2. Doerfler, D., Deslippe, J., Williams, S., Oliker, L., Cook, B., Kurth, T., Lobet, M., Malas, T., Vay, J.-L., Vincenti, H.: Applying the roofline performance model to the Intel Xeon Phi knights landing processor. In: Taufer, M., Mohr, B., Kunkel, J.M. (eds.) ISC High Performance 2016. LNCS, vol. 9945, pp. 339–353. Springer, Cham (2016). doi:10.1007/978-3-319-46079-6_24
3. Frigo, M., Strumpen, V.: The memory behavior of cache oblivious stencil computations. J. Supercomput. **39**(2), 93–112 (2007)
4. Grosser, T., Cohen, A., Holewinski, J., Sadayappan, P., Verdoolaege, S.: Hybrid hexagonal/classical tiling for gpus. In: Proceedings of Annual IEEE/ACM International Symposium on Code Generation and Optimization, CGO 2014, pp. 66:66–66:75. ACM, New York (2014). http://doi.acm.org/10.1145/2544137.2544160
5. Henretty, T., Veras, R., Franchetti, F., Pouchet, L.N., Ramanujam, J., Sadayappan, P.: A stencil compiler for short-vector simd architectures. In: Proceedings of the 27th International ACM Conference on International Conference on Supercomputing, ICS 2013, pp. 13–24. ACM, New York (2013). http://doi.acm.org/10.1145/2464996.2467268
6. Levchenko, V., Perepelkina, A., Zakirov, A.: Diamondtorre algorithm for high-performance wave modeling. Computation **4**(3), 29 (2016). http://www.mdpi.com/2079-3197/4/3/29

7. Levchenko, V.: Asynchronous parallel algorithms as a way to archive effectiveness of computations. J. Inf. Technol. Comput. Syst. (1), 68 (2005). (in Russian)
8. McCalpin, J., Wonnacott, D.: Time skewing: a value-based approach to optimizing for memory locality. Technical report (1999). http://www.haverford.edu/cmsc/davew/cache-opt/cache-opt.html
9. Muranushi, T., Makino, J., Hosono, N., Inoue, H., Nishizawa, S., Tomita, H., Nitadori, K., Iwasawa, M., Maruyama, Y., Yashiro, H., Nakamura, Y., Hotta, H.: Automatic generation of efficient codes from mathematical descriptions of stencil computation. In: Proceedings of the 5th International Workshop on Functional High-Performance Computing, FHPC 2016. Association for Computing Machinery (ACM) (2016). https://doi.org/10.1145/2F2975991.2975994
10. Nguyen, A., Satish, N., Chhugani, J., Kim, C., Dubey, P.: 3.5DD blocking optimization for stencil computations on modern CPUs and GPUs. In: Proceedings of the 2010 ACM/IEEE International Conference for High Performance Computing, Networking, Storage and Analysis, SC 2010, pp. 1–13 (2010). http://dx.doi.org/10.1109/SC.2010.2
11. Williams, S., Waterman, A., Patterson, D.A.: Roofline: an insightful visual performance model for multicore architectures. Commun. ACM 52(4), 65–76 (2009). http://dblp.uni-trier.de/db/journals/cacm/cacm52.html#WilliamsWP09
12. Wolfe, M.: More iteration space tiling. In: Proceedings of the 1989 ACM/IEEE Conference on Supercomputing, Supercomputing 1989. ACM, New York (1989). http://doi.acm.org/10.1145/76263.76337
13. Yount, C., Duran, A.: Effective use of large high-bandwidth memory caches in hpc stencil computation via temporal wave-front tiling. In: Proceedings of the 7th International Workshop on Performance Modeling, Benchmarking and Simulation of High Performance Computing Systems, PMBS 2016, pp. 65–75. IEEE Press, Piscataway (2016). https://doi.org/10.1109/PMBS.2016.12
14. Zakirov, A., Levchenko, V.D., Perepelkina, A., Yasunari, Z.: High performance fdtd code implementation for gpgpu supercomputers. Keldysh Institute Preprints (44), 22 pages (2016). http://library.keldysh.ru/preprint.asp?id=2016-44

A Parallel Locally-Adaptive 3D Model on Cartesian Nested-Type Grids

Igor Menshov[1,2(✉)] and Viktor Sheverdin[2]

[1] Keldysh Institute for Applied Mathematics, Russian Academy of Sciences, Moscow, Russia
menshov@kiam.ru
[2] VNIIA, Moscow, Russia

Abstract. The paper addresses the 3D extension of the Cartesian multilevel nested-type grid methodology and its software implementation in an application library written in C++ object-oriented language with the application program interface OpenMP for parallelizing calculations on shared memory. The library accounts for the specifics of multithread calculations of 3D problems on Cartesian grids, which makes it possible to substantially minimize the loaded memory via non-storing the grid information. The loop order over cells is represented by a special list that remarkably simplifies parallel realization with the OpenMP directives. Test results show high effectiveness of dynamical local adaptation of Cartesian grids, and increasing of this effectiveness while the number of adaptation levels becomes larger.

Keywords: Locally-adaptive nested-type Cartesian grid · Shared memory parallel calculation · Gas dynamics equations

1 Introduction

Dynamic adaptation of the computational grid via local subdividing cells with the aim to increase accuracy of numerical solutions has been being studied for already more than thirty years [1–6]. However, with the development of modern supercomputers, such grid technology assumes additional specifics that require to design new approaches. The algorithm implementation on high performance parallel multi-core computer systems presents several difficulties and special aspects in organizing calculations and data transfers. For example, using GPUs leads to necessity of searching simple computational algorithms with small local computational stencils.

Cartesian grids are most appropriate for multi-core parallel computer systems due to simplicity in programming directional splitting methods, regular stencils of memory access, simple grid parameters as face normal, volume, face area, and so on. On the other hand, using structured regular Cartesian grids to obtain high quality results requires so many computational cells that the problem execution with even a modern supercomputer may take inadmissibly much time.

Commonly, high spatial resolution is required only in small subdomains where the solution has steep gradients. Using a large grid to be fine in all the computational domain seems to be not rational; it is more reasonable to employ grids with the option of refining

© Springer International Publishing AG 2017
V. Malyshkin (Ed.): PaCT 2017, LNCS 10421, pp. 136–142, 2017.
DOI: 10.1007/978-3-319-62932-2_12

grid elements (by dividing the cell in smaller subcells) only locally in accordance with some criteria. In this way we obtain a grid that is referred to as locally adaptive (LA) of the nested-type structure.

The technology of LA grids causes several problems related with its software implementation. The first is the data format on LA grids. This format must maintain the homogeneity of memory access, minimize resources used for calculation, and be sufficiently complete to realize basic solver functions (as, e.g., searching neighboring elements). Parallelizing algorithms on LA grids is another problem. For shared memory systems, issues of optimal data design and cell loop order require special consideration. Dynamic balancing work between nodes is very crucial for the distributed memory systems. The geometry representation and treatment of the boundary conditions on LA Cartesian grids is also an important issue.

The present paper addresses the above issues of Cartesian LA grids. We describe an application library written in C++ object-oriented language with the application program interface OpenMP for parallelizing calculations on the shared memory. The library accounts for the specifics of multithread calculations of 3D problems on Cartesian grids, which makes it possible to substantially minimize the loaded memory of the computer via non-storing the grid information. The loop order over cells is represented by a special list that remarkably simplifies parallel realization with the OpenMP directives. To treat the geometry, we implement local adaptation of the grid to the geometry with subsequent treatment of the boundary conditions by means of the free boundary method [7, 8].

2   Representation of Cartesian LA Grid in Computer Memory

We use an octree data format to represent the discrete solution on the Cartesian LA grid. A computational cell is allowed to split into eight equal subcells (child cells) by dividing the cell in two along each coordinate direction. As the base grid, a Cartesian structured grid of a size of $M \times N \times K$ cells is taken; the zero-level of adaptation is assigned to base cells.

Each cell of the LA grid is completely described by its level of adaptation lvl (indicates the number of divisions) and by the triple index (i, j, k) that shows the cell location in the virtual structured grid corresponding to lvl level (when all cells of the base grid recursively divided lvl times). Thus, the grid cells are represented by the octree network. Each node of this tree has a flag that shows whether the corresponding cell is divided or not. If the flag equals 1, the cell has 8 child cells and stores 8 pointers to these cells. The leaf cells (with no more subdivisions) have flags equal 0. These cells are computational, which are assigned by pointers to vertex coordinates and the solution parameters. The adaptation is ruled by the 1:2 principle that admits neighborhood of only cells with difference in the level of adaptation not more than 1.

The octree format with the 1:2 limitation has useful properties that allows us to effectively perform the operation of searching neighbors to a current cell. These properties are as follows. The parent cell for a cell (lvl, i, j, k) has corresponding coordinates $(lvl - 1, [i/2],[j/2],[k/2])$, where $[./.]$) is integer division, and index starts from 0,

$i = 0, M - 1; j = 0, N - 1; k = 0, K - 1$. Let the virtual coordinates (i, j, k) at the level lvl of a cell be written binary so that the number of digits equals lvl. Then, the integers in the s position in these binary notations of the (i, j, k) index indicates the position of the s ancestor ($s < lvl$) in the $s - 1$ ancestor. For example, if the integers are $(0,0,0)$, then the s ancestor is the bottom-left-front subcell of the $s - 1$ ancestor, $(1,0,0)$ corresponds to the upper-left-front cell, $(0,1,0)$ is the bottom-right-front subcell, $(1,0,1)$ is the upper-left-rear subcell, and so on.

To this end, ordinary sweep over cells of the LA grid is performed as follows. We do program loop over cells of the base grid. If the current cell is found to be leave (with no subdivisions), we move to another cell. Otherwise, we recursively sweep the child cells from level to level until the leaf cell (with the flag 0) is found. To do the sweep over neighbors, we have to find all leaf neighbors for a current leaf cell. Let us consider a current cell that has coordinates (lvl, i, j, k), and try to find leaf neighbors adjoin the cell from the right (i-direction). We take the cell with coordinates $(lvl, i + 1, j, k)$, and search such a cell in the octree network by using the above properties. This searching results in the following 3 situations: (1) the cell $(lvl, i + 1, j, k)$ is found and it is leaf (flag = 0), (2) the cell $(lvl, i + 1, j, k)$ is found and its flag equals 1, and (3) the cell $(lvl, i + 1, j, k)$ is not found because the searching procedure terminates at a cell of the level $lvl - 1$. The case (1) corresponds the neighbor cell of the same level as the current cell, in (2) we have 4 neighbors that are child cells to the $(lvl, i + 1, j, k)$ cell, and in (3) we have one neighbor cell found at a lower $lvl - 1$ level.

3   Discrete Model

We consider numerical solution of the non-stationary 3D Euler equations that describe compressible fluid flows. The discrete model on Cartesian LA grids is developed on the base of the numerical method on a structured Cartesian grid [8, 9]. Spatial discretization of the system of governing equations is performed with the finite volume method, which determines the cell-centered solution through the numerical fluxes of mass, momentum, and energy across the cell faces bordering the cell. The Godunov method [10] is implemented to calculate the numerical flux in terms of the exact solution to the Riemann problem stated at the cell interface. To increase the accuracy of the numerical scheme, a high-order subcell reconstruction of the solution is used, which provides interpolated solution values on the both sides of the cell interface. In the present discrete model, we use a linear subcell reconstruction based on the MUSCL (Monotone Upstream-Centered Scheme for Conservation Laws) interpolation [11]. Time integration of the semi-discrete system of equations is performed with the explicit two-stage predictor-corrector scheme that is stable under the CFL condition on the time step. Details of the scheme can be seen in [8].

Treatment of the boundary conditions on the Cartesian LA grid that does not fit the geometry of flow domain is carried out with the method of free boundaries [7, 8]. A key point of this method is an alternative mathematical statement of the problem which allows us to replace the solution of the boundary value problem in a part of the space with the solution of an initial value problem in the whole space. We modify the original system of Euler equations by adding in the right-hand side a vector-function that we refer to as compensating flux. The modified equations are then solved in the whole computational domain that encompasses the geometry. The compensating flux is defined so that the solution to the modified problem off the geometry exactly matches the solution to the original boundary value problem.

Extending the numerical method to the case of unstructured Cartesian LA grids we need, first of all, reconsider calculation of numerical fluxes at cell interfaces. This calculation procedure must take into account possibility of non-conformal (not face-to-face) neighborhood of grid cells when the cell face is adjoined by 4 cells of the lower adaptation level or the neighborhood is a cell of the upper level. In the first case, the numerical flux to the current cell is determined by means of averaging of 4 numerical fluxes calculated between the current cell and the 4 neighboring cells. In doing so, the current cell solution is interpolated at the sub-faces centers with the MUSCL interpolation. If the neighborhood is one level higher cell then the current cell, the neighbor cell solution is MUSCL-interpolated at the face center of the current cell, and the flux is calculated in the standard way as in the base scheme.

4   Application Library for Cartesian LA Grid Operation

The discrete model described above is realized in an application library. This library operates with 3D Cartesian LA nested-type grids, and encompasses functions of grid generation, grid refinement and coarsening, searching neighborhoods, and also others which are required to generate the numerical code. There are two types of grid functions in the library, which conventionally referred to as geometrical and physical. The geometrical grid function serves for generating geometry grid elements by means of surfaces determined parametrically or with polygons. The physical grid function represents the vector of numerical solution on the grid considered.

The library is organized with templates, which make it possible to easy set up the size of physical vectors. To do this, we need only assign the number of components of the solution vector, the numerical flux, and the number of adaptation coefficients: **TPhysGrid<5,4,2>**. Refinement of a grid element is performed with the function **Divide()**, coarsening (merging child cells into one) with **Merge()**. Searching neighborhoods is implemented with the function **FoundNeights()**. There are several other functions and procedures used in the numerical code that will be described below.

The software library is implemented in five basic parts: **BasicLib, MathLib, GeometryLib, GridLib, DrawLib**. Each of them is a dynamically linked application library in C++. The **BasicLib** library includes basic components for operating with strings, arrays, lists, pointers, and other similar structures. It encompasses base function and subroutines which are used in other libraries.

The **MathLib** library consists of subroutines for work with mathematical objects such as matrix, vector, normal, basis, angle, etc. **GeometryLib** includes classes and functions for work with 2D surfaces in 3D space. Basically, we use triangulation for representing the surface, but representation with simple primitive is also available. Subroutines for manipulating 3D locally adaptive grids are assembled in **GridLib** library. **DrawLib** serves for visualizing geometry, grid, and results of calculations by means of OpenGL.

The LA Cartesian grid represents a multilevel structure of recursively embedded Cartesian subgrids. The structure of a grid cell can be written in pseudo-language as

```
class Cell {
    Cell *child[N];
    int level;
    int xNum, yNum, zNum;
    enum type;
    void *additionalData;
};
```

where (*level, xNum, yNum, zNum*) are coordinates of the cell (Sect. 2). Each cell have different data depending on its type. For example, fluid cells involve the vector of physical parameters, border cells must include the data of boundary conditions, and cut cells must have also the data about volume of fluid and area of the geometry inside the cell, and the coordinates of the outer unit normal to accomplish discretization of the equations. One way to realize this in the code is to store all the data uniformly for all types of cell, but this might result in consuming large memory resources. Therefore, we store these data dynamically as pointers to the type *void*, for example. We use the mechanism of virtual functions in the library. In the pseudo code the additional data are defined as

The above description roughly represents set-up and functioning of the library. For searching and sweeping the neighborhoods, dynamical cell refinement and coarsening, the structures of dynamic storage management such as array, list, wordindex, set are implemented. Mathematical objects as matrix and vectors are used to describe geometrical and solution parameters, the objects polygon, surface, primitives (sphere, parallelepiped, cylinder) - to represent the geometry, there are special function included in the library to visualize computed results. More detail description of all these structures is given in [12].

5   Numerical Results

In this section we present results of numerical experiments aimed at verification of the method and testing the application library. The problem of propagation of the 3D blast wave in a closed domain bounded by rigid walls is considered as the test problem. The initial data for this calculation correspond to an initial stage of the self-similar 1D blast wave problem [13] (for details of calculations see [12]). The Intel Core I5-6600, 3.3 GHz, 3.2 Gb Quad-Core processor is used for these calculations.

Figure 1 shows numerical results. One can see that the grid adaptation well captures the shock front in a narrow zone with enhanced grid resolution. On the other hand the grid becomes rather coarse off these zones. The effect of shock interference is clear seen on side walls of the computational domain (left figure). The top view (right figure) demonstrates the shock reflection process.

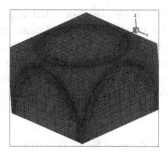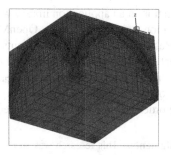

Fig. 1. The grid and density distribution, time = 1, 7 levels of adaptation: top view (left), bottom view (right).

Effectiveness of the adaptive grid can be expressed via the parameter of adaptation effectivity $P = N_c/8^L$, where N_c is the total number of computational cells of the Cartesian LA grid, and L is the number of levels of adaptation. Figure 2 shows plots of P versus number of time steps T. The rate of effectiveness in average amounts to 18.6% for $L = 6$, and 8.7% for $L = 7$.

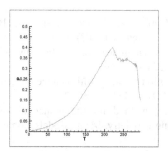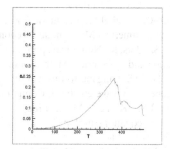

Fig. 2. Effectiveness (P) vs time (T): 6 levels (left), 7 levels (right).

Table 1 demonstrates effectiveness of the OpenMP parallelization. The speedup parameter A_n shows the ratio of the computational times in *1CPU* and *nCPUs* calculations, respectively. Effectiveness of parallelization is defined as $E_n = (1 + A_n)/n$.

Table 1. Effectiveness of the OpenMP parallelization; 6 levels of adaptation.

CPU	Time, c	Speedup, A_n	Effectiveness, E_n
1	5724	–	100%
2	3597.5	59.1%	79.6%
4	2957.8	93.5%	48.4%

6   Conclusions

A numerical method of using locally adaptive nested-type Cartesian grids in numerical simulations of 3D gas dynamics problems has been presented along with its software implementation in an application library written in C++ object-oriented language with the application program interface OpenMP for parallelizing calculations on shared memory. Results of numerical experiments have shown high effectiveness of dynamical grid adaptation with the use of multilevel nested-type Cartesian grids, which becomes more profound as the number of adaptation levels increases because local grid refining in vicinity of two-dimensional manifolds results in the increase of computational cells that is one order less with comparison to equivalent global grid refinement.

Acknowledgments. This research was supported by the grant No 17-71-30014 from Russian Scientific Fund.

References

1. Bramkamp, F., Lamby, P.H., Mueller, S.: An adaptive multiscale finite volume solver for unsteady and steady state flow computations. J. Comput. Phys. **197**(2), 460–490 (2004)
2. Harten, A.: Multiresolution algorithms for the numerical solution of hyperbolic conservation laws. Comm. Pure Appl. Math. **48**(12), 1305–1342 (1995)
3. Harten, A.: Multiresolution representation of data: a general framework. SIAM J. Numer. Anal **33**(3), 1205–1256 (1996)
4. Zumbusch, G.: Parallel Multilevel Methods: Adaptive Mesh Refinement and Load Balancing. Advances in Numerical Mathematics. Teubner, Wiesbaden (2003)
5. Osher, S., Sanders, R.: Numerical approximations to nonlinear conservation laws with locally varying time and space grids. Math. Comp. **41**, 321–336 (1983)
6. Vasilyev, O.V.: Solving multi-dimensional evolution problems with localized structures using second generation wavelets. Int. J. Comp. Fluid Dyn. **17**, 151–168 (2003)
7. Menshov, I.S., Kornev, M.A.: Free_boundary method for the numerical solution of gas_dynamic equations in domains with varying geometry. Math. Models Comput. Simul. **6**(6), 612–621 (2014)
8. Menshov, I.S., Pavlukhin, P.V.: Efficient parallel shock-capturing method for aerodynamics simulations on body-unfitted cartesian grids. Comput. Math. Math. Phys. **56**(9), 1651–1664 (2016)
9. Menshov, I.S., Pavlukhin, P.V.: Highly scalable implementation of an implicit matrix-free solver for gas dynamics on GPU-accelerated clusters. J. Supercomput. **73**, 631–638 (2017)
10. Godunov, S.K.: Difference method for computing discontinuous solutions of fluid dynamics equations. Mat. Sb. **47**(3), 271–306 (1959)
11. Van Leer, B.: Towards the ultimate conservative difference scheme: V. A second-order sequel to Godunov's method. J. Comput. Phys. **32**, 101–136 (1979)
12. Menshov, I.S., Nikitin, V.S., Sheverdin, V.V.: Parallel three-dimensional LAD model on Cartesian grids of nested structure. Keldysh Inst. Prepr. **118**, 1–32 (2016)
13. Sedov, L.I.: Propagation of strong shock waves. J. Appl. Math. Mech. **10**, 241–250 (1946)

Auto-Vectorization of Loops on Intel 64 and Intel Xeon Phi: Analysis and Evaluation

Olga V. Moldovanova$^{(\boxtimes)}$ and Mikhail G. Kurnosov

Rzhanov Institute of Semiconductor Physics,
Siberian Branch of Russian Academy of Sciences, Novosibirsk, Russia
{ovm,mkurnosov}@isp.nsc.ru

Abstract. This paper evaluates auto-vectorizing capabilities of modern optimizing compilers Intel C/C++, GCC C/C++, LLVM/Clang and PGI C/C++ on Intel 64 and Intel Xeon Phi architectures. We use the Extended Test Suite for Vectorizing Compilers consisting of 151 loops. In this work, we estimate speedup by running the loops in scalar and vector modes for different data types and determine loop classes which the compilers used in the study fail to vectorize. We use the dual CPU system (NUMA, 2 x Intel Xeon E5-2620v4, Intel Broadwell microarchitecture) with the Intel Xeon Phi 3120A co-processor for our experiments.

1 Introduction

Modern high-performance computer systems are multiarchitectural systems and implement several levels of parallelism: process level parallelism, thread level parallelism, instruction level parallelism, and data level parallelism. Processor vendors pay great attention to the development of vector extensions. In particular, Fujitsu announced in its future version of the exascale K Computer system a transition to processors with the ARMv8-A architecture, which implements scalable vector extensions. And Intel extensively develops AVX-512 vector extension. That is why problem definitions and works on automatic vectorizing compilers have given the new stage in development in recent decades.

In this work, we studied the effectiveness of auto-vectorizing capabilities of modern compilers: Intel C/C++, GCC C/C++, LLVM/Clang, PGI C/C++. The main goal is to identify classes of problem loops.

Since there was no information about vectorizing methods implemented in the commercial compilers, the evaluation was implemented by the "black box" method. We used the Extended Test Suite for Vectorizing Compilers [1–4] as a benchmark for our experiments. We determined classes of typical loops that the compilers used in this study failed to vectorize and evaluated them.

The rest of this paper is organized as follows: Sect. 2 discusses the main issues that explain effectiveness of vectorization; Sect. 3 describes the benchmark we used; Sect. 4 presents results of our experiments; and finally Sect. 5 concludes.

This work is supported by Russian Foundation for Basic Research (projects 15-07-00048, 16-07-00712).

© Springer International Publishing AG 2017
V. Malyshkin (Ed.): PaCT 2017, LNCS 10421, pp. 143–150, 2017.
DOI: 10.1007/978-3-319-62932-2_13

2   Vector Instruction Sets

Instruction sets of almost all modern processor architectures include vector extensions. Processors implementing vector extensions contain one or several vector arithmetic logic units (ALU) functioning in parallel and several vector registers.

The main application of the vector extensions consists in decreasing of time of one-dimensional arrays processing. As a rule, a speedup achieved using the vector extensions is primarily determined by the number of array elements that can be loaded into a vector register. To achieve a maximum speedup during vector processing it is necessary to consider the microarchitectural system parameters. One of the most important of them is an alignment of array initial addresses. Effectiveness decreasing can also be caused by a mixed usage of SSE and AVX vector extensions [5].

When vector instructions are used, the achieved speedup can exceed the expected one due to the processor overhead decreases and parallel execution of vector instructions by several vector ALUs. Thus, an efficiently vectorized program overloads subsystems of a superscalar pipelined processor in a less degree. This is the reason of less processor energy consumption during execution of a vectorized program as compared to its scalar version [6].

Application developers have different opportunities to use vector instructions: inline assembler, intrinsics, SIMD directives of compilers (OpenMP and OpenACC standards), automatic vectorizing compilers.

In this work, we study the last approach, since it does not require large code modification and provides its portability between different processor architectures.

3   Benchmark

We used the Extended Test Suite for Vectorizing Compilers (ETSVC) [2] as a benchmark containing main loop classes, typical for scientific applications in C language. The original package version was developed in the late 1980s by the J. Dongarra's group and contained 122 loops in Fortran to test the analysis capabilities of automatic vectorizing compilers for vector computer systems [3, 4]. In 2011 the D. Padua's group translated the TSVC suite into C and added to it new loops [1]. The extended version of the package contains 151 loops. The loops are divided into categories: dependence analysis (36 loops), vectorization (52 loops), idiom recognition (27 loops), language completeness (23 loops). Besides that, the test suite contains 13 "control" loops, trivial loops that are expected to be vectorized by every vectorizing compiler.

The loops operate on one- and two-dimensional 16-byte aligned global arrays. The one-dimensional arrays contain $125 \cdot 1024/\texttt{sizeof(TYPE)}$ elements of the given type TYPE, and the two-dimensional ones contain 256 elements by each dimension.

Each loop is contained in a separate function. In the init function an array is initialized by individual for this test values before loop execution. The outer loop is used to increase the test execution time (for statistics issues). A call to an empty dummy function is used in each iteration of the outer loop so that, in case where the inner loop is invariant with respect to the outer loop, the compiler is still required to execute each iteration rather than just recognizing that the calculation needs to be done only once [4]. After execution of the loop is complete, a checksum is computed by using elements of the resulting array and is displayed.

4   Results of Experiments

We used two systems for our experiments. The first system was a server based on two Intel Xeon E5-2620 v4 CPUs (Intel 64 architecture, Broadwell microarchitecture, 8 cores, Hyper-Threading was on, AVX 2.0 support), 64 GB RAM DDR4, GNU/Linux CentOS 7.3 × 86-64 operating system (linux 3.10.0-514.2.2.el7 kernel). The second system was Intel Xeon Phi 3120A co-processor (Knights Corner microarchitecture, 57 cores, AVX-512 support, 6 GB RAM, MPSS 3.8) installed in the server.

The compilers evaluated in these experiments were Intel C/C++ Compiler 17.0; GCC C/C++ 6.3.0; LLVM/Clang 3.9.1; and PGI C/C++ 16.10. The vectorized version of the ETSVC benchmark was compiled with the command line options shown in Table 1 (column 2). To generate the scalar version of the test suite the optimization options were used with the disabled compilers vectorizer (column 3, Table 1).

32-byte aligned global arrays were used for the Intel Xeon processor, and 64-byte aligned global arrays were used for the Intel Xeon Phi processor. We used arrays with elements of double, float, int and short data types for our evaluation.

Table 1. Compilers options

Compiler	Compilers options	Disabling vectorizer
Intel C/C++ 17.0	-O3 -xHost -qopt-report3 -qopt-report-phase=vec,loop -qopt-report-embed	-no-vec
GCC C/C++ 6.3.0	-O3 -ffast-math -fivopts -march=native -fopt-info-vec -fopt-info-vec-missed -fno-tree-vectorize	-fno-tree-vectorize
LLVM/Clang 3.9.1	-O3 -ffast-math -fvectorize -Rpass=loop-vectorize -Rpass-missed=loop-vectorize -Rpass-analysis=loop-vectorize	-fno-vectorize
PGI C/C++ 16.10	-O3 -Mvect -Minfo=loop,vect -Mneginfo=loop,vect	-Mnovect

The following results were obtained for the `double` data type on the Intel 64 architecture (Intel Xeon Broadwell processor). The Intel C/C++ vectorized 95 loops in total, 7 from which were vectorized by it alone. For GCC C/C++ the total amount of vectorized loops was 79. But herewith there was no loop that was vectorized only by this compiler. The PGI C/C++ vectorized the largest number of loops, 100, 13 from them were vectorized by it alone. The minimum number of loops was vectorized by the LLVM/Clang compiler, 52, 4 from which were vectorized only by it. The number of loops unvectorized by any compiler was equal to 28.

The similar results were obtained for arrays with elements of the `float` and `int` types by all compilers. The consistent results were obtained for the `short` type when Intel C/C++, GCC C/C++ and LLVM/Clang were used. The exception to this rule was the PGI C/C++ compiler that vectorized no loops processing data of this type.

Figure 1 shows the results of loop vectorization for the `double` data type on the Intel 64 architecture. Abbreviated notations of the vectorization results are shown in the table cells. They were obtained from vectorization reports of compilers for all 151 loops. The full form of these notations is shown in Table 2. The similar results were obtained for other data types.

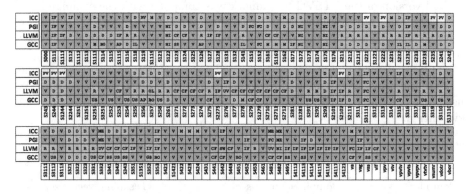

Fig. 1. Results of loops vectorization (Intel 64 architecture, `double` data type)

In the "Dependence analysis" category 9 loops were not vectorized by any compiler for the `double` data type. The compilers used in this study failed to vectorize loops with linear data dependences, induction variables together with conditional and unconditional branches, loop nesting and variable values of lower and/or upper loop bounds and/or iteration step.

In the "Vectorization" category the compilers failed to vectorize 11 loops. These loops required transformations as follows: loop fission, loop interchange, node splitting (to avoid cycles in data dependence graphs and output and anti-dependences [7]) and array expansions. Among causes of problems were interdependence of iteration counts of nested loops; linear data dependences in a loop body; conditional and unconditional branches in a loop body.

Table 2. Abbreviated notations of vectorization results

V	Loop is vectorized
PV	Partial loop is vectorized
RV	Remainder is not vectorized
IF	Vectorization is possible but seems inefficient
D	Vector dependence prevents vectorization (supposed data dependence in a loop)
M	Loop is multiversioned (multiple loop versions are generated)
BO	Bad operation or unsupported loop bound (`sinf` or `cosf` function is used)
AP	Complicated access pattern (e.g., value of iteration count is more than 1)
R	Value that could not be identified as reduction is used outside the loop (induction variables are present in a loop)
IL	Inner-loop count not invariant (iteration count of inner loop depends on iteration count of outer loop)
NI	Number of iterations cannot be computed (lower and/or upper loop bounds are set by function's arguments)
CF	Control flow cannot be substituted for a select (conditional branches inside a loop)
SS	Loop is not suitable for scatter store (in case of packing a two-dimensional array into a one-dimensional array)
ME	Loop with multiple exits cannot be vectorized (`break` or `exit` are present inside a loop)
FC	Loop contains function calls or data references that cannot be analyzed
OL	Value cannot be used outside the loop (scalar expansion or mixed usage of one- and two-dimensional arrays in one loop)
UV	Loop control flow is not understood by vectorizer (conditional branches inside a loop)
SW	Loop contains a `switch` statement
US	Unsupported use in statement (scalar expansion, wraparound variables recognition)
GS	No grouped stores in basic block (unrolled scalar product)

The following idioms (6 loops) from the "Idiom recognition" category were not vectorized by the compilers used: 1st and 2nd order recurrences, array searching, loop rerolling (for loops that were unrolled by hand before vectorization [8]) and reduction with function calls.

The "Language completeness" category contains 2 loops unvectorized by any compiler. The problem of both loops consisted in breaking loop computations (`exit` in the first case and `break` in the second case). Compiler vectorizers could not analyze control flow in these loops.

A median value and maximum speedups of vectorized loops are shown in Figs. 2 and 3. The maximum speedup obtained on the Intel 64 architecture by

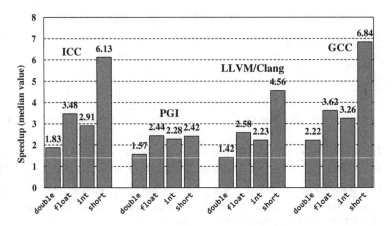

Fig. 2. Median value of speedup for vectorized loops on Intel Xeon E5-2620 v4 CPU

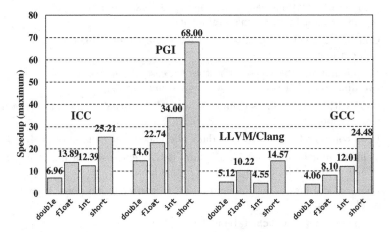

Fig. 3. Maximum speedup for vectorized loops on Intel Xeon E5-2620 v4 CPU

the Intel C/C++ was 6.96 for the `double` data type, 13.89 for the `float` data type, 2.39 for `int` and 25.21 for `short int`. The maximum speedup obtained by GCC C/C++ was equal to 4.06, 8.1, 12.01 and 24.48 for types `double`, `float`, `int` and `short int`, correspondingly. The LLVM/Clang obtained results as follows: 5.12 (`double`), 10.22 (`float`), 4.55 (`int`) and 14.57 (`short int`). For PGI C/C++ these values were 14.6, 22.74, 34.0 and 68.0, correspondingly. The speedup is the ratio of the running time of the scalar code over the running time of the vectorized code.

As our evaluation showed maximum speedups for Intel C/C++, GCC C/C++ and LLVM/Clang correspond to the loops executing reduction operations with elements of one-dimensional arrays of all data types. For PGI C/C++ maximum speedup was achieved for the loop calculating an identity matrix for the `double` and `float` data types. And for `int` and `short` this value was obtained in the loop calculating product reduction. However, the speedup value 68.0 for the `short` data type can be explained by the fact that calculations in a loop are not executed at all because of the compiler optimization.

On the Intel Xeon Phi architecture we studied vectorizing capabilities of the Intel C/C++ Compiler 17.0. The `-mmic` command line option was used instead of the `-xHost` during compilation. The compiler could vectorize 99 loops processing data of the `double` type and 102 of the `float` type. Supposed data dependences (28 loops for the `double` type and 27 for the `float` type) were the main reason of loop vectorization failing. 12 loops were partially vectorized for both data types. Similar results were obtained for the `int` and `short` types. In this case the maximum speedup for the `double` type was 13.7, for `float` – 19.43, `int` – 30.84, and `short` – 46.3. For `float` and `short` maximum speedups were obtained for loops executing reduction operations for elements of one-dimensional arrays. For the `double` data type `sinf` and `cosf` functions were used in a loop. In the case with `int` it was a "control" loop `vbor` calculating a scalar product of six one-dimensional arrays.

5 Conclusion

In this work we studied auto-vectorizing capabilities of modern optimizing compilers Intel C/C++, GCC C/C++, LLVM/Clang, PGI C/C++ on the Intel 64 and Intel Xeon Phi architectures. Our study shows that the compilers evaluated could vectorize 39–77 % of the total number of loops in the ETSVC package. The best results were shown by the Intel C/C++, and the worst ones – by the LLVM/Clang. The compilers failed to vectorize loops containing conditional and unconditional branches, function calls, induction variables, variable loop bounds and iteration count, as well as such idioms as 1st or 2nd order recurrences, search loops and loop rerolling.

The future work will consist of evaluation and development of efficient vectorizing methods for the obtained class of challenging loops, applicability analysis of JIT compilation [9] and profile-guided optimization.

References

1. Maleki, S., Gao, Y., Garzaran, M.J., Wong, T., Padua, D.A.: An evaluation of vectorizing compilers. In: Proceedings of the International Conference on Parallel Architectures and Compilation Techniques, pp. 372–382 (2011)
2. Extended Test Suite for Vectorizing Compilers. http://polaris.cs.uiuc.edu/~maleki1/TSVC.tar.gz
3. Callahan, D., Dongarra, J., Levine, D.: Vectorizing compilers: a test suite and results. In: Proceedings of the ACM/IEEE Conference on Supercomputing, pp. 98–105 (1988)
4. Levine, D., Callahan, D., Dongarra, J.: A comparative study of automatic vectorizing compilers. J. Parallel Comput. **17**, 1223–1244 (1991)

5. Konsor, P.: Avoiding AVX-SSE transition penalties. https://software.intel.com/en-us/articles/avoiding-avx-sse-transition-penalties
6. Jibaja, I., Jensen, P., Hu, N., Haghighat, M., McCutchan, J., Gohman, D., Blackburn, S., McKinley, K.: Vector parallelism in JavaScript: language and compiler support for SIMD. In: Proceedings of the International Conference on Parallel Architecture and Compilation, Techniques, pp. 407–418 (2015)
7. Program Vectorization: Theory, Methods, Implementation (1991)
8. Metzger, R.C., Wen, Z.: Automatic Algorithm Recognition and Replacement: A New Approach to Program Optimization. MIT Press, Cambridge (2000)
9. Rohou, E., Williams, K., Yuste, D.: Vectorization technology to improve interpreter performance. ACM Trans. Archit. Code Optim. **9**(4), 26: 1–26: 22 (2013)

Parallel Algorithms for an Implicit CFD Solver on Tree-Based Grids

Pavel Pavlukhin[1,2(✉)] and Igor Menshov[1]

[1] Keldysh Institute of Applied Mathematics, Moscow 125047, Russia
{pavelpavlukhin,menshov}@kiam.ru
[2] Research and Development Institute "Kvant", Moscow 125438, Russia

Abstract. Parallel implementation of the implicit LU-SGS solver is considered. It leads to the graph coloring problem. A novel recursive graph coloring algorithm has been proposed that requires only three colors on 2:1 balanced quadtree-based meshes. The algorithm has been shown to allow simple parallel implementations, including GPU architectures, and is fully coherent with local grid coarsing/refining procedures resulting in highly effective co-execution with local grid adaptation.

Keywords: CFD · CUDA · LU-SGS · Implicit schemes · Parallel algorithms · Tree-based grids · AMR

1 Introduction

Highly scalable implementations of CFD solvers for large massively-parallel computing systems is a big challenge. It involves several problems one of which is discretization of the computational domain. In case of complex geometries unstructured meshes are commonly used. However dynamic load balancing for this type of grids leads to large communication overheads and therefore is hard to realize.

Conversely, Cartesian grid methods well fit requirements of highly scalable parallel algorithms. To treat geometries on Cartesian grids we can employ the free boundary method (FBM) [1]. This method allows us to solve the problem in domains with complex geometry on Cartesian grids with only minor modifications of the baseline method without geometry. The modification is just addition of special terms (referred to as compensating fluxes) to the right-hand side of the gas dynamics equations:

$$\frac{\partial \boldsymbol{q}}{\partial t} + \frac{\partial \boldsymbol{f}_i}{\partial x_i} = -\boldsymbol{F}_w. \tag{1}$$

Implicit time integration scheme is applied to these equations after spatial discretization by the finite volume method. The resulting system of discrete equations is then solved with Newtonian iterations represented as a linear system with a sparse block matrix, $A\delta q = R$, where δq is the iterative residual. Solution

© Springer International Publishing AG 2017
V. Malyshkin (Ed.): PaCT 2017, LNCS 10421, pp. 151–158, 2017.
DOI: 10.1007/978-3-319-62932-2_14

for this linear system is found by the matrix-free Lower-Upper Symmetric Gauss-Seidel (LU-SGS) approximate factorization method [2–4]. The matrix A is split into diagonal, low-, and upper-triangle parts, $A = D + L + U$, and then is replaced by approximate factorization $A \approx (D + L)D^{-1}(D + U)$ what finally leads to solving the following systems (by corresponding forward and backward sweeps):

$$\begin{cases} (D + L)\delta \boldsymbol{q}_i^* = -\boldsymbol{R}_i^{n+1} \\ (D + U)\delta \boldsymbol{q}_i = D\delta \boldsymbol{q}_i^*. \end{cases} \tag{2}$$

Cartesian grids limitations and possible substitutions for FBM are discussed in Sect. 2. LU-SGS parallelization on octree-based meshes (as alternative to Cartesian ones) and involved graph coloring problem are considered in Sects. 3 and 4, respectively. Parallel implementation details of the proposed recursive coloring algorithm and its properties are described in Sect. 5. Conclusions are presented in final Section.

2   Choosing an Adaptive Mesh Refinement Approach

For better geometry representation in computational aerodynamics, it is necessary to increase grid resolution near solid inclusions in the computational domain. This is even more required when we need to resolve boundary layers on the surfaces of solid inclusions for taking into account viscous effects. At regimes with high Reynolds numbers, fluid flow forms very thin boundary layer near the solid body where the effect of viscosity is significant, and high grid resolution must be used for tracking this layer. Consequently, it is worth to employ one of the adaptive mesh refinement (AMR) aproaches. These approaches can be divided into three types which are considered in what follows.

The first is the block-structured AMR which is based on coarsening the grid by cells merge in appropriate domains. Since such patches of different resolutions can be overlapped, the block-structured AMR increases numerical complexity and leads to difficulties in extending to high-order discretizations to complex geometries. Moreover, this type of AMR is hard to implement on massively-parallel architectures like multi-GPU clusters. For example, strong scaling of the GPU-aware framework described in [5] is limited by 8 GPU on 6.4 million zone problem.

The second type is unstructured AMR (for example, [6]). This approach can treat geometries of different complexities. A problem associated with this type of AMR is maintaining element quality in the process of mesh coarsening and refining. Another problem is dynamic load balancing in computing systems with distributed memory which generally involves considerable communication overhead.

Block-structured and unstructured AMR methods, except difficulties mentioned above, are not appropriate for using with the FBM. These types of AMR are supposed to work with conforming (to geometry domain surfaces) grids while the FBM works with non-conforming grids consisted of simple, rectangular computational cells. Therefore, natural choice for the free boundary method is tree-based AMR. For this AMR type, each grid cell of the initial base Cartesian grid can be recursively divided in 8 subcells (4 subcells in the 2D case). The resulting grid can be represented as the octree graph (each internal node in such tree has exactly eight children, Fig. 1). With a space-filling curve connecting elements of this graph, the data on such adapted grid can be arranged in line which enables simple partitioning and dynamic load balancing.

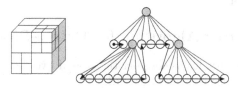

Fig. 1. Octree-based grid (left) and corresponding graph representation with space-filling curve (right).

3   LU-SGS Method on Octree-Based Grids

Besides developing of GPU-aware AMR handling, there is one problem which is directly related to the solver used in free boundary method. If an explicit scheme is used then parallelization is quite simple since all grid cells can be computed simultaneously. The weakness of this type of schemes is stability restriction: smaller cells in the adaptive mesh limit the global time step resulting in unreasonable computational costs. Conversely, the implicit scheme based on the LU-SGS method [2–4] haven't this disadvantage but their implementation on massively-parallel architectures is complicated since data dependency between neighbor grid cells (separated with common face) arises [7,8]. The solution algorithm of (2) in the LU-SGS is represented as forward and backward sweeps over all grid cells, and data dependency has local nature and is only defined by the sweep order over all geometrical neighbors relative to the current cell. In other words, different computations are performed depending on the neighbor position in relation to the current cell ("before" or "after"). The sweep order can be chosen in different ways based not only on cell geometry neighborhood. Hence, executing simultaneous calculations in the grid cells is realized if only these cells are not geometrical neighbors, and therefore developing a parallel algorithm for the LU-SGS method leads to the problem of graph coloring. The problem is to color the grid cells so that any two neighbor grid cells (with the common face) would have different colors. In case of structured, for example, Cartesian grids, there is a simple solution which requires only two colors. This leads to the "chessboard" cells sweep when calculations are fist performed for the all "black" cells (simultaneously), and then over all "white" cells (simultaneously).

There are several graph coloring algorithms that can be applied to color cells of the unstructured grid. However most of these algorithms have drawbacks. The widely used greedy Multi-Coloring algorithm is a strictly sequential. Parallel coloring algorithms produce partitions with overestimated number of colors and have limited scalability [9,10]. On the other hand, octree-based grids are not quite unstructured, rather they mimic Cartesian structured grids because of partially ordered recursive subcell nature. Because of this reason, definite motivation appears to develop a new graph coloring algorithm which exploits the tree-like structure of the AMR grid. The description of such an algorithm is given in the next section.

4   Graph Coloring Algorithm for Octree-Based Grids

For the sake of simplicity, let us consider the case of two-dimensional grids. The extension to the 3D case is straightforward.

Among variety of octree/quadtree-based grids, it is worth to distinguish those which meet the so called 2:1 balance property. This property says that each grid cell has to have no more than two neighbors over any its face (Fig. 2). Such meshes considerably simplify cell-to-cell solver interface, and many numerical methods essentially utilize this property. Therefore we develop the coloring algorithm for AMR Cartesian grids with the 2:1 balance property. Hereafter this property is assumed unless mentioned otherwise.

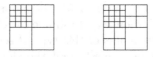

Fig. 2. 2:1 balance status for a 2D quadtree mesh: unbalanced (left) and face balanced (right).

What is the minimal number of colors which must be used to partition the cells of the quadtree-based Cartesian grid? Obviously, 2-color "chessboard" partitioning used for Cartesian meshes becomes insufficient and consequently at least 3 colors are needed. The four color theorem states that for 2-dimensional unstructured grids no more than 4 colors are needed. In general, reduction of the number of colors used to partition grid cells increases effectiveness of multithreaded computations on GPU since more cells of the same color can be computed simultaneously. The algorithm we propose satisfies this favorable property and involves only 3 colors referenced to as "black", "white", and "red".

First, we introduce the algorithm in step-by-step way, and then generalize its description. As it was mentioned above, specifics of the tree-like grid structure is exploited. The construction is recursive. The algorithm starts with coloring the cells of the top level (according to tree graph). These cells form the regular Cartesian grid, and therefore the 2-color "chessboard" scheme is applied (Fig. 3(a)).

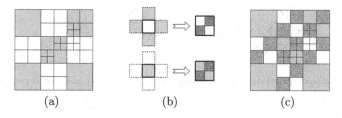

Fig. 3. The graph coloring algorithm: (a) 1st tree layer coloring; (b) recoloring rules; (b) 2nd tree layer coloring. (Color figure online)

Next, for all cells having four recursive subcells the following recoloring operation is performed. As can be seen, there are two templates of neighborhood at the current step for which specific recoloring rules are applied as shown in Fig. 3(b). In other words, depending on the color of the current cell and its neighbors corresponding color conversion is executed. The resulting grid coloring is shown on Fig. 3(c).

Then, all cells with child subcells at the successive third layer are considered. Four more templates of neighborhood appear in this step (in addition to 2 previous ones) which are associated with corresponding recoloring rules as shown in Fig. 4(a). After applying these rules, we obtain the grid coloring which is shown in Fig. 4(b).

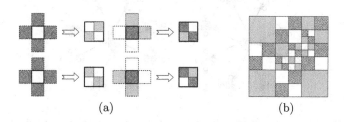

Fig. 4. The graph coloring algorithm: (a) additional recoloring rules; (b) 3rd tree layer coloring.

The next recursive steps (up to the latest tree layer) are performed in the same manner which actually can produce nine additional templates of neighborhood at most so the total number of different templates is fifteen. It is not difficult to see that all these templates can be represented by means of two color-generalized templates shown in (Fig. 5) and referred to as $T1$ (two-color) and $T2$ (three-color). Here "$\times$", "$\circ$" and "$\diamond$" mean mutually exclusive values of the three colors. It can be also noted that all the recoloring rules are accordingly expressed in only two generalized ones as in Fig. 5.

Nontrivial (and actually most substantial) statement is that the described above algorithm provides the solution to the graph coloring problem for any 2-dimensional quadtree-based, 2:1 balanced grid. To prove this, let us suppose that

Fig. 5. Generalized templates of neighborhood ($T1$ and $T2$) and corresponding recoloring rules.

at a certain recursive step each cell of the grid matches the generic template $T1$ or $T2$, and the mentioned above recoloring procedure is being applied. Hence, there are two kind of neighborhood appear (with corresponding templates) across each face depending on the cell template which are shown in Fig. 6. Each grid cell either remains the same or is divided into four recolored (in accordance to the above rules) subcells so that all possible situations of neighboring cells (after recoloring) can be displayed as in Fig. 7. From this figure, it is not difficult to see that the following statements are true (at the end of the current step):

1. There are no conflicts between any neighboring cells, i.e., they are always of different colors.
2. All templates of neighborhood match $T1$ or $T2$.

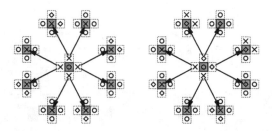

Fig. 6. All possible neighbors combinations with related templates of neighborhood at the beginning of the recursive algorithm step (generic form).

Consequently, in the next recursive step of the algorithm the grid will be properly colored and each cell can be recolored by the same old rules. This proves full correctness of the proposed coloring algorithm for quadtree-based grids.

5   Parallel Implementations of the Graph Coloring Algorithm

As we could see in the above considerations, each grid cell at a current step of the algorithm can be recolored independently of others, providing that current colors of its neighbors are available. Therefore, on systems with shared memory

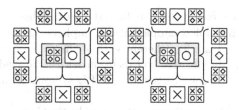

Fig. 7. All possible combinations of any grid cell (in grey rectangle) with its neighbors at the end of recursive algorithm step (generic form).

like GPU and multi-core CPU, all cells are recolored simultaneously with only synchronization between recursive steps. In the case of implementation for GPU, the algorithm is represented as consecutive running of kernels which perform recoloring at corresponding tree layer.

For systems with distributed memory one needs to perform exchange data in border cells of grid partitions. However, such communications are reduced to minimum since two of four cell neighbors are only needed for choosing proper recoloring rule. As a matter of fact, the recoloring rule applied (Fig. 5) depends only on the type of the generic template of neighborhood. To recognize this template, we need only the two neighbors which form the right angle with the current cell. If these neighbors are of the same color, then the template is $T1$ else $T2$ (Fig. 5). Since quadtree-based grids partitioning based on space-filling curves generally leads to simply connected regions (partitions) with vast majority of inner cells, the number of border cells for which the template of neighborhood is identified only by local neighbors (within partition) is dominating. Hence, only a small fraction of border cells requires neighbors from other partitions. Thus, the implementation of the algorithm for distributed memory parallel systems is expected to have good scalability because only negligible data exchange between partitions will need to do before each recursive step of the algorithm.

A very important property of the algorithm proposed is its full coherence with grid local adaptation procedures. In other words, recoloring is performed locally only in having changed cells after grid coarsing/refining with the rules of Fig. 5. In coarsed cells, reverted recoloring is applied since the rules (Fig. 5) actually state one-to-one correspondence between cell and its subcells. After any grid modification, updated coloring will be fully valid. The only required condition for the coloring algorithm is the 2:1 grid balancing property. The algorithm is fully conjugated with dynamic grid adaptation and possesses low overhead due to parallel recoloring performing only in modified cells.

6   Conclusions

It has been shown that octree-based adaptive Cartesian grids are most suitable for further development of the free boundary method. Parallelization of the implicit LU-SGS solver with the free boundary method leads to the graph coloring problem. A novel recursive graph coloring algorithm has been proposed that requires only three colors on 2:1 balanced quadtree-based meshes. The algorithm has been shown to allow simple parallel implementations, in particular for GPU architectures, and is fully coherent with local grid coarsing/refining procedures resulting in highly effective co-execution with local grid adaptation. The implementation of this algorithm for multi-GPU systems is in progress.

Acknowledgement. This research was supported by the Grant No. 17-71-30014 from the Russian Scientific Fund.

References

1. Menshov, I.S., Pavlukhin, P.V.: Efficient parallel shock-capturing method for aerodynamics simulations on body-unfitted cartesian grids. Comput. Math. Math. Phys. **56**(9), 1651–1664 (2016)
2. Menshov, I., Nakamura, Y.: Hybrid explicit-implicit, unconditionally stable scheme for unsteady compressible flows. AIAA J. **42**(3), 551–559 (2004)
3. Jameson, A., Turkel, E.: Implicit schemes and LU decomposition. Math. Comp. **37**, 385–397 (1981)
4. Menshov, I., Nakamura, Y.: On implicit Godunov's method with exactly linearized numerical flux. Comput. Fluids **29**(6), 595–616 (2000)
5. Beckingsale, D., Gaudin, W., Herdman, A., Jarvis, S.: Resident block-structured adaptive mesh refinement on thousands of graphics processing units. In: 2015 44th International Conference on Parallel Processing Parallel Processing (ICPP), pp. 61–70. IEEE (2015)
6. Lawlor, O.S., Chakravorty, S., Wilmarth, T.L., Choudhury, N., Dooley, I., Zheng, G., Kal, L.V.: ParFUM: a parallel framework for unstructured meshes for scalable dynamic physics applications. Eng. Comput. **22**(3–4), 215–235 (2006)
7. Pavlukhin, P.: Parallel LU-SGS numerical method implementation for gas dynamics problems on GPU-accelerated computer systems. Vestn. Lobachevsky State Univ. Nizhni Novgorod **1**, 213–218 (2013)
8. Pavlukhin, P., Menshov, I.: On implementation high-scalable CFD solvers for hybrid clusters with massively-parallel architectures. In: Malyshkin, V. (ed.) PaCT 2015. LNCS, vol. 9251, pp. 436–444. Springer, Cham (2015). doi:10.1007/978-3-319-21909-7_42
9. Bozdag, D., Gebremedhin, A.H., Manne, F., Boman, E.G., Catalyurek, U.V.: A framework for scalable greedy coloring on distributed-memory parallel computers. J. Parallel Distrib. Comput. **68**(4), 515–535 (2008)
10. Boman, E.G., Bozdağ, D., Catalyurek, U., Gebremedhin, A.H., Manne, F.: A scalable parallel graph coloring algorithm for distributed memory computers. In: Cunha, J.C., Medeiros, P.D. (eds.) Euro-Par 2005. LNCS, vol. 3648, pp. 241–251. Springer, Heidelberg (2005). doi:10.1007/11549468_29

Software Implementation of Mathematical Model of Thermodynamic Processes in a Steam Turbine on High-Performance System

Aleksandr Sukhinov[1], Aleksandr Chistyakov[1(✉)], Alla Nikitina[1],
Irina Yakovenko[2], Vladimir Parshukov[3], Nikolay Efimov[3], Vadim Kopitsa[3],
and Dmitriy Stepovoy[4]

[1] Don State Technical University, Rostov-on-Don, Russia
sukhinov@gmail.com, cheese_05@mail.ru, nikitina.vm@gmail.com
[2] Taganrog University, Named After A.P. Chekov – Branch of Rostov State
University of Economics, Taganrog, Russia
[3] RPE "Donskie Technologii" Ltd., Novocherkassk, Russia
[4] Azov-Black Sea Engineering Institute, Don State Agrarian University,
Zernograd, Russia

Abstract. The aim of this paper is the development of the mathematical model of thermal processes in steam turbine based on the modern information technologies and computational methods, with help of which the accuracy of calculations of thermal modes. The practical significance of the paper are: the model of thermal processes in steam turbine is proposed and implemented, the information about the temperature modes of the steam turbine is derived, limits and prospects of the proposed mathematical model is defined. The thermal processes in the turbine are characterized by a strong non-uniformity of the heat flow, which has significantly influence to the reliability and efficiency of the facility. As a rule, it the influence of these parameters on the geometry is not considered in the designing of the system that results in premature wear of the machine. The developed model takes into account the complex geometry of the steam turbine, does not require the significant changes in the processing of the design features and can be used to calculate the thermal processes other construction such as turbines. Software solution was developed for two-dimensional simulation of thermal processes in steam turbine that takes into account the occupancy control volumes.

Keywords: Steam turbine · Thermal conductivity · Mathematical model · Computational experiments

The paper is performed under an agreement with the Ministry of Education and Science of the Russian Federation No. 14.579.21.0123 about the granting subsidies from 10.27.2015. The theme: The development of high-efficiently steam turbine technology for processing liquid and solid organic wastes in the energy production for the small-scale distributed energy, the federal target program Research and development in priority areas of Russian scientific and technological complex for 2014–2020. The unique identifier of the applied scientific researches (project) RFMEFI57915X0123.

© Springer International Publishing AG 2017
V. Malyshkin (Ed.): PaCT 2017, LNCS 10421, pp. 159–171, 2017.
DOI: 10.1007/978-3-319-62932-2_15

1 Introduction

The question about the optimization problem of installation and exploitation of steam turbines is the actual. The strict requirements are presented with the development of modern technology and industry needs to operate the turbines associated with the reliability and efficiency of their operation. A large number of existing turbines is practically close to the elaboration its resource. So, the introduction of more modern units is required. The fundamentals of the theory of heat transfer and analysis results of transfer processes are required to assess the reliability and efficiency of the facility. Therefore, these data should be taken into account in designing of steam turbines. Thermal systems modeling include the problems of optimal control of thermal modes, due to we can choose the best from different implementations. The optimization of thermal modes is reduced to the solution of heat conduction problem. Mathematical modeling of thermal processes in technogenic systems is relevant at the present. Due to it, we can check the correctness of engineering ideas and correct errors at the stage of designing by the simple, low-cost means. The developed mathematical model is represented by the scheme model – algorithm – program, and must contain the structure, characteristic features of the process and described by equation system or functional relations [1]. After the design stage, it is necessary to determine the real values of temperature at significant points of the steam facility and analyze the compliance with required values.

2 Problem Statement

Thermal processes in turbine $\bar{G}$ was described by the heat conduction equation:

$$c\rho \frac{\partial T}{\partial t} = \frac{\partial}{\partial x}\left(\lambda \frac{\partial T}{\partial x}\right) + \frac{\partial}{\partial y}\left(\lambda \frac{\partial T}{\partial y}\right) + \frac{\partial}{\partial z}\left(\lambda \frac{\partial T}{\partial z}\right) + q_v, \tag{1}$$

which in the case of axial symmetry can be written as:

$$\rho c r T'_t = r\left(\lambda T'_x\right)'_x + \left(\lambda r T'_r\right)'_r + rf. \tag{2}$$

In the system (1) and (2) T is the temperature, $°K$; λ is the conductivity of water; ρ is the metal density; c is the heat capacity of metal; r is the polar radius; q_v is the source function.

We will consider the Eq. (2) with the boundary conditions of third kind:

$$T'_n(x,\, r,\, t) = \alpha_n T + \beta_n, \tag{3}$$

where n is the normal vector to the $\bar{G}$.

An initial condition were added to (2):

$$T(x, r, 0) = T_0(x, r),\ (x, r) \in \bar{G}. \tag{4}$$

We will describe an algorithm for defining the coefficient of the water and water steam thermal conductivity.

3 Thermal Conductivity Coefficient

The equation for determining the conductivity coefficient of water and water steam in the international practice has the following form:

$$\lambda = \lambda_0(\tau) + \lambda_1(\delta) + \lambda_2(\tau, \delta), \tag{5}$$

where λ is the thermal conductivity, W/(m·K); $\tau = T/T^*$; T is an absolute temperature, °K (ITS - 90); $T^* = 647.256$ °K; $\delta = \rho/\rho^*$; ρ is a density, kg/m³; $\rho^* = 317.7$ kg/m³. The water vapor thermal conductivity in ideal gas state is determined by the equation:

$$\lambda_0(\tau) = \tau^{0.5} \sum_{k:=0}^{3} a_k \tau^k,$$

where $a_0 = 0.0102811$; $a_1 = 0.0299621$; $a_2 = 0.0156146$; $a_3 = -0.00422464$. The function $\lambda_1(\delta)$ is defined as:

$$\lambda_1(\delta) = b_0 + b_1\delta + b_2 \exp\left\{B_1(\delta + B_2)^2\right\},$$

where $b_0 = -0.397070$; $b_1 = 0.400302$; $b_2 = 1.06000$; $B_1 = -0.171587$; $B_2 = 2.392190$, and the function $\lambda_2(\tau, \delta)$ has the form:

$$\lambda_2(\tau, \delta) = \left(\frac{d_1}{10} + d_2\right) \delta^{9/5} \exp\left[C_1(1 - \delta^{14/5})\right]$$

$$+ d_3 S \delta^Q \exp\left[\left(\frac{Q}{1+Q}\right)(1 - \delta^{1+Q})\right] + d_4 \exp\left(C_2\tau^{3/2} + \frac{C_3}{\delta^5}\right),$$

where Q and S are functions of argument $\Delta\tau = |\tau - 1| + C_4$:

$$Q = 2 + C_5/\Delta\tau^{0.6}; S = \begin{cases} 1/\Delta\tau & for\ \tau \geq 1; \\ C_6/\Delta\tau^{0.6} & for\ \tau < 1. \end{cases}$$

Coefficients d_i and C_i have the following values:
$d_1 = 0.0701309$; $d_2 = 0.0118520$; $d_3 = 0.00169937$; $d_4 = -1.0200$; $C_1 = 0.642857$; $C_2 = -4.11717$; $C_3 = -6.17937$; $C_4 = 0.00308976$; $C_5 = 0.0822994$; $C_6 = 10.0932$.

The Eq. (5) is applicable with the following values of temperatures and pressures: $p \leq 100$ MPa for 0 °C $\leq T \leq 500$ °C; $p \leq 70$ MPa for 500 °C $< T \leq 650$ °C; $p \leq 40$ MPa for 650 °C $< T \leq 800$ °C.

The error values in liquid at temperatures of 25–200 °C and pressures up to 5 MPa is equaled to the 1.5% in the calculations, at higher temperatures up to 300 °C – 2%. The error is equaled to the 1.5% for water vapor at temperatures up to 550 °C at a pressure of 0.1 MPa, at pressures up to 40 MPa – 3%.

The Eq. (5), in comparison with the theoretical conclusions, determines the not infinite, and the final coefficient of conductivity at the critical point, which does not allow estimating the error value near its critical point.

4   Discrete Model

The estimated domain inscribed in a rectangle. A uniform mesh is introduced for the numerical realization of the discrete mathematical model of the problem in the form:

$$w_h = \{t^n = nh_t; \; x_i = ih_x, \; r_j = jh_r; \;\; n = \overline{0, N_t}; \; i = \overline{0, N_x}; j = \overline{0, N_r};$$

$$N_t h_t = l_t; \; N_x h_x = l_x; N_r h_r = l_r\}, \tag{6}$$

where h_t is the time step; h_x, h_r are space steps; N_t is the upper time bound; N_x, N_r are space bounds.

To improve the discrete "smoothness" solution we assume that the cell are not completely filled. The domain Ω_{xr} is the filled part of the domain $D_{xr} : \{x \in [x_{i-1/2}, x_{i+1/2}], r \in [r_{j-1/2}, r_{j+1/2}]\}$. In addition, we introduce the notation for the following domains:

$$D_1 : \{x \in [x_i, x_{i+1/2}], r \in [r_{j-1/2}, r_{j+1/2}]\};$$

$$D_2 : \{x \in [x_{i-1/2}, x_i], r \in [r_{j-1/2}, r_{j+1/2}]\};$$

$$D_3 : \{x \in [x_{i-1/2}, x_{i+1/2}], r \in [r_j, r_{j+1/2}]\};$$

$$D_4 : \{x \in [x_{i-1/2}, x_{i+1/2}], r \in [r_{j-1/2}, r_j]\}.$$

The occupancy coefficients q_0, q_1, q_2, q_3, q_4 for the domains $D_{xr}, D_1, D_2, D_3, D_4$ are introduced as the following: $q_0 = S_{D_{xr}}/S_{\Omega_{xr}}$, $q_i = S_{D_i}/S_{\Omega_i}$, $i = \overline{1,4}$, where S is an area of the corresponding domain part, Ω_i is a filled part of the domain D_i.

The discrete analogue of the heat equation written in cylindrical coordinates with boundary conditions of the third kind has the form:

$$(q_0)_{i,j}\, \rho_{i,j} c_{i,j} r_j \frac{\hat{T}_{i,j} - T_{i,j}}{h_t} = (q_1)_{i,j}\, \lambda_{i+1/2,j} r_j \frac{\bar{T}_{i+1,j} - \bar{T}_{i,j}}{h_x^2}$$

$$- (q_2)_{i,j}\, \lambda_{i-1/2,j} r_j \frac{\bar{T}_{i,j} - \bar{T}_{i-1,j}}{h_x^2} - \left|(q_1)_{i,j} - (q_2)_{i,j}\right| \lambda_{i,j} r_j \frac{\alpha_x \bar{T}_{i,j} + \beta_x}{h_x}$$

$$+ (q_3)_{i,j}\, \lambda_{i,j+1/2} r_{j+1/2} \frac{\bar{T}_{i,j+1} - \bar{T}_{i,j}}{h_r^2} - (q_4)_{i,j}\, \lambda_{i,j-1/2} r_{j-1/2} \frac{\bar{T}_{i,j} - \bar{T}_{i,j-1}}{h_r^2}$$

$$- \left|(q_3)_{i,j} - (q_4)_{i,j}\right| \lambda_{i,j} r_j \frac{\alpha_r \bar{T}_{i,j} + \beta_r}{h_r} + (q_0)_{i,j}\, r_j f_{i,j}, \;\; i = \overline{1,4}, \tag{7}$$

where $\hat{T} \equiv T^{n+1}$, $T \equiv T^n$, $\bar{T} = \sigma\hat{T} + (1-\sigma)T$, $\sigma \in [0,1]$ is the scheme weight.

5   Discrete Model Research

Let's research the discrete model (7). Error magnitude orders were defined for the proposed approximations:

$$\lambda_{i+1/2,j}\frac{T_{i+1,j}-T_{i,j}}{h_x^2}-\lambda_{i-1/2,j}\frac{T_{i,j}-T_{i-1,j}}{h_x^2}=\left(\lambda_{i,j}\left(T_{i,j}\right)_x'\right)_x'+O\left(h_x^2\right),$$

$$\lambda_{i,j+1/2}r_{j+1/2}\frac{T_{i,j+1}-T_{i,j}}{h_r^2}-\lambda_{i,j-1/2}r_{j-1/2}\frac{T_{i,j}-T_{i,j-1}}{h_r^2}$$

$$=\left(\lambda_{i,j}r_j\left(T_{i,j}\right)_r'\right)_r'+O\left(h_r^2\right). \tag{8}$$

The estimation for the discrete heat equation:

$$\left\|T^{n+1}\right\|_c\leq\left\|T^0\right\|_c+\tau\sum_{k=0}^{n}\left\|f^k\right\|_c$$

$$+\left\|\frac{\left|q_1\left(P\right)-q_2\left(P\right)\right|\frac{\beta_x}{h_x}+\left|q_3\left(P\right)-q_4\left(P\right)\right|\frac{\beta_r}{h_r}}{\left|q_1\left(P\right)-q_2\left(P\right)\right|\frac{\alpha_x}{h_x}+\left|q_3\left(P\right)-q_4\left(P\right)\right|\frac{\alpha_r}{h_r}}\right\|_c. \tag{9}$$

The verification of the conservatism of the scheme (7):

$$\sum_{i,j}\left(q_0\right)_{i,j}c_{i,j}\rho_{i,j}r_j\hat{T}_{i,j}=\sum_{i,j}\left(q_0\right)_{i,j}c_{i,j}\rho_{i,j}r_jT_{i,j}$$

$$+\tau\sum_{i,j}\left(q_0\right)_{i,j}r_jf_{i,j}-\tau\sum_{i,j}r_j\lambda_{i,j}\left(\left|\left(q_1\right)_{i,j}-\left(q_2\right)_{i,j}\right|\frac{\alpha_x\bar{T}_{i,j}+\beta_x}{h_x}\right)$$

$$+\tau\sum_{i,j}r_j\lambda_{i,j}\left(\left|\left(q_3\right)_{i,j}-\left(q_4\right)_{i,j}\right|\frac{\alpha_r\bar{T}_{i,j}+\beta_r}{h_r}\right), \tag{10}$$

where $i\in\left[1,N_x-1\right],j\in\left[1,N_r-1\right]$.

The verification of balance relations for discrete model (7) showed that the quantity of heat at the next time layer is equaled to the total quantity of heat and the thermal energy emitted by the internal and boundary sources (drains).

6   Weight Scheme Optimization

The error estimation for the numerical solution of problem (1)–(3):

$$\phi=\max_{\chi\in[0,h_\tau]}\left|\left(\left(1-\frac{\chi}{1+\chi\sigma}\right)-e^{-\chi}\right)\frac{1+\chi\sigma}{\chi}\right|, \tag{11}$$

where $\chi=\frac{\lambda_i}{\lambda_{\max}}h_\tau$, $h_\tau=\frac{h_t}{\lambda_{\max}}$, $t=\lambda_{\max}\tau$, λ_i are eigenvalues of operator, σ is the scheme weight.

Table 1. Optimal proportion values of error ϕ, grid step h_τ and weight σ.

ϕ	h_τ	σ	ϕ	h_τ	σ
0.0001	0.08468	0.5058	0.006	0.7203	0.5478
0.0003	0.148	0.5102	0.008	0.8476	0.5558
0.0005	0.1924	0.5132	0.01	0.9642	0.563
0.0008	0.2452	0.5167	0.04	2.374	0.6373
0.001	0.2754	0.5188	0.06	3.333	0.6748
0.002	0.3964	0.5268	0.07	3.882	0.692
0.003	0.4925	0.5332	0.08	4.508	0.7084
0.005	0.6508	0.5434	0.1	6.166	0.7397

The optimum parameter σ^* is defined from the condition:

$$\sigma^* = \arg\min \left[\max_{\chi \in (0, h_\tau]} \left| \left(1 - \frac{\chi}{1 + \chi\sigma} \right) - e^{-\chi} \frac{1 + \chi\sigma}{\chi} \right| \right]. \qquad (12)$$

Values of the optimal weights depending on the time step variable are given in the Table 1.

We obtained that the use of optimized schemes for the solution of the problem (2)–(4) are reduced the computational labor expenditures in 2.5–3 times.

7   Solution Method of Grid Equations

We describe a solution method of grid equations occurred in the discretization (7) of difference equations. For this, we consider the problem of solution of the operator equation:

$$Ax = f, A : H \to H, \qquad (13)$$

where A is the linear, self-adjoint ($A = A^*$), positive definite operator ($A > 0$). We use the implicit two-layer iterative process for the solution of the problem (13):

$$B \frac{x^{m+1} - x^m}{\tau} + Ax^m = f, B : H \to H. \qquad (14)$$

In the Eq. (14) m is an iteration number, $\tau > 0$ is an iterative parameter and B is a preconditioner. We assume the additive decomposition of the operator in the construction of B:

$$A = A_1 + A_2, A_1^* = A_2. \qquad (15)$$

In view in the Eq. (14): $(Ay, A) = 2(A_1 y, y) = 2(A_2 y, y)$. Therefore, $A_1 > 0$, $A_2 > 0$ in the Eq. (14). Let in the Eq. (13):

$$B = (D + \omega R_1) D^{-1} (D + \omega R_2), \qquad (16)$$

where A is an operator of grid equation, D is a diagonal part of operator A, ω is an iterative parameter, R_1, R_2 are upper and lower triangular parts of an operator A.

Since $A = A^* > 0$, this gives $B = B^* > 0$ with the Eq. (15). The relations (14)–(16) are defined the modified alternating triangular iterative method (MATM) [2–4] of the problem solution (13). The algorithm of the adaptive modified alternating triangular iterative method of steepest descent is in the form [5,6]:

$$r^m = Ax^m - f, B(\omega_m)w^m = r^m, \widehat{\omega}_m = 2\,\|w^m\|\,/\,\|Aw^m\|,$$

$$\tau_{m+1} = \widehat{\omega}_m + 2\,\|w^m\|^2\,/(Aw^m, w^m), x^{m+1} = x^m - \tau_{m+1}w^m, \omega_{m+1} = \widehat{\omega}_m,$$

where x^m is a solution vector; w^m is a correction vector; r^m is a residual vector; f is a right part of the grid equation.

8   Parallel Implementation of the Modified Alternating Triangular Iterative Method

The scheme of two-layer iterative modified alternating triangular method [7,8] is in the form:

$$x^{m+1} = x^m - \tau_{m+1}w^m, (D + \omega R_1)\,D^{-1}\,(D + \omega R_2)\,w^m = r^m, r^m = Ax^m - f.$$

For the methods of the domain decomposition in one direction were used for parallel implementation of adaptive MATM method. The most laborious calculation from the point of view the development of parallel software implementation is the calculation of the amendment vector, which is performed in two steps [9–11]:

$$(1)\,(D + \omega R_1)\,y^m = r^m; (2)\,(D + \omega R_2)\,w^m = Dy^m.$$

At the first step, the elements of the auxiliary vector y^m are calculated bottom up, and then, knowing it, at the second step the elements of the correction vector w^m are calculated top-down.

The schemes of the calculation of the auxiliary vector and correction vector are given in Fig. 1 (the arrows indicate the directions of calculation and transfer between processors of multiprocessor computer system).

The adaptive modified alternating triangular method (MATM) of minimal corrections was used for solving the heat transport problem (2)–(4) on the multiprocessor computer system (MCS). The decomposition methods of grid domains were used for computational laborious problems diffusion-convection in parallel implementation, taking into account the architecture and parameters of MCS. Maximum performance of MCS is 18.8 teraflops. The 512 uniform 16-core HP ProLiant BL685c Blade servers are used as computational nodes, each of which is equipped the four Quad-core AMD Opteron 8356 2.3 GHz processors and the operative memory in volume of 32 GB. The time costs for performing the one iteration of the MATM method on various grids and values of acceleration and efficiency for different numbers of computational cores are given in Table 2 [12–17].

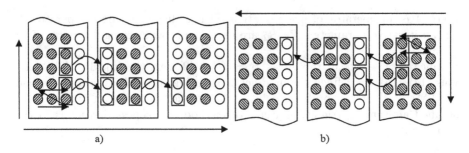

Fig. 1. Calculation schemes: (a) the auxiliary vector y^m; (b) the correction vector w^m.

Table 2. The acceleration and efficiency of MATM parallel version.

		100×100	200×200	500×500	1000×1000	2000×2000	5000×5000
1	Time	0.001183	0.010633	0.026031	0.10584	0.381988	3.700073
	Acceleration	1	1	1	1	1	1
	Efficiency	1	1	1	1	1	1
4	Time	0.000232	0.00106	0.005755	0.026683	0.132585	1.2655
	Acceleration	5.099	10.031	4.523	3.967	2.881	2.924
	Efficiency	1.275	**2.508**	1.131	0.992	0.72	0.731
16	Time	0.000231	0.000407	0.001869	0.013105	0.085056	0.472151
	Acceleration	5.121	**26.125**	13.928	8.076	4.491	7.837
	Efficiency	0.32	1.633	0.87	0.505	0.281	0.49
64	Time	0.000642	0.000748	0.001557	0.004189	0.026844	0.182296
	Acceleration	1.843	14.215	16.719	25.266	14.23	20.297
	Efficiency	0.029	0.222	0.261	0.395	0.222	0.317
128	Time	-	0.001612	0.002064	0.003442	0.016437	0.076545
	Acceleration	-	6.596	12.612	**30.75**	23.24	48.338
	Efficiency	-	0.052	0.099	0.24	0.182	0.378
512	Time	-	-	-	0.009793	0.012362	0.058805
	Acceleration	-	-	-	10.808	**30.9**	**62.921**
	Efficiency	-	-	-	0.021	0.06	0.123

According to the Table 2, the acceleration takes the highest value at the certain value of calculators and the decreases at further increasing of the number of cores for each of the computational grids. This is due to the time costs for data exchange between the calculaters [18–21].

The initial parameters of the turbine were the values of temperatures and conductivity of the environment and in the internal turbine cameras. Environmental temperature was specified of 20 °C, the thermal conductivity coefficient 0.022 W/(m·K). For the first camera the temperature was specified of 500 °C, the thermal conductivity coefficient 0.099 W/(m·K). For the second cameras temperature was specified of 337.7 °C, the thermal conductivity coefficient 0.0555 W/(m·K). For the third cameras the temperature was specified of 200.4 °C, the thermal conductivity coefficient 0.0361 W/(m·K). For the fourth cameras the temperature was specified of 104 °C, the thermal conductivity coefficient 0.0257 W/(m·K). The initial surface temperature is equal to 20 °C, the thermal conductivity coefficient 92 W/(m·K).

The initial data for modeling were: steps at the spatial coordinates hx = 0.0164 m, hy = 0.0164 m, and the time step ht = 1 s; the time interval lt = 3600; the optimal scheme weight, corresponding to the given time step, sigma = 0.563.

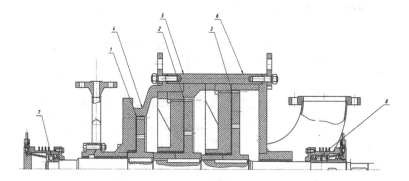

Fig. 2. Thermal field in the area of the left bearing.

The numerical experiment described the thermal modes in the main functional points of the steam turbine: 1–3–points, located on the impellers and turbine blades; 4–6–on the surface material; 7–8–on bearings. The location shows of points, in which the temperature field was determined, is given in Fig. 2. The results of numerical experiments of modeling the thermal processes in the steam turbine are given in Figs. 3, 4 and 5.

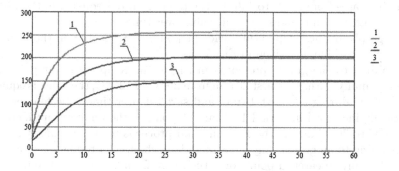

Fig. 3. Thermal field at points on the impellers of the turbine.

The graphs of the temperature fields at points 1–3, located on the working wheels of steam turbines, are given in Fig. 3. The calculation period was equaled to the 1 h.

According to the graph, presented in Fig. 3, the temperature at point 1 is changing rapidly during the first 15 min from 25 °C to 250 °C and then saves the achieved value. The temperature increases from 25 °C to 200 °C during the period 1–25 min at points 2 and 3 and 150 °C respectively, and after saves the value over time.

The graphs of the temperature fields at points 4–6, located on the surface of the steam turbine, are given in Fig. 4.

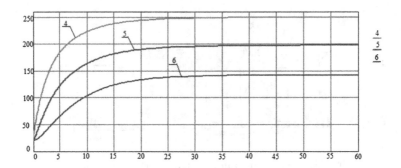

Fig. 4. Thermal field at points on the turbine surface.

According to the graph, presented in Fig. 4, the temperature at point 4 is changing rapidly during the first 15 min from 25 °C to 240 °C, and then from 15 min to 25 min the value is equaled to the 250 °C and saved it over time. the temperature is abruptly equaled to the value from 25 °C to 190 °C at point 5 during 1–20 min, then increased to 200 °C during from 20 min to 35 min and proceeded to the stationary state. At point 6 in the first 20 min, the temperature rapidly increases from 25 °C to 140 °C, the temperature is equaled to the 140 °C within the next 10 min and saved this value over time.

The graphs of the temperature fields at points 7–8, located on the bearings of the steam turbine, are given in Fig. 5.

According to the graph, presented in Fig. 5, the temperature at point 7 is changing rapidly during the first 20 min from 25 °C to 120 °C, the value is equaled to the 130 °C during the next 15 min and saved it over time. The temperature is gradually increased in period 1–35 min at point 8, the temperature increases from 25 °C to 70 °C during the 1–35 min and proceeds to the stationary state.

According to all graphs, any leaps or sudden changes of temperature are not observed in any researched points over the range from 20 min to 60 min.

The thermal field on the surface of the steam turbine and the left and right bearings through 10, 20 and 30 min after starting respectively is given in Fig. 6.

The data of the above graphs (Figs. 3, 4 and 5) is corresponded to the heat pattern shown in Fig. 6.

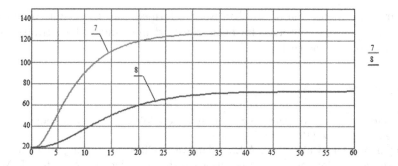

Fig. 5. Thermal field at points on the turbine bearings.

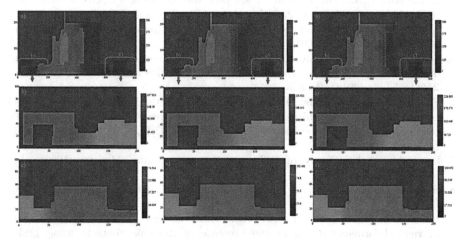

Fig. 6. Thermal field in the turbine surface and in areas of the left and right bearings through 10, 20 and 30 min after starting.

9   Conclusion

The heat transport mathematical model in the steam turbine was proposed in this paper. The finite-difference analogues of the diffusion transport operator were obtained in the polar coordinate system. The occupancy degree of the cells was taken into account in the construction of difference schemes. The schemes with optimal weights were used to discretize the model. The use of optimized schemes are reduced the computational labour input in 2.5–3 times. It is shown that the proposed approximations have the second order of error in the spatial variable. The stability of the proposed difference schemes are checked on the basis of discrete maximum principle. The condition of applicability of the discrete maximum principle was obtained, and the stability of the proposed difference schemes was proved for the initial data, boundary conditions and right part. The main balance ratios were checked for mathematical model describing the thermal processes on the surface of the steam turbine.

Software implementation of problem was performed, and the results of numerical calculations were obtained for heat transport modeling of heat transport on the surface of the gas turbine.

References

1. Samarsky, A.A., Nikolaev, E.S.: Methods of Solving Grid Equations, p. 588. Science, Moscow (1978). (in Russian)
2. Sukhinov, A.I., Chistyakov, A.E.: Adaptive modified alternating triangular iterative method for solving grid equations with non-selfadjoint operator. J. Math. Model. **24**(1), 3–20 (2012). (in Russian)
3. Konovalov, A.N.: The method of steepest descent with adaptive alternately-triangular preamplification. J. Differ. Equ. **40**(7), 953 (2004). (in Russian)
4. Konovalov, A.N.: The theory of alternating-triangular iterative method. J. Siberian Math. J. **43**(3), 552 (2002). (in Russian)
5. Sukhinov, A.I., Chistyakov, A.E., Shishenya, A.V.: Error estimate of the solution of the diffusion equation on the basis of the schemes with weights. J. Math. Model. **25**(11), 53–64 (2013). (in Russian)
6. Sukhinov, A.I., Chistyakov, A.E., Fomenko, N.A.: Methods of constructing difference schemes for the problem of diffusion-convection-reaction, taking into account the occupancy level of the control cells. J. Izv. SFedU Eng. Sci. **4**, 87–98 (2013). (in Russian)
7. Samarskiy, A.A.: Theory of Difference Schemes. Nauka, Moscow (1989). (in Russian)
8. Samarskiy, A.A., Gulin, A.V.: Numerical Methods. Nauka, Moscow (1989). (in Russian)
9. Beklemishev, K.A., Petrov, I.B., Favorsky, A.V.: Numerical simulation of processes in rigid deformable media in the presence of dynamic contacts using grid-characteristic method. J. Math. Model. **25**(11), 3–16 (2013). (in Russian)
10. Petrov, I.B., Favorsky, A.V., Sannikov, A.V., Kvasov, I.E.: Grid-characteristic method using high order interpolation on tetrahedral hierarchical meshes with a multiple time step. J. Math. Model. **25**(2), 42–52 (2013). (in Russian)
11. Sukhinov, A.I., Chistyakov, A.E., Semenyakina, A.A., Nikitina, A.V.: Parallel realization of the tasks of the transport of substances and recovery of the bottom surface on the basis of schemes of high order of accuracy. J. Comput. Meth. Program. New Comput. Technol. **16**(2), 256–267 (2015). (in Russian)
12. Nikitina, A.V., Semenyakina, A.A., Chistyakov, A.E., Protsenko, E.A., Yakovenko, I.V.: Application of schemes of increased order of accuracy for solving problems of biological kinetics in a multiprocessor computer system. J. Fundam. Res. **12–3**, 500–504 (2015). (in Russian)
13. Chistyakov, A.E., Hachunts, D.S., Nikitina, A.V., Protsenko, E.A., Kuznetsova, I.: Parallel library of iterative methods of the SLAE solvers for problem of convection-diffusion-based decomposition in one spatial direction. J. Mod. Prob. Sci. Educ. **1**(1), 1786 (2015). (in Russian)
14. Sukhinov, A.I., Nikitina, A.V., Semenyakina, A.A., Protsenko, E.A.: Complex programs and algorithms to calculate sediment transport and multi-component suspensions on a multiprocessor computer system. J. Eng. J. Don **38**(4), 52 (2015). (in Russian)

15. Nikitina, A.V., Abramenko, Y.A., Chistyakov, A.E.: Mathematical modeling of oil spill in shallow waters. J. Inform. Comput. Sci. Eng. Educ. **3**(23), 49–55 (2015). (in Russian)

16. Chistyakov, A.E., Nikitina, A.V., Ougolnitsky, G.A., Puchkin, V.M., Semenov, I.S., Sukhinov, A.I., Usov, A.B.: A differential game model of preventing fish kills in shallow waterbodies. J. Game Theory Appl. **17**, 37–48 (2015)

17. Sukhinov, A.I., Nikitina, A.V., Semenyakina, A.A., Chistyakov, A.E.: A set of models, explicit regularized schemes of high order of accuracy and programs for predictive modeling of consequences of emergency oil spill. In: Proceedings of the International Scientific Conference Parallel Computational Technologies (PCT 2016), pp. 308–319 (2016). (in Russian)

18. Chistyakov, A.E., Nikitina, A.V., Sumbaev, V.V.: Solution of the Poisson problem based on the multigrid method. J. Herald Comput. Inf. Technol. **8**(146), 3–7 (2016). (in Russian)

19. Nikitina, A.V., Semenyakina, A.A., Chistyakov, A.E.: Parallel implementation of the tasks of diffusion-convection-based schemes of high order of accuracy. J. Herald Comput. Inf. Technol. **7**(145), 3–8 (2016). (in Russian)

20. Sukhinov, A.I., Chistyakov, A.E., Semenyakina, A.A., Nikitina, A.V.: Numerical modeling of an ecological condition of the Sea of Azov with application of schemes of the raised accuracy order on the multiprocessor computing system. J. Comput. Res. Model. **8**(1), 151–168 (2016). (in Russian)

21. Sukhinov, A.I., Nikitina, A.V., Semenyakina, A.A., Chistyakov, A.E.: Complex of models, explicit regularized schemes of high-order of accuracy and applications for predictive modeling of after-math of emergency oil spill. In: 10th Annual International Scientific Conference on Parallel Computing Technologies, PCT 2016, Arkhangelsk, Russian Federation, 29–31 March 2016, Code 121197. CEUR Workshop Proceedings, vol. 1576, pp. 308–319 (2016). (in Russian)

Predictive Modeling of Suffocation in Shallow Waters on a Multiprocessor Computer System

Aleksandr Sukhinov[1], Alla Nikitina[1(✉)], Aleksandr Chistyakov[1],
Vladimir Sumbaev[1,2], Maksim Abramov[1], and Alena Semenyakina[3]

[1] Don State Technical University, Rostov-on-Don, Russia
sukhinov@gmail.com, nikitina.vm@gmail.com, cheese_05@mail.ru,
maxim-abramov@yandex.ru
[2] South Federal University, Rostov-on-Don, Russia
valdec4813@mail.ru
[3] Kalyaev Scientific Research Institute of Multiprocessor Computer Systems,
Southern Federal University, Taganrog, Russia
j.a.s.s.y@mail.ru

Abstract. The model of the algal bloom, causing suffocations in shallow waters takes into account the follows: the transport of water environment; microturbulent diffusion; gravitational sedimentation of pollutants and plankton; nonlinear interaction of plankton populations; biogenic, temperature and oxygen regimes; influence of salinity. The computational accuracy is significantly increased and computational time is decreased at using schemes of high order of accuracy for discretization of the model. The practical significance is the software implementation of the proposed model, the limits and prospects of it practical use are defined. Experimental software was developed based on multiprocessor computer system and intended for mathematical modeling of possible progress scenarios of shallow waters ecosystems on the example of the Azov Sea in the case of suffocation. We used decomposition methods of grid domains in parallel implementation for computationally laborious convection-diffusion problems, taking into account the architecture and parameters of multiprocessor computer system. The advantage of the developed software is also the use of hydrodynamical model including the motion equations in the three coordinate directions.

Keywords: Multiprocessor computer system · Water bloom · Mathematical model · Suffocation · Phytoplankton · Computational experiments

1 Introduction

Shallow waters like the Azov Sea are suffered the great anthropogenic influence. However, most of them is the unique ecological systems of fish productivity.

This paper was partially supported by the grant No. 17-11-01286 of the Russian Science Foundation, the program of fundamental researches of the Presidium of RAS No. 43 Fundamental problems of mathematical modeling, and partial financial support of RFFR for projects No. 15-01-08619, No. 15-07-08626, No. 15-07-08408.

V. Malyshkin (Ed.): PaCT 2017, LNCS 10421, pp. 172–180, 2017.
DOI: 10.1007/978-3-319-62932-2_16

The biogenic matters are entered in the shallow waters with the river flows which causing the growth of the algae – water bloom. The suffocation periodically occurs in shallow waters in summer. Because there is the significant decrease of dissolved oxygen in them, consumed in the decomposition of organic matter, due to the high temperature. The fish is suffering the oxygen starvation and the mass dying of suffocation.

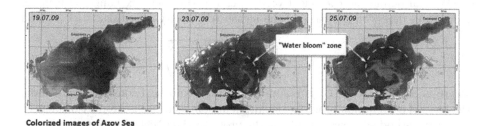

Colorized images of Azov Sea
(spectral channels: 0.620-0.670 mcm, 0.545-0.565 mcm, 0.459-0.479 mcm)

Fig. 1. The wide areas of the water bloom in the Azov Sea.

The results of satellite monitoring of the Earth are used in this paper to control the quality modeling of processes of hydrodynamics and biological kinetics [1,2]. The satellite monitoring data of the Azov Sea, obtained by SRC Planeta, are given in Fig. 1 [3]. The analysis of satellite data reveals the water areas of suffocations.

2   Hydrodynamic Mathematical Model

The Navier-Stokes motion equations are initial equations of hydrodynamics of shallow water:

$$u'_t + uu'_x + vu'_y + wu'_z = -\frac{1}{\rho}p'_x + (\mu u'_x)'_x + (\mu u'_y)'_y + (\nu u'_z)'_z + 2\Omega(v\sin\theta - w\cos\theta),$$

$$v'_t + uv'_x + vv'_y + wv'_z = -\frac{1}{\rho}p'_y + (\mu v'_x)'_x + (\mu v'_y)'_y + (\nu v'_z)'_z - 2\Omega u\sin\theta, \quad (1)$$

$$w'_t + uw'_x + vw'_y + ww'_z = -\frac{1}{\rho}p'_z + (\mu w'_x)'_x + (\mu w'_y)'_y + (\nu w'_z)'_z + 2\Omega u\cos\theta + g\left(\rho_0/\rho - 1\right);$$

– continuity equation was written for the case of variable density:

$$\rho'_t + (\rho u)'_x + (\rho v)'_y + (\rho w)'_z = 0, \quad (2)$$

where $u = \{u, v, w\}$ are velocity vector components; p is an excess pressure above the undisturbed fluid hydrostatic pressure; ρ is a density; Ω is an Earth's angular velocity; θ is an angle between the angular velocity vector and the vertical vector; μ, ν are horizontal and vertical components of turbulent exchange coefficient. Tangential stress components for bottom are in the form:

$$\tau_x = \rho C_p \left(|u|\right) u \left|u\right|, \tau_y = \rho C_p \left(|u|\right) v \left|u\right|.$$

We can define the coefficient of the vertical turbulent exchange with inhomogeneous depth on the basic of the measured velocity pulsation:

$$\nu = C_s^2 \Delta^2 \frac{1}{2} \sqrt{\left(\frac{\partial u}{\partial z}\right)^2 + \left(\frac{\partial v}{\partial z}\right)^2}, \qquad (3)$$

where Δ is a grid scale; C_s is a non-dimensional empirical constant, defined on the basis of attenuation process calculation of homogeneous isotropic turbulence.

Grid method was used for solving the problem (1)–(2) [4]. The approximation of equations by time variable was performed on the basis of splitting schemes into physical processes [5–7] in the form of the pressure correction method.

3   Mathematical Model of Water Bloom Processes of Shallow Waters

The spatially heterogeneous model of water bloom (WB) is described by equations:

$$S_{i,t} + u\frac{\partial S_i}{\partial x} + v\frac{\partial S_i}{\partial y} + (w - w_{gi})\frac{\partial S_i}{\partial x} = \mu_i \Delta S_i + \frac{\partial}{\partial z}\left(\nu_i \frac{\partial S_i}{\partial z}\right) + \psi_i. \quad (4)$$

(4) are equations of changes the concentration of impurities, index i indicates the substance type, S_i is the concentration of i-th impurity, $i = \overline{1,6}$; 1 is the total organic nitrogen (N); 2 are phosphates (PO_4); 3 is a phytoplankton; 4 is a zooplankton; 5 is a dissolved oxygen (O_2); 6 is a hydrogen sulfide (H_2S); u, v, w are components of water flow velocity vector; ψ_i is a chemical-biological source (drain) or a summand that describes the aggregation (clumping-declumping) if the corresponding component is a suspension.

The WB model takes into account the transport of water flow; microturbulent diffusion; gravitational sedimentation of pollutants and plankton; nonlinear interaction of planktonic populations; nutrient, temperature and oxygen regimes; influence of salinity.

Computational domain $\bar{G}$ is a closed area, limited by the undisturbed water surface Σ_0, bottom $\Sigma_H = \Sigma_H(x, y)$, and the cylindrical surface, the undisturbed surface σ for $0 < t \leq T_0$. $\Sigma = \Sigma_0 \cup \Sigma_H \cup \sigma$ – the sectionally smooth boundary of the domain G [8–10].

We consider the system (4) with the following boundary conditions:

$$\begin{aligned} &S_i = 0 \text{ on } \sigma, \text{ if } U_n < 0; \frac{\partial S_i}{\partial n} = 0 \text{ on } \sigma, \text{ if } U_n \geq 0; \\ &S'_{i,z} = \phi(S_i) \text{ on } \Sigma_0; S'_{i,z} = -\varepsilon_i S_i \text{ on } \Sigma_H, \end{aligned} \qquad (5)$$

where ε_i is the absorption coefficient of the i-th component by the bottom material.

We has to add the following initial conditions to (4):

$$S_i|_{t=0} = S_{i0}(x, y, z), \ i = \overline{1, 6}. \tag{6}$$

Water flow velocity fields, calculated according to the model (1)–(2), are used as input data for the model (4)–(6). The discretization of models (1)–(2), (4)–(6) was performed on the basis of the high-resolution schemes which are described in [11].

4   Parallel Implementation of the Modified Alternating Triangular Method (MATM)

We describe the parallel algorithms, which are used for solving the problems (1)–(2), (4)–(6), with different types of domain decomposition.

Algorithm 1. Each processor is received its computational domain after the partition of the initial computational domain into two coordinate directions. The adjacent domains overlap by two layers of nodes in the perpendicular direction to the plane of the partition.

The residual vector and it uniform norm are calculated after that as each processor will receive the information for its part of the domain. Then, each processor determines the maximum element in module of the residual vector and transmits its value to all remaining calculators. Now receiving the maximum element on each processor is enough to calculate the uniform norm of the residual vector.

The parallel algorithm for calculating the correction vector is in the form:

$$(D + \omega_m R_1)D^{-1}(D + \omega_m R_2)w^m = r^m,$$

where R_1 is the lower-triangular matrix, and R_2 is the upper-triangular matrix. We should solve consistently the next two equations for calculating the correction vector:

$$(D + \omega_m R_1)y^m = r^m, (D + \omega_m R_2)w^m = Dy^m.$$

At first, the vector y^m is calculated, and the calculation is started in the lower left corner. Then, the correction vector w^m is calculated from the upper right corner. The calculation scheme of the vector y^m is given in Fig. 2 (the transferring elements after the calculation of two layers by the first processor is presented).

In the first step of calculating the first processor work on with the top layer. Then the transfer of overlapping elements is occurred to the adjacent processors. In the next step the first processor work on with the second layer, and its neighbors – the first. The transfer of elements after calculating two layers by the first processor is given in Fig. 2. In the scheme for the calculation of the vector y^m only the first processor does not require additional information and can independently work on with its part of the domain. Other processors are waiting the results from the previous processor, while it transfers the calculated values of the grid functions for the grid nodes, located in the preceding positions of this line. The process continues until all the layers will be calculated. Similarly, we can solve the systems of linear algebraic equations (SLAE) with the upper-triangular matrix for calculating the correction vector. Further, the scalar products are calculated, and the transition is proceeded to the next iteration layer.

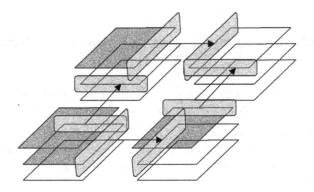

Fig. 2. Scheme of calculation the vector y^m.

We constructed the theoretical estimate of the time. It's required to perform the MATM step for SLAE with seven-diagonal matrix with using decomposition in two spatial directions on a cluster of distributed calculations.

We considered the case of the problem solution with the rectangular domain. The domain has a complex shape in the case of real water. At the same time the real acceleration is less than its theoretical estimation. The dependence of the acceleration, obtained in the theoretical estimates, can be used as the upper estimate of the acceleration for parallel implementation of the MATM algorithm by the domain decomposition in two spatial directions.

We describe the domain decomposition in two spatial directions with using the *k-means* algorithm.

Algorithm 2. The algorithm of *k-means* method.

(1) The initial centers of subdomains are selected with using maximum algorithm.
(2) All calculated nodes are divided into m Voronoi's cells by the method of the nearest neighbor, i.e. the current calculated grid node $x \in X_c$, where X_c is a subdomain, which is chosen according to the condition $\|x - s_c\| = \min_{1 \leq i \leq m} \|x - s_i\|$, where the s_c is the center of the subdomain X_c.
(3) New centers are calculated by the formula: $s_c^{(k+1)} = \frac{1}{\left|X_i^{(k)}\right|} \sum_{x \in X_i^{(k)}} x$.

(4) The condition of the stop is checked $s_c^{(k+1)} = s_c^{(k)}$, $k = 1, ..., m$. If the condition of the stop is not performed, then the transition is proceeded to the item 2 of the algorithm.

The result of the *k-means* method for model domains is given in Fig. 3 (arrows are indicated exchanges between subdomains). All points in the boundary of each subdomains are required to data exchange in the computational process. The Jarvis's algorithm was used for this aim (the task of constructing the convex hull). The list of the neighboring subdomains for each subdomain was created, and an algorithm was developed for data transfer between subdomains.

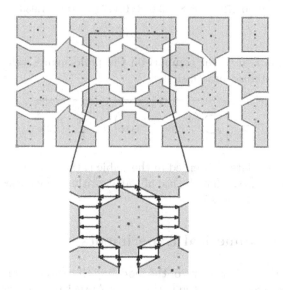

Fig. 3. Domain decomposition.

The comparison of the developed parallel algorithms 1 and 2 for the solution (1)–(2), (4)–(6) was performed. The results are given in the Table 1.

In Table 1: n is the number of processors; $t_{(k)}, S_{(k)}, E_{(k)}$ are the processing time, the acceleration and efficiency of the k-th algorithm; $S_{(k)}^t, E_{(k)}^t$ are the theoretical estimates of the efficiency and acceleration of the k-th algorithm, $k = \{1, 2\}$.

According to the Table 1 we can conclude that the developed algorithms based on the decomposition method in two spatial directions and *k-means* method can be effectively used for solving hydrodynamics problems in the case the sufficiently large number of computational nodes.

Table 1. Comparison of acceleration and efficiency of algorithms

n	$t_{(1)}$	$S_{(1)}^t$	$S_{(1)}$	$t_{(2)}$	$E_{(2)}^t$	$E_{(2)}$
1	7.491	1.0	1.0	6.073	1.0	1.0
2	4.152	1.654	1.804	3.121	1.181	1.946
4	2.549	3.256	2.938	1.811	2.326	3.354
8	1.450	6.318	5.165	0.997	4.513	6.093
16	0.882	11.928	8.489	0.619	8.520	9.805
32	0.458	21.482	16.352	0.317	15.344	19.147
64	0.266	35.955	28.184	0.184	25.682	33.018
128	0.172	54.618	43.668	0.117	39.013	51.933

The estimation is used for comparison the performance values of the algorithms 1 and 2, obtained practically:

$$\delta = \sqrt{\sum_{k=1}^{n}\left(E_{(2)k} - E_{(1)k}\right)^2} \Big/ \sqrt{\sum_{k=1}^{n} E_{(2)k}^2}. \qquad (7)$$

On the basis the data, presented in the Table 1, the comparison of the developed algorithms is shown that the use of the algorithm 2 increased the efficiency for the problem (1)–(2) on 15%.

5   Results of Numerical Experiments

A series of numerical experiments of the modeling the water bloom processes was performed in the Azov Sea for the period from April 1 to October 31, 2013. The results of the numerical experiment for reconstruction of the suffocation caused by the phytoplankton bloom in July 2013 is given in Fig. 4.

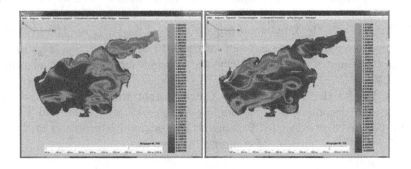

Fig. 4. Phytoplankton concentration change in Azov Sea.

The comparison of the developed software complex that implements the designed scenarios for the changing of ecological situation in the Azov Sea using the numerical realization of model plankton evolution problems of the biological kinetics use with the similar works in the mathematical modeling of hydro-biological processes.

The verification criterion of the developed models (1)–(2), (4)–(6) was an estimate of the error modeling taking into account the available field data measurements at the same time, calculated according to the formula:

$$\delta = \sqrt{\sum_{k=1}^{n} (S_{k\ nat} - S_k)^2} / \sqrt{\sum_{k=1}^{n} S_{k\ nat}^2},$$

where $S_{k\ nat}$ – the value of the harmful algae concentration, obtained through field measurements; S_k – the value of the harmful algae concentration, calculated by the model (1)–(2). The concentrations of pollutants and plankton calculated for different wind situations were taken into consideration, if the relative error did not exceed 30%.

The analysis of the same software complexes for shallow waters has shown that the accuracy of the predictive changes in pollutant concentrations of plankton in shallow waters has been increased at 10–15% depending on the chosen problem of the biological kinetics.

6   Conclusion

The model of hydrodynamics and water bloom were proposed in this paper. They used for the reconstruction of suffocation, occurred on July 16, 2013 in the south-eastern part of the Azov Sea. The numerical implementation of the developed models was performed on the multiprocessor computer system with distributed memory. The theoretical values of the acceleration and efficiency of parallel algorithms was calculated. The developed experimental software is designed for mathematical modeling of possible scenarios of development of ecosystems of shallow waters on the example of Azov-Black Sea basin. The decomposition methods of grid domains were used in parallel implementation for computationally laborious convection-diffusion problems, taking into account the architecture and parameters of multiprocessor computer system. The maximum acceleration value was achieved with using 128 computational nodes and equaled to 43 times. Two algorithms, including the algorithm, had been developed in the parallel algorithm implementation for solving the problem on the MCS and the data distribution between the processors. Using k-$means$ method, the algorithm efficiency of the problem was increased at 15% compared with the algorithm, based on a standard partition the computational domain.

References

1. Samarsky, A.A., Nikolaev, E.S.: Methods of Solving Grid Equations, p. 588. Science, Moscow (1978). (in Russian)
2. Sukhinov, A.I., Chistyakov, A.E.: Adaptive modified alternating triangular iterative method for solving grid equations with non-selfadjoint operator. Math. Model. **24**(1), 3–20 (2012). (in Russian)
3. State Research Center Planeta. http://planet.iitp.ru/english/index_eng.htm
4. Samarskiy, A.A.: Theory of Difference Schemes. Nauka, Moscow (1989). (in Russian)
5. Konovalov, A.N.: The method of steepest descent with adaptive alternately-triangular preamplification. Differ. Equ. **40**(7), 953 (2004). (in Russian)
6. Konovalov, A.N.: The theory of alternating-triangular iterative method. Siberian Math. J. **43**(3), 552 (2002). (in Russian)
7. Sukhinov, A.I., Chistyakov, A.E., Shishenya, A.V.: Error estimate of the solution of the diffusion equation on the basis of the schemes with weights. Math. Model. **25**(11), 53–64 (2013). (in Russian)
8. Sukhinov, A.I., Chistyakov, A.E., Semenyakina, A.A., Nikitina, A.V.: Parallel realization of the tasks of the transport of substances and recovery of the bottom surface on the basis of high-resolution schemes. Comput. Meth. Program. New Comput. Technol. **16**(2), 256–267 (2015). (in Russian)
9. Chistyakov, A.E., Hachunts, D.S., Nikitina, A.V., Protsenko, E.A., Kuznetsova, I.: Parallel library of iterative methods of the SLAE solvers for problem of convection-diffusion-based decomposition in one spatial direction. Mod. Probl. Sci. Educ. **1**(1), 1786 (2015). (in Russian)
10. Sukhinov, A.I., Nikitina, A.V., Semenyakina, A.A., Protsenko, E.A.: Complex programs and algorithms to calculate sediment transport and multi-component suspensions on a multiprocessor computer system. Eng. J. Don **38**(4), 52 (2015). (in Russian)
11. Sukhinov, A.I., Nikitina, A.V., Semenyakina, A.A., Chistyakov, A.E.: A set of models, explicit regularized high-resolution schemes and programs for predictive modeling of consequences of emergency oil spill. In: Proceedings of the International Scientific Conference Parallel Computational Technologies (PCT 2016), pp. 308–319 (2016). (in Russian)

Cellular Automata and Discrete Event Systems

Finite and Infinite Computations and a Classification of Two-Dimensional Cellular Automata Using Infinite Computations

Louis D'Alotto[1,2(✉)]

[1] Department of Mathematics and Computer Science,
York College/City University of New York, Jamaica, New York 11451, USA
[2] The Doctoral Program in Computer Science, CUNY Graduate Center,
New York, USA
ldalotto@gc.cuny.edu

Abstract. This paper proposes an application of the Infinite Unit Axiom and *grossone*, introduced by Yaroslav Sergeyev (see [19–23]), to the development and classification of two-dimensional cellular automata. This application establishes, by the application of grossone, a new and more precise nonarchimedean metric on the space of definition for two-dimensional cellular automata, whereby the accuracy of computations is increased. Using this new metric, open disks are defined and the number of points in each disk computed. The forward dynamics of a cellular automaton map are also studied by defined sets. It is also shown that using the Infinite Unit Axiom, the number of configurations that follow a given configuration, under the forward iterations of the cellular automaton map, can now be computed and hence a classification scheme developed based on this computation.

Keywords: Cellular automata · Infinite Unit Axiom · Grossone · Nonarchimedean metric · Dynamical systems

1 Introduction

Cellular automata, originally developed by von Neuman and Ulam in the 1940's to model biological systems, are discrete dynamical systems that are known for their strong modeling and self-organizational dynamical properties (for examples of some of these properties see [1, 3, 4, 8, 26–28, 30]). Cellular automata are defined on an infinite lattice and can be defined for all dimensions. In the one-dimensional case the integer lattice $\mathbb{Z}$ is used. In the two-dimensional case, $\mathbb{Z} \times \mathbb{Z}$. An example of a two-dimensional cellular automaton is John Conway's ever popular "Game of Life" (for a complete description of "The Game of Life", including some of the more interesting structures that emerge, see [2], Chap. 25). Probably the most interesting aspect about cellular automata is that which seems to conflict our physical systems. While physical systems tend to maximal entropy, even starting

© Springer International Publishing AG 2017
V. Malyshkin (Ed.): PaCT 2017, LNCS 10421, pp. 183–195, 2017.
DOI: 10.1007/978-3-319-62932-2_17

with complete disorder, forward evolution of cellular automata can generate highly organized structure.

As with all dynamical systems, it is important and interesting to understand their long term behavior under forward time evolution and achieve an understanding or hopefully a classification of the system. The concept of classifying cellular automata was initiated by Stephen Wolfram in the early 1980's, see [29, 30]. Wolfram classified one-dimensional cellular automata through numerous computer simulations. He actually noticed that if an initial configuration (sequence) was chosen at random then the probability is high that the cellular automaton rule will fall within one of four classes. Later, R. Gilman (see [10]) produced his measure theoretic/probabilistic classification of one-dimensional cellular automata and partitioned them into three classes. This was a more rigorous classification of cellular automata and based on the probability of choosing a configuration that will stay arbitrarily close to a given initial configuration under forward iteration of the map. To accomplish this Gilman used a metric that considers the central window where two configurations agree and continue to agree upon forward iterations. For an overview and a comparison of classifications of one-dimensional cellular automata see [7, 15]. However, This paper is concerned with the classification of two-dimensional cellular automata. Gilman and Wolfram's results have not been formally extended to the two-dimensional case, however, presented herein is a new approach to a classification of two-dimensional cellular automata.

2   The Infinite Unit Axiom

The new methodology of computation, initiated by Sergeyev (see [19–22]), provides a new way of computing with infinities and infinitesimals. Indeed, Sergeyev uses concepts and observations from physics (and other sciences) to set the basis for this new methodology. This basis is philosophically founded on three postulates:

Postulate 1. *"We postulate the existence of infinite and infinitesimal objects but accept that human beings and machines are able to execute only a finite number of operations."*

Postulate 2. *"We shall not tell what are the mathematical objects we deal with. Instead, we shall construct more powerful tools that will allow us to improve our capacities to observe and to describe properties of mathematical objects."*

Postulate 3. *"We adopt the principle: 'The part is less than the whole', and apply it to all numbers, be they finite, infinite, or infinitesimal, and to all sets and processes, finite or infinite."*

These postulates set the basis for a new way of looking at and measuring mathematical objects. The postulates are actually important philosophical realizations that we live in a finite world (i.e. that we, and machines, are incapable of infinite or infinitesimal computations). All the postulates are important in the application presented herein, however Postulate 1 has a ready illustration. In this paper we will deal with counting and hence representing infinite quantities and measuring (by way of a metric) extremely small or infinitesimal quantities. Postulate 2 also has a ready consequence herein. In the classification presented in this paper, more powerful numeral representations will be constructed that actually improve our capacity to observe, and describe, mathematical objects and quantities. Postulate 3 culminates in the actual classification scheme presented in this paper. Indeed, the cellular automata classification presented here is developed by partitioning the entire space into three classes. It is interesting to note that the order of Postulates 1 – 3 seem to dictate the exposition and order of results of this paper. It is important to note that the Postulates should not be conceived as axioms in this new axiomatic system but rather set the methodological basis for the new system (See [22], Sect. 2 and also [14], for a more rigorous discussion of the Postulates and Axioms).

The *Infinite Unit Axiom* is formally stated in three parts below. This axiom involves the idea of an infinite unit from finite to infinite. The infinite unit of measure is expressed by the numeral ①, called *grossone*, and represents the number of elements in the set $\mathbb{N}$ of natural numbers.

1. *Infinity*: For any finite natural number n, it follows that $n < ①$.
2. *Identity*: The following involve the identity elements 0 and 1
   (a) $0 \cdot ① = ① \cdot 0 = 0$
   (b) $① - ① = 0$
   (c) $\frac{①}{①} = 1$
   (d) $①^0 = 1$
   (e) $1^① = 1$
3. *Divisibility*: For any finite natural number n, the numbers

$$①, \frac{①}{2}, \frac{①}{3}, ..., \frac{①}{n}, ...$$

are the number of elements of the nth part of $\mathbb{N}$, see [22].

An important aspect of ① that will be used extensively in this paper is the numeric representation of $①^{-i}$ for $i > 0$ (note that i can be infinite as well). These numbers are called *infinitesimals*. The simplest infinitesimal is $①^{-1} = \frac{1}{①}$. It is noted that $①^{-1}$ is the multiplicative inverse element for ①. That is, $①^{-1} \cdot ① = ① \cdot ①^{-1} = 1$. It is also important (and essential in this paper) to note that all infinitesimals are not equal to 0. In particular, $\frac{1}{①} > 0$. In [19,21] this is also shown as a limiting process. That is,

$$\lim_{n \to ①} \frac{1}{n} = \frac{1}{①} \neq 0.$$

As noted above, the set of natural numbers is represented by

$$\mathbb{N} = \{1, 2, 3, ..., ① - 2, ① - 1, ①\}$$

and the set of integers, with the new grossone methodology, is represented by

$$\mathbb{Z} = \{-①, -①+1, -①+2, ..., -3, -2, -1, 0, 1, 2, 3, ..., ①-2, ①-1, ①\}$$

However, since we will be working with the set $S^{\mathbb{Z}}$ as the domain of definition for cellular automata maps, we will need to make use of the set of extended natural numbers by applying the arithmetical operations with grossone and other infinite numbers (see [19, 22] for a complete description on the formation of these sets).

$$\hat{\mathbb{N}} = \{1, 2, 3, ..., ①-2, ①-1, ①, ①+1, ..., ①^n, ..., 2^{①}, ..., ①^{①}, ...\}$$

Where

$$1 < 2 < 3 < ... < ①-1 < ① < ①+1 <$$
$$... < ①^{10} < ... < 2^{①} < ... < ①^{①} < ...$$

and hence the infinitesimals

$$0 < ... < \frac{1}{①^{①}} < ... < \frac{1}{2^{①}} < ... < \frac{1}{①^{10}} < ... < \frac{1}{①} < ...$$

The extended natural numbers will be used to represent the number of elements in a set and their reciprocals used for infinitesimal quantities. The sequence of forward iterates of an automaton map will only go up to ①, as the maximum number of elements in a sequence cannot be more than grossone, see [22] for a complete discussion. Cellular automata are important models of computation, namely parallel computation. However, the theory of grossone has already been successfully applied to studying other models of computation, see [24, 25].

Herein it is important to note the number of elements in a set, especially an infinite set.

Theorem 1. *The number of elements in the set $\mathbb{Z}$ of integers is $2①+1$*

Proof. See [22].

Theorem 2. *The number of elements in the set $\mathbb{Z} \times \mathbb{Z}$ is $|\mathbb{Z} \times \mathbb{Z}| = (2①+1)(2①+1)$.*

Proof. The number of elements in the set $\mathbb{Z}$ of integers is $|\mathbb{Z}| = 2①+1$, see [22]. For any ordered pair (a, b), with a and b both belonging to the set $\mathbb{Z}$, there are $2①+1$ possibilities for a and $2①+1$ possibilities for b. Hence the product $(2①+1)(2①+1) = 4①^2 + 4① + 1$ for the total number of possibilities.

Theorem 3. *The number of elements in the set $\mathbb{N} \times \mathbb{Z}$ is $|\mathbb{N} \times \mathbb{Z}| = ①(2①+1) = 2①^2 + ①$.*

Proof. The proof is similar to Theorem 2 and hence omitted.

3   Two-Dimensional Cellular Automata

Let $\mathbf{S}$ be a finite alphabet of size s such that $2 \leq s$ and let $\mathbf{X} = \mathbf{S}^{\mathbb{Z} \times \mathbb{Z}}$, i.e. the set of all maps from the two-dimensional lattice $\mathbb{Z} \times \mathbb{Z}$ to the set $\mathbf{S}$. That is, for $x \in \mathbf{X}$, $x : \mathbb{Z} \times \mathbb{Z} \to \mathbf{S}$. Two-dimensional cellular automata are induced by arbitrary (local) maps:

$$F : \mathbf{S}^{(2r+1)^2} \longrightarrow \mathbf{S}$$

We will call these local maps local rules or block maps. Let $\mathbb{N}$ denote the set of natural numbers, the value $r \in \mathbb{N} \cup \{0\}$ is called the range of the map. The automaton map f induced by F is defined by $f(x) = y$ with

$$y(i,j) = F[x(i-r,j-r), ..., x(i+r,i-r), x(i-r,j-r+1), ..., x(i+r,j-r+1), ...,$$

$$x(i-r,j+r), ..., x(i+r,j+r)]$$

To illustrate the importance of discrete time steps in the forward evolution of the automaton, we will use the following formula, where t represents time.

$$y(i,j)_{t+1} = F[x(i-r,j-r)_t, ..., x(i+r,i-r)_t, x(i-r,j-r+1)_t, ..., x(i+r,j-r+1)_t, ...,$$

$$x(i-r,j+r)_t, ..., x(i+r,j+r)_t]$$

This is usually called the Moore neighborhood ($r = 1$), or the extended Moore neighborhood ($r > 1$) in the literature. The restriction of $x \in X$ to a non-empty region $[m, n] \times [p, q]$ of $\mathbb{Z} \times \mathbb{Z}$, where $-\text{①} \leq m \leq n \leq \text{①}$ and $-\text{①} \leq p \leq q \leq \text{①}$ is called a *configuration*. Configurations are written $x([m, n] \times [p, q])$.

Denote by R_n the square region in $\mathbb{Z} \times \mathbb{Z}$ bounded by n. The notation $f|_{R_n}$ denotes the restriction of f to the region R_n. Define:

$$\rho(f,g) = \begin{cases} \prod_{(i,j) \in R_n} \lambda_{i,j} & \text{if } f|_{R_n} = g|_{R_n} \text{ but } f|_{R_{n+1}} \neq g|_{R_{n+1}} \\ 1 & \text{if } f(0,0) \neq g(0,0) \end{cases}$$

Where λ is any real-valued function defined on $\mathbf{S}$ and taking values in the open interval $(0, 1)$, i.e. $\lambda : \mathbf{S} \to (0, 1)$ where $\lambda_{i,j} = \lambda(f(i,j))$ for each $f(i,j) \in \mathbf{S}$ and not infinitesimal, hence each $0 < \lambda_{i,j} < 1$. The metric is defined for $f, g \in X$ as follows:

$$d(f,g) = \begin{cases} 0 & \text{if } f = g \\ \rho(f,g) & \text{otherwise} \end{cases}$$

The metric just defined will be called the *two-dimensional Kolmogorov metric* and satisfies the nonarchimedean (ultra metric) property,

$$d(x,y) \leq max\{d(x,z), d(z,y)\}.$$

An example of the use of this metric is given in the following example.

Example 1. Given the alphabet $S = \{0, 1\}$, the following configuration x consisting of all 1's, and for simplicity choose $\lambda(1) = \lambda(0) = 1/2$

$$
\begin{array}{ccc}
(-\mathbb{1},\mathbb{1}) & & (\mathbb{1},\mathbb{1}) \\
\overbrace{} & & \overbrace{}
\end{array}
$$

$$
\begin{array}{ccccccccc}
\vdots & \vdots & \vdots\vdots & \vdots & \vdots\vdots & \vdots & \vdots \\
1 & \ldots 1\,1 & 1 & 1\,1 & \ldots & 1 \\
1 & \ldots 1\,1 & 1 & 1\,1 & \ldots & 1 \\
(-\mathbb{1}, 0) \longrightarrow 1 & \ldots 1\,1 & \langle 1 \rangle & 1\,1 & \ldots & 1 \longleftarrow (\mathbb{1}, 0) \\
1 & \ldots 1\,1 & 1 & 1\,1 & \ldots & 1 \\
1 & \ldots 1\,1 & 1 & 1\,1 & \ldots & 1 \\
\vdots & \vdots & \vdots\vdots & \vdots & \vdots\vdots & \vdots & \vdots
\end{array}
$$

$$
\begin{array}{ccc}
(-\mathbb{1},-\mathbb{1}) & & (\mathbb{1},-\mathbb{1})
\end{array}
$$

The brackets $\langle$ and $\rangle$ represent the $(0, 0)$ position. Then the configuration below, call it y, is identical to the one above, except for the 0 in the $(2, 1)$ position.

$$
\begin{array}{ccc}
(-\mathbb{1},\mathbb{1}) & & (\mathbb{1},\mathbb{1}) \\
\overbrace{} & & \overbrace{}
\end{array}
$$

$$
\begin{array}{ccccccccc}
\vdots & \vdots & \vdots\vdots & \vdots & \vdots\vdots & \vdots & \vdots \\
1 & \ldots 1\,1 & 1 & 1\,1 & \ldots & 1 \\
1 & \ldots 1\,1 & 1 & 1\,0 & \ldots & 1 \\
(-\mathbb{1}, 0) \longrightarrow 1 & \ldots 1\,1 & \langle 1 \rangle & 1\,1 & \ldots & 1 \longleftarrow (\mathbb{1}, 0) \\
1 & \ldots 1\,1 & 1 & 1\,1 & \ldots & 1 \\
1 & \ldots 1\,1 & 1 & 1\,1 & \ldots & 1 \\
\vdots & \vdots & \vdots\vdots & \vdots & \vdots\vdots & \vdots & \vdots
\end{array}
$$

$$
\begin{array}{ccc}
(-\mathbb{1},-\mathbb{1}) & & (\mathbb{1},-\mathbb{1})
\end{array}
$$

Hence the center region is denoted R_1 and we can compute the distance of the two configurations as follows.

$$
\rho(x, y) = \prod_{(i,j) \in R_1} \lambda_{i,j} = \left(\frac{1}{2}\right)^9 = \frac{1}{512} = d(x, y)
$$

Under the usual product topology, a two-dimensional *cylinder* is a set $C(i, j, w) = \{x \in X | x([i, j] \times [i, j]) = w\}$, where $|w| = (j - i + 1)^2$. We define the open disk of radius ε around x to be $C_n(x) = C(-n, n, x([-n, n] \times [-n, n]))$. Here, it is important to note, $\varepsilon > 0$ and that ε can be infinitesimal. It should be clarified that ε must be computed with respect to the metric defined above but first with the respective values of λ chosen. As the following example illustrates.

Example 2. Given the alphabet $S = \{0, 1\}$ and $\lambda(0) = \lambda(1) = 1/2$, then the disk centered at x and of radius $\varepsilon = 1/512$ is denoted by $C_1(x)$. We also take the convention, once the λ values are fixed, to denote $C_{1/512}(x)$ as the disk of radius

1/512. For instance, if x is the configuration of all 1's and given the λ values $\lambda(0) = \lambda(1) = 1/2$, The open disk $C_{1/512}(x)$ is illustrated.

$$(-①,①) \qquad\qquad (①,①)$$

$$
\begin{array}{ccccccccc}
\vdots & \vdots & \vdots\vdots & \vdots & \vdots\vdots & \vdots & \vdots \\
1 & \ldots 1\,1 & 1 & 1\,1 & \ldots 1 \\
1 & \ldots 1\,1 & 1 & 1\,1 & \ldots 1 \\
(-①,0) \longrightarrow 1 & \ldots 1\,1 & \langle 1\rangle & 1\,1 & \ldots 1 & \longleftarrow (①,0) \\
1 & \ldots 1\,1 & 1 & 1\,1 & \ldots 1 \\
1 & \ldots 1\,1 & 1 & 1\,1 & \ldots 1 \\
\vdots & \vdots & \vdots\vdots & \vdots & \vdots\vdots & \vdots & \vdots
\end{array}
$$

$$(-①,-①) \qquad\qquad (①,-①)$$

The brackets $\langle$ and $\rangle$ represent the $(0,0)$ position. Then any other configuration in the disk $C_{1/512}(x)$ would have to be of the form with the center Moore neighborhood consisting of all 1's.

$$(-①,①) \qquad\qquad (①,①)$$

$$
\begin{array}{ccccccccc}
\vdots & \vdots & \vdots\vdots & \vdots & \vdots\vdots & \vdots & \vdots \\
* & \ldots *\,* & * & *\,* & \ldots * \\
* & \ldots *\,1 & 1 & 1\,* & \ldots * \\
(-①,0) \longrightarrow * & \cdots *\,1 & \langle 1\rangle & 1\,* & \cdots * & \longleftarrow (①,0) \\
* & \ldots *\,1 & 1 & 1\,* & \ldots * \\
* & \ldots *\,* & * & *\,* & \ldots * \\
\vdots & \vdots & \vdots\vdots & \vdots & \vdots\vdots & \vdots & \vdots
\end{array}
$$

$$(-①,-①) \qquad\qquad (①,-①)$$

where $*$ is a "wildcard" and can represent either a 0 or 1.

Since the metric is nonarchimedean, given any two disks $C_\varepsilon(f)$, $C_\alpha(y)$, either $C_\varepsilon(f) \cap C_\alpha(y) = \emptyset$ or one contains the other. In this topology, the C_ε sets are also closed. For fixed $\varepsilon > 0$, the relation $f \sim y$ if $d(f,y) \leq \varepsilon$ is an equivalence relation with equivalence classes $\{C_\varepsilon(f)\}$.

It should be noted, with the given definitions and the Infinite Unit Axiom, it is possible to define an open disk of infinitesimal radius. A disk of infinitesimal radius is an open disk around an infinite square configuration. For example, the disk $C_{①^{-2}}(x)$ is a disk of such radius.

Theorem 4. *Given the space $S^{\mathbb{Z} \times \mathbb{Z}}$ of two-dimensional bi-infinite configurations, the number of elements $x \in S^{\mathbb{Z} \times \mathbb{Z}}$ is equal to*

$$|S|^{(4①^2 + 4① + 1)}$$

Proof. By Theorem 2 there are $(2① + 1)(2① + 1)$ elements (or places) in the two-dimensional lattice $\mathbb{Z} \times \mathbb{Z}$ and each lattice point can hold a value from the finite alphabet S. Hence there are

$$|S|^{(2①+1)(2①+1)} = |S|^{(4①^2+4①+1)}$$

distinct configurations.

Corollary 1. *The open disk $C_n(x)$, for finite or infinite n, around x contains*

$$|S|^{(4①^2+4①-4n^2-4n)} \ \text{elements.}$$

Proof. An open disk $C_n(x)$ around x must have a fixed square center where a side equals $2n + 1$. The number of possible configurations outside this square center must be computed. Above the square there are $|S|^{(2①+1)(①-n)}$ possible configurations. Below the square, the same. To the right of the square, there are $|S|^{(2n+1)(①-n)}$ possible configurations and the same to the left of the center square. Hence the total number of possible configurations (elements in the open disk $C_n(x)$) are given by the following computation.

$$|S|^{(2①+1)(①-n)} \cdot |S|^{(2①+1)(①-n)} \cdot |S|^{(2n+1)(①-n)} \cdot |S|^{(2n+1)(①-n)}$$

$$= |S|^{2(2①+1)(①-n)} \cdot |S|^{2(2n+1)(①-n)}$$

$$= |S|^{(4①^2+4①-4n^2-4n)}$$

Example 3. For $n = ①-1$, $C_{①-1}(x)$ is a disk of infinitesimal radius and contains

$$|S|^{(4①^2+4①-4(①-1)^2-4(①-1))} = |S|^{8①}$$

points.

As shown in the previous example, disks of infinitesimal radius contain, although still infinite, many fewer points than disks of finite radius. This is in contrast to the one-dimensional case (see [6]) where there are only finitely many elements in a disk of infinitesimal radius.

The study of dynamical systems, in this case discrete dynamical systems, endeavors to understand the forward evolution (or forward iterations) of the system map, in this case the automaton rule. For $t \in \mathbb{N} \cup \{0\}$, $f^t(x)$ is used to represent the tth iterate of the automaton map f. That is,

$$f^t(x) = f \circ f \circ f \cdots \circ f(x)$$

where $0 \le t \le ①$.

To understand the dynamics of two-dimensional cellular automata it is necessary to study the forward iterates of configurations that equal or match those of a given configuration, call it "x", on a given region of $\mathbb{Z} \times \mathbb{Z}$. Here the relation

$x \sim y$ iff $\forall i \in \mathbb{N} \cup \{0\}$, $(f^i(y))([m,n] \times [p,q]) = (f^i(x))([m,n] \times [p,q])$ forms an equivalence relation with equivalence classes denoted by $B_{m,n,p,q}(x)$. That is,

$$B_{m,n,p,q}(x) = \{y \mid (f^i(y))([m,n] \times [p,q]) = (f^i(x))([m,n] \times [p,q]) \ \forall i \in \mathbb{N} \cup \{0\}\}.$$

$B_{m,n,p,q}(x)$ is the set of y for which $(f^i(y))([m,n] \times [p,q]) = (f^i(x))([m,n] \times [p,q])$, for $m \leq 0 \leq n$ and $p \leq 0 \leq q$, under forward iterations of the cellular automaton function. That is, $\forall i \in \mathbb{N}_0$. Recall, $(f^i(y))([m,n] \times [p,q])$ represents configurations and that the cellular automaton function, f is first applied to the entire configuration x (or y), and then restricted to the region $[m,n] \times [p,q]$. Note that m and/or p can equal $-① + k$ and n and/or q can equal $① - k$, for some finite integer $k \geq 0$. In those cases the configurations are left-sided, right-sided or both sided infinite. Hence elements in the $B_{m,n,p,q}(x)$ classes will agree with $x([m,n] \times [p,q])$ and all forward iterations of $x([m,n] \times [p,q])$ under the automaton map f. This will form the effect of an infinite vertical rectangular prism, not necessarily symmetric, around the central window.

The dynamical analysis of cellular automata presented herein is based on counting the number of elements in the entire domain space, X. Hence, in this section we will use ① to count the number of elements in the class $B_{m,n,p,q}(x)$ whose forward iterates match those of x in some window containing the center and develop a simple classification of two-dimensional cellular automata based on this count. Similar to the one-dimensional case, two-dimensional cellular automata rules are thus partitioned into three classes.

Definition 1. *Define the classes of two-dimensional cellular automata, f, as follows:*

1. *$f \in \mathcal{A}$ if there is a $B_{m,n,p,q}(x)$ that contains at least $|S|^{4(①^2+①)-k}$ elements, for some finite integer $k \geq 0$.*
2. *$f \in \mathcal{B}$ if there is a $B_{m,n,p,q}(x)$ that contains at least $|S|^{\alpha ①^2 + \beta ① - k}$ elements, for some finite integer $k \geq 0$, $0 < \alpha \leq 4$, α not infinitesimal and $0 < \beta < 4$, but f does not belong to class $\mathcal{A}$.*
3. *$f \in \mathcal{C}$ otherwise.*

Class $\mathcal{C}$ is the most chaotic class of automata. Indeed, in this class there may only be finitely many elements or simple infinitely many elements in any $B_{m,n}(x)$ class. Hence, beginning with an initial configuration, most other configurations will diverge away from the initial configuration. Automata in class $\mathcal{A}$ are the least chaotic and most elements will equal an initial configuration upon repeated applications (iterations) of the automata rule on the infinite strip. The following theorem shows the relationship between an open disk and the number of configurations in a $B_{m,n,p,q}(x)$ class.

Theorem 5. *If there exists a $B_{m,n,p,q}(x)$, for cellular automaton f, that contains an open disk of non-infinitesimal radius, then $f \in \mathcal{A}$.*

Proof. If there is a $B_{m,n,p,q}(x)$, for cellular automaton f that contains an open disk $C_n(x)$ of non-infinitesimal radius, then $C_n(x)$ contains $|S|^{(4①^2+4①-4n^2-4n)}$

elements. Therefore $B_{m,n,p,q}(x)$ contains at least $|S|^{(4\textcircled{1}^2+4\textcircled{1}-4n^2-4n)}$ elements. Since n is finite, take finite $k = 4n^2 - 4n$ and by Definition 1 the theorem is proved.

The following example shows a class $\mathcal{A}$ two-dimensional automaton of range $r = 1$.

Example 4. For simplicity we use the binary alphabet. Let $S = \{0, 1\}$, and define the two-dimensional automaton function, F, on the Moore neighborhood as follows.

$$F(a, b, c, d, e, f, g, h, i) = \begin{cases} 1 \ if \ \ a = b = c = d = e = f = g = h = i = 1 \\ 0 \ otherwise \end{cases}$$

That is, all configurations go to 0 except the configuration of all 1's. Hence it is easily seen there is a $B_{m,n,p,q}(x)$, except in the case x is the configuration of all 1's, that contains an open disk. Therefore by Theorem 5 this automaton is of class $\mathcal{A}$.

The next example is a cellular automaton map that belongs to class $\mathcal{B}$ and shows the new computational power of the Infinite Unit Axiom and grossone.

Example 5. Again, for simplicity, we use the binary alphabet $S = \{0, 1\}$. We can define the cellular automaton on either the Von Neuman or Moore neighborhood and use coordinates.

$$\sigma(x(i, j)) = x(i + 1, j).$$

This is the simple horizontal left shift map and illustrated by the following.

```
 ⋮  ⋮  ⋮ ⋮ ⋮ ⋮ ⋮  … ⋮
 1  0  1 1 0 1 1  … 1
 0  0  1 0 1 1 1  … 1
 0 ⟨1⟩ 1 1 0 1 1  … 1
 1  1  1 0 1 1 1  … 1
 0  1  0 0 1 1 1  … 1
 ⋮  ⋮  ⋮ ⋮ ⋮ ⋮ ⋮  … ⋮
```

where the brackets ⟨ and ⟩ represent the $(0, 0)$ position. The next iteration yields,

```
 ⋮  ⋮  ⋮ ⋮ ⋮ ⋮ ⋮  … ⋮
 0  1  1 0 1 1 1  … 1
 0  1  0 1 1 1 1  … 1
 1 ⟨1⟩ 1 0 1 1 1  … 1
 1  1  0 1 1 1 1  … 1
 1  0  0 1 1 1 1  … 1
 ⋮  ⋮  ⋮ ⋮ ⋮ ⋮ ⋮  … ⋮
```

Hence any other configuration, y, in $B_{m,n,p,q}(x)$ would have to agree on the right of the center square out to ①. Hence there are at most

$$|S|^{①\cdot(2①+1)} \cdot |S|^{①(2①+1)} \cdot |S|^{①} = |S|^{4①^2+3①}$$

and clearly by Definition 1 the left shift, $\sigma(x(i,j)) = x(i+1,j)$, belongs in $\mathcal{B}$.

4   Discussion and Conclusion

In this paper a classification scheme for two-dimensional cellular automata, based on the Infinite Unit Axiom and grossone, has been presented. The entire domain space of two-dimensional automata, $X = S^{\mathbb{Z}\times\mathbb{Z}}$, contains $|S|^{(2①+1)(2①+1)}$ configurations. This puts an upper bound representation on the number of elements in the entire space, hence we sub-divided the space into three components and used this to build a classification on the number of configurations whose forward evolution, under a cellular automaton, equal those (on a central window) of a given initial configuration.

This classification is based on a numeric representation of counting elements in a set. Automata in class $\mathcal{A}$ are the least chaotic, having a very large number of configurations equaling those of a given configuration, on some central window (given the definition of the metric, it is allowable to say "staying close together" upon forward iterations), upon forward iterations of the automaton map. Automata in class $\mathcal{B}$, such as the left shift automaton, are more chaotic than those in class $\mathcal{A}$. However, it seems that they can still be described without too much complexity. Automata in class $\mathcal{C}$ are more difficult to find and are the most chaotic in the respect that there are relatively very few other configurations that will follow and stay close to a given. Indeed, the number of configurations that equal a given initial configuration, upon forward iterations, is much less than the other classes and may be simple infinite (either ①, or ①², ..., or ①^n, or some part thereof), finite or a single configuration. Conway's Game of Life has been shown to be capable of universal computation. Due to the nature of universal computation, some of these automata can fall into class $\mathcal{C}$. It is left as an open problem to prove or disprove this. It is noted that the presented classification would be stronger if there was an algorithm to determine membership in the different classes and it is also posed as an open problem.

References

1. Baetens, J.M., Gravner, J.: Stability of cellular automata trajectories revisited: branching walks and Lyapunov profiles. J. Nonlinear Sci. **26**, 1329–1367 (2016)
2. Berlekamp, E.R., Conway, J.H., Guy, R.K.: Winning Ways for Your Mathematical Plays, vol. 4, 2nd edn. A. K. Peters, Wellesley (2004)
3. Calidonna, C.R., Naddeo, A., Trunfio, G.A., Di Gregorio, S.: From classical infinite space-time CA to a hybrid CA model for natural sciences modeling. Appl. Math. Comput. **218**(16), 8137–8150 (2012)

4. Chopard, B., Droz, M.: Cellular Automata Modeling of Physical Systems. Cambridge University Press, Cambridge (1998)
5. D'Alotto, L.: Cellular automata using infinite computations. Appl. Math. Comput. **218**(16), 8077–8082 (2012)
6. D'Alotto, L.: A classification of one-dimensional cellular automata using infinite computations. Appl. Math. Comput. **255**, 15–24 (2014). http://dx.doi.org/10.1016/j.amc.2014.06.087
7. D'Alotto, L., Pizzuti, C.: Characterization of one-dimensional cellular automata rules through topological network features. In: Numerical Computations Theory and Algorithms 2016, AIP Conference Proceedings, vol. 1776, pp. 090048-1–090048-4 (2016)
8. D'Ambrosio, D., Filippone, G., Marocco, D., Rongo, R., Spataro, W.: Efficient application of GPGPU for lava flow hazard mapping. J. Supercomput. **65**(2), 630–644 (2013)
9. De Cosmis, S., De Leone, R.: The use of grossone in mathematical programming and operations research. Appl. Math. Comput. **218**(16), 8029–8038 (2012)
10. Gilman, R.: Classes of linear automata. Ergod. Theor. Dyn. Syst. **7**, 105–118 (1987)
11. Hedlund, G.A.: Edomorphisms and automorphisms of the shift dynamical system. Math. Syst. Theor. **3**, 51–59 (1969)
12. Iudin, D.I., Sergeyev, Y.D., Hayakawa, M.: Interpretation of percolation in terms of infinity computations. Appl. Math. Comput. **218**(16), 8099–8111 (2012)
13. Lolli, G.: Infinitesimals and infinities in the history of mathematics: a brief survey. Appl. Math. Comput. **218**(16), 7979–7988 (2012)
14. Lolli, G., Metamathematical Investigations on the Theory of Grossone, Preprint, Applied Mathematics and Computation. Elsevier (submitted and accepted for publication)
15. Mart'nez, G.J.: A note on elementary cellular automata classification. J. Cell. Automata **8**, 233–259 (2013)
16. Margenstern, M.: Using grossone to count the number of elements of infinite sets and the connection with bijections. p-Adic Numbers Ultrametric Anal. Appl. **3**(3), 196–204 (2011)
17. Margenstern, M.: An application of grossone to the study of a family of tilings of the hyperbolic plane. Appl. Math. Comput. **218**(16), 8005–8018 (2012)
18. Narici, L., Beckenstein, E., Bachman, G.: Functional Analysis and Valuation Theory. Marcel Dekker Inc., New York (1971)
19. Sergeyev, Y.D.: Arithmetic of Infinity. Edizioni Orizzonti Meridionali, Italy (2003)
20. Sergeyev, Y.D.: Numerical Point of view on calculus for functions assuming finite, infinite, and infinitesimal values over finite, infinite, and infinitesimal domains. Nonlinear Anal. Ser. A Theor. Methods Appl. **71**(12), e1688–e1707 (2009)
21. Sergeyev, Y.D.: Numerical computations with infinite and infinitesimal numbers: theory and applications. In: Sorokin, A., Pardalos, P.M. (eds.) Dynamics of Information Systems: Algorithmic Approaches, pp. 1–66. Springer, New York (2013)
22. Sergeyev, Y.D.: A new applied approach for executing computations with infinite and infinitesimal quantities. Informatica **19**(4), 567–596 (2008)
23. Sergeyev, Y.D.: Measuring fractals by infinite and infinitesimal numbers. Math. Methods Phys. Methods Simul. Sci. Technol. **1**(1), 217–237 (2008)
24. Sergeyev, Y.D., Garro, A.: Observability of turing machines: a refinement of the theory of computation. Informatica **21**(3), 425–454 (2010)
25. Sergeyev, Y.D., Garro, A.: Single-tape and multi-tape turing machines through the lens of grossone methodology. J. Supercomput. **65**(2), 645–663 (2013)

26. Sirakoulis, G.C., Krafyllidis, I., Spataro, W.: A computational intelligent oxidation process model and its VLSI implementation. In: International Conference on Scientific Computing Proceedings, pp. 329–335 (2009)
27. Trunfio, G.A.: Predicting wildfire spreading through a hexagonal cellular automata model. In: Sloot, P.M.A., Chopard, B., Hoekstra, A.G. (eds.) ACRI 2004. LNCS, vol. 3305, pp. 385–394. Springer, Heidelberg (2004). doi:10.1007/978-3-540-30479-1_40
28. Trunfio, G.A., D'Ambrosio, D., Rongo, R., Spataro, W., Di Gregorio, S.: A new algorithm for simulating wilfire spread through cellular automata. ACM Trans. Model. Comput. Simul. **22**, 1–26 (2011)
29. Wolfram, S.: Statistical mechanics of cellular automata. Rev. Mod. Phys. **55**(3), 601–644 (1983)
30. Wolfram, S.: A New Kind of Science. Wolfram Media Inc., Champaign (2002)
31. Wolfram, S.: Universality and complexity in cellular automata. Phys. D **10**, 1–35 (1984)
32. Zhigljavsky, A.: Computing sums of conditionally convergent and divergent series using the concept of grossone. Appl. Math. Comput. **218**(16), 8064–8076 (2012)

Multiple-Precision Residue-Based Arithmetic Library for Parallel CPU-GPU Architectures: Data Types and Features

Konstantin Isupov[✉], Alexander Kuvaev, Mikhail Popov,
and Anton Zaviyalov

Department of Electronic Computing Machines, Vyatka State University,
Kirov 610000, Russia
{ks_isupov,mv_popov}@vyatsu.ru, kyvaevy@gmail.com,
antonzaviyalov@gmail.com

Abstract. In this paper a new software library for multiple-precision (integer and floating-point) and extended-range computations is considered. The library is targeted at heterogeneous CPU-GPU architectures. The use of residue number system (RNS), enabling effective parallelization of arithmetic operations, lies in the basis of library multiple-precision modules. The paper deals with the supported number formats and the library features. An algorithm for the selection of an RNS moduli set for a given precision of computations are also presented.

Keywords: Multiple-precision computations · Extended-range computations · Parallel processing · GPGPU · Residue number system

1 Introduction

Numerous recent scientific applications have involved processing multiple-precision numbers, i.e. numbers where precision exceeds IEEE 754 double format. Such applications include, for example, satellite collision simulation without the availability of complete data about their trajectories [9], the use of complete elliptic integrals for solving the Hertzian elliptical contact problems [8], and applying inverse Laplace transforms for solutions of fractional order differential equations [3]. Due to the large amount of computations, performance is considered to be one of the most important non-functional requirements in such applications. This leads to the necessity of using modern parallel computing systems, where multicore processors (CPUs) are often used alongside with graphics processing units (GPUs). Due to its massively parallel architecture and relatively low power consumption, modern GPUs are very powerful and cost-effective for resource-intensive computing. The NVIDIA CUDA platform allows software developers to simply and effectively use GPU for general purpose computing.

At the same time, the maximum performance of multiple-precision computations on modern massively parallel architectures needs the algorithms which

© Springer International Publishing AG 2017
V. Malyshkin (Ed.): PaCT 2017, LNCS 10421, pp. 196–204, 2017.
DOI: 10.1007/978-3-319-62932-2_18

would allow effective parallelization of arithmetic operations. However this cannot be achieved by the most common way to represent multiple-precision numbers, namely a dense representation in a fixed base [2,5].

In this regard, the attention of researchers has recently been attracted to the alternative ways of representing multiple-precision numbers. The use of residue number system (RNS) is one of such ways [15,16]. Due to the non-positional nature of RNS, operations on multiple-precision numbers can be split into several reduced-precision operations executed in parallel. This enables the achievement of the maximum efficiency while using the resources of parallel architectures.

The article focuses on the implementation concept of a new RNS-based software library MPRES (Multiple-Precision Residue-based Arithmetic Library) for multiple-precision computations on CPUs and CUDA compatible GPUs.

2   Background of RNS

An RNS is defined by a set of moduli $\{m_1, m_2, \ldots, m_n\}$ that are coprime integers. In RNS, an integer X is represented by an n-tuple $\langle x_1, x_2, \ldots, x_n \rangle$, where $x_i = |X|_{m_i}$ is the least non-negative remainder of X divided by m_i. The digit x_i is called the residue of X (mod m_i). Such a representation is unique for any integer $X \in [0, M-1]$, where $M = \prod_{i=1}^{n} m_i$ is the dynamic range of an RNS.

RNS is traditionally used as the basis for high-performance hardware in such applications as digital signal processing [1] and cryptography [6]. At the same time, the development of general purpose parallel architectures enables efficient use of RNS in software, particularly for the implementation of parallel multiple-precision arithmetic. In RNS, multiple-precision operations such as addition, subtraction and multiplication are naturally divided into groups of reduced-precision operations on residues, performed in parallel and without carry propagation.

However, the effective use of RNS is limited by the complexity of operations, requiring estimation of the number magnitudes, such as comparison and overflow detection. In large dynamic ranges, the classical method to perform these operations, which is based on the Chinese Remainder Theorem, becomes slow. An alternative method, targeted at large dynamic ranges, is based on computation of interval floating-point characteristic (IFC) of the RNS number [10]. IFC is denoted by $I(X/M) = [\underline{X/M}, \overline{X/M}]$ and represents an interval with floating-point bounds, which localize the value of X, scaled with respect to M, that is: $\underline{X/M} \leq X/M \leq \overline{X/M}$. IFC bounds are machine-precision numbers and are computed using directed roundings. The average-case complexity of serial and parallel IFC computation is a linear and logarithmic function of the size of the moduli set, respectively. IFC computation requires only small integer operations and standard floating-point operations (no residue-to-binary conversion is required). The modern computing platforms, including CPU, GPU, FPGA and ASIC, allow for the efficient execution of both these types of operations, while compatibility with the IEEE 754 standard allows to estimate the IFC accuracy.

3   Moduli Set Selection

The choice of an optimal moduli set significantly influences calculation efficiency in the RNS. For example, the well-known three-moduli set $\{2^n - 1, 2^n, 2^n + 1\}$ provides fast implementations of many operations in RNS, such as scaling, sign detection and magnitude comparison [4, 17]. However, this moduli set is not suitable when large dynamic ranges are required.

In the RNS with the moduli set $\{m_1, m_2, \ldots, m_n\}$, the numbers range from 0 to $M - 1$. Therefore, satisfying the inequality $\log_2 M \geq p$ means that all p-bit nonnegative integers are representable in this RNS. The following is a moduli set generation algorithm, which can be used to implement multiple-precision arithmetic in RNS.

Algorithm 1. Moduli Set Generation

Require: p (precision), k (bit size of the most significant modulus m_1)
Ensure: $n, \{m_1, m_2, \ldots, m_n\}$ such that $\log_2 \prod_{i=1}^{n} m_i \geq p$
1: $i \leftarrow 1$; $M \leftarrow m_1 \leftarrow 2^k - 1$
2: **while** $\log_2 M < p$ **do**
3:     $A \leftarrow \{a \in \mathbb{Z}^+ \mid a < m_i \wedge a \pmod 2 \equiv 1 \wedge \gcd(a, m_j) = 1,$ for all $1 \leq j \leq i\}$
4:     **if** $A \neq \varnothing$ **then**
5:         $i \leftarrow i + 1$
6:         $m_i \leftarrow \max A$
7:         $M \leftarrow M \cdot m_i$
8:     **else** *Raise exception*: impossible to generate a moduli set for the given p and k
9:     **end if**
10: **end while**
11: $n \leftarrow i$
12: **return** $n, \{m_1, m_2, \ldots, m_n\}$

An alternative algorithm, which performs generation of the moduli set starting from the least significant modulus, is presented in [13]. The above presented algorithm is preferable from the viewpoint of reducing the size of the moduli set and memory overhead for the software implementation of the computations.

4   Extended-Range and Multiple-Precision Data Formats

The MPRES library supports three data types, which are based on the following numeric formats: (1) extended-range floating-point format; (2) multiple-precision modular integer format; (3) modular-positional floating-point format. Operations on numbers in these formats makes up the arithmetic level of the library (Fig. 1). The following is a consideration of each of these formats.

Extended-Range Floating-Point. This format (hereafter ER-format) is intended to represent machine-precision floats with extended exponent. It does not increase the accuracy and precision of computations, but in MPRES it is involved

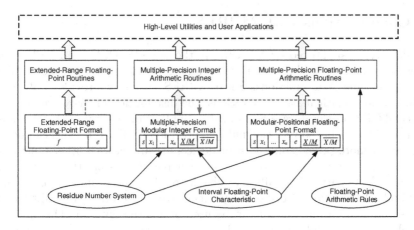

Fig. 1. Arithmetic level of the MPRES library

in the formation of two other multiple-precision formats. ER-format is constructed by pairing a machine integer e with an ordinary machine floating-point number f. This pair is considered as the number $f \times B^e$, where B is a predetermined constant that is a power of the floating-point base (radix) [7].

Multiple-Precision Modular Integer. This format (hereafter MI-format) is intended to represent arbitrary precision integers. A number x in this format is represented by a sign $s \in \{0, 1\}$, a significand in RNS X and an IFC $I(X/M)$ of the significand. If $x \geq 0$, then $s = 0$, otherwise $s = 1$. Significand X expresses the absolute value of the number and is represented by the tuple of residues $\langle x_1, x_2, \ldots, x_n \rangle$ with respect to the moduli set $\{m_1, m_2, \ldots, m_n\}$. The residues are represented by machine integers. The significand is interpreted as an integer in the interval $[0, M - 1]$. The precision of x in bits, represented in the described way, is defined by the value $\lfloor \log_2 M \rfloor$. IFC bounds $\underline{X/M}$ and $\overline{X/M}$ are represented in the ER-format thus enabling the use of IFC without any restrictions for M. Notation $x \rightarrow \{s, X, I(X/M)\}$, which expresses the value $x = (-1)^s \times \left| \sum_{i=1}^{n} x_i \left| M_i^{-1} \right|_{m_i} M_i \right|_M$, is used for numbers represented in the MI-format. Here $M_i = M/m_i$ and $\left| M_i^{-1} \right|_{m_i}$ are RNS constants.

Modular-Positional Floating-Point. This format (hereafter MF-format) extends the above-described MI-format by adding the unbiased exponent e, which is a signed integer. The format name results from the fact that the RNS significand represents the modular part of a number, and the binary exponent — the positional part. Notation $x \rightarrow \{s, X, e, I(X/M)\}$, which expresses the value $x = (-1)^s \times \left| \sum_{i=1}^{n} x_i \left| M_i^{-1} \right|_{m_i} M_i \right|_M \times 2^e$, is used for numbers in the MF-format.

5   Computations with MPRES

The data types supported by MPRES are the same for CPUs and GPUs. This allows to exchange data between the host and the device by calling the standard CUDA Runtime API functions. Since MPRES currently provides only basic arithmetic and mathematical operations, the distribution of tasks among CPU cores and GPU accelerators is performed by application programmer. Depending on the problem being solved, the multiple-precision CUDA-functions can be run from one or more threads on a host or device, in parallel with other functions.

The RNS moduli, IFC accuracy and other parameters are the same for the CPU and GPU parts of the library. Thanks to this, the results obtained from CPU and GPU are found to be in agreement with each other. In addition, the results obtained on one system can be used on another system. This makes it possible to more flexibly distribute the computations among CPUs and GPUs.

Functions for converting numbers from standard data types, (e.g. long, double) to the data types supported by MPRES are implemented for both CPU and GPU, so if source data are represented with standard data types, they can be transferred to the GPU memory without any additional conversion. There are also special functions (only for CPU) that provide interaction with the well-known multiple-precision libraries, GMP and MPFR. Functions that convert computational results from MI- and MF-formats to binary system require multiple-precision binary arithmetic and are implemented only for CPU.

Extended-Range Computations. The algorithms for handling extended-range numbers are well-known and easy to implement [7]. The following CPU- and CUDA-functions are implemented in MPRES for computations in the ER-format: addition, subtraction, multiplication and division, supporting four IEEE 754 rounding modes; floor and ceiling functions, computation of the fractional part, as well as comparison functions; a number of mathematical functions: exp, fact, pow, sqrt, sin, cos, ceil, floor, etc. Some of the algorithms implemented in MPRES are described in [14]. CUDA-implementation of extended-range arithmetic enables the use of the GPU for the tasks, associated with processing extremely large or small quantities. In particular, the implemented routines are used to calculate normalized associated Legendre polynomials with very high degrees [14].

Multiple-Precision Computations. The algorithms for multiple-precision computations in MF- and MI-formats are described in [11,12], respectively. In general, an algorithm for performing a multiple-precision arithmetic operation in MPRES consists of the following steps:

1. Magnitudes of the significands of the operands are preliminary estimated by means of IFCs analysis.
2. The exponent (for the MF-format) and the sign of the result are determined.
3. The multiple-precision significand is computed in RNS. All of the computations are performed separately for each modulus.
4. The IFC of the result is calculated.

Currently MPRES supports four basic arithmetic operations over MF- and MI-formats. The representation of multiple-precision significands in RNS makes it possible to eliminate the carry propagation. This simplifies the execution of arithmetic operations and allows to process all digits of significands in parallel.

IFC that provides information about the magnitude of the significand is used to control the overflow of the range, determine the sign of the difference, compare numbers, determine the need for rounding, and for other operations. Since IFC is included in the multiple-precision representation and stored in memory, it eliminates the need for its full recalculation after performing each operation on the significand. In general, computations with IFCs are performed using interval arithmetic. After performing a lot of computations, the IFC widens, thereby becoming less informative. Therefore, it needs to be refreshed periodically, i.e., recalculated on the basis of the residues of the significand.

For CPU computing, processing of the significand and IFC is vectorized using SIMD extensions. In the future, support for multi-threaded OpenMP implementations is planned, which may be expedient at very high precision of computations and large size of the RNS moduli set.

For GPU computing, two sets of CUDA functions are implemented:

1. Serial __device__ functions which can be called from the main CUDA kernel and used on any CUDA compatible GPU cards.
2. Parallel __global__ functions designed to be run on devices that support CUDA dynamic parallelism (Fig. 2). In such functions, n child threads are running, where n is the size of the moduli set. Each i-th child thread performs operations modulo m_i. After performing the arithmetic operation, the control flow returns back to the to the calling function. It is worth noting that __global__ MPRES functions can be called both from GPU and CPU.

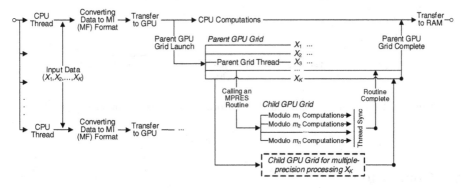

Fig. 2. Parallel multiple-precision calculations with MPRES

6   Memory Overhead Evaluation

The use of MI- and MF-formats is associated with memory overhead, which consists of two parts:

1. Overheads, associated with the necessity of storing IFC, which is not involved in the formation of a number value. These overheads are fixed and do not depend on the precision of computations.
2. Overheads, associated with the fact that the RNS moduli must be coprime numbers, and their product M is not a power of two. These overheads depend on the precision of computations.

Figure 3 shows the calculation results for the size of a number represented in the MI-format and total memory overhead. In order to generate the moduli set Algorithm 1 was used. The bit size of each modulus was assumed to be 16 bits, and it was assumed that the IFC bounds are ER-numbers with 64-bit significands and 16-bit exponents (thus, storing one IFC requires 80 bits). The results show that the memory overhead, associated with the use of MI-format, is insignificant when the precision is more than 256 bits. Memory overhead assessment for the MF-format will be very similar to the presented estimates.

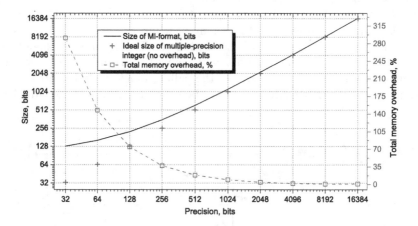

Fig. 3. The size of a number in the MI-format and total memory overhead

7   Conclusion

This article describes data types and main features of a new software library for multiple-precision and extended-range computations on hybrid CPU-GPU systems. The use of RNS to represent multiple-precision significands enables effective parallelization of arithmetic operations without any carry propagation.

By changing the RNS moduli set it is possible to set arbitrary precision of computations. All the library functions are presented in the CPU- and CUDA-implementations, which can be used in parallel. All procedures are thread-safe, thus allowing parallelization of the computation process on the application level.

Currently, a limitation of the integer arithmetic in the MI-format lies in the inability to dynamically increase precision of computations. This causes the need to analyze the whole computational process and assign a large number of RNS moduli, excluding the range overflow. In order to overcome this drawback, efficient RNS base extension algorithms are required.

Acknowledgement. This work is supported by the Russian Foundation for Basic Research (project no. 16-37-60003 mol_a_dk) and FASIE UMNIK grant.

References

1. Albicocco, P., Cardarilli, G., Nannarelli, A., Re, M.: Twenty years of research on RNS for DSP: lessons learned and future perspectives. In: Proceedings of 14th International Symposium on Integrated Circuits (ISIC), Singapore, pp. 436–439, December 2014
2. Brent, R., Zimmermann, P.: Modern Computer Arithmetic. Cambridge University Press, New York (2010)
3. Brzeziński, D.W., Ostalczyk, P.: Numerical calculations accuracy comparison of the inverse Laplace transform algorithms for solutions of fractional order differential equations. Nonlinear Dyn. **84**(1), 65–77 (2016)
4. Chang, C.H., Low, J.Y.S.: Simple, fast, and exact RNS scaler for the three-moduli set $\{2^n - 1, 2^n, 2^n + 1\}$. IEEE Trans. Circ. Syst. I Regul. Pap. **58**(11), 2686–2697 (2011)
5. Defour, D., de Dinechin, F.: Software carry-safe for fast multiple-precision algorithms. In: Proceedings of 1st International Congress of Mathematical Software, Beijing, China, pp. 29–39, August 2002
6. Esmaeildoust, M., Schinianakis, D., Javashi, H., Stouraitis, T., Navi, K.: Efficient RNS implementation of elliptic curve point multiplication over GF(p). IEEE Trans. Very Large Scale Integr. (VLSI) Syst. **21**(8), 1545–1549 (2013)
7. Hauser, J.R.: Handling floating-point exceptions in numeric programs. ACM Trans. Program. Lang. Syst. **18**(2), 139–174 (1996)
8. He, K., Zhou, X., Lin, Q.: High accuracy complete elliptic integrals for solving the Hertzian elliptical contact problems. Comput. Math. Appl. **73**(1), 122–128 (2017)
9. Hemenway, B., Lu, S., Ostrovsky, R., Welser IV, W.: High-precision secure computation of satellite collision probabilities. In: Zikas, V., De Prisco, R. (eds.) SCN 2016. LNCS, vol. 9841, pp. 169–187. Springer, Cham (2016). doi:10.1007/978-3-319-44618-9_9
10. Isupov, K., Knyazkov, V.: Non-modular Computations in Residue Number Systems Using Interval Floating-Point Characteristics. Deposited in VINITI, No. 61-B2015 (2015). (in Russian)
11. Isupov, K., Knyazkov, V.: Parallel multiple-precision arithmetic based on residue number system. Program Syst. Theor. Appl. **7**(1), 61–97 (2016). (in Russian)
12. Isupov, K., Knyazkov, V.: RNS-based data representation for handling multiple-precision integers on parallel architectures. In: Proceedings of the 2016 International Conference on Engineering and Telecommunication (EnT 2016), Moscow, pp. 76–79, November 2016

13. Isupov, K., Knyazkov, V., Kuvaev, A., Popov, M.: Development of high-precision arithmetic package for supercomputers with graphics processing units. Programmnaya Ingeneria **7**(9), 387–394 (2016). (in Russian)
14. Isupov, K., Knyazkov, V., Kuvaev, A., Popov, M.: Parallel computation of normalized legendre polynomials using graphics processors. In: Voevodin, V. (ed.) Russian Supercomputing Days 2016. CCIS, vol. 687. Springer International Publishing, Cham (2017)
15. Mohan, P.V.A.: Residue Number Systems: Theory and Applications. Birkhäuser, Basel (2016)
16. Szabo, N.S., Tanaka, R.I.: Residue Arithmetic and its Application to Computer Technology. McGraw-Hill, New York (1967)
17. Tomczak, T.: Fast sign detection for RNS $\{2^n - 1, 2^n, 2^n + 1\}$. IEEE Trans. Circ. Syst. I Regul. Pap. **55**(6), 1502–1511 (2008)

Parallel Implementation of Cellular Automaton Model of the Carbon Corrosion Under the Influence of the Electrochemical Oxidation

A.E. Kireeva[1(✉)], K.K. Sabelfeld[1,2], N.V. Maltseva[3], and E.N. Gribov[2,3]

[1] Institute of Computational Mathematics and Mathematical Geophysics SB RAS,
Pr. Lavrentjeva 6, Novosibirsk, Russia
`kireeva@ssd.sscc.ru, karl@osmf.sscc.ru`
[2] Novosibirsk State University, Pirogova str., 2, Novosibirsk, Russia
`gribov@catalysis.ru`
[3] Boreskov Institute of Catalysis, pr. Lavrentieva, 5, Novosibirsk, Russia
`maltseva.n.v@catalysis.ru`

Abstract. In the paper we present a cellular automaton model of electrochemical oxidation of the carbon. A two-dimensional sample of the electro-conductive carbon black "Ketjenblack ES DJ 600" is simulated. In the model the sample consists of a ring-formed granules of carbon. The carbon granules under the influence of the electrochemical oxidation are destroyed through a few successive stages. The rates of these oxidation stages are chosen to fit the simulation result with the experiment. In result of a computer simulation of carbon electrochemical oxidation the portions of surface atoms and atoms with different degree of oxidation were calculated and compared with the experimental data. In addition, a parallel implementation of the cellular automaton simulating the carbon corrosion is developed and efficiency of the parallel code is analyzed.

Keywords: Cellular automaton · Parallel implementation · Domain decomposition · Electrochemical oxidation · Carbon corrosion

1 Introduction

The cellular automata (CA) approach is useful method for simulation of nonlinear spatially inhomogeneous phenomena in physics and chemistry. Cellular automaton is a discrete dynamical system consisting of a set of cells [1]. Cells have states corresponding to the elements of the system under study. The states are changed with the time according to the local rules imitating the system behavior. Locality of the rules allows to describe complex dynamically changing spacial structures. Such a problem arises in simulation of a degradation of particles of some substance decomposing during the chemical reaction. For example, the carbon corrosion under the influence of the electrochemical oxidation is a

Supported by Russian Science Foundation under Grant 14-11-00083.

V. Malyshkin (Ed.): PaCT 2017, LNCS 10421, pp. 205–214, 2017.
DOI: 10.1007/978-3-319-62932-2_19

problem with boundary constantly changing during the system evolution. This problem cannot be solved by the conventional finite-difference or finite-element methods. However, the carbon corrosion can be effectively simulated by the CA approach. CA algorithms can also be treated as the Monte Carlo method with discrete space and time. The application of a Monte Carlo method to the simulation of chemical reactions is well developed (e.g., see [2–4]).

The study of the proton exchange membrane fuel cells is attracting increasing attention since they are considered as clean power sources with high energy efficiency, suitable for many applications including automobile engines [5]. The carbon supported platinum catalyst is typically used for anode and cathode in the fuel cells. Conductive carbon black "Vulcan" and "Ketjenblack" are currently widely used as a support for catalyst [6]. One of the main problems in the fuel cells commercialization is the degradation of Pt/C catalysts [6–8], which resulted mainly from the low corrosion stability of carbon support [9,10]. Carbon corrosion leads to detachment of large pieces of the support materials on which Pt is loaded. The mathematical models of carbon and Pt electrochemical oxidation in the fuel cell are developed [11–13] and investigated. However, these models are based on the ordinary differential equations or one-dimensional partial differential equations [14]. The main disadvantages of these models is that many reaction steps are considered which can hardly be verified by other methods.

Recently, the model of carbon corrosion was suggested and the electrochemical stabilities of different carbon supports were analyzed [15,16]. In this paper a two-dimensional CA model is developed based on the mechanism of carbon corrosion supposed by our group, which can be described by gradual carbon surface degradation through successive electrooxidation stages. The advantage of CA approach for simulation of the carbon corrosion is the possibility to study in details the spatial distribution of carbon atoms with different oxidation degree progressing in time.

Simulation of electrochemical oxidation of carbon sample of real physical size is a highly challenging problem because a huge number of atoms for a long time is desired to simulate. Moreover, since stochastic processes are involved, the averaging over an ensemble of initial distributions of atoms is required to obtain reliable values of integral characteristics. Therefore in this paper, a parallel implementation of CA model of the carbon corrosion is developed, and its efficiency is estimated. Using this parallel code, an evolution of CA simulating the electrochemical oxidation of carbon is obtained for chemically meaningful values of the parameters. The calculated values of the integral characteristics are compared with the chemical experimental data.

2   The Cellular Automaton Model of Carbon Corrosion

The corrosion experiments were performed in three-electrode electrochemical cell which can roughly be represented as in Fig. 1. The carbon sample (2) was deposited on polished glass carbon rod (1) and served as working electrode. The thickness of a carbon layer ($\sim 1\ \mu m$) was significantly lower than the diameter of

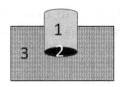

Fig. 1. Scheme of the chemical experiments: 1 - glass carbon electrode, 2 - carbon sample film, 3 - electrolyte (0.1 M HClO$_4$ aqueous solution)

sample (7 mm), so in the mathematical model carbon sample can be represented as a thin layer.

The glass carbon rod is non-porous electroconductive material, which degradation can be neglected. The porous carbon sample is fixed on the surface of glass carbon rod and immersed in 0.1 M HClO$_4$ solution as a background electrolyte. The detailed experimental description can be found in [15,17]. The electrooxidation of carbon occurs throughout the surface according to overall equation:

$$C + 2H_2O = CO_2 + 4H^+ + 4e^- \tag{1}$$

Commercial carbon black "Ketjenblack EC 600 DJ" is well studied in [8,9], therefore, this material is chosen for investigation in the current paper. The following estimations were suggested using the data obtained by low temperature nitrogen adsorption and the transmission electron microscopy [8–10,18]. The "Ketjenblack" consists of hollow nanospheres-granules of carbon atoms. The surface area is $S_{BET} = 1420$ m^2/g, the total pore volume is $V_{pores} = 3$ cm^3/g, the average diameter of carbon grains is ~ 30 nm. The percentage of surface atoms, including atoms within the spheres, to the total number of atoms is $P_{surf} = 36\%$. The porosity of "Ketjenblack" is $Por = 84\%$.

The mechanism of corrosion proposed in this work [15] is based on a simple suggestion that in the first approximation the carbon oxidation rate depends on the number of covalent bonds with oxygen only. So we denote "C" as pure surface carbon, "COH" as surface carbon having a single bond, "COOH" as surface carbon with two and three bonds. The carbon corrosion proceeds through the following oxidation stages:

$$\text{"C"} + H_2O \xrightarrow{k_1} \text{"COH"} + H^+,$$

$$\text{"COH"} + H_2O \xrightarrow{k_2} \text{"COOH"} + 2H^+, \tag{2}$$

$$\text{"COOH"} + H_2O \xrightarrow{k_3} CO_2(\text{gas}) + 2H^+$$

where "C" denotes the carbon atom, "COH" and "COOH" are oxidized carbon atoms, CO$_2$ is a carbon dioxide that is formed after the final destruction of the carbon atom and desorbs into the gas. The coefficients k_1, k_2, k_3 are oxidation probabilities for each stage.

Based on this mechanism and definition of cellular automaton (CA) [19], the CA model of carbon corrosion can be determined by the following notion:

$\aleph = \langle A, X, \Theta, \mu \rangle$, where $A = \{C_0, C, COH, COOH, \emptyset\}$ is an alphabet of admissible in the model states. Symbol "C_0" denotes a carbon atom inside the sample volume, "C" is an outer atom on the sample surface, "COH" and "$COOH$" are surface carbon atoms with different oxidation degree, symbol "$\emptyset$" corresponds to a place without any carbon atom. The set of names "X" defines the set of coordinates of atoms, here $X = \{m = (i,j) : i = 0, ..., N, \ j = 0, ..., M\}$ is a two-dimensional Cartesian lattice. The set of cells with names $m \in X$ is called a cellular array.

The set of rules updating states of cells are defined by the operator Θ. Here, Θ is a sequential composition of two operators Θ_{oxid} and Θ_{surf} [20].

The operator Θ_{oxid} simulates the oxidation stages (2) and being a sequential composition of three local operators:

$$\theta_1(m) : \{(C, m)\} \xrightarrow{k_1} \{(COH, m)\},$$
$$\theta_2(m) : \{(COH, m)\} \xrightarrow{k_2} \{(COOH, m)\}, \tag{3}$$
$$\theta_3(m) : \{(COOH, m)\} \xrightarrow{k_3} \{(\emptyset, m)\}.$$

The CA model assumes that the water is everywhere in large quantities, so the water atoms are not simulated, it is believed that they are always available. The application of Θ_{oxid} to cell m consists in a choosing the local operator θ_i, whose left-hand side coincides with the state of the cell m, and replacement this state by the state of the right side of the selected operator θ_i with probability k_i.

The operator Θ_{surf} finds the new surface carbon atoms, i.e. inner carbon atoms that after application of Θ_{oxid} have become an outer one.

$$\theta_{surf}(m) : \{(C_0, m), (\emptyset, \varphi(m))\} \rightarrow \{(C, m)\},$$
$$\text{where } \varphi(m) = \varphi(i,j), \ \varphi(i,j) \in T_4(i,j), \tag{4}$$
$$T_4(i,j) = \{(i-1,j),(i+1,j),(i,j-1),(i,j+1)\},$$

where $\varphi(m)$ is a neighbor of cell m selected by the template $T_4(m)$, being a cross with a center in the cell m. The operator Θ_{surf} replaces the cell state "inner atom" C_0 to state "outer atom" C, if the cell has at least one neighboring cell without any carbon atom ($\emptyset$).

The mode μ of application of the operator Θ to cells $m \in X$ is the synchronous. This mode prescribes the operator to be applied to all cells of a cellular array, all being updated simultaneously. The application of the operator Θ to all cells $m \in X$ is called an iteration.

Based on the geometry of the fuel cell (Fig. 1), in the CA model it is supposed that the carbon is fixed from above. The carbon pieces unconnected with the upper atoms are considered as detached and hence disappear. Therefore, after each iteration it is necessary to find and remove the detached carbon atoms. To this end, all atoms connected with upper atoms are marked by "one scan connected component labeling technique" [21]. The atoms not marked as connected are removed, i.e., states of these cells are replaced by $\emptyset$.

The initial state of cellular array in the CA model is constructed based on the characteristics of the "Ketjenblack", consisting of the hollow nanospheres-granules of carbon atoms. In the two-dimensional CA model these granules are

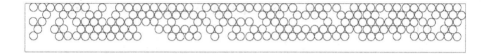

Fig. 2. The initial state of cellular array generated according to the carbon black "Ketjenblack" characteristics

represented by rings formed from cells with states corresponding to the carbon atoms. A ring is formed by cells lying between two circles: outer circle with radius R_{out} and nested inner circle with radius R_{in}. A cell belongs to a ring if the distance r from the center of the ring to the center of this cell satisfies the condition: $R_{in} \leq r \leq R_{out}$. The radii of the outer R_{out} and inner R_{in} circles are selected in such a way that the percentage of cells with state "C" (surface atom) is equal to $P_{surf} = 36\%$. The rings with the calculated radii are randomly distributed in the cellular array until the ratio of the number of cells with states "C_0" and "C" to the carbon sample size is equal to 0.16. The rings may overlap no more than 5%. In Fig. 2 one of the generated initial states is presented. The obtained sample consists of 5 rows by 50 columns of carbon granules, each being a ring with $R_{out} = 29.7$ and $R_{in} = 25.2$.

During the simulation the following characteristics are calculated after each iteration: the number of surface carbon atoms N_C, the number of oxidized carbon atoms N_{COH}, N_{COOH} and the total number of surface atoms $N_b = N_C + N_{COH} + N_{COOH}$. These characteristics are computed as the number of cells with corresponding state.

The initial states of cellular array are random, and the local operators θ_i are probabilistic, so the carbon corrosion should be considered as a stochastic process. Therefore, to obtain reliable values of statistical characteristics, an averaging over a large ensemble of initial states is required.

For the initial state shown on Fig. 2, CA simulation of carbon corrosion is performed for values of probabilities $k_1 = 0.01$, $k_2 = k_3 = 0.005$. The values of N_C, N_{COH}, N_{COOH} and N_b averaged over 500 different initial values of the random number generator are presented in Fig. 3.

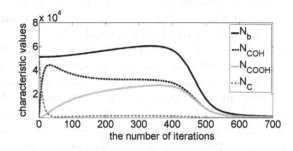

Fig. 3. The averaged values of characteristics obtained by the CA model ℵ of the carbon corrosion

3   Parallel Implementation of CA Model of Carbon Corrosion

To simulate oxidation of a large carbon sample a parallel implementation of the CA model $\aleph$ is developed. Parallel implementation of the CA model of carbon corrosion is performed using MPI library by a decomposition of a cellular array into subdomains. The cellular array is divided horizontally into n domains (Ω_k, $k = 1, \ldots, n$) according to the number of available computational cores. Due to synchronous mode, for application of the operator Θ it is sufficient to perform data exchange once per iteration. Here, data exchange means the exchange of boundary column values of domains of neighboring MPI processes.

Labeling of all atoms connected with the upper ones may require n data exchanges, because the whole carbon sample can be attached to the top only in the cellular array part, belonging to the domain Ω_k. In this case, in other domains atoms connected, possibly through several other domains, with labeled atoms of the domain Ω_k should be found. The search of connected components is performed until the sum of the new labeled atoms in all domains is equal to zero.

Thus, on each iteration each MPI process applies the operator Θ to all cells of its domain. Then it finds the connected components including needed data exchanges, removes detached atoms. Finally, it exchanges new values of boundary cells with neighboring processes. In addition, each process calculates the characteristic values for its domain's cells and summarizes the values obtained for the same iteration for different initial values of the random number generator. After all iterations, the process with rank equal to 0 summarizes the characteristic values calculated by all processes.

To estimate the efficiency of the parallel implementation of the CA model of carbon corrosion, computing experiments have been performed for the initial states generated for 5×50 carbon granules with $R_{out} = 29.7$, $R_{in} = 25.2$, for probability values $k_1 = 0.01$, $k_2 = k_3 = 0.005$ and for 50 different initial values of the random number generator. In the calculations, the cluster "MVS-10P" of the Joint Supercomputer Center of the Russian Academy of Sciences[1] is used. Each computational node of "MVS-10P" consists of two processors with 8 cores.

Figure 4 presents a computation time $T(n)$ obtained for using different number n of cores, speed-up $S(n) = T(1)/T(n)$, and a strong scaling efficiency $Q(n) = T(1)/(T(n) \cdot n)$ of the parallel implementation of $\aleph$.

The computation time of the parallel code decreases and the speed-up increases when using up to 64 cores. Further increasing of the core numbers leads to the time growth and sharp drop of the speed-up. However, the efficiency of the $\aleph$ parallel implementation is significantly reduced already within a single node, i.e. for 16 cores. This can be explained by the fact that within a single cluster node, when the number of MPI processes increase, memory access conflicts occur. The calculations performed by the operator Θ for each cell are very simple, so the memory access time is the main limiting factor.

---

[1] The website of the JSCC RAS is http://www.jscc.ru/.

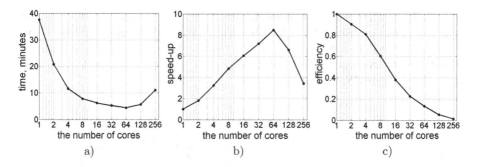

Fig. 4. The values of characteristics of the ℵ parallel implementation: (a) computation time, (b) speed-up and (c) efficiency

The drop of the efficiency and the speed-up when using several nodes ($n > 16$) is associated with a large number of data exchanges on each iteration. After every recalculation of the cellular array states the connected components labeling is performed. It requires $l \geq 1$ interprocess data exchanges in addition to the boundary exchanges.

Despite the low efficiency of the parallel code, it allows us to accelerate the calculation in 8 times for 64 cores and can be used to simulate the corrosion of a large carbon sample.

4   Results of Simulation of a Large Carbon Sample

In the chemical experiments [17] the oxidation stability of "Ketjenblack" carbon have been estimated by changes of an electrochemical capacity upon electrochemical potential cycling. According to [17], the specific capacity C_Σ (F/m^2) of the sample obtained by cyclic voltammetry method at scan rate of 50 mV/s depends linearly on the number of surface carbon (N_C) and oxidized carbon (N_O) atoms:

$$C_\Sigma = N_O \cdot C_O + N_C \cdot C_C, \tag{5}$$

where $C_O = 0.799$ F/m^2 and $C_C = 0.012$ F/m^2 are specific capacity of a single oxidized and "pure" carbon atoms on the surface.

To compare CA simulation results with the experimental data the specific capacity (5) is calculated for the values N_O and N_C obtained using the parallel implementation of the CA model ℵ. The carbon corrosion is simulated for the sample consisting of 5×500 carbon granules with $R_{out} = 29.7$ and $R_{in} = 25.2$. It corresponds to the cellular array size $|X| = 296 \times 29699$ cells. The probability values are taken equal to $k_1 = 0.1$, $k_2 = 0.005$, $k_3 = 0.00005$.

In Fig. 5 the result of CA simulation and the capacity obtained by the chemical experiment are presented. The portion of "pure" and oxidized carbon atoms is computed by ℵ. The portion of atoms in question is calculated as ratio of the number of cells with state corresponding to this atom to the cellular array size: $\rho_C = N_C/|X|$, $\rho_{COH} = N_{COH}/|X|$, $\rho_{COOH} = N_{COOH}/|X|$, $\rho_b = N_b/|X|$.

The graph shown in Fig. 5b is taken from [17]. It presents the specific capacity for the samples of "Ketjenblack" carbon (KB) and KB activated at 600 °C in air during 10 (KB-10), 20 (KB-20) and 30 (KB-30) min. The capacity calculated using the CA model $\aleph$ is qualitatively similar to that experimentally obtained for the initially oxidized sample "KB-20".

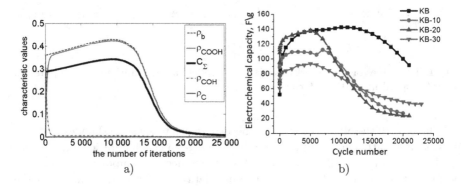

a)   b)

Fig. 5. The characteristic values computed by the CA model $\aleph$ (a) and the specific capacity measured by the chemical experiment (b)

Figure 6 shows a part of the carbon sample obtained by the CA model with the parameter values given above. At the initial time the carbon sample consists of "pure" atoms. Then surface carbon atoms are gradually oxidized. The shape of the sample remains practically unchanged up to 10000 iterations. When almost all atoms are converted to "COOH", the process of their destruction begins and, consequently, the degradation of the granules happens. The qualitative coincidence of the experimental and calculation results indicates the feasibility of the suggested corrosion model.

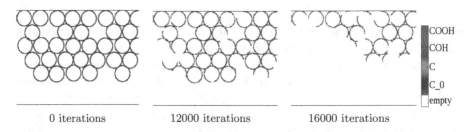

0 iterations   12000 iterations   16000 iterations

Fig. 6. A part of the carbon sample obtained by the CA model for $k_1 = 0.1$, $k_2 = 0.005$, $k_3 = 0.00005$

5   Conclusion

Two-dimensional CA model of the carbon corrosion is developed and investigated. The model allows us to directly observe the time evolution of the spatial structures of "pure" and oxidized carbon atoms and calculate integral characteristics of the sample: portion of different atoms and estimation of capacity. Choosing the modeling parameters: the probability values, initial porosity of the sample, and portion of surface atoms, we have obtained the specific capacity qualitatively similar to the one measured experimentally. For the better quantitative description of the carbon corrosion, we plan to extend the developed CA model to the three-dimensional case.

To simulate the large carbon sample, the parallel implementation of the CA model of carbon oxidation has been performed using the domain decomposition technique. The parallel code allows us to accelerate the calculation 8 times for 64 cores on the cluster "MVS-10P" JSCC of RAS, compared to the sequential code. However, the parallel code efficiency is significantly reduced for several cluster nodes. Therefore, in the future work it is planned to improve the parallel implementation of the CA model ℵ by modification of the parallel connected components labeling algorithm.

References

1. Toffoli, T., Margolus, N.: Cellular Automata Machines: A New Environment for Modeling, p. 259. MIT Press, USA (1987)
2. Sabelfeld, K.K., Brandt, O., Kaganer, V.M.: Stochastic model for the fluctuation-limited reaction-diffusion kinetics in inhomogeneous media based on the nonlinear Smoluchowski equations. J. Math. Chem. **53**(2), 651–669 (2015)
3. Karasev, V.V., Onischuk, A.A., Glotov, O.G., Baklanov, A.M., Maryasov, A.G., Zarko, V.E., Panlov, V.N., Levykin, A.I., Sabelfeld, K.K.: Formation of charged aggregates of Al_2O_3 nanoparticles by combustion of aluminum droplets in air. Combust. Flame **138**, 40–54 (2004)
4. Gillespie, D.T.: A diffusional bimolecular propensity function. J. Chem. Phys. **131**(16), 164109-1–164109-13 (2009)
5. DOE The US Department of Energy (DOE). Energy Efficiency and Renewable Energy. http://www.eere.energy.gov/hydrogenandfuelcells/mypp/pdfs/fuel_Cells. pdf and the US DRIVE Fuel Cell Technical Team Technology Roadmap. www. uscar.org/guest/teams/17/Fuel-Cell-Tech-Team
6. Li, L., Hu, L., Li, J., Wei, Z.: Enhanced stability of Pt nanoparticle electrocatalysts for fuel cells. Nano Res. **8**(2), 418–440 (2015)
7. Capelo, A., de Esteves, M.A., de Sá, A.I., Silva, R.A., Cangueiro, L., Almeida, A., et al.: Stability and durability under potential cycling of Pt/C catalyst with new surface-functionalized carbon support. Int. J. Hydrog. Energy **41**(30), 12962–12975 (2016)
8. Gribov, E.N., Kuznetzov, A.N., Golovin, V.A., Voropaev, I.N., Romanenko, A.V., Okunev, A.G.: Degradation of Pt/C catalysts in start-stop cycling tests. Russian J. Electrochem. **50**(7), 700–711 (2014)

9. Gribov, E.N., Kuznetsov, A.N., Voropaev, I.N., Golovin, V.A., Simonov, P.A., Romanenko, A.V., et al.: Analysis of the corrosion kinetic of Pt/C catalysts prepared on different carbon supports under the Start-Stop cycling. Electrocatalysis **7**, 159–73 (2016)

10. Shrestha, S., Liu, Y., Mustain, W.E.: Electrocatalytic activity and stability of Pt clusters on state-of-the-art supports: a review. Catal. Rev. Sci. Eng. **53**, 256–336 (2011)

11. Meyers, J.P., Darling, R.M.: Model of carbon corrosion in PEM fuel cells. J. Electrochem. Soc. **153**(8), A1432–A1442 (2006)

12. Pandy, A., Yang, Z., Gummalla, M., Atrazhev, V.V., Kuzminyh, N., Vadim, I.S., Burlatsky, S.F.: A carbon corrosion model to evaluate the effect of steady state and transient operation of a polymer electrolyte membrane fuel cell. J. Electrochem. Soci. **160**(9), F972–F979 (2013). arXiv:1401.4285 [physics.chem-ph]. doi:10.1149/2.036309jes

13. Chen, J., Siegel, J.B., Matsuura, T., Stefanopoulou, A.G.: Carbon corrosion in PEM fuel cell dead-ended anode operations. J. Electrochem. Soc. **158**(9), B1164–B1174 (2011)

14. Gallagher, K.G., Fuller, T.F.: Kinetic model of the electrochemical oxidation of graphitic carbon in acidic environments. Phys. Chem. Chem. Phys. **11**, 11557–11567 (2009)

15. Gribov, E.N., Maltseva, N.V., Golovin, V.A., Okunev, A.G.: A simple method for estimating the electrochemical stability of the carbon materials. Int. J. Hydrog. Energy **41**, 18207–18213 (2016)

16. Golovin, V.A., Maltseva, N.V., Gribov, E.N., Okunev, A.G.: New nitrogen-containing carbon supports with improved corrosion resistance for proton exchange membrane fuel cells. Int. J. Hydrog. Energy (in press). doi:10.1016/j.ijhydene.2017.02.117

17. Maltseva, N.V., Golovin, V.A., Chikunova, Y., Gribov, E.N.: Influence of the number of surface oxygen on the electrochemical capacity and stability of high surface Ketjen Black ES 600 DJ. Submitted in Russ. J. Electrochem

18. Meier, J.C., Katsounaros, I., Galeano, C., Bongard, H.J., Topalov, A.A., Kostka, A., et al.: Stability investigations of electrocatalysts on the nanoscale. Energy Environ. Sci. **5**, 9319–9330 (2012)

19. Bandman, O.L.: Mapping physical phenomena onto CA-models, AUTOMATA-2008. In: Adamatzky, A., Alonso-Sanz, R., Lawniczak, A., Martinez, G.J., Morita, K., Worsch, T. (eds.) Theory and Applications of Cellular Automata, pp. 381–397. Luniver Press, UK (2008)

20. Bandman, O.L.: Cellular automata composition techniques for spatial dynamics simulation. In: Hoekstra, A.G., et al. (eds.) Simulating Complex Systems by Cellular Automata. Understanding Complex Systems, Berlin, pp. 81–115 (2010)

21. Abubaker, A., Qahwaji, R., Ipson, S., Saleh, M.: One scan connected component labeling technique, signal processing and communications. In: IEEE International Conference on ICSPC 2007, pp. 1283–1286 (2007)

A Fine-Grained Parallel Particle Swarm Optimization on Many-core and Multi-core Architectures

Nadia Nedjah[1(✉)], Rogério de Moraes Calazan[2],
and Luiza de Macedo Mourelle[3]

[1] Department of Electronics Engineering and Telecommunications,
State University of Rio de Janeiro, Rio de Janeiro, Brazil
`nadia@eng.uerj.br`
[2] Center of Electronics, Communications and Information Technology,
Brazilian Navy, Rio de Janeiro, Brazil
`rgc.moraes@gmail.br`
[3] Department of Systems Engineering and Computation,
State University of Rio de Janeiro, Rio de Janeiro, Brazil
`ldmm@eng.uerj.br`

Abstract. Particle Swarm Optimization (PSO) is a stochastic meta-heuristics yet very robust. Real-world optimizations require a high computational effort to converge to a viable solution. In general, parallel PSO implementations provide good performance, but this depends on the parallelization strategy as well as the number and/or characteristics of the exploited processors. In this paper, we propose a fine-grained paralellization strategy that focuses on the work done w.r.t. each of the problem dimensions and does it in parallel. Moreover, all particles act in parallel. This strategy is useful in computationally demanding optimization problems wherein the objective function has a very large number of dimensions. We map the computation onto three different parallel high-performance multiprocessor architectures, which are based on many and multi-core architectures. The performance of the proposed strategy is evaluated for four well-known benchmarks with high-dimension and different complexity. The obtained speedups are very promising.

1 Introduction

Parallel processing is a strategy used in computing to solve complex computational problems faster by splitting them into sub-tasks that will be allocated on multiple processors to run concurrently [9]. These processors communicate so there is synchronization or information exchange. The methodology for designing parallel algorithms comprises four distinct stages [7]: partitioning, communication, aggregation and mapping.

A multi-core processor is typically a single computing machine composed of up of 2 to 8 independent processor cores in the same silicon circuit die connected through an on-chip bus. All included cores communicate with each other, with

© Springer International Publishing AG 2017
V. Malyshkin (Ed.): PaCT 2017, LNCS 10421, pp. 215–224, 2017.
DOI: 10.1007/978-3-319-62932-2_20

the memory and I/O peripherals via this internal bus. The multi-core processor executes multiple threads concurrently, typically to boost performance in compute intensive processes. However as more cores are added to the processor, the information traffic that flows along the on-chip bus, increases as all the data must travel through the same path. This limits the benefits of a multi-core processor.

A many-core also known as a massively multi-core processors are simply multi-core processors with an especially high number of cores, ranging from 10 to 100 cores. Of course, in this multi-processors communication infrastructure between the included cores must be upgraded to a sophisticated interconnection network to cope with the amount of data exchanged by the cores. Furthermore, cores are coupled with private and local memories to reduce data traffic on the main interconnection network.

Particle Swarm Optimization (PSO) was introduced by Kennedy and Eberhart [8] and is based on collective behavior, social influence and learning. It imitates the social behavior of a flock of birds. If one element of the group discovers a way where there is easy to find food, the other group members tend instantly, to follow same way. Many successful applications of PSO have been reported, in which this algorithm has shown many advantages over other algorithms based on swarm intelligence, mainly due to its robustness, efficiency and simplicity. Moreover, it usually requires less computational effort when compared to other stochastic algorithms [6,11]. The PSO algorithm maintains a swarm of particles, each of which represents a potential solution. In analogy with evolutionary computation, a *swarm* can be identified as the population, while a *particle* with an individual. In general terms, the particle flows through a multi-dimensional search space, and the corresponding position is adjusted according to its own experience and that of its neighbors [6].

Several works show that PSO implementation on dedicated hardware [3,4] and GPUs [1,12–14] provide a better performance than CPU-based implementations. However, these implementations take advantage of the parallelization only within the loop of particles processing and also the stopping condition used in those works is always based on the total number of iterations. It is worth noting that the number of iterations required to reach a good solution is problem-dependent. Few iterations may terminate the search prematurely and large number of iterations has the consequence of unnecessary added computational effort. In contrast, the purpose of this paper is to implement a massively parallel algorithm of PSO in many and multi-core and compare with a serial implementation using the stopping condition that depends on the acceptability of the solution that has been found so far. We investigate the impact of fine-grained parallelism on the convergence time of high-dimension optimization problems and also analyze the efficiency of the process to reach the solution.

In order to take full advantage of the massively parallel in multi-cores architectures, in this paper, we explore a new strategy for fine-grained parallelism, which focuses on the work done for each of the problem dimension and does it in parallel. Throughout this paper, the strategy is termed Fine-Grained Parallel PSO — FGP-PSO. We implement the strategy using OpenMP and OpenMP with MPI. Both implementations are executed on a cluster of multi-core processors.

This paper is organized as follows: First, in Sect. 2, we sketch briefly the PSO process and the sequential version of the algorithm; Then, in Sect. 3, we describe the FGP-PSO strategy. Thereafter, in Sect. 4, we present some issues for the implementation of the FGP-PSO on a shared memory multi-core processor architecture using OpenMP; Subsequently, in Sect. 5, we report on the implementation of FGP-PSO on a cluster of multi-core shared memory processor combining the use of OpenMP and MPI; Note that the implementation of this strategy on many-core architectures (GPUs) is detailed in [2], and its performance is used in the comparison presented in this paper. It is also noteworthy to point out that interested reader can find answers to how communications are performed and how the data structures are mapped on the GPU in that publication. There follows, in Sect. 6, a thorough analysis of the obtained results for a set of benchmark functions is given; Finally, in Sect. 7, we draw some conclusions and point out some directions for future work.

2   Particle Swarm Optimization

The main steps of the PSO algorithm are described in Algorithm 1. Note that, in this specification, the computations are planned to be executed sequentially. In this algorithm, each particle has a *velocity* and an *adaptive direction* [8] that determine its next movement within the search space. The particle is also endowed with a memory that makes it able to remember the best previous position it passed by. In Algorithm 1, as well as in the remainder of this paper, we denote by $Pbest[i]$ the best fitness particle i has achieved so far and $Pbestx[i]$ the coordinates of the position that yielded it. In the same way, we denote $Sbest[i]$ the swarm best fitness particle i and its neighbors have achieved so far and $Sbestx[i]$ the coordinates of the corresponding position in the search space.

The PSO uses a set of particles, where each is a potential solution to the problem, having position coordinates in a space of d-dimensional search. Thus, each particle has a position vector with the corresponding fitness, a vector keeping the coordinates of the best position reached by the particle so far and one field to fitness and another to better fitness. To update the position of each particle i of the PSO algorithm is a set velocity for each dimension j of this position. The velocity is the element that promotes the ability of movement of the particles.

In the implemented variation of the PSO algorithm, a ring topology is used as a social network topology where smaller neighborhoods are defined for each particle. The social component, denominated *local best*, reflects the information exchanged within the neighborhood of the particle [6]. The Local Best PSO is less susceptible to being trapped into a local minimum and also the ring topology used improves performance. The velocity is the component that promotes the capacity of particle locomotion and can be computed as described in [6,8], wherein ω is called *inertia weight*, r_1 and r_2 are random numbers in $[0,1]$,

Algorithm 1. Local Best PSO

1: **for** $i = 1$ *to* η **do**
2:     **initialize** the position and velocity of particle i
3: **end for**
4: **repeat**
5:     **for** $i = 1 \rightarrow \eta$ **do**
6:         **compute** $fitness[i]$
7:         **if** $fitness[i] \leq Pbest[i]$ **then**
8:             **update** $Pbestx[i]$ using position of particle i
9:         **end if**
10:        **if** $Pbest[i] \leq Sbest[i]$ **then**
11:            **update** $Sbestx[i]$ using the $Pbestx[i]$
12:        **end if**
13:    **end for**
14:    **for** $i = 1$ *to* η **do**
15:        **update** velocity and position of particle i
16:    **end for**
17: **until** stopping criterion
18: **return** $Sbest[i]$ and $Sbestx[i]$

ϕ_1 and ϕ_2 are positive constants, y_{ij} is the particle best position *Pbest* found by particle i so far, regarding dimension j, and l_{ij} is the local best position *Lbest* found by all the particles in the neighborhood of particle i, regarding dimension j. The position of each particle is updated at each iteration [6]. Note that $x_{i,j}^{(t+1)}$ is the current position and $x_{i,j}^{(t)}$ is the previous position.

3   The Fine-Grained Parallel PSO

Parallelization of the PSO computation can be done in many ways. The most evident approach consists of doing the particle's work in parallel. Another approach could be dividing the optimization of a problem that involves d dimensions into k sub-problems of d/k dimensions each. Each sub-problem is optimized by a group of particles that form a sub-swarm of the overall acting swarm. A sub-swarm is responsible of optimizing the original problem only with respect to the corresponding d/k dimensions. Within a given sub-swarm, the computation done by the particles can thus be performed in parallel. Note that this parallelization strategy can be seen as coarse-grained [2]. In contrast, the proposed parallelization approach FGP-PSO considers the fact that in some computationally demanding optimization problems, the objective function has a large number of dimensions. Here, we are talking about more than 512 different dimensions and can even reach more than a thousand different dimensions. In this approach, the parallelism is fine-grained as it is associated with the problem dimensions. For instance, if the number of the problem dimensions is 1024, using the simple strategy that does the particle work in parallel would require 1,024 iterations to compute the fitness value. In contrast, the proposed approach requires only one single iteration to yield the fitness value for a given dimension, as all 1024 computations are done in parallel, followed by 10 iterations to compute the intermediate fitness values and thus obtaining the final fitness value. This process is called fitness *reduction*, and will be explained in more details later in this section.

4   FGP-PSO on Shared Memory Multi-core Processors

Algorithm 2 presents an overview of the implementation of FGP-PSO using OpenMP. The objective function to be optimized has d dimensions. The swarm is composed of n particles. In this approach, a particle is mapped to a block of threads wherein the computation with respect to a given dimension is handled by a thread of the block. Given the considered ring neighborhood topology, the procedure used to update the neighborhood best, namely *Lbest*.

Algorithm 2. FGP-PSO implemented using OpenMP

1: **Let** nt = number of threads
2: #pragma omp parallel
3: **Begin parallel region**
4: tid := omp_get_thread_num(); srand($seed + tid$)
5: #pragma omp for schedule(static)
6: **for** $i := 0 \rightarrow n$ **do**
7:    **initialize** particle i
8: **end for**
9: **repeat**
10:    #pragma omp for schedule(static)
11:    **for** $i := 0 \rightarrow n$ **do**
12:       #pragma omp parallel for schedule(static)
13:       **update** v_{ij} e x_{ij}; **compute** $fitness_{ij}$; **update** $Pbest_i$
14:    **end for**
15:    #pragma omp for schedule(static)
16:    **for** $i := 0 \rightarrow n$ **do**
17:       **update** $Lbest()$
18:       **if** ($Lbest_i < Best_{tid}$) **then**
19:          **update** $Best_{tid}$
20:       **end if**
21:    **end for**
22:    **if** $tid = Master$ **then**
23:       **for** $t := 0 \rightarrow nt - 1$ **do**
24:          **get** the smallest value in $Best_t$
25:       **end for**
26:    **end if**
27:    **synchronize** *threads*
28: **until** stopping condition
29: **End parallel region**
30: **return** *result* and *position*

After the beginning of the parallel region, the threads initialize the velocity and coordinates of the particles. Then, each thread initializes the context of vector, at the position indicated by the thread index, using the coordinated of particle 0 of the respective sub-swarm. After this procedure, the threads are synchronized in order to prevent copying of uninitialized context vector for the local memory of the thread.

Additionally, a second parallel region is created to divide the dimensions into groups to be run by the threads of this region. This new parallel region is implemented using nested parallelism [5] OpenMP. In nested parallelism, if a thread that belongs to a group of threads running a parallel region comes across directive to create another parallel region, the thread creates a new group and becomes the master thread of this new group. The procedures for fitness calculation, and velocity and position update were implemented with nested parallelism. The updates regarding *Pbest* and *Lbest* were kept only with parallelism in the loop of the particles since the creation of parallel regions for upgrading the positions did not improve the performance of the algorithm.

5   FGP-PSO on Clusters of Multi-core Shared Memory Processors

Algorithm 3 sketches the FGP-PSO algorithm implemented in OpenMP with MPICH (OpenMPI). Each particle is mapped as a MPI process while the loop regarding the dimensions is parallelized via the OpenMP *for* constructor. The pseudo-random number generators are initialized in the same manner as before. In order to compute the fitness value, the OpenMP reduction operation is used for synthesize the partial values of fitness obtained by each thread with respect to the considered dimension. At the end of dimensions loop, the threads, which executed this region, will compute the value of the result to the thread with index equal to 0 (*i.e.* identifying the master thread) using the informed operator depending on the objective function. Then, *Pbest* is updated and its value and position are sent by the master thread to process $(rank + 1) \mod n$ and $(rank - 1) \mod n$. After the update of *Lbest*, all the processes send messages including the value and the position of their respective *Lbest* to the master process.

The master process then checks whether, among the obtained results, there some result that satisfies the stopping condition. The other threads remain waiting in the synchronization barrier (line 28 of Algorithm 3). If so, the master process sets the exit flag and sends a message to all processes (line 26 of Algorithm 3). With this flag enabled, the processes halt the optimization and the master process returns the achieved fitness value and corresponding position.

6   Performance Results

In order to evaluate the performance of the parallelization strategy, two alternative implementations were explored: *(i)* the first implementation is based on

---

Algorithm 3. FGP-PSO implemented using OpenMPI

 1: **let** $p =$ number of processes (particles)
 2: **let** $d =$ number of dimensions
 3: **MPI_Init()**
 4: #pragma omp parallel
 5: **Begin** of an OpenMP parallel region
 6: **initialize** $particle_{rank}$
 7: **repeat**
 8:    #pragma omp for schedule(static)
 9:    **for** $j := 0 \rightarrow d$ **do**
10:       **update** x_j e v_j
11:    **end for**
12:    #pragma omp for schedule(static) reduction($operator : result$)
13:    **for** $j := 0 \rightarrow d$ **do**
14:       **compute** $fitness_j$
15:    **end for**
16:    $fitness := result$
17:    **if** $tid = Master$ **then**
18:       **update** $Pbest$; **send** $Pbest$ to process $(rank + 1)$ mod n; **update** $Lbest$
19:       **send** $Lbest$ to process $Master$
20:       **if** $rank = Master$ **then**
21:          **if** $Lbest_{rank} \leq error$ **then**
22:             **activate** the exit $flag$
23:          **end if**
24:       **end if**
25:       **Master** sends exit $flag$ to all process
26:    **end if**
27:    **synchronize** $threads$
28: **until** $flag$ activated
29: **End** parallel region OpenMP
30: **MPI_Finalize()**
31: **return** result & corresponding position

OpenMP; *(ii)* the second one uses both OpenMP and MPI. The proposed multi-core implementations were run on an SGI Octane III cluster. Each cluster node has two 2.4 GHz Intel Xeon processors, which include 4 HT cores each, hence a total of 16 cores. The results about a GPU-based implementations are taken from [10]. The many-core implementation was run on an NVIDIA GeForce GTX 460 GPU to run the CUDA implementation. The GPU includes 7 SMs with 48 CUDA cores of 1.3 GHz each, hence a total of 336 cores. Two classical benchmark functions, as listed in Table 1, were used to evaluate the performance of the proposed implementations.

We run the compared implementations using the same configuration parameters as defined in Table 2. These are the number of dimensions in the original problem d and the total number of particles n. All charts horizontal axes are represented in terms of $n \times d$ as computed in the last row of Table 2. The optimization processes were repeated 50 times with different seeds to guarantee the

Table 1. Objective functions used as benchmarks

Function	Domain	f^*
$f_2(x) = 418.9829\delta - \sum\limits_{i=1}^{\delta} x_i sin\left(\sqrt{|x_i|}\right)$	$[-500, 500]^\delta$	0.0
$f_4(x) = \sum\limits_{i=1}^{\delta} (x_i^2 - 10cos(2\pi x_i) + 10)$	$[-5.12, 5.12]^\delta$	0.0

robustness of the obtained results. The PSO basic parameters were set up as follows: inertial coefficient ω initialized at 0.99 and decreased linearly as far as 0.2; $vmax$ computed using δ of 0.2 and the stopping condition adopted was either achieving at most an error of 10^{-4} or 6,000 iterations.

Table 2. Arrangements of particles and dimensions for the evaluations

Case	C_1	C_2	C_3	C_4	C_5	C_6	C_7	C_8
d	2	4	8	16	32	64	128	256
n	8	16	32	64	128	256	512	1024
$n \times d$	4,096	16,384	65,536	26,2144	4,096	16,384	65,536	26,2144

The charts of Fig. 1 show the convergence times of benchmark functions using the serial implementation *vs.* the three implementations of the FGP-PSO proposed parallelization strategy.

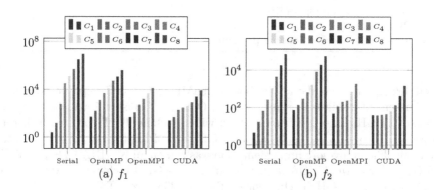

Fig. 1. Optimization time comparison for the benchmark functions

In order to make an assessment of the impact of the FGP-PSO parallelization strategy and the three alternative implementations on different parallel architectures, we compare the execution time of the proposed implementation to that obtained by a the sequential implementation of the PSO. The charts of

Fig. 2 show the speedups achieved during the minimization of benchmark functions using the three compared implementations of FGP-PSO. In all cases, the speedup is computed with respect to the sequential implementation of the PSO running on a single core of the Xeon processor.

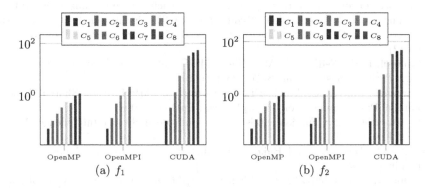

(a) f_1 (b) f_2

Fig. 2. Speedup comparison for the benchmark functions

When compared with the PSO serial implementation, the FGP-PSO OpenMP implementation achieved almost no speedup for the benchmark functions. The OpenMP with MPI implementation achieved speedups of up to 2.09× for function f_1 and 2.4× for function f_2. In contrast, the CUDA implementation occasioned speedups of up to 53.81× for function f_1 and 48.03× for function f_2. It can easily be observed that, in general, the speedup yielded by the alternative implementations increases with the complexity of the optimized problem.

7 Conclusion

This paper investigates the implementation of a fine-grained strategy to parallelize the particle swarm optimization algorithm, aiming an efficient implementation on multi-core and many-core architectures. The parallelization strategy divides the optimization workload among the swarm particles that run in parallel. Moreover, the workload of a given particle is distributed among many threads that execute the required computation regarding one given dimension of the problem.

As a interesting future work, we intend to use the parallel implementations as proposed in this work to solve an engineering problem to prove its viability and efficiency in real-world applications.

References

1. Cádenas-Montes, M., Vega-Rodríguez, M.A., Rodríguez-Vázquez, J.J., Gómez-Iglesias, A.: Accelerating particle swarm algorithm with GPGPU. In: Proceedings of the 19th Euromicro International Conference on Parallel, Distributed and Network-Based Processing, pp. 560–564. IEEE Press (2011)

2. Calazan, R.M., Nedjah, N., Mourelle, L.M.: Swarm grid: a proposal for high performance of parallel particle swarm optimization using GPGPU. In: Proceedings of the 4th International Symposium of IEEE Circuits and Systems in Latin America (LASCAS 2013), Cuzco, Peru. IEEE Computer Press, Los Alamitos (2013)

3. Calazan, R.M., Nedjah, N., Mourelle, L.M.: A massively parallel reconfigurable co-processor for computationally demanding particle swarm optimization. In: Proceedings of the 3rd International Symposium of IEEE Circuits and Systems in Latin America (LASCAS 2012), Cancun, Mexico. IEEE Computer Press, Los Alamitos (2012)

4. Calazan, R.M., Nedjah, N., Mourelle, L.M.: Parallel co-processor for PSO. Int. J. High Perform. Syst. Archit. **3**(4), 233–240 (2011)

5. Chapman, B., Jost, G., Van Der Pas, R.: Using OpenMP: Portable Shared Memory Parallel Programming, vol. 10. MIT Press, England (2008)

6. Engelbrecht, A.P.: Fundamentals of Computational Swarm Intelligence. Wiley, New Jersey (2005)

7. Foster, I.: Designing and Building Parallel Programs, vol. 95. Addison-Wesley, Reading (1995)

8. Kennedy, J., Eberhart, R.: Particle swarm optimization. In: Proceedings of IEEE International Conference on Neural Network, pp. 1942–1948. IEEE Press, Australia (1995)

9. Kirk, D.J., Hwu, W.: Programming Massively Parallel Processors. Morgan Kaufmann, San Francisco (2010)

10. Nedjah, N., Calazan, R.M., Mourelle, L.M.: Particle, dimension and cooperation-oriented PSO parallelization strategies for efficient high-dimension problem optimizations on graphics processing units. Comput. J. Sect. C: Comput. Intell. Mach. Learn. Data Anal. (2015). doi:10.1093/comjnl/bxu153

11. Nedjah, N., Coelho, L.S., Mourelle, L.M.: Multi-Objective Swarm Intelligent Systems – Theory & Experiences. Springer, Berlin (2010)

12. Papadakis, S.E., Bakrtzis, A.G.: A GPU accelerated PSO with application to economic dispatch problem. In: 16th International Conference on Intelligent System Application to Power Systems (ISAP), pp. 1–6. IEEE Press (2011)

13. Veronese, L., Krohling, R.A.: Swarm's flight: accelerating the particles using C-CUDA. In: 11th IEEE Congress on Evolutionary Computation, pp. 3264–3270. IEEE Press, Trondheim (2009)

14. Zhou, Y., Tan, Y.: GPU-based parallel particle swarm optimization. In: 11th IEEE Congress on Evolutionary Computation (CEC 2009), pp. 1493–1500. IEEE Press, Trondheim (2009)

The Implementation of Cellular Automata Interference of Two Waves in LuNA Fragmented Programming System

V.P. Markova[1,2,3] and M.B. Ostapkevich[1,2(✉)] (iD)

[1] The Institute of Computational Mathematics and Mathematical Geophysics SB RAS, Novosibirsk, Russia
{markova,ostap}@ssd.sscc.ru
[2] The Novosibirsk State Technical University, Novosibirsk, Russia
[3] The Novosibirsk National Research State University, Novosibirsk, Russia

Abstract. In this paper, a parallel implementation of the cellular-automata interference algorithm for two waves using the fragmented programming technology and LuNA system based on it is proposed. The technology is based on a strategy of data flow control. Unlike existing systems and technologies, LuNA provides a unified technology for implementing parallel programs on a heterogeneous multicomputer. The LuNA program contains a description of data fragments, computational fragments, and information dependencies between them. In the work, the LuNA program was executed on a computational cluster with homogeneous nodes. The results of comparison of the LuNA and MPI implementations showed that the execution time of the LuNA program exceeded that of the MPI program. This is due to the peculiarities of algorithms used for the distribution, search and transfer of data and computation fragments between the nodes of a cluster. The complexity of writing the LuNA program is much lower than for the MPI program.

Keywords: Parallel programming · Fragmented programming · Graph of information dependencies · LuNA system · Cellular automata

1 Introduction

With the advent of computers with heterogeneous nodes, the parallelization became complicated, because different computing devices in such a node have different architectures and each of them is programmed using a separate interface (technology). The implementation of an efficient program for such computers is done in two ways. The first way is to use MPI for inter-node parallelism, and a set of versatile technologies, such as OpenMP, OpenCL, CUDA, and HLS for intra-node parallelism.

The second way implies use of a single technology for implementation of parallel programs on a heterogeneous multicomputer. The examples of such technologies are StreamIt [1, 2] and Lift [3]. Within the framework of the second approach, the

---

The work is supported by the projects of Presidium RAS 14.1, 15.4.

© Springer International Publishing AG 2017
V. Malyshkin (Ed.): PaCT 2017, LNCS 10421, pp. 225–231, 2017.
DOI: 10.1007/978-3-319-62932-2_21

technology of fragmented programming [4], which is based on a strategy of data flow, was implemented at ICMMG SB RAS. The LuNA (Language for Numerical Algorithms) programming system was built on its basis.

The implementations of cellular automata wave interference are examined in the paper. The first implementation is built using MPI, while the second one is based on LuNA. The comparison between the two implementations is drawn and the advantages of LuNA technologies are outlined.

2   The Main Definitions and Characteristics of the LuNA System

The LuNA system is a tool for building parallel programs based on fragmented programming technology. The LuNA system consists of a fragmented program compiler and a subsystem of fragmented program execution (executive subsystem).

The basic concepts in the system are data fragments, code fragments and fragments of computations.

A *data fragment* is a set of a given size of neighboring sites in a cellular array.

A *code fragment* is a function that takes the values of some input data fragments and computes the values of the output data fragments from them. When constructing fragmented programs, two types of code fragments are used: atomic and structured. Atomic fragments of code in the system are represented as functions of C programs. Structured code fragments contain an assembly of fragments of computations.

A *fragment of computations* is a call to a code fragment with the specified names of all the input and output data fragments.

The construction of the LuNA program by an existing sequential C/C++ program consists of the following steps.

- In a given sequential C/C++ program in the code sections that are responsible for the computation are extracted. In these sections of the code, the processing of arbitrary data objects is replaced by the processing of data fragments, and the sections themselves are formalized as atomic code fragments.
- Structured code fragments and information dependencies between computational fragments and data fragments are described in the LuNA language. Unlike the MPI program, in the fragmented program (this is a synonym for the LuNA program), there is no need to specify the strict order of computations, control the allocation of resources, and program inter-node communications (see Table 1).
- The instructions are inserted that specify the release of memory occupied by data fragments that are no longer used.
- The sizes of data fragments are determined, in which the execution time is minimal.

The programming experience in the LuNA system allows us to formulate its advantages in comparison with MPI.

- The absence of rigid binding of data fragments in LuNA to MPI processes allows the LuNA program to adapt to available resources of the computer system. Currently it is possible when the program is starting. Later is will also be possible during its execution, when the set of available resources changes.

- A rather small size of data fragments provides the balance of computational load both on processes in different nodes of the computer system, and in separate threads within nodes in automatic mode. In the MPI program, the load-balancing task is either not solved at all, or is solved by the programmer.
- The description of information dependencies between fragments in LuNA is a simpler task for the programmer than a description of the precise sequence of actions in the MPI program, especially with a large number of them. The consequence of this simplification is the reduction of the number of errors in the program associated with the wrong sequence of actions. Such errors in parallel programs are difficult to identify and correct.

Table 1. The implementation of system functions in MPI and LuNA

Function	MPI	LuNA
Order of computations	Hardcoded by the programmer, fixed at the moment of execution	The final order is determined already at execution, when there is information about the computational fragments for which all the initial data is ready and collected in one node and which can be started
Management of resources in a node	Implemented by the programmer	Implemented by the LuNA system
Data distribution to nodes	Implemented by the programmer	Executed by the system on the basis of information about available resources of the computer system
Inter-node communications	Explicitly coded by the programmer	Activated by the execution subsystem based on the location information of the requested data
Load balancing	Not implemented at all or implemented by the programmer	Implemented by the LuNA system

3   The Implementation of LuNA-Program for CA Interference Algorithm

The single-particle nondeterministic Lattice Gas Cellular Automata (HPP1rp) [5] is used for wave interference simulation. In contrast to the classical methods of modeling, HPP1rp Cellular Automata (CA) method considers a physical phenomenon as a set of hypothetical particles. They move on the lattice space with the speed of a finite set of discrete velocities according to certain rules. These rules present phenomenon on micro-level based on general laws of physics.

3.1 Cellular Automata

Cellular Automata is defined on a 2D square lattice, periodically wrapped around. Each node is connected to four neighbors by unit lattice links e_i, $i = 1,2,3,4$ (e_1 – up, e_2 – right, e_3 – down, e_4 – left). Each lattice node with name r is assigned to a cell with the same name r. Two types of hypothetical particles (the moving and the rest particles) are located at each cell at discrete time. The *moving* particles are indistinguishable: each particle has unit mass and unit velocity. It is moving along one of the four links and located at the cells at discrete time. Not more than one particle is to be found at a given time and cell, moving in a given direction (exclusion principle). The *rest* particles have velocity equal to zero and mass 2. The set of particles in the cell determines its *state* $s(r)$. The vector $s(r)$ consists of 5 elements. The value of the first four elements of the vector indicates the presence ($s_i(r) = 1$) or absence ($s_i(r) = 0$) of a *moving* particle with velocity in a cell r the last element of the vector indicates the presence or absence of a *rest* particle. A pair (s,r) is called a *cell*. The total sum of particle masses in a cell is called the *model density*. For example, if a cell has the state $s = (10101)$, this implies that it contains one mass-2 rest particle and two moving particles with velocities in the directions e_1 and e_3, the model density of this cell being 4. A set of cells in which all cells have unique names forms a *cellular array*. A set of states of all cells of the array at a moment of time is called *the global state* of the automaton. The change in the global states of the automaton describes the *evolution* of CA.

The cellular automaton HPP1rp operates synchronously: all the cells of the automaton change their states simultaneously at each time step. The step consists of 2 phases: collision and propagation.

In the *collision* phase, particles at each cell collide with each other in such a way that the total particle mass and the total momentum are conserved at each cell. The collision function creates or destroys moving particles with unit speed in the cell at the time and depends only on its initial state at a given time. For example, a cell in the state (00101) (**5**) in the collision phase changes it to 3 states: (01010) (**10**) with probability $p_{5 \to 10}$ (Fig. 1a), (10000) (**16**) with probability $p_{5 \to 10}$ (Fig. 1b) and remains in its state (00101) (**5**) with probability $p_{5 \to 5}$ (Fig. 1c).

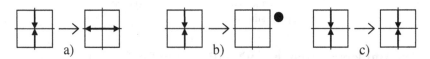

Fig. 1. An example of a collision rule: (a) transition $5 \to 10$, (b) transition $5 \to 16$, (c) transition $5 \to 5$.

In the *propagation* phase, moving particles from each cell are shifted to the nearest neighbors with unit velocity c_i.

The set of probabilities of the transitions of the HPP1rp cells from one state to another forms a matrix of transitions P of order $2^5 \times 2^5$. As a result of executing the iterative step, automaton HPP1rp goes from one global state to another. It is shown in [6] that if the matrix of transitions that implements the collision phase satisfies the semi-detailed balance condition, then the evolution of the HPP1rp CA describes the dynamics of the wave process.

3.2   Simulation of Interference of Two Waves

The interference of two waves is modeled by the evolution of a cellular automaton of size 1400 × 1200 and the density of the cells of the medium 3.3. Two circular sources of periodic waves are at a distance of four wavelengths (an integer number of wavelengths is the condition for the existence of constructive interference [7]). Two similar interference patterns of two waves that are obtained in different ways and at different modeling steps are shown in Fig. 2. The first one is simulated in the traditional way at the time step 100 (Fig. 2a), while the second is simulated in the cell-automaton way at the 975th step of the evolution of the cellular automaton (Fig. 2b).

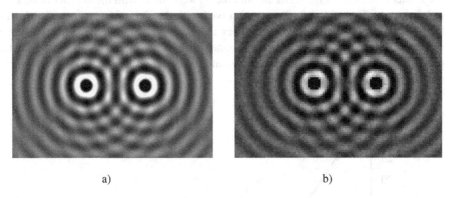

a)   b)

Fig. 2.   Interference of two waves: (a) simulated in the traditional way, (b) simulated in HPP1rp model.

3.3   Description of the Fragmented Algorithm of CA Interference

In the implementation, the original cellular array is divided into data fragments with the linear topology.

The main data fragments in the implementation are:

- a[t][x] – fragments that store the initial state of the cellular array at the simulation step t,
- b[t][x] – fragments that store the state of the cellular array after computing the particle collision at the simulation step t.
- bu[t][x], bd[t][x] – fragments for storage of shadow edges.

The upper line of the fragment b[t][x] is duplicated in the fragment bu[t][x]. The lower line of the fragment b[t][x] is duplicated in the fragment bd[t][x]. The introduction of data fragments bu[t][x], bd[t][x] allows to reduce the volume of transfers between nodes, since their size is much smaller than that of the fragment b[t][x].

Index x determines the position of the fragment in space. The value of x is 0 for the fragment containing the uppermost rows of the cellular array. x for the fragment containing the lowest rows of the array is equal to the number of fragments to which the array is divided minus one.

The fragmented implementation of the CA interference algorithm is represented by code fragments:

- init - initialization of the cellular array,
- collision - computation of the collision rule of particles in cells,
- propagation - the computation of the propagation of particles between cells,
- print - saving the simulation result to a file.

The LuNA program implements CA interference in a cellular array of the 4096 × 4096 size. The dependence of the execution time of this program on the size of fragments is shown at Fig. 3. All measurements were made with the number of nodes 1 and 8 threads. The minimum time is observed with the size of fragments 256. As the size of fragments decreases, their number increases and the overhead costs of their processing by the system grow. As the fragment size increases, the number of fragments processed by a thread becomes too small, and the system poorly balances the computational load between threads.

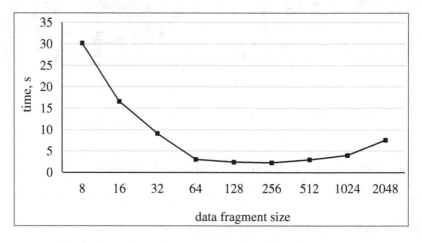

Fig. 3. Dependence of program execution time on fragment size

The first experiments and the analysis of the time complexity of the LuNA program have shown that an improvement in the time complexity can be achieved by optimizing the algorithms of distribution, searching and transferring data fragments between the nodes of the multicomputers and memory allocation control algorithms for data fragments within the nodes.

4   Conclusion

Experience in the implementation of programs has shown that, although the current version of LuNA has a bigger execution time than MPI, the LuNA technology has a number of significant advantages over MPI.

For tasks with uneven computational load on nodes, using LuNA's built-in dynamic balancing gives similar performance results compared to manual implementation of balancing in the MPI program. LuNA-programs do not have such dependence on the architecture and available resources of the computer, as MPI-programs. Therefore, LuNA programs have significantly higher portability.

The absence of the need to define the order of computations and program inter-node communications greatly simplifies programming in LuNA as compared to MPI. Same features of LuNA eliminate the appearance of some types of errors inherent in programming in MPI. The description of information dependencies between fragments has a local character, in contrast to specifying the order of computations in MPI. This simplifies the debugging of LuNA programs. All this together reduces the development time of parallel implementations of numerical algorithms in the LuNA system in comparison with MPI.

References

1. StreamIt Project Homepage. http://groups.csail.mit.edu/cag/streamit/. Accessed 3 Jan 2017
2. Thies, W., Karczmarek, M., Amarasinghe, S.: StreamIt: a language for streaming applications. In: Horspool, R.N. (ed.) CC 2002. LNCS, vol. 2304, pp. 179–196. Springer, Heidelberg (2002). doi:10.1007/3-540-45937-5_14
3. Steuwer, M., Remmelg, T., Dubach, C: Lift: a functional data-parallel IR for high-performance GPU code generation. In: Proceedings of the 2017 International Symposium on Code Generation and Optimization, CGO 2017, pp. 74–85 (2017)
4. Malyshkin, V.: Active knowledge, LuNA and literacy for oncoming centuries. In: Bodei, C., Ferrari, G.-L., Priami, C. (eds.) Programming Languages with Applications to Biology and Security. LNCS, vol. 9465, pp. 292–303. Springer, Cham (2015). doi: 10.1007/978-3-319-25527-9_19
5. Zhang, M., Cule, D., Shafai, L., Bridges, G., Simons, N.: Computing electromagnetic fields in inhomogeneous media using lattice gas automata. In: Proceedings of 1998 Symposium on Antenna Technology and Applied Electromagnetics, Ottawa, Canada, 14–16 August 1988
6. Markova, V.: Designing a collision matrix for a cellular automaton with rest particles for simulation of wave processes. Bull. Nov. Comput. Center Comput. Sci. **36**, 47–56 (2014). NCC Publisher, Novosibirsk
7. Conditions for interference, http://physics.bu.edu/~duffy/sc545_notes09/interference_conditions.html. Accessed 3 Jan 2017

A New Class of the Smallest Four-State Partial FSSP Solutions for One-Dimensional Ring Cellular Automata

Hiroshi Umeo$^{(\boxtimes)}$ and Naoki Kamikawa

University of Osaka Electro-Communication,
Hastu-cho, 18-8, Neyagawa-shi, Osaka 572-8530, Japan
{umeo,naoki}@osakac.ac.jp

Abstract. The synchronization in cellular automata has been known as the firing squad synchronization problem (FSSP) since its development, where the FSSP gives a finite-state protocol for synchronizing a large scale of cellular automata. A quest for smaller state FSSP solutions has been an interesting problem for a long time. Umeo, Kamikawa and Yunès [9] answered partially by introducing a concept of partial FSSP solutions and proposing a full list of the smallest four-state *symmetric* powers-of-2 FSSP protocols that can synchronize any one-dimensional (1D) ring cellular automata of length $n = 2^k$ for any positive integer $k \geq 1$. Afterwards, Ng [7] also added a list of *asymmetric* FSSP partial solutions, thus completing the four-state powers-of-2 FSSP partial solutions. The number *four* is the smallest one in the class of FSSP protocols proposed so far. A question remained is that "are there any other four-state partial solutions?". In this paper, we answer to the question by proposing a new class of the smallest four-state FSSP protocols that can synchronize any 1D ring of length $n = 2^k - 1$ for any positive integer $k \geq 2$. We show that the class includes a rich variety of FSSP protocols that consists of 39 *symmetric* solutions and 132 *asymmetric* ones, ranging from minimum-time to linear-time in synchronization steps. In addition, we make an investigation into several interesting properties of these partial solutions such as swapping general states, a duality between them, inclusion of powers-of-2 solutions, reflected solutions and so on.

1 Introduction

We study a synchronization problem that gives a finite-state protocol for synchronizing a large scale of cellular automata. The synchronization in cellular automata has been known as the firing squad synchronization problem (FSSP) since its development, in which it was originally proposed by J. Myhill in Moore [6] to synchronize some/all parts of self-reproducing cellular automata. The FSSP has been studied extensively for more than fifty years in [1–12].

The minimum-time (i.e., $(2n - 2)$-step) FSSP algorithm was developed first by Goto [4] for synchronizing any one-dimensional (1D) array of length $n \geq 2$. The algorithm needed many thousands of internal states for its realization.

© Springer International Publishing AG 2017
V. Malyshkin (Ed.): PaCT 2017, LNCS 10421, pp. 232–245, 2017.
DOI: 10.1007/978-3-319-62932-2_22

Afterwards, Waksman [11], Balzer [1], Gerken [3] and Mazoyer [5] also developed a minimum-time FSSP algorithm and reduced the number of states realizing the algorithm, each with 16, 8, 7 and 6 states.

On the other hand, Balzer [1], Sanders [8] and Berthiaume et al. [2] have shown that there exists no four-state synchronization algorithm. Thus, an existence or non-existence of five-state FSSP protocol has been an open problem for a long time. Umeo and Yanagihara [10] gave the first 5-state FSSP solution that can synchronize any array of length $n = 2^k (k \geq 1)$ in 3n-3 steps. Umeo, Kamikawa and Yunès [9] answered partially by introducing a concept of *partial versus full* FSSP solutions and proposing a full list of the smallest four-state symmetric powers-of-2 FSSP partial protocols that can synchronize any 1D ring cellular automata of length $n = 2^k$ for any positive integer $k \geq 1$. Afterwards, Ng [7] also added a list of asymmetric FSSP partial solutions, thus completing the four-state powers-of-2 FSSP partial solutions. A question remained is that "are there any other four-state partial solutions?".

In this paper, we answer to the question by proposing a new class of the smallest four-state FSSP protocols that can synchronize any 1D ring of length $n = 2^k - 1$ for any positive integer $k \geq 2$. We show that the class includes a rich variety of FSSP protocols that consists of 39 symmetric solutions and 132 asymmetric ones, ranging from minimum-time to linear-time in synchronization steps. In addition, we make an investigation into several interesting properties of these partial solutions such as swapping general states, a duality between them, inclusion of powers-of-2 solutions, reflected solutions and so on.

In Sect. 2 we give a description of the 1D FSSP on rings and review some basic results on ring FSSP algorithms. Sections 3 and 4 present a new class of the symmetric and asymmetric partial solutions for rings. Section 5 gives a summary and discussions of the paper.

2   Firing Squad Synchronization Problem on Rings

2.1   Definition of the FSSP on Rings

The FSSP on rings is formalized in terms of the model of cellular automata. Figure 1 shows a 1D ring cellular automaton consisting of n cells, denoted by C_i, where $1 \leq i \leq n$. All cells are identical finite state automata. The ring operates in lock-step mode such that the next state of each cell is determined by both its own present state and the present states of its right and left neighbors. All cells (*soldiers*), except one cell, are initially in the *quiescent* state at time $t = 0$ and have the property whereby the next state of a quiescent cell having quiescent neighbors is the quiescent state. At time $t = 0$ the cell C_1 (*general*) is in the *fire-when-ready* state, which is an initiation signal to the ring.

The FSSP is stated as follows: given a ring of n identical cellular automata, including a *general* cell which is activated at time $t = 0$, we want to give the description (state set and next-state transition function) of the automata so that, *at some future time*, all of the cells will *simultaneously* and, *for the first time*, enter a special *firing* state. The set of states and the next-state transition

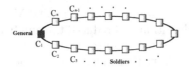

Fig. 1. One-dimensional (1D) ring cellular automaton.

function must be independent of n. Without loss of generality, we assume $n \geq 2$. The tricky part of the problem is that the same kind of soldier having a fixed number of states must be synchronized, regardless of the length n of the ring.

A formal definition of the FSSP on ring is as follows: a cellular automaton $\mathcal{M}$ is a pair $\mathcal{M} = (\mathcal{Q}, \delta)$, where

1. $\mathcal{Q}$ is a finite set of states with three distinguished states G, Q, and F. G is an initial general state, Q is a quiescent state, and F is a firing state, respectively.
2. δ is a next state function such that $\delta : \mathcal{Q}^3 \rightarrow \mathcal{Q}$.
3. The quiescent state Q must satisfy the following conditions: $\delta(\mathtt{Q}, \mathtt{Q}, \mathtt{Q}) = \mathtt{Q}$.

A ring cellular automaton $\mathcal{M}_n$ of length n, consisting of n copies of $\mathcal{M}$, is a 1D ring whose positions are numbered from 1 to n. Each $\mathcal{M}$ is referred to as a cell and denoted by C_i, where $1 \leq i \leq n$. We denote a state of C_i at time (step) t by $\mathtt{S}_i^t$, where $t \geq 0, 1 \leq i \leq n$. A *configuration* of $\mathcal{M}_n$ at time t is a function $\mathcal{C}^t : [1, n] \rightarrow \mathcal{Q}$ and denoted as $\mathtt{S}_1^t \mathtt{S}_2^t \ldots. \mathtt{S}_n^t$. A *computation* of $\mathcal{M}_n$ is a sequence of configurations of $\mathcal{M}_n, \mathcal{C}^0, \mathcal{C}^1, \mathcal{C}^2, \ldots., \mathcal{C}^t, \ldots$, where $\mathcal{C}^0$ is a given initial configuration. The configuration at time $t+1$, $\mathcal{C}^{t+1}$, is computed by synchronous applications of the next transition function δ to each cell of $\mathcal{M}_n$ in $\mathcal{C}^t$ such that: $\mathtt{S}_1^{t+1} = \delta(\mathtt{S}_{n-1}^t, \mathtt{S}_1^t, \mathtt{S}_2^t)$, $\mathtt{S}_i^{t+1} = \delta(\mathtt{S}_{i-1}^t, \mathtt{S}_i^t, \mathtt{S}_{i+1}^t)$, for any i, $2 \leq i \leq n-1$, and $\mathtt{S}_n^{t+1} = \delta(\mathtt{S}_{n-1}^t, \mathtt{S}_n^t, \mathtt{S}_1^t)$.

A *synchronized configuration* of $\mathcal{M}_n$ at time t is a configuration $\mathcal{C}^t$, $\mathtt{S}_i^t = \mathtt{F}$, for any $1 \leq i \leq n$.

The FSSP is to obtain an $\mathcal{M}$ such that, for any $n \geq 2$,

1. A synchronized configuration at time $t = T(n)$, $\mathcal{C}^{T(n)} = \overbrace{\mathtt{F}, \cdots, \mathtt{F}}^{n}$ can be computed from an initial configuration $\mathcal{C}^0 = \mathtt{G} \overbrace{\mathtt{Q}, \cdots, \mathtt{Q}}^{n-1}$.
2. For any t, i such that $1 \leq t \leq T(n) - 1$, $1 \leq i \leq n, \mathtt{S}_i^t \neq \mathtt{F}$.

2.2   Full vs. Partial Solutions

One has to note that any solution in the original FSSP problem is to synchronize any array of length $n \geq 2$. We call it **full** solution. Berthiaume et al. [2] presented an eight-state full solution for the ring. On the other hand, Umeo, Kamikawa, and Yunès [9] and Ng [7] constructed a rich variety of 4-state protocols that can synchronize some infinite set of rings, but not all. We call such protocol **partial** solution. Here we summarize recent developments on small state solutions in the ring FSSP.

Theorem 1. Berthiaume, Bittner, Perkovic, Settle, and Simon [2] *(Time Lower Bound) The minimum time in which the ring FSSP could occur is no earlier than n steps for any ring of length n.*

Theorem 2. Berthiaume, Bittner, Perkovic, Settle, and Simon [2] *There is no 3-state full solution to the ring FSSP.*

Theorem 3. Berthiaume, Bittner, Perkovic, Settle, and Simon [2] *There is no 4-state, symmetric, minimal-time full solution to the ring FSSP.*

Theorem 4. Umeo, Kamikawa, and Yunès [10] *There exist **17 symmetric** 4-state partial solutions to the ring FSSP for the ring of length $n = 2^k$ for any positive integer $k \geq 1$.*

Theorem 5. Ng [7] *There exist **80 asymmetric** 4-state partial solutions to the ring FSSP for the ring of length $n = 2^k$ for any positive integer $k \geq 1$.*

2.3 A Quest for Four-State Partial Solutions for Rings

- **Four-state ring cellular automata**
  Let $\mathcal{M}$ be a four-state ring cellular automaton $\mathcal{M} = \{\mathcal{Q}, \delta\}$, where $\mathcal{Q}$ is an internal state set $\mathcal{Q} = \{\mathtt{A}, \mathtt{F}, \mathtt{G}, \mathtt{Q}\}$ and δ is a transition function such that $\delta : \mathcal{Q}^3 \rightarrow \mathcal{Q}$. Without loss of generality, we assume that Q is a quiescent state with a property $\delta(\mathtt{Q}, \mathtt{Q}, \mathtt{Q}) = \mathtt{Q}$, G is a general state, A is an auxiliary state and F is the *firing* state, respectively. The initial configuration is $\mathtt{G}\overbrace{\mathtt{QQ}, ..., \mathtt{Q}}^{n-1}$ for $n \geq 2$. We say that an FSSP solution is *symmetric* if its transition table has a property such that $\delta(x, y, z) = \delta(z, y, x)$, for any state x, y, z in $\mathcal{Q}$. Otherwise, the FSSP solution is called *asymmetric* one.
- **A computer investigation into four-state FSSP solutions for rings**
  Figure 2 is a four-state transition table, where a symbol • shows a possible state in $\mathcal{Q} = \{\mathtt{A}, \mathtt{F}, \mathtt{G}, \mathtt{Q}\}$. Note that we have totally 4^{26} possible transition rules. We make a computer investigation into the transition rule set that might yield possible FSSP solutions. Our strategy is based on a backtracking searching. A similar technique was employed in Ng [7]. Due to the space available we omit the details of the backtracking searching strategy. The outline of those solutions will be described in the next section.

Q		Right State		
		Q	G	A
Left State	Q	Q	•	•
	G	•	•	•
	A	•	•	•

G		Right State		
		Q	G	A
Left State	Q	•	•	•
	G	•	•	•
	A	•	•	•

A		Right State		
		Q	G	A
Left State	Q	•	•	•
	G	•	•	•
	A	•	•	•

Fig. 2. Four-state transition table.

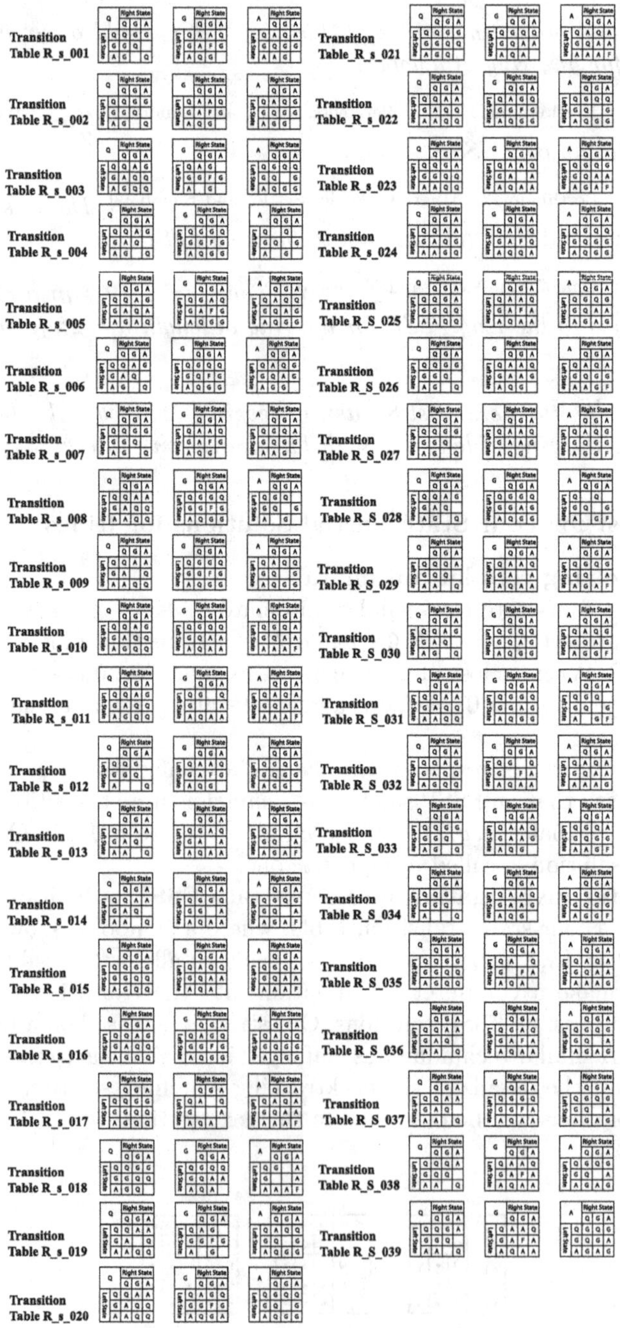

Fig. 3. Transition tables for 39 minimum-time, nearly minimum-time and non-minimum-time symmetric solutions.

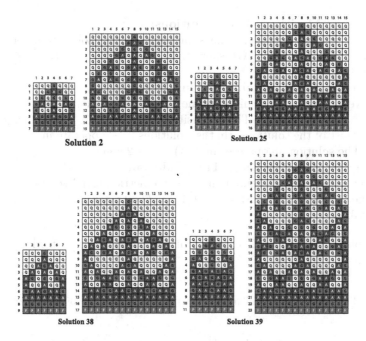

Fig. 4. Snapshots on 7 and 15 cells for symmetric solutions 2, 25, 38, and 39.

3   Four-State Symmetric Partial Solutions

In this section we will establish the following theorem with a help of computer investigation.

Theorem 6. *There exist **39 symmetric** 4-state partial solutions to the ring FSSP for the ring of length $n = 2^k - 1$ for any positive integer $k \geq 2$.*

Let $R_{S\_i}, 1 \leq i \leq 39$ be a transition table for symmetric solutions obtained in this paper. We refer to the ith symmetric transition table as symmetric solution i, where $1 \leq i \leq 39$. The details are as follows:

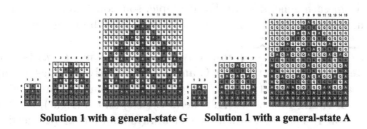

Solution 1 with a general-state G   Solution 1 with a general-state A

Fig. 5. Synchronized configurations on 3, 7, and 15 cells with a general-state G (left) and A (right), respectively, for the Solution 1.

- **Symmetric Minimum-Time Solutions:**
  We have got 24 minimum-time symmetric partial solutions operating in exactly $T(n) = n$ steps. We show their transition rules $R_{S\_i}, 1 \leq i \leq 24$ in Fig. 3.
- **Symmetric Nearly Minimum-Time Solutions:**
  We have got 14 nearly minimum-time symmetric partial solutions operating in $T(n) = n + O(1)$ steps. Their transition rules $R_{S\_i}, 25 \leq i \leq 38$ are given in Fig. 3. Most of the solutions, that is, solutions 25–37 operate in $T(n) = n+1$ steps. The solution 38 operates in $T(n) = n + 2$ steps.
- **Symmetric Non-Minimum-Time Solution:**
  It is seen that one non-minimum-time symmetric partial solution 39 exists. Its time complexity is $T(n) = (3n + 1)/2$. The transition rule R_{S_39} is given in Fig. 3.

In Table 1 we give the time complexity and number of transition rules for each symmetric solution.

Table 1. Time complexity and number of transition rules for 39 symmetric partial solutions.

Symmetric partial solutions	Time complexity	# of Transition rules
R_{S_i}, $1 \leq i \leq 24$	$T(n) = n$	$21 - 27$
R_{S_i}, $25 \leq i \leq 37$	$T(n) = n + 1$	$22 - 27$
R_{S_38}	$T(n) = n + 2$	24
R_{S_39}	$T(n) = (3n + 1)/2$	25

Here we give some snapshots on 7 and 15 cells for minimum-time, nearly minimum-time and non-minimum-time FSSP solutions, respectively, in Fig. 4.

Now, we give several interesting observations obtained for the rule set.

Observation 1 (Swapping General States). It is noted that some solutions have a property that both of the states G and A can be an initial general state without introducing any additional transition rules and yield successful synchronizations from each general state. For example, solution 1 can synchronize any ring of length $n = 2^k - 1, k \geq 2$ in $T(n) = n$ steps from both an initial configuration G $\overbrace{Q, \cdots, Q}^{n-1}$ and A $\overbrace{Q, \cdots, Q}^{n-1}$, respectively. Let $T_G(n)$ and $T_A(n)$ be synchronization steps staring from the state G and A, respectively, for rings of length n. Then, we have $T_G(n) = T_A(n) = n$. In Fig. 5 we show some synchronized configurations on 3, 7, and 15 cells with a general G (left) and A (right), respectively, for the solution 1. The observation doesn't always hold for all symmetric rules. For example, the solution 3 can synchronize any ring of length $n = 2^k - 1, k \geq 2$ in $T(n) = n$ steps from the general state G, but not from the state A.

The Observation 1 yields the following *duality* relation among the four-state rule sets.

Observation 2 (Duality). Let x and y be any four-state FSSP solution for rings and x is obtained from y by swapping the states G and A in y and vice versa. We say that the two rules x and y are dual concerning the states G and A. The relation is denoted as $x \leftrightarrows y$. For example, we have:

$$\mathrm{R_{S\_1}} \leftrightarrows \mathrm{R_{S\_14}}, \mathrm{R_{S\_2}} \leftrightarrows \mathrm{R_{S\_13}}.$$

4   Asymmetric Solutions

In this section we will establish the following theorem with a help of computer investigation.

Theorem 7. *There exist **132 asymmetric** 4-state partial solutions to the ring FSSP for the ring of length $n = 2^k - 1$ for any positive integer $k \geq 2$.*

Let $\mathrm{R_{AS\_i}}, 1 \leq i \leq 132$ be a transition table for asymmetric solutions obtained in this paper. We refer to the ith asymmetric transition table as asymmetric solution i, where $1 \leq i \leq 132$. Their breakdown is as follows:

- **Asymmetric Minimum-Time Solutions:**
  We have got 60 minimum-time asymmetric partial solutions operating in exactly $T(n) = n$ steps. Their transition rule sets $\mathrm{R_{AS\_i}}, 1 \leq i \leq 60$, are given in Figs. 6 and 7.
- **Asymmetric Nearly Minimum-Time Solutions:**
  We have got 56 nearly minimum-time asymmetric partial solutions operating in $T(n) = n + O(1)$ steps. Transition rule sets $\mathrm{R_{AS\_i}}, 61 \leq i \leq 116$, shown in Figs. 7 and 8, are the nearly minimum-time solutions obtained.
- **Asymmetric Non-Minimum-Time Solutions:**
  We have got 16 non-minimum-time asymmetric partial solutions operating in non-minimum-steps. Their transition rules are denoted by $\mathrm{R_{AS\_i}}, 117 \leq i \leq 132$. Figure 8 shows those transition rules. Each solution in $\mathrm{R_{S\_i}}, 117 \leq i \leq 124$ operates in $T(n) = 3n/2 \pm O(1)$ steps, respectively. Each solution with the rule set $\mathrm{R_{S\_125}}$ and $\mathrm{R_{S\_130}}$ operates in $T(n) = 2n + O(1)$ steps, respectively.

In Table 2 we give the time complexity and number of transition rules for each asymmetric solution.

Here we give some snapshots on 7 and 15 cells for minimum-time, nearly minimum-time and non-minimum-time FSSP solutions, respectively, in Fig. 9.

Observation 3 (Swapping General States). It is noted that some asymetric solutions have a property that both of the states G and A can be an initial general state without introducing any additional transition rules and yield successful synchronizations from each general state. For example, asymetric solution 1 can synchronize any ring of length $n = 2^k - 1, k \geq 2$ in $T(n) = n$ steps from both an initial configuration $\mathrm{G}\overbrace{\mathrm{Q},\cdots,\mathrm{Q}}^{n-1}$ and $\mathrm{A}\overbrace{\mathrm{Q},\cdots,\mathrm{Q}}^{n-1}$, respectively and we have $T_G(n) = T_A(n) = n$.

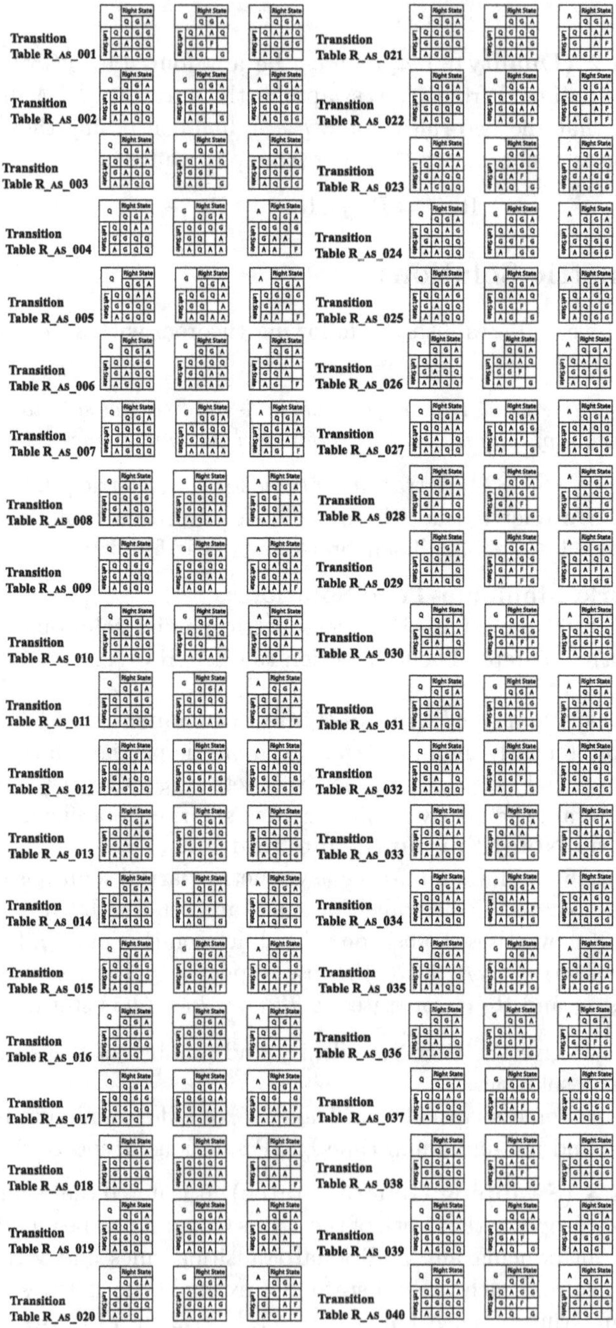

Fig. 6. Transition tables $R_{AS\_i}, 1 \le i \le 40$ for minimum-time asymmetric solutions.

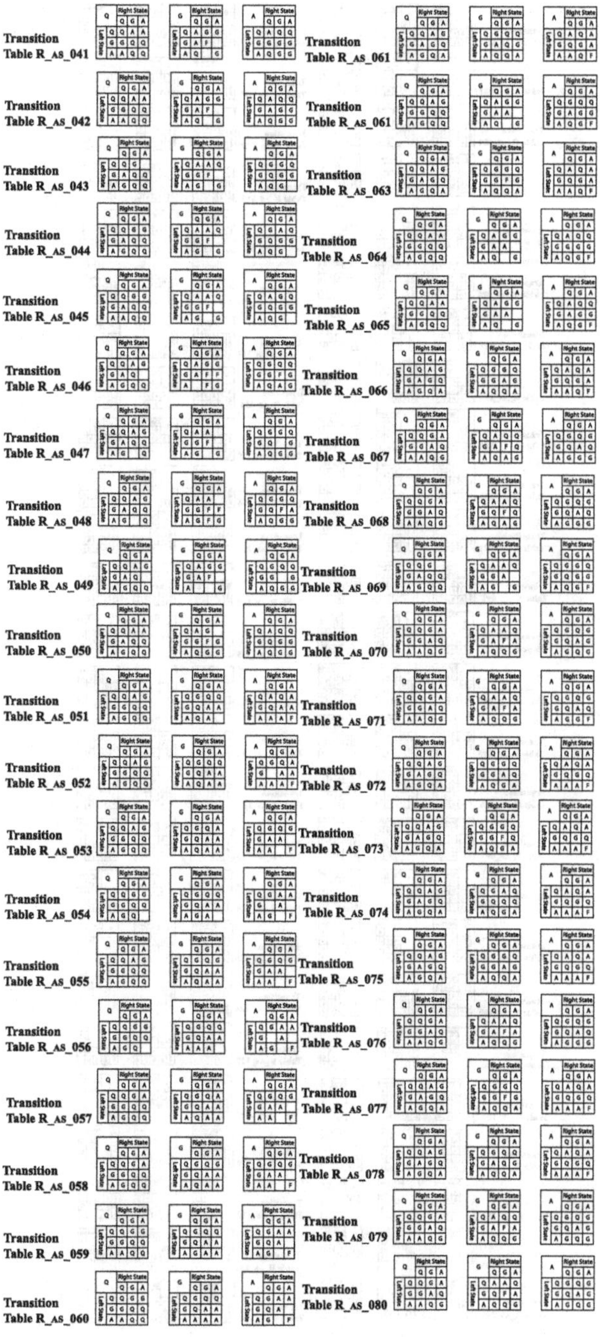

Fig. 7. Transition tables $R_{AS\_i}, 41 \leq i \leq 80$ for minimum-time and nearly-minimum-time asymmetric solutions.

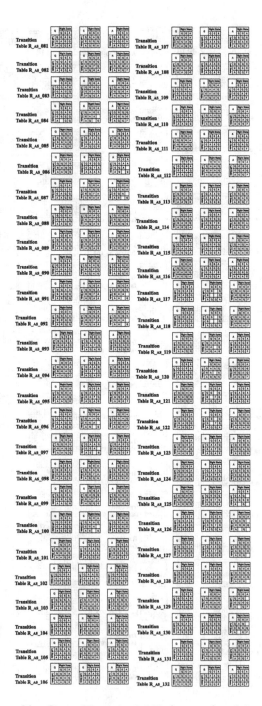

Fig. 8. Transition tables $R_{AS-i}, 81 \leq i \leq 132$ for nearly-minimum-time and non-minimum-time asymmetric solutions.

Table 2. Time complexity and number of transition rules for 132 asymmetric solutions.

Asymmetric partial solutions	Time complexity	# of Transition rules
R_{AS_i}, $1 \leq i \leq 60$	$T(n) = n$	22 – 26
R_{AS_i}, $61 \leq i \leq 116$	$T(n) = n + O(1)$	25 – 27
R_{AS_i}, $117 \leq i \leq 124$	$T(n) = 3n/2 \pm O(1)$	24 – 27
R_{AS_i}, $125 \leq i \leq 132$	$T(n) = 2n + O(1)$	24 – 27

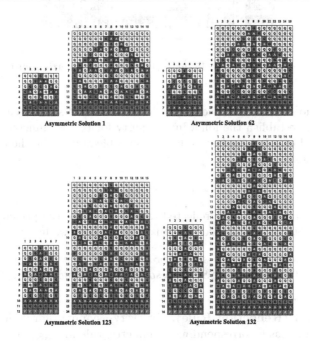

Fig. 9. Snapshots on 7 and 15 cells for asymmetric solutions 1, 62, 123, and 132.

Observation 4 (Duality). A duality relation exists among the asymetric solutions. For example, we have:

$$R_{AS\_1} \leftrightarrows R_{AS\_4}, R_{AS\_2} \leftrightarrows R_{AS\_57}.$$

Observation 5 (Inclusion of Powers-of-2 Rule). It is noted that some solutions can synchronize not only rings of length $2^k - 1, k \geq 2$ but also rings of length $2^k, k \geq 1$. For example, solution 130 can synchronize any ring of length $n = 2^k - 1, k \geq 2$ in $T(n) = 2n + 1$ steps and simultaneously the solution can synchronize any ring of length $n = 2^k, k \geq 1$ in $T(n) = 2n - 1$ steps. See the snapshots given in Fig. 10 on 7, 8, 15, and 16 cells for the solution 130. A relatively large number of solutions includes powers-of-2 solutions as a proper subset of rules.

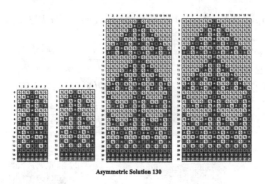

Fig. 10. Snapshots on 7, 8, 15, and 16 cells for asymmetric solutions 130.

Now we show a one to one correspondence between 4-state asymmetric solutions. First, we establish the following property for the asymmetric FSSP solution for rings. Let x be any k-state transition table and x^R be the k-state table defined such that:

$$x^R(i,j) = x(j,i), \text{ for any } 1 \le i, j \le k.$$

The transition table x^R is the reflected table concerning the principal diagonal of the table x, which is obtained by transposition. Now we have:

Theorem 8. *Let x be any k-state FSSP ring solution with time complexity $T_x(n)$. Then, x^R is also an FSSP ring solution with time complexity $T_{x^R}(n) = T_x(n)$.*

Observation 6 (Reflection Rule). For every asymmetric rule in R_{AS_i}, $1 \le i \le 132$, the rule has a corresponding asymmetric rule in R_{AS_i}, $1 \le i \le 132$.

For example, R_{AS_1} is the reflected rule of R_{AS_40} and vice versa.

5   Summary and Discussions

A quest for smaller state FSSP solutions has been an interesting problem for a long time. We have answered to the question by proposing a new class of the smallest four-state FSSP protocols that can synchronize any 1D ring of length $n = 2^k - 1$ for any positive integer $k \ge 2$. We show that the class includes a rich variety of FSSP protocols that consists of 39 symmetric solutions and 132 asymmetric ones, ranging from minimum-time to linear-time in synchronization steps. Some interesting properties in the structure of 4-state partial solutions have been discussed. We strongly believe that no smallest solutions exist other than the ones proposed for length 2^k rings in Umeo, Kamikawa and Yunès [9] and Ng [7] and for rings of length $2^k - 1$ in this paper. A question "how many 4-state partial solutions exist for arrays (open-rings)?" remains open. We think that there would be a large number of smallest 4-state partial solutions for arrays.

Its number would be larger than several thousands. The structure of the 4-state partial array synchronizers is far more complex than the 4-state partial ring synchronizers.

References

1. Balzer, R.: An 8-state minimal time solution to the firing squad synchronization problem. Inf. Control **10**, 22–42 (1967)
2. Berthiaume, A., Bittner, T., Perković, L., Settle, A., Simon, J.: Bounding the firing synchronization problem on a ring. Theor. Comput. Sci. **320**, 213–228 (2004)
3. Gerken, H.D.: Über Synchronisationsprobleme bei Zellularautomaten, pp. 1–50. Diplomarbeit, Institut für Theoretische Informatik, Technische Universität Braunschweig (1987)
4. Goto, E.: A Minimal Time Solution of the Firing Squad Problem. Dittoed course notes for Applied Mathematics 298 (with an illustration in color). Harvard University, Cambridge (1962)
5. Mazoyer, J.: A six-state minimal time solution to the firing squad synchronization problem. Theor. Comput. Sci. **50**, 183–238 (1987)
6. Moore, E.F.: The firing squad synchronization problem. In: Moore, E.F. (ed.) Sequential Machines, Selected Papers, pp. 213–214. Addison-Wesley, Reading MA (1964)
7. Ng, W.L.: Partial Solutions for the Firing Squad Synchronization Problem on Rings, pp. 1–363. ProQuest publications, Ann Arbor (2011)
8. Sanders, P.: Massively parallel search for transition-tables of polyautomata. In: Jesshope, C., Jossifov, V., Wilhelmi, W. (eds.). Proceeding of the VI International Workshop on Parallel Processing by Cellular Automata and Arrays, Akademie, pp. 99–108 (1994)
9. Umeo, H., Kamikawa, N., Yunès, J.-B.: A family of smallest symmetrical four-state firing squad synchronization protocols for ring arrays. Parallel Process. Lett. **19**(2), 299–313 (2009)
10. Umeo, H., Yanagihara, T.: A smallest five-state solution to the firing squad synchronization problem. In: Durand-Lose, J., Margenstern, M. (eds.) MCU 2007. LNCS, vol. 4664, pp. 291–302. Springer, Heidelberg (2007). doi:10.1007/978-3-540-74593-8_25
11. Waksman, A.: An optimum solution to the firing squad synchronization problem. Inf. Control **9**, 66–78 (1966)
12. Yunès, J.B.: A 4-states algebraic solution to linear cellular automata synchronization. Inf. Process. Lett. **19**(2), 71–75 (2008)

Properties of the Conservative Parallel Discrete Event Simulation Algorithm

Liliia Ziganurova[1,2(✉)] and Lev Shchur[1,2,3]

[1] Scientific Center in Chernogolovka,
142432 Chernogolovka, Moscow Region, Russia
ziganurova@gmail.com, levshchur@gmail.com
[2] National Research University Higher School of Economics,
101000 Moscow, Russia
[3] Landau Institute for Theoretical Physics,
142432 Chernogolovka, Moscow Region, Russia

Abstract. We address question of synchronisation in parallel discrete event simulation (PDES) algorithms. We study synchronisation in conservative PDES model adding long-range connections between processing elements. We investigate how fraction of the random long-range connections in the synchronisation scheme influences the simulation time profile of PDES. We found that small fraction of random distant connections enhance synchronisation, namely, the width of the local virtual times remains constant with increasing number of processing elements. At the same time the conservative algorithm of PDES on small-world networks remains free from deadlocks. We compare our results with the case-study simulations.

Keywords: Parallel discrete event simulation · PDES · Conservative algorithm · Small-world

1 Introduction

Modern high performance systems consist with hundreds of thousands of nodes, which in turn may have many CPUs, cores, and numerical accelerators. The development of hardware architecture influences on the development environments (programming models, frameworks, compilers, libraries, etc.). They now need to deal with a high level of parallelism and can solve difficulties arising from system heterogeneity [1].

In the paper we discuss synchronisation in one of the methods of large-scale simulation known as parallel discrete event simulation (PDES) [2]. The method is widely used in physics and computer science, as well as in economics, engineering, and society. The first ideas come about 40 years ago in order to overcome limitation of memory/time resources, and revising of the method is still important nowadays. PDES has a property of good scalability with the physical system size (number of objects) as well as with the hardware size (number of nodes, cores, and level of the hyper-threadings).

© Springer International Publishing AG 2017
V. Malyshkin (Ed.): PaCT 2017, LNCS 10421, pp. 246–253, 2017.
DOI: 10.1007/978-3-319-62932-2_23

PDES allows to run one single discrete event simulation task on the number of processing elements (PE), which physically can be nodes, or CPUs, or cores, or threads depending on the particular system architecture. The system being simulated is divided into subsystems which are mapped onto programming objects, or logical processes (LPs). Logical process is a sequential subprogram executed by some PE.

The system changes its state at some discrete moments of time, which are usually Poisson arrivals. The changes are called *discrete events*. The events generates messages with timestamps which are saved in the output queue and sent to other LPs. Received messages are stored by LPs in their input queues. LPs during the simulation maintain a loop, sequentially taking the event with the lowest timestamp from the input queue, executing it, and communicating with other LPs if necessary. The communication between LPs goes exclusively via time-stamped messages. It is important that LPs do not use any shared memory. Synchronisation process is done locally by the analysis of the values of timestamps of messages in the queue. When LP processes an event, it updates its own local virtual time (LVT) to the time of the processed event. Each LP evolves independently in time and there is no global synchronisation in the simulation process. Since the dynamic is asynchronous some synchronisation protocol is required. There are three groups of such protocols: conservative, optimistic and FaS [2–4].

In the paper we investigate the performance and scalability properties of the conservative PDES algorithm. In conservative algorithm it is assumed that all dependencies between LPs must be checked before every portion of computations in order to preserve causality of the computations. The performance of conservative algorithm depends on the communication network: the more dependencies in the system the lower speed of the computation. We study the influence of long-range communication links on the synchronisation and performance of PDES conservative algorithm. We build a simplified model of the evolution of LVT profile. The model allows to measure local time variance and average speed of the utilisation of processing times by LPs. The observables are then mapped onto synchronisation aspects of PDES scheme.

The paper is organised as follows. In the next section we describe a background of the problem. Section 3 provides detailed information about the model under consideration. The results of our simulation are given in Sect. 4. The discussion and further work are presented in Sect. 5.

2   Models of Evolution of LVT Profile in PDES

In this section we describe the general approach to investigation of synchronisation in PDES algorithms and review main results in this area.

Conservative PDES model on regular networks. Model of evolution of times in PDES conservative algorithm is proposed in [5]. Authors consider communication scheme with only nearest-neighbour interactions, which is equivalent to one-dimensional system with periodic boundary conditions. It was found in simulation that evolution of time profile reminds the evolution of a growing surface which is known from literature in physics does belong to Kardar-Parisi-Zhang-like kinetic roughening [6]. This analogy provides a cross-disciplinary application of well-known concepts from non-equilibrium statistical physics to our problem. More details on the relation of PDES algorithms with physical models can be found in [7,8].

Synchronisation of the parallel schemes can be described using this analogy. *Efficiency* of parallel implementation can be defined as a fraction of the non-idle processing elements. This fraction exactly coincides with the density of local minima in the growing model. It is shown in [5] that in the worst-case scenario the efficiency of the PDES algorithm remains nonzero as the number of PEs goes to infinity.

Freeze-and-Shift PDES model. The conservative PDES is proved to be free from deadlock, and efficiency is about 1/4, on average, i.e., at least one PE out of four is working at any given time. In [4] an alternative synchronisation algorithm of PDES was proposed. It is based on i) the extension of the PDES concept on the hierarchical hardware architecture, including multi-core and hyper-threadings and ii) using analogy of the evolution of time-profile interface with the physical models of surface interface growth. In the late case classification of the boundary conditions leads to the classification of possible PDES algorithms. Authors give a way to increase utilisation by giving each node a large portion of LPs which are processed by threads running on the same CPU. The LPs within one CPU communicate conservatively, whereas LPs from different CPUs communicate according to either conservative, or optimistic scheme, or scheme with fixed LVT on the boundary LPs (those which can communicate with LP in the neighbouring CPUs). In the last case CPUs do not communicate with other CPUs for some time interval window (the frozen part of the algorithm), and after that time the message exchange is implemented as part of the memory shifting between CPUs (the shift part of the algorithm). The algorithm is therefore called Freeze-and-Shift (FaS).

Optimistic PDES models on regular networks. Model of evolution of time profile in optimistic PDES algorithm is introduced in [9]. Dynamical behaviour of the optimistic PDES model is quite different from the conservative PDES model. The optimistic model corresponds to another surface growing model and demonstrates features of the roughening transition and directed percolation [10].

Conservative PDES models on small-world like networks. All models described above consider PDES algorithms with short-range connections. The idea of studying the model with other type of communication topology is proposed in [11]. Authors investigate the behaviour of the local virtual time profile on a small-world like network [12]. The network is build as a regular one-dimensional lattice with additional long-range connections randomly wired above it. The links are used dynamically, i.e. at each time step additional synchronisation check between distant neighbours is made with some probability p. Small value of the p significantly improves synchronisation. Variance between local times becomes finite, while utilisation decreases just slightly.

In present paper we revise the approach of [11], considering more realistic topology, and compare with the result of [11]. In addition we compare our results with the case-study [13].

3   Model Definition

We build a model of *evolution of LVTs profile* for analysis of the desynchronisation processes in conservative PDES algorithm. The PEs are said to be synchronised when the differences between LVTs stay finite with the simulation process. The efficiency of the algorithm is measured as a number of PEs working at a given moment of time. It reflects the load of the processing elements (CPU, core, or thread). When the efficiency of the algorithm is strictly greater than zero, one can say that the algorithm is deadlock free. These properties of the synchronisation algorithm can be extracted from the analysis of the LVT profile.

The model is constructed on the *small-world communication networks* [12]. Long-range connections in addition to short-range reflect real systems properties. For example, computer networks, social networks, electric power grid, and network of brain neurones are known to be small-world networks. Additional communication links between distant nodes also enhance synchronisation of simulation.

First we build a communication topology. For simplicity of the model we assume that each PE does process only one LP. Nodes in the communication graph represent PEs, and edges represent dependencies between them. Each PE has its local variable τ, which is the value of LVT. The set of all LVTs is stored as an array. Two observables are computed: the efficiency $\langle u \rangle$ and the profile width $\langle w^2 \rangle$. We run the simulation program with different set of parameters N and p. Finally we take an average over multiple samples.

Topology. Denote number of processing elements as N. We build a small-world topology of PEs using the parameter p – the fraction of random long-range connections. First we connect all PEs into a regular one-dimensional lattice (equivalent to the ring) and then add pN random long-range connections. Each edge is chosen only once. The result is a communication graph of N nodes and $N(1 + p)$ edges stored as adjacency list.

Initialisation. We set the parameter p of the network to some value from 0.002 to 0.01, build a topology, and set all local times to zero: $\tau_i(0) = 0$, $i = 1..N$.

Simulation. We are interested in evolution of LVTs profile in *conservative* algorithm. We assume that only those PEs, whose current time is lower than the time of their neighbours (i.e. the PEs which it is connected with), may proceed with computations. These PEs are called *active*. Such scheme guarantees that causality will be preserved [2].

In our model we implement an ideal scheme of message passing. At each time step t every LP broadcasts the message with the time stamp equal to its LVT to all LPs connected with it. We assume that time needed for message distribution is negligibly small, so there is no difference between sending and receiving time. The PEs who have received only messages with higher time stamp than their LVT, may proceed. These PEs have minimal LVT among their neighbours.

At each simulation step t we find PEs with the lowest LVTs among their neighbouring nodes and increment local time of those PEs by an exponentially distributed random amount:

$$\tau_i(t+1) = \begin{cases} \tau_i(t) + \eta & \text{if } \tau_i(t) \leq \tau_K(t), \\ \tau_i(t) & \text{otherwise,} \end{cases} \tag{1}$$

where η is a random value drawn from the Poisson distribution, K is a set of all PEs which are connected to i-th PE by local or long-range communication links, and $i = 1..N$.

After updating array of LVTs the observables are computed, and PDES goes to the next simulation cycle.

Observables. We compute two essential features of the model: the efficiency $\langle u \rangle$, which is equivalent to the average utilisation of the algorithm, and the width of the profile $\langle w^2 \rangle$, i.e. the variance of local virtual times, which is associated to the desynchronisation of PEs.

1. The *efficiency* (utilisation) of the algorithm is equivalent to the density of local minima of the LVT profile. The efficiency shows how many PEs is working at a given moment of time. In basic conservative scheme on a ring topology (when each PE is connected with exactly two neighbours) the efficiency is approximately $1/4$. The figure is derived analytically from the observation of all possible combination of LVTs of neighbouring PEs. Numerical result is equal to $0.24641(7)$ [5]. The number shows that only approximately one quarter of all PEs are working at a given moment of time and other three quarters are idling.
   We calculate the efficiency of the algorithm $\langle u \rangle$ as an average fraction of active PEs at each time step:

$$\langle u \rangle = \frac{\langle \overline{N_{activePE}} \rangle}{N}. \tag{2}$$

   The underlined average is taken over all time steps and $\langle \cdot \rangle$ states for the average over independent 1500 runs with fixed parameters N and p.
2. The *width* (variance) of the LVT profile shows the average spread between local virtual times. If the width remains constant during the simulation, then PEs are well synchronised. The increasing of the profile width corresponds to the growing of the desynchronisation with simulation time.
   The width of the LVT profile is calculated according the formula below:

$$\langle w^2(N,t) \rangle = \langle \frac{1}{N} \sum_{i=1}^{N} [\tau_i(t) - \overline{\tau}(t)]^2 \rangle, \tag{3}$$

where $\overline{\tau}(t) = \frac{1}{N} \sum_{i=1}^{N} \tau_i(t)$ is the mean value of the time profile.

In our simulation program we use random number generation library RNGAVXLIB [14]. We run program on the Manticore cluster using MVA-PICH2 [15].

4   Results

We are interested in scalability properties of the synchronisation of conservative PDES model with the long-range connections. We perform simulation of conservative PDES model on the ring topology with long-range connections. We simulate systems of size N (number of PEs) varying from $N = 10^3$ to $N = 10^5$, and for number of values of fraction p for the long-range connections. Note that $p = 0$ corresponds to the basic conservative model with only short-range connections studied in [5].

The main results are: (1) efficiency of the algorithm remains finite and slightly reduces with adding long-range connections p; (2) profile width for any p grows with the system size; (3) profile width saturates for system sizes larger than 10^4; (4) degree of desynchronisation depends logarithmically with p.

The efficiency. We observe that for any system size the average density of local minima $\langle u(t) \rangle$ monotonically decreases as a function of time and approaches a constant. The constant depends on the fraction of random connections p and system size N. For small p the utilisation of events reduces slightly. The small-world-synchronised simulation scheme maintains an average rate greater than zero. For example, for $p = 0.01$ it is $\langle u \rangle = 0.22137(7)$, while for basic conservative scheme $\langle u_0 \rangle = 0.24641(7)$ [5].

The efficiency $\langle u \rangle$ has nonlinear dependence on the parameter p. It is possible to fit utilisation dependence on p by expression:

$$\langle u(p) \rangle = u_0 - A(N)p^{B(N)}. \tag{4}$$

The coefficient A and the exponent B depend on the system size, and can be fit using logarithmic or exponential dependencies. Using logarithmic fit we obtain $A = 0.078(3) + \frac{0.345(9)}{\log N}$ and $B = 0.092(3) + \frac{1.26(1)}{\log N}$. Using power-law fit we obtain $A = 0.08(2) + \frac{0.253(5)}{N^{0.12(3)}}$ and $B = 0.24(1) + \frac{1.14(5)}{N^{0.21(1)}}$. We could not choose which fit is better.

Finally, we found that as N goes to infinity, $\langle u(p) \rangle = u_0 - 0.078(3)p^{0.092(3)}$ if we approximate A and B with logarithm, or $\langle u(p) \rangle = u_0 - 0.08(2)p^{0.24(1)}$ in the case of the power-law approximation for the coefficients A and the exponent B.

The width. We observed that the profile width grows as $\langle w^2(t) \rangle \sim t^{2\beta}$ and saturates at some time t_* reaching the value $\langle w_\infty^2 \rangle$. We measured the growth exponent β for each combination of the parameters N and p. For large systems ($N > 10^4$) exponent β becomes almost constant. The asymptotic value of β is found to behave logarithmically with p

$$\beta = -0.137(4) - 0.162(1)\ln(p). \tag{5}$$

We remind that without long-range connections ($p = 0$) it is $\beta = 1/3$.

It is interesting, that during the simulation on more then ten thousand PEs the desynchronisation of PEs will grow equally fast for systems of any sizes.

For large systems synchronisation depends only on the amount of long-range connections. The profile width approaches a constant value $\langle w_\infty^2 \rangle$ with growing system size. In contrary, in the basic short-range conservative scheme ($p = 0$) the width is increasing with the system size as $\langle w_\infty^2 \rangle \sim N^{2\alpha}$, $\alpha = 0.49(1)$. The topology with additional distant connections allows the simulation of large systems to preserve the same degree of synchronisation.

Since the small-world conservative PDES scheme progresses with positive rate and the profile width becomes *finite* in the limit of infinitely many PEs, one can say that the conservative algorithm with long-range connections is fully scalable.

5   Discussion and Future Work

In the paper we present analysis of the synchronisation in conservative PDES algorithm on the small-world networks.

Paper [13] presents detailed results of the case-study simulations of different models. Two optimistic simulators were used: ROSS [16] and WARPED2 [17]. Simulation results of three models were reported: traffic model, wireless network model, and epidemic model. Average utilisation $\langle u \rangle$ varied from 0.47 for epidemic model to 0.0043 for traffic model, and down to $5 \cdot 10^{-5}$ for wireless network model.

We guess that the results of case-study [13] can be explained in part by the concept of small-world network with varying parameter p. To answer this question in details it is necessary to perform case-studies of the mentioned models measuring quantities, which can be mapped on the parameters of our model.

In addition, it is interesting to investigate properties of the *optimistic* algorithm of PDES on small-world networks provided with the comparison with the results of case-studies.

Acknowledgements. This work is supported by grant 14-21-00158 of the Russian Science Foundation.

References

1. Bailey, D.H., David, H., Dongarra, J., Gao, G., Hoisie, A., Hollingsworth, J., Jefferson, D., Kamath, C., Malony, A., Quinian, D.: Performance Technologies for Peta-Scale Systems: A White Paper Prepared by the Performance Evaluation Research Center and Collaborators. White paper, Lawrence Berkeley National Laboratories (2003)
2. Fujimoto, R.M.: Parallel discrete event simulation. Commun. ACM **33**, 30–53 (1990). doi:10.1145/84537.84545
3. Jefferson, D.R.: Virtual time. ACM Trans. Program. Lang. Syst. **7**, 404–425 (1985). doi:10.1145/3916.3988
4. Shchur, L.N., Novotny, M.A.: Evolution of time horizons in parallel and grid simulations. Phys. Rev. E **70**, 026703 (2004). doi:10.1103/PhysRevE.70.026703

5. Korniss, G., Toroczkai, Z., Novotny, M.A., Rikvold, P.A.: From massively parallel algorithms and fluctuating time horizons to nonequilibrium surface growth. Phys. Rev. Lett. **84**, 1351 (2000). doi:10.1103/PhysRevLett.84.1351
6. Kardar, M., Parisi, G., Zhang, Y.C.: Dynamic scaling of growing interfaces. Phys. Rev. Lett. **56**, 889 (1986). doi:10.1103/PhysRevLett.56.889
7. Shchur, L.N., Shchur, L.V.: Relation of parallel discrete event simulation algorithms with physical models. J. Phys: Conf. Ser. **640**, 012065 (2015). doi:10.1088/1742-6596/640/1/012065
8. Shchur, L., Shchur, L.: Parallel discrete event simulation as a paradigm for large scale modeling experiments. Selected Papers of the XVII International Conference on Data Analytics and Management in Data Intensive Domains (DAMDID/RCDL 2015), Obninsk, Russia, 13–16 October 2015, pp. 107–113, (2015). http://ceur-ws.org/Vol-1536/
9. Ziganurova, L., Novotny, M.A., Shchur, L.N.: Model for the evolution of the time profile in optimistic parallel discrete event simulations. J. Phys. Conf. Ser. **681**, 012047 (2016). doi:10.1088/1742-6596/681/1/012047
10. Alon, U., Evans, M.R., Hinrichsen, H., Mukamel, D.: Roughening transition in a one-dimensional growth process. Phys. Rev. Lett. **76**, 2746 (1996). doi:10.1103/PhysRevLett.76.2746
11. Guclu, H., Korniss, G., Novotny, M.A., Toroczkai, Z., Racz, Z.: Synchronization landscapes in small-world-connected computer networks. Phys. Rev. E **73**, 066115 (2006). doi:10.1103/PhysRevE.73.066115
12. Watts, D.J., Strogatz, S.H.: Collective dynamics of small-world networks. Nature **393**, 440–442 (1998). doi:10.1038/30918
13. Wilsey, P.A.: Some properties of events executed in discrete-event simulation models. In: Proceedings of the 2016 annual ACM Conference on SIGSIM Principles of Advanced Discrete Simulation, pp. 165–176. ACM, New York (2016). doi:10.1145/2901378.2901400
14. Guskova, M.S., Barash, L.Y., Shchur, L.N.: RNGAVXLIB: Program library for random number generation: AVX realization. Comput. Phys. Commun. **200**, 402–405 (2016). doi:10.1016/j.cpc.2015.11.001
15. MVAPICH: MPI over InfiniBand, Omni-Path, Ethernet/iWARP, and RoCE. http://mvapich.cse.ohio-state.edu
16. Carothers, C.D., Bauer, D., Pearce, S.: ROSS: A high-performance, low-memory, modular Time Warp system. J. Parallel Distrib. Comput. **62**, 1648–1669 (2002). doi:10.1016/S0743-7315(02)00004-7
17. Weber, D.: Time warp simulation on multi-core processors and clusters. Master's thesis, University of Cincinnati, Cincinnati, OH (2016)

Organization of Parallel Computation

Combining Parallelization with Overlaps and Optimization of Cache Memory Usage

S.G. Ammaev, L.R. Gervich$^{(\boxtimes)}$, and B.Y. Steinberg

Southern Federal University, Rostov-on-Don, Russian Federation
ammsaid@mail.ru, lgervith@gmail.com, borsteinb@mail.ru

Abstract. This paper allows L. Lamport hyperplane method modified for improvement of the temporal data locality. Gauss-Seidel algorithm optimized by modified hyperplane method is faster than non-optimized in 2.5 times. This algorithm was paralleled by the technique of data placement with overlaps and we have got the speedup in 28 times on 16 processors in comparison with the non-optimized sequential algorithm.

Keywords: Hyperplane method · Temporal data locality · Data placement with overlaps · Optimization · Tiling

1 Introduction

The hyperplane method was published by L. Lamport 40 years ago [1]. This method was subjected to many modifications and generalizations [2] etc. Sometimes, when there are no loops can be paralleled, the hyperplane method allows parallelization of the perfect nested loop. Points from hyperplanes of iteration space of nested loop can be executed concurrently. Many generations of computers were changed in 40 years of method existence. The execution time of data processing was accelerated on 30% on average every year, but the execution time of memory access on 9% [12]. It brought to the development of memory hierarchy in processors, that is necessary to consider for fast program development [13]. For many cases, the direct application of hyperplane method doesn't accelerate program due to ineffective using of cache memory.

The hyperplane method is applied for parallelization. However, the modified hyperplane method can improve data locality in iterative methods.

Improving the locality of data is an important approach for accelerating programs. There are two types of techniques for improving data locality: data access optimizations and data layout optimizations [3]. Data access optimizations change the order in which iterations in a loop nest are executed. Loop fusion and loop tiling [4] is data access optimizations [3]. There are more complex techniques of data access optimization, such as chunking [5]. Data locality is also considered in articles [6–10].

Data layout optimizations reorder the placement of original data in memory. Such methods as array padding, array transpose is data layout optimization. One of the effective data layout optimization is diamond tiling. Diamond tiling

© Springer International Publishing AG 2017
V. Malyshkin (Ed.): PaCT 2017, LNCS 10421, pp. 257–264, 2017.
DOI: 10.1007/978-3-319-62932-2_24

was applied to the wave equation together with vector optimizations and GPU parallelization in [11].

2 The Iteration Space of Nested Loop and Hyperplane Method

The perfect nest consisting of n loops has the following form.

```
for(I1=L1; I1<=R1; ++I1)
    for(I2=L2; I2<=R2; ++I2)

        for(In=Ln; In<=Rn; ++In)
        {
            LOOPBODY (I1,I2,...,In)
        }
```

The set of all values of index vector $I = (I1, I2, \cdots, In)$ named the iteration space of nested loop. If loop bounds are affinity-dependent on higher loop index, the iteration space is a convex polyhedron. If loop bounds are constant, the iteration space is multidimensional rectangular cuboid. Each point of iteration space corresponds to a certain value of loop index vector $I = (I1, I2, \cdots, In)$. By the execution of point of iteration space, we consider the execution of the nested loop body for index values, that corresponds to this point of iteration space.

The idea of the hyperplane method is to divide the set of iteration space points into subsets, placed on certain parallel hyperplanes. Herewith, the order of computation of iteration space points is transformed to new nested loop. In the new nested loop, the outer loop computes hyperplanes, and inner loops compute points of each hyperplane. It is assumed that loops computing hyperplane points are computed parallel.

3 Using the Hyperplane Method for Temporal Data Locality in Iterative Algorithm

The hyperplane method can accelerate even without parallelization by improving temporal data locality. We will present it on the Gauss-Seidel iterative method for solving the Laplace's equation of Dirichlet problem.

The Gauss-Seidel method consists in computing element u(i,j) of the array, that is the original function, at numerical solving by the formula:

```
for (k=L1; k<=R1; ++k)
    for (i=L2; i<=R2; ++i)
        for (j=Ln; j<=Rn; ++j)
            u(i, j) = ( u(i+1, j)+ u(i-1, j)+ u(i, j+1)+ u(i, j-1) ) / 4
```

The classic hyperplane method is applied to set of two innermost loops. It leads to deceleration on modern computers, even at parallelization without using transformation to blocking computations [14]. In a case of an iterative algorithm for solving two-dimensional Dirichlet problem, the iteration space is three-dimensional. Information dependencies in this iteration space are parallel to the coordinate axes. The new nested loop has the following form.

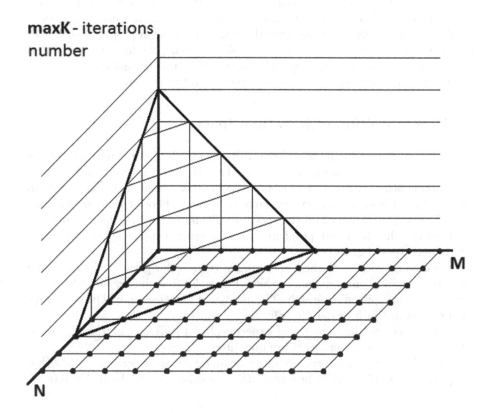

Fig. 1. Hyperplane, that is orthogonal to the vector (1,1,1)

```
D = N + M - 1;
P = D + maxK - 1;
for (p = 1; p <= P; ++p) {
    L = min( min(D,maxK) , min(p , P - p + 1) );
    for (l = L; l > 0; --l) {
        N1 = max(p-maxK, 0) + 1;
        R = min(min(D - N1 + 1, N1), min(N,M));
        i = N1 - min(N1,M);
        j = min(N1,M) + 1;
        for (r = 1; r <= R; ++r) {
```

```
        ++i;
        --j;
        x[i][j] = ( x[i-1][j] + x[i+1][j] + x[i][j-1] + x[i][j+1] ) / 4.0;
      }
    }
}
```

The idea of the method described in this section is to begin computations of the next iteration without finishing of computation of the current iteration. It allows to place computations with the same data near in program, that is to improve temporal data locality. It is equivalent to using the hyperplane method to the nest of all three loops. Herewith, the iteration space is partitioned to set of points, placed on hyperplanes, that are orthogonal to the vector (1,1,1) (Fig. 1).

4   Decomposition of Iteration Space

It was found that at small sizes of the width or length of the matrix we get the gain in execution speed for hyperplane method. However, the standard order of execution is faster for large sizes of both the matrix parameters. This is due to the fact that if both parameters are large then elements of hyperplane diagonal don't fit in cache memory and the speed of calculation falls on this diagonal.

To solve the described problem iteration space can be split into three-dimensional tiles (Fig. 2).

It should be noted that these tiles are not rectangular parallelepipeds. These tiles are straight prisms lying on the side. For each three-dimensional tile hyperplane method is applied separately. The number of iterations should be no greater than the prism width of the tile (Fig. 2).

There were executed numerical experiments for different widths of prisms. The results of these experiments are presented in Table 1.

Numerical experiments were executed on an Intel Core i7 processor - 4700HQ; Frequency: 2400 MHz; L3 cache size: 6 MB. The dimension of the grid is 10000 *

Table 1. Dependence of the speedup of computations on the width of tiles (prisms)

Prism width	8 iterations	Speedup	16 iterations	Speedup	32 iterations	Speedup
20	1.7 s	2.53	3.2 s	2.66	-	-
30	1.6 s	2.65	3.2 s	2.66	-	-
40	1.6 s	2.65	3.1 s	2.74	6.1 s	2.79
50	1.7 s	2.53	3.1 s	2.74	6.2 s	2.76
80	1.8 s	2.43	3.3 s	2.58	6.7 s	2.55
200	1.8 s	2.43	3.5 s	2.42	6.9 s	2.47
300	2.1 s	2.09	4.3 s	1.97	8.7 s	1.97
900	2.4 s	1.80	5.1 s	1.66	10.2 s	1.68

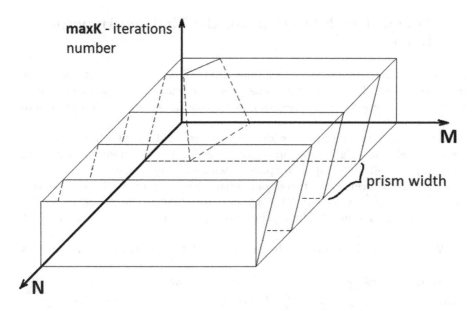

Fig. 2. Decomposition of iteration space into prisms

10000. The width of the strip is measured by the number of grid points, the time is measured in milliseconds. Hyperplanes in each three-dimensional figure are considered orthogonal to the vector $(1, 1, 1)$. The normal calculation time for 8 iterations is 4.3 s, for 16 iterations – 8.5 s, for 32 iterations – 17.1 s.

It can be seen from the table that this method gives an acceleration of about 2.6 times. The presented approach can be applied to iterative algorithms for the numerical solution of other problems with other schemes in the basis (seven-point templates, etc.).

5   Combining Parallelization and Usage of the Hyperplane Method for Temporary Data Localization in Iterative Algorithms

The next step is to increase speedup obtained by data localization by speedup obtained by parallelization. The points of the iteration space lying on different long figures can not always be executed in parallel. To apply parallel computations tiles considered in the previous section are divided into smaller pieces. For parallelization on a cluster, the standard block allocation is not suitable since such parallelization will allow using hyperplane method for only one step of the iterative process. This is due to the fact that the tiles used (Fig. 2) are not rectangular parallelepipeds. If we want to execute several steps of modified hyperplane method independently, then we need to duplicate elements lying near the boundaries of the tile in neighboring processor elements. Therefore, placement with overlaps [15,16] may be appropriate here.

6   Placement with Overlaps for the Modified Hyperplane Method

Placement with overlaps is a technique that allows reducing interprocessor exchanges. This placement allows interprocessor transfers not after each iteration, but after every k iterations. It assumes that distributed data divided into tiles.

Placement with overlaps provides the storage of data lying near the border of tile also in neighboring processors (Fig. 3). This technique can give an acceleration of 30–40% [15] in comparison with standard placement.

Consider the grid of the two-dimensional Dirichlet problem for the Laplace equation. We map processor topology into two-dimensional topology, so the processor rank is represented by two indexes (rank_p, rank_q), $0 <= rank\_p < p, 0 <= rank\_q < q$.

We will place the nodes of the original grid to the processor elements as follows:

$$u(i,j) \in (rank\_p, rank\_q), i = [max(\frac{N}{p} \cdot (rank\_p - 1) - m, 0), min(\frac{N}{p} \cdot (rank\_p) + m, N)), j = [max(\frac{N}{q} \cdot (rank\_q - 1) - m, 0), min(\frac{N}{q} \cdot (rank\_q) + m, N)),$$

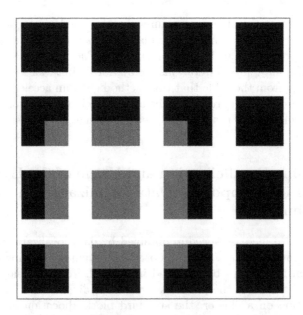

Fig. 3. The set of elements of a two-dimensional array is distributed on 16 nodes of a supercomputer. White lines divide the array into disjoint regions, each of which belongs to its cluster node in a standard location. Gray color shaded part of the array belonging to one of the nodes of the supercomputer when placed with overlapping. Other cluster nodes have similar parts of the array. Some parts of the array belong to only one cluster node, some to two, some to four.

where N is the dimension of the grid and m is the overlap size.

Then represent each tile in three-dimensional space, postponing a new dimension – the iterations of the Gauss-Seidel method. For each tile we start the hyperplane method for the number of iterations equal to m. On iterations multiples of m we make interprocessor exchanges.

Numerical experiments (Table 2) were executed on the cluster «Angara-K1» [17]. Note that without a use of overlaps the algorithm is processed in parallel extremely inefficiently and works slower than sequential.

Table 2. Results of parallelizing Gauss-Seidel method using hyperplane method and placement with overlaps (the dimension of the grid = 10000 * 10000, count of iterations = 32, the overlap size = 8)

Method	Processors count	Time (sec)	Speedup
Standard method	1	25.2	-
Modified hyperplane method	1	10.2	2.47
Parallel hyperplane method	2	6.3	3.98
	4	3.5	7.3
	8	1.7	14.5
	16	0.9	28.1

7   Conclusion

We showed that the modified hyperplane method can improve temporal data locality. It allowed to speed up Gauss-Seidel method in 2.5 times. However, straightforward parallelization for the modified hyperplane method is not suitable due to data dependencies between elements of different tiles. Therefore, we applied the technique of data placement with overlaps.

Through combining of data placement with overlaps and modified hyperplane method we got speedup that bigger than linear in 1.8 times (Table 2) in comparison with the standard algorithm.

References

1. Lamport, L.: The parallel execution of DO loops. Commun. ACM **17**(2), 83–93 (1974)
2. Fernandez, A., Llaberia, J.M., Valero-Garcia, M.: Loop transformation using nonunimodal matrices. IEEE Trans. Parallel Distrib. Syst. **6**(8), 832–840 (1995)
3. Kowarschik, M., Weiß, C.: An overview of cache optimization techniques and cache-aware numerical algorithms. In: Meyer, U., Sanders, P., Sibeyn, J. (eds.) Algorithms for Memory Hierarchies. LNCS, vol. 2625, pp. 213–232. Springer, Heidelberg (2003). doi:10.1007/3-540-36574-5_10

4. Wolfe, M.: More iteration space tiling. In: Proceedings of the 1989 ACM/IEEE conference on Supercomputing (Supercomputing 1989), pp. 655–664. ACM, New York (1989). https://doi.org/10.1145/76263.76337

5. Bastoul, C., Feautrier, P.: Improving data locality by chunking. In: Hedin, G. (ed.) CC 2003. LNCS, vol. 2622, pp. 320–334. Springer, Heidelberg (2003). doi:10.1007/3-540-36579-6_23

6. Likhoded, N.A.: Generalized tiling. Doklady NAN Belarusi, T. 55, N. 1, pp. 16–21 (2011). (in Russian)

7. Yurushkin, M.V.: Double block placement of data in RAM for solving the problem of matrix multiplication. Programmnaya inzheneriya, pp. 132–139 (2016). (in Russian)

8. Gervich, L.R., Steinberg, B.Y., Yurushkin, M.V.: Development of parallel programs with optimizing the use of memory structures. 120 p. Southern Federal University, Rostov-on-Don (2014). (in Russian)

9. Lam, S.M.: A data locality optimizing algorithm. In: Proceedings of the ACM SIGPLAN 1991 Conference on Programming Language Design and Implementation, pp. 30–44. ACM, New York (1991). ISBN:0-89791-428-7

10. Goto, K.: Anatomy of high-performance matrix multiplication. ACM Trans. Math. Softw. **34**(3), 1–25 (2008)

11. Perepelkina, A.Y., Levchenko, V.D.: DiamondTorre algorithm for high-performance wave modeling. Keldysh Institute preprints, vol. 018, 20 p. (2015)

12. Graham, S.L., Snir, M., Patterson, C.A.: Getting Up To Speed: The Future Of Supercomputing, p. 289. National Academies Press, Washington (2005)

13. Abu-Khalil, J., Guda, S., Steinberg, B.: Porting Parallel Programs Without Loss of Efficiency. Open Syst. DBMS J. 23(4) (2015)

14. Steinberg, B.J., Abu-Khalil, J.M., Adigeyev, M.G., Bout, A.A., Kermanov, A.V., Pshenichnyy, E.A., Ramanchauskayte, G.V., Kroshkina, A.P., Gutnikov, A.V., Ponomareva, N.S., Panich, A.E., Shkurat, T.P.: A package of fast tools for genomic sequence analysis. Int. J. Math. Models Methods Appl. Sci. **10**, 42–50 (2016). ISSN:1998-0140

15. Gervich, L.R., Kravchenko, E.N., Steinberg, B.Y., Yurushkin, M.V.: Automatic program parallelization with block data distribution. Sib. Zh. Vychisl. Mat. **18**(1), 41–53 (2015)

16. Gervich, L.R., Steinberg, B.Y., Yurushkin, M.V.: ExaScale Systems Programming. Open Syst. J. 21(8) (2013)

17. Simonov, A.S.: High-speed Angara network: opportunities and prospects. PaVT 2016. http://omega.sp.susu.ru/books/conference/PaVT2016/talks/Simonov.pdf. (in Russian)

Defining Order of Execution in Aspect Programming Language

Sergey Arykov[(⊠)]

Institute of Computational Mathematics and Mathematical Geophysics,
Siberian Branch of Russian Academy of Sciences, Novosibirsk, Russia
arykov@sscc.ru

Abstract. A fragmented approach to parallel programming and its implementation in the Aspect programming language are considered. Approach to define order of execution of computation fragments in Aspect language is described and illustrated by the example of matrix LU decomposition task.

Keywords: Parallel programming · Technology of fragmented programming · Aspect programming language · Control schemes

1 Introduction

Attempts to increase the level of parallel programming, shifting most of the technical problems to the programming system, are constantly being made both in Russia and abroad. However, despite the abundance of projects, MPI and OpenMP (together with specialized frameworks for specific architectures) remain the main tools for developing effective applied parallel programs, and the problem of simple parallel programs development and automatic use of new hardware capabilities is not solved, which is because of a very high complexity of the task of increasing the level of parallel programming.

The goal of the research is to develop a parallel programming system in which it is possible to smoothly change the degree of nonprocedural representation of algorithms through the explicit definition of information and control dependencies and provide the user with higher level of programming environment.

2 A Fragmented Approach to Parallel Programming

The essence of the fragmented approach [1] is to represent an algorithm and its implementing program as a set of *data fragments* and *code fragments*. In the course of execution, the fragmented structure of a program is kept.

Each code fragment is supplied with a set of *input data fragments* (formal parameters) used to compute *output data fragments*. The substitution of data fragments as parameters into a code fragment is referred to as *applying* a code fragment to data fragments (the same code fragment may be applied to different data fragments). The code fragment with its input and output data fragments constitutes a *computation*

© Springer International Publishing AG 2017
V. Malyshkin (Ed.): PaCT 2017, LNCS 10421, pp. 265–271, 2017.
DOI: 10.1007/978-3-319-62932-2_25

fragment. On the set of computation fragments, a partial order (*control scheme*) is defined. The resulting program is created from such computation fragments, with fragmentation of the program kept during the program execution.

Execution of a fragmented program is the execution of computation fragments in any order that does not contradict to the defined control scheme. Each computation fragment receives its resources during setting on execution, creates a new process of the program and can migrate from one processor to another.

Consider an example of LU decomposition algorithm. The source matrix (a matrix that should be factorized) is built out of the data fragments (sub-matrixes) and four code fragments. The first code fragment will process the data located on the main diagonal; the second code fragment – the data fragments to the right of the main diagonal; the third code fragment – the data fragments below the main diagonal and the fourth code fragment will calculate the rest data fragments.

Computations are performed through iterations. The first iteration is shown on Fig. 1(left matrix). Each block represents data fragment and marked inside with the name of computation fragment that will process that data fragment. Arrows indicate dependencies between computation fragments.

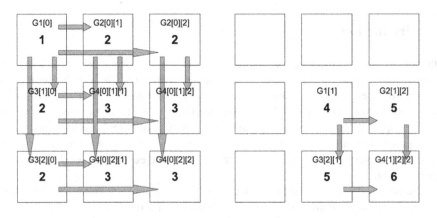

Fig. 1. Fragmentation of the matrix LU decomposition task

On the first iteration, the data fragment with number 1 will be computed. After that, the data fragments to the right and below with number 2 can be computed simultaneously. Finally, the internal matrix (data fragments with number 3) should be recomputed. The next iteration will be applied only to the internal matrix recursively, as shown on Fig. 1(right matrix).

It is important to notice that there is no necessity to wait until all computation fragments on current iteration will be finished before moving to the next iteration. For example, as soon as G4[0][1][1] is finished (iteration 1), it's possible to start execution of G1[1] (iteration 2). In other words, the fragmented algorithm of LU decomposition has a high degree of nonprocedurality.

3 Declarative Language Aspect

3.1 Language Fundamentals

The Aspect language [2, 3] allows representing algorithms in fragmented form with high degree of nonprocedurality by defining dependencies between different computation fragments in declarative form. Computation fragments that do not have explicitly specified dependencies are considered independent. Defining dependencies between computation fragments is the main purpose of the language, it does not define any computations – an imperative language (like C or C++) can be used to do that.

The Aspect language was created for numeric computation and therefore it is focused on regular data structures and its main data type is array. Another key peculiarity of the language is an ability to make several assignments to the same variable (partial distribution of resources).

A typical Aspect program consists from the following sections: *program* – defines name of the (sub)program; *preface* – constants declaration; *data fragments* – data fragments declaration; *code fragments* – code fragments declaration; *task data* – assemble program data from data fragments; *task computation* – defining computation fragments by applying code fragments to data fragments; *task control* – defining dependencies between computation fragments (and thus order of execution), i.e. control scheme.

A detailed description of Aspect programming language is out of the scope of this paper, but the meaning of each section reflects fragmented approach directly and self-illustrating in the example of matrix LU decomposition program (see text below).

3.2 Defining Control Scheme in Aspect Language

The most interesting thing in Aspect language is how the control scheme is defined, i.e. "task control" section. The construction of control scheme is based on two key features: strict partial order and explicit support of massive control schemes. By default, all computation fragments can be executed simultaneously. That can be changed by defining dependencies between computation fragments in task control section of Aspect program (static control scheme).

The task control section consists from lines divided by semicolon. Each line defines dependencies using the following syntax:

$$ \text{id1}\{[\text{expression}]\} < \text{id2}\{[\text{expression}]\}, \{\text{id3}\{[\text{expression}]\}, \ldots\} $$

where *id1* is the name of computation fragment that should be executed before computation fragments with names *id2*, *id3*, etc. and symbol '<' reflects strict order relation. For example,

```
task control
  A < B, C;
```

means the execution of computation fragment *A* should be completed before computation fragments *B* or *C* can start execution. Symbol '<' can be used to define dependencies of two types (Aspect does not distinguish these types of dependencies from each other):

- data dependency, when 'A < B' means A computes some data required by B to proceed;
- control dependency, when 'A < B' means B doesn't need data from A, but have to be executed after A anyway because of other reason (resource allocation, performance optimization, etc.).

Expressions are used to define massive control scheme. The format of allowed expression is <index name> {+ <const>}. For example,

```
task control
    S1[i] < S2[i];
```

means each i-th computation fragment of code fragment S2 can be executed only after i-th computation fragment of code fragment S1 has finished execution, and

```
task control
    S[p][q] < S[p][q+1];
    S[p][q] < S[p+1][q];
```

means each (p,q) computation fragment S depends from its left and up neighbor.

By default, the massive control scheme applies to all range of definition of the index where both computation fragments in '<' relation are defined. That can be changed using keyword *where* followed by logical expression for the index. In that case, massive control scheme will be applicable only to those index values where logical expression is true, e.g.

```
task control
    S1[i] < S2[i+1] where {i%2 == 0};
```

will apply only to even-valued index numbers.

Aspect language also permits to use logical operators to construct massive control scheme. In the example below

```
task control
    (S1[i] & S2[i]) < S3[i];
```

i-th computation fragment S3 can be executed only when both S1[i] and S2[i] have been finished. Similarly, in the control scheme

```
task control
    (S1[i] | S2[i]) < S3[i];
```

i-th computation fragment S3 can be executed when either S1[i] or S2[i] or both have been finished.

3.3 Example of Aspect Program

The main properties of Aspect language will be illustrated on task of matrix LU decomposition. The solution of that task is shown below.

```
program LUDecomposition
preface {
  const int m = 3; const int n = 4;
};
data fragments
  double Matrix[m][m];
code fragments
  F1(in Matrix A; out Matrix B) {
    for(int i=0; i<m; ++i)
      for(int j=i+1; j<m; ++j) {
        B[j][i] = A[j][i]/A[i][i];
        for(int k=i+1; k<m; ++k)
          B[j][k] = A[j][k] - A[j][i]*A[i][k];
      }
  };
  F2(in Matrix A, Matrix B; out Matrix C) {...};
  F3(in Matrix A, Matrix B; out Matrix C) {...};
  F4(in Matrix A, Matrix B, Matrix C; out Matrix D){...};
task data
  Matrix A[n][n];
task computations
  G1[i]: F1(A[i][i], A[i][i]) where i: 0..n-1 priority 0;
  G2[i][j]: F2(A[i][i], A[i][j], A[i][j])
    where i: 0..n-2, j: i+1..n-1;
  G3[i][j]: F3(A[j][j], A[i][j], A[i][j])
    where i: j+1..n-1, j: 0..n-2;
  G4[k][i][j]: F4(A[i][k], A[k][j], A[i][j], A[i][j])
    where k: 0..n-1, i: k+1..n-1, j: k+1..n-1;
task control
  G1[i] < G2[i][], G3[][i];
  G2[i][j] < G4[i][][j];
  G3[i][j] < G4[j][i][];
  G4[i-1][i][i] < G1[i];
  G4[i-1][i][j] < G2[i][j];
  G4[j-1][i][j] < G3[i][j];
  G4[k][i][j] < G4[k+1][i][j];
end
```

The meaning of most constructions of the program is obvious. Computations inside the code fragments *F1–F4* are defined using C++ language. In the section *task computations* application of the code fragments *F1–F4* to the task data are defined. Each of the indices i, j, k goes through all values in the range defined after the index name; each combination of (i, j) or (k, i, j) creates a separate computation fragment. In the section *task control,* order of execution of different computation fragments (control scheme) is defined according to syntax described in Sect. 3.2.

Program code for code fragments *F2*, *F3* and *F4* was omitted to reduce the length of the program. It is similar to code of *F1* with minor changes.

4   Results of Experiments

As a test platform, the computer with the following configuration was used: HP Pro-Liant DL580 G5 (4 Intel Xeon X7350/256 GB RAM/Cent OS 5.3 64 bit/Intel C++ Compiler).

Test task is matrix LU decomposition. The size of the matrix is 5040×5040; all entries are real numbers with double precision. The results are shown in Table 1 (the time of computing is given in seconds).

Table 1. LU decomposition program performance

Approach/Number of cores	1	2	4	8	16
Non-fragmented (C++/OpenMP)	267,14	139,10	90,53	67,27	67,29
Fragmented (Aspect)	31,01	15,6	7,93	4,13	2,35

The performance and scalability of fragmented approach is much better than the same parameters of non-fragmented approach. That is because effective use of cache memory and high degree of nonprocedurality.

5   Related Works

The closest work to Aspect is the Bars language [4]. Like Aspect, the Bars has advanced capabilities of constructing control schemes, it separates definition of control scheme from computations and allows writing expressions in a non-procedural form. Massive operations can be implemented by infiltrating a unary operation into the data of arbitrary structure (for example, arrays), and massive control schemes – by binary infiltrate of C-formulas, but the mechanism for the formation of computational structures for such infiltration is not worked out. Currently, the project is not developing; a working version for modern supercomputers is also missing.

Another attempt to create a fragmented programming system is LuNA project [5, 6], which is very high level implementation of fragmented approach to parallel programming for distributed memory systems.

A large body of associated research is carried out in the field of producing high-performance libraries for linear algebra, where the blocked algorithms that are friendly to the cash-memory of processors are developed. Plasma project [7, 8] uses the same fragmented programming approach, but introduce another term – "tiled" algorithms. While the concept is the same, they do not provide full value programming system, but program each "tiled" algorithm manually, because their goal is to create a new high-performance successor of LAPACK.

6 Conclusion

The research of control schemes in numeric computations programs has important practical application. When it becomes clear how many control schemes are exist in that area and what is their essence, it will be possible to improve parallel programming to a higher-level using high-quality control schemes implementation embedded in parallel programming systems. Aspect programming system is a ready-made instrument for such type of research.

References

1. Kireev, S., Malyshkin, V.: Fragmentation of numerical algorithms for parallel subroutines library. J. Supercomput. **57**, 161–171 (2011). doi:10.1007/s11227-010-0385-3
2. Arykov, S., Malyshkin, V.: Asynchronous language and system of numerical algorithms fragmented programming. In: Malyshkin, V. (ed.) PaCT 2009. LNCS, vol. 5698, pp. 1–7. Springer, Heidelberg (2009). doi:10.1007/978-3-642-03275-2_1
3. Arykov, S.: Asynchronous model of computation controlled by strict partial order. In: 10th Annual International Scientific Conference on Parallel Computing Technologies (PCT-2016). CEUR Workshop Proceedings, vol. 1576, pp. 54–67. CEUR-WS (2016)
4. Bystrov, A., Dudorov, N., Kotov, V.: About the core language. In: Languages and Programming Systems (in Russian), pp. 85–106. CC SB AS USSR, Novosibirsk (1979)
5. Malyshkin, V., Perepelkin, V.: The PIC implementation in LuNA system of fragmented programming. J. Supercomput. **69**, 89–97 (2014). doi:10.1007/s11227-014-1216-8
6. Malyshkin, V., Perepelkin, V., Schukin, G.: Scalable distributed data allocation in LuNA fragmented programming system. J. Supercomput. **73**, 726–732 (2017). doi:10.1007/s11227-016-1781-0
7. Buttari, A., Langou, J., Kurzak, J., Dongarra, J.: A class of parallel tiled linear algebra algorithms for multicore architectures. Parallel Comput. **35**, 38–53 (2009). doi:10.1016/j.parco.2008.10.002. Elsevier, Amsterdam
8. YarKhan, A., Kurzak, J., Luszczek, P., Dongarra, J.: Porting the PLASMA numerical library to the OpenMP standard. Int. J. Parallel Prog. **45**, 1–22 (2016). doi:10.1007/s10766-016-0441-6

Automated GPU Support in LuNA Fragmented Programming System

Belyaev Nikolay[1] and Vladislav Perepelkin[1,2(✉)]

[1] Institute of Computational Mathematics and Mathematical Geophysics SB RAS,
Novosibirsk, Russia
bl0ckzer01@gmail.com, perepelkin@ssd.sscc.ru
[2] National Research University of Novosibirsk, Novosibirsk, Russia

Abstract. The paper is devoted to the problem of reduction of complexity of development of numerical parallel programs for distributed memory computers with hybrid (CPU+GPU) computing nodes. The basic idea is to employ a high-level representation of an application algorithm to allow its automated execution on multicomputers with hybrid nodes without a programmer having to do low-level programming. LuNA is a programming system for numerical algorithms, which implements the idea, but only for CPU. In the paper we propose a LuNA language extension, as well as necessary run-time algorithms to support GPU utilization. For that a user only has to provide a limited number of computational GPU procedures using CUDA, while the system will take care of such associated low-level problems, as jobs scheduling, CPU-GPU data transfer, network communications and others. The algorithms developed and implemented take advantage of concerning informational dependencies of an application and support automated tuning to available hardware configuration and application input data.

Keywords: Hybrid multicomputers · GPGPU · Parallel programming automation · Fragmented programming · LuNA system

1 Introduction

When implementing large-scale numerical models on a supercomputer one can significantly improve performance by utilizing both CPUs and GPUs available. Unfortunately, development of such a program is often problematic due to necessity to distribute computational load between CPUs and GPUs, organize data transfer and computations' synchronization. The distribution depends on relative performance of CPUs and GPUs, RAM available, network topology and other architectural peculiarities of given hardware. Implementation of such a distribution is usually troublesome and requires skills in system parallel programming, thus impeding numerical programs development.

Despite the fact that such system programming skills are not expected from application programmers, their involvement is still necessary, because efficient workload distribution problem is far from being solved in general case. In particular, it requires an understanding of application's data and computations structure and sometimes even

© Springer International Publishing AG 2017
V. Malyshkin (Ed.): PaCT 2017, LNCS 10421, pp. 272–277, 2017.
DOI: 10.1007/978-3-319-62932-2_26

understanding of peculiarities of the numerical model implemented (see [1] for an example).

Automation of construction of numerical parallel programs, which efficiently utilize available hardware, is a powerful way to hide the data distribution programming problem from application programmers, thus simplifying numerical programs development. Nowadays there are different systems and tools, aimed at simplifying GPU utilization.

OpenCL [2], for example, is an open standard and a library to support "kernel" development, which can be executed on CPU, GPU or FPGA. OpenCL employs a C-like language to define a kernel. Computational device is selected automatically, based on static analysis and profiling [3]. OpenCL is still a low-level programming tool, where a programmer has to program control manually. OpenCL also does not concern data locality of the application.

OpenACC [4] offers compiler directives to denote "GPU parts" and an API (Application Programmer Interface) to invoke them or transfer data. OpenACC does not concern application data locality, does not balance workload and only supports shared memory systems. DVMH [5] is similar to OpenACC, it allows tuning workload distribution for hybrid multicomputers, but does not provide dynamic load balancing.

Charm++ [6] is a platform-independent programming system with a compiler and a run-time system. Charm++ program consists of "chares", which can execute simultaneously and interact with each other. A chare can be assigned to GPU or CPU by a run-time system depending on the strategy, chosen by a programmer.

It can be concluded, that different systems provide some automation of GPU usage, but either for a particular case, or at cost of a significant involvement of the programmer. This is caused by peculiarities of models these systems employ.

A programming system LuNA [7] is being developed in Institute of Computational Mathematics and Mathematical Geophysics SB RAS. LuNA is aimed at automation of numerical parallel programs construction and consists of LuNA language, compiler and a run-time system. LuNA system was chosen for this work, because it is designed for automation of tuning program to hardware resources, which makes is useful to examine algorithms of CPU-GPU load distribution algorithms. This paper is devoted to an attempt to provide automated GPU support for LuNA system.

2 LuNA-Program

In LuNA an application program is represented as a set of computational fragments (CF) and a set of data fragments (DF). Each DF is an aggregated immutable piece of data (say, a subdomain of a numerical mesh at given time step or iteration). Each CF is an operation on DFs, which takes a number of DFs as inputs and produces values of a number of output DFs. Each CF is implemented by a conventional sequential procedure without "side-effects". LuNA-program consists of two parts: a number of sequential procedures in C++ and a description of sets of CFs and DFs in LuNA language. LuNA compiler translates programs into an internal representation, executable by LuNA run-time system.

3   CPU and GPU Workload Distribution Algorithm with Automatic Data Refragmentation

The problem of workload distribution is formulated as follows. For each CF a device (CPU or GPU) must be assigned to be executed on. The goal is to reduce overall application execution time, mainly by providing load balance of available devices and by saving CPU-GPU data transfer "bottleneck."

The proposed algorithm is based on the Rope-of Beads [8] (RoB) algorithm, employed in LuNA system. In the RoB algorithm each CF and DF has a number n assigned ($0 \leq n < L$), where L is a parameter of the algorithm. Number n is called coordinate on the $[0; L)$ segment. More than one fragment can share the same coordinate. The segment $[0; L)$ is split into a number of sub-segments, one sub-segment for each computational node. All the fragments, mapped to a sub-segment of a node, are considered to be assigned to the node. Dynamic load balancing is possible through resplitting the segment, causing CFs and DFs to migrate if their assignment has changed.

In the proposed algorithm an additional split within one node is proposed. A subsegment of a node is split into three new parts. The fist part corresponds to CPU(s) of the node (the "CPU part"), the last part corresponds to GPU (the "GPU part"), and the middle part corresponds to fragments, stored in CPU memory, but executed on GPU (the "drag-through part"). For the drag-through part once a CF has to be executed, its input data are copied to GPU, then the CF is executed, and its output DFs are transferred to CPU, releasing occupied GPU memory (this allows running on GPU more CFs than fit in its memory). DFs of GPU and CPU parts never leave their devices in order to optimize CPU-GPU connection usage.

The reasoning behind this splitting a sub-segment into three parts is the following. To save GPU-CPU traffic, each of the devices should have its part of computations (CPU and GPU parts). In order to achieve load balance GPU may require more workload, than its memory can hold. To handle this case the drag-through part of the sub-segment is introduced. The drag-through workload occupies CPU memory, but is transferred in smaller portions to GPU for computations (trading off CPU-GPU bandwidth against CPU or GPU idle time). Although it is unclear, what proportions of parts 1, 2 and 3 would be the best for certain application and hardware, the optimum can be searched for (in particular, one or two parts can degenerate).

To reduce run-time system overhead, bound with number of fragments, static data refragmentation is suggested to be combined with the proposed algorithm. The refragmentation is performed as follows: All the GPU fragments (GPU part) are merged into one, all the CPU fragments (CPU part) are resplit into a number of fragments equal (or proportional) to the number of CPU cores, and the drag-through part is resplit into a number of fragments (portion size), which is a parameter of the proposed algorithm. Such refragmentation requires, that involved C++ procedures fit the "merge requirements", i.e. processed domain size must be a parameter of the procedures, which is annotated in code by the programmer.

The proposed algorithm concerns informational dependencies and data structure of the application algorithm (this is inherited from the original RoB algorithm). It can be tuned to properties of an application and hardware configuration using the parameters of the proposed algorithm. The parameters can be defined automatically on the basis of static analysis, hardware benchmarking and/or application profiling, but this is out of scope of the paper. The drawback of the proposed algorithm is the 1D refragmentable decomposition requirement.

4   Testing

The proposed algorithm was implemented as a part of LuNA programming system. To study performance characteristics of the algorithm a number of tests was performed. All the tests were conducted on single computing node with 2 × Xeon 5670 (3 GHz) CPUs and GPU Nvidia Tesla M 2090. The application tested is a model finite scheme solver, where the number of computations per single data unit is a parameter. This parameter (called *load*) is used to represent different application classes with different volume of computations per data, which is one of the key properties of an algorithm.

The first test is devoted to finding an optimal CPU/GPU workload proportion (CPU workload percentage is the X axis). The drag-through parameter is degenerated to zero. It can be seen in Fig. 1, the optimal time is achieved when both CPU and GPU are used, despite the fact, that such execution requires extra CPU-GPU communications, as compared to CPU-only or GPU-only execution. Note, that the optimal proportion is different for different *load* value.

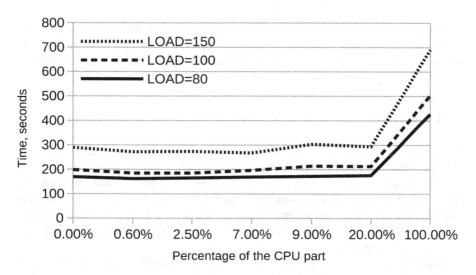

Fig. 1. Program execution time dependency on the amount of computations, assigned to CPU, for different computation-per-data intensity

The second test is devoted to obtaining optimal value of the drag-through parameter. The X axis corresponds to different value of the parameter. It can be seen from Fig. 2 that optimal drag-through parameter is non-zero, which is an evidence of usefulness of the dragging-through part of the proposed algorithm. It also can be seen, that the optimal value of the parameter depends on the *load* parameter. It means, that different applications would require different value of the drag-through parameter.

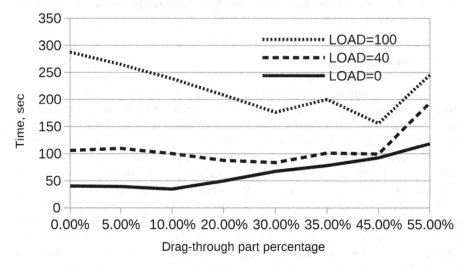

Fig. 2. Program execution time dependency on amount of the "middle-part" DFs for different computation-per-data intensity.

It is worth mentioning, that during the testing the absolute performance achieved is close to that of manually developed programs, which means that the conclusions made are essential to the proposed algorithm, and are not significantly affected by foreign factors, such as LuNA run-time system overhead.

5   Conclusion

An algorithm to distribute workload to CPUs and GPUs of a multicomputer is proposed. The algorithm possesses parameters, capable of tuning to application and hardware peculiarities to reduce program execution time. The algorithm was implemented as a part of LuNA system and performance tests were performed. The tests showed that the algorithm proposed allows automated efficient usage of hybrid (GPU+CPU) computing nodes of a multicomputer. The tests also showed, that the parameters of the proposed algorithm are essential.

Future work supposes solution of the problem of automatic (or at least automated) definition of parameters of the algorithm to allow LuNA tune to given hardware configuration and application peculiarities automatically.

References

1. Kraeva, M.A., Malyshkin, V.E.: Assembly technology for parallel realization of numerical models on mimd-multicomputers. Int. J. Futur. Gener. Comput. Syst. **17**(6), 755–765 (2001). Elsevier Science
2. https://www.khronos.org/opencl/ accessed May 2017
3. Wen, Y., Wang, Z., O'Boyle, M.F.P.: Smart multi-task scheduling for OpenCL programs on CPU/GPU heterogeneous platforms. In: 21st International Conference on High Performance Computing (HiPC), pp. 1–10 (2014)
4. http://www.openacc.org/ accessed May 2017
5. Bakhtin, V.A., Chetverushkin, B.N., Krukov, V.A., Shilnikov, E.V.: Extension of the DVM parallel programming model for clusters with heterogeneous nodes. Doklady Math. **84**(3), 879–881 (2011). Moscow: Pleiades Publishing Ltd
6. http://charm.cs.illinois.edu/research/charm accessed May 2017
7. Malyshkin, V.E., Perepelkin, V.A.: LuNA fragmented programming system, main functions and peculiarities of run-time subsystem. In: Malyshkin, V. (ed.) PaCT 2011. LNCS, vol. 6873, pp. 53–61. Springer, Heidelberg (2011). doi:10.1007/978-3-642-23178-0_5
8. Malyshkin, V.E., Perepelkin, V.A., Schukin, G.A.: Distributed algorithm of data allocation in the fragmented programming system LuNA. In: Malyshkin, V. (ed.) PaCT 2015. LNCS, vol. 9251, pp. 80–85. Springer, Cham (2015). doi:10.1007/978-3-319-21909-7_8

Automation Development Framework of Scalable Scientific Web Applications Based on Subject Domain Knowledge

Igor V. Bychkov, Gennady A. Oparin, Vera G. Bogdanova,
Anton A. Pashinin, and Sergey A. Gorsky[(✉)]

Matrosov Institute for Systems Dynamics and Control Theory,
Siberian Branch of Russian Academy of Sciences, Irkutsk, Russia
{oparin,bvg}@icc.ru, apcrol@gmail.com, gorskysergey@mail.ru

Abstract. Currently high-performance computing technologies using computational capabilities for solving scientific, are actively improving. The purpose of our research is the development of toolkit for construction and execution of scientific service-oriented application in heterogeneous distributed computing environment (HDCE). These tools provide the access for subject domain experts to the high-capacity computing resource, using these resources without extensive knowledge of computing architecture and low-level software, and the parallel execution of the user application on the base of the service-oriented technology and multi-agent control. We describe an architecture and functional capabilities of automated toolkit for the service-oriented application creation based on applied programs package, and multi-agent control of this application parallel running in HDCE. We demonstrate an example of the creation of the web-application for parametric feedback synthesis of linear dynamic object by these tools. The offered technology allows simplifying service creation and provides new qualitative opportunities of controlling parallel high-performance computations.

Keywords: Scalable application · Service · Parametric synthesis of control law

1 Introduction

One of current trends in high performance computing (HPC) is to apply its possibilities for complicated scientific problems solving in different subject domains. Therefore, there is relevant the creation of toolkit intended for program environments creation provided the access to HPC resources for subject domain experts, and using these resources without extensive knowledge of computing architecture and low-level software tools. There is also a tendency of combining the service-oriented approach that provides access to resources via the Internet, and the multi-agent technology for managing these resources.

The computation works automation issues are relevant for many subject domains. This paper discusses the automation of problem solving from researching the dynamic and design of control systems for moving objects, in particular, the problem of structurally parametric synthesis of linear control systems (LCS). Using HPC requires

V. Malyshkin (Ed.): PaCT 2017, LNCS 10421, pp. 278–288, 2017.
DOI: 10.1007/978-3-319-62932-2_27

automatizing complex of scientific and technological works, performed on the level of mathematical models, analytical and numerical methods of its research. The foundation of this complex involves three closely connected conceptions – mathematical models, approach methods of these models and methodology of main research object achievement. The integration of these conceptions constitutes the basis of algorithmic knowledge of computation works. In the process of dynamics research and LCS designing multivariate calculations are of great need. By multivariate calculations, we mean computational experiments, where the structure and parameter values of model, methods, and methodology vary according to research strategy and tactics, and the search of the acceptable solution. In previous decades, new tendency was formed in programing, namely the creation of applied programs packages, representing a set of interrelated programs of functional filling and control tools. These packages provide the solution of certain class problems described with terms of subject domain glossary. The represented research purpose is the creation of toolkit, which automatizes converting an applied program package into scalable service-oriented application, and controls the parallel computation schemes execution of this application in HDCE. In the context of parallel computing process, we consider the scalability as increasing the speedup while saving the stable level of the efficiency with increasing the processors number. Offered toolkit is also horizontally scalable. Integration with additional computational node, activated into HDCE, for using its resources is realized by installing on this node configurable agent-manager, which functional capabilities are examined in the description of the toolkit architecture.

We offer such toolkit, namely High-performing computing Service-Oriented Multi-agent System (HPCSOMAS) Framework, which was developed in accordance with the following requirements:

- Providing automation tools for creating service-oriented applications based on an applied programs package;
- Availability of tools for describing knowledge about the subject domain;
- Providing the access to high-performance resources;
- Organizing multi-agent control of parallel execution in HDCE with distribution of computing resources on the user application level;
- Orientation towards two categories of users, subject domain experts and developers.

This approach suggests developing by these tools specialized multi-agent systems (MAS). MAS agents are represented in form of services and perform decentralized resource management on the base of cooperation. Such approach provides new qualitative opportunities of controlling high-performance computations. The paper explores the HPCSOMAS architecture and the technology of developing HPCSOMAS-based web-application. We described such application for the parametric synthesis of the stabilized feedback law for various classes of linear dynamic objects.

2 Related Data

Currently controlled dynamic systems are actively developing and improving. Despite the large number of existing methods, the problem of analyzing and synthesizing different classes of such systems is still a challenging issue. Particularly, binary dynamic systems (Boolean networks) are of great interest in theory and practice. Stability of control systems is a mandatory requirement for such systems. Synthesis of the control law for such objects is often associated with significant computational difficulties. The use of modern computing tools greatly speeds up the synthesis process, reducing the amount of the researcher routine work by dozens of times. Program tools for solving tasks of controlling complex dynamic systems have been developing for a long time [1]. The system MATLAB [2] is very well known, however this system does not provide an expert with a necessary level of automation of computation work, and is designed mostly for the extensive tests of various algorithms and methods rather than effective solution of complex engineering problems of controller design [3]. Therefore program tools for automated solution of problems referred to above, oriented on different categories of users are being developed along with this system. These tools provide ready-for-use applications for typical project solutions, and possibilities for creating new applications. As an example, we can consider the system, presented in [4]. However, issues, connected with using high-capacity resources for solving complicated problems referred to above and organization of control of their parallel solution in HDCE, still have not been resolved. In the review [5] some classes of problems of linear control theory, the non-convexity and NP-hardness of which create difficulties in the solution search are described. Thus, in the work [6] the NP-hardness of the problem of State Output Feedback (SOF) is shown. A large number of topical problems in contemporary control theory, which require high-capacity resources, are connected with the problem of parametric synthesis of linear controller for linear dynamic objects.

In recent years, problem-oriented environments based on HPC and service-oriented technology are actively developing and improving [7, 8]. These researches concerned with development of workflow systems, performed by Russian [9] and foreign [10–12] scientists. Various aspects of the implementation of management in such systems actively explores [13, 14]. There are successful solutions [15] in the field of multi-agent control of distributed computing, but the orientation of many MAS to operate in a particular HDCE reduces the efficiency of their operation in environments with different characteristics. Currently, developers do not have the necessary high-level tool for the mass creation and application of MAS [16].

The systems for transparent user access to distributed computing resources, and high-level tools for creating services is also actively improving. Thus, for example, in [17] the toolkit for compounding computing resources into a single service and providing users an access to this service through web-application was described. However, this toolkit do not provide tools of automation for designing of scalable web-applications and managing multivariate computations. In [18] the web-service for execution of such computations based on Everest platform are described. However, additional means for description of the execution logic of web-application may be used only by programming way. In the work [8] some challenges unresolved by such systems as grid middleware, scientific workflow systems, web service toolkits, gateways and platforms are discussed, and the Everest platform architecture and its basic components are overviewed. Despite the variety of tools of this kind, a number of questions is still a challenging issue. There are such questions as decentralized control of resource distribution, more preferable in partitioned environments for providing scalability [19], the employment of tools for subject domain specification, and declarative language tools for the execution logic description. Therefore, the development of automation tools integrating creating agents in the form of services, the composition of services with description of execution logic, and multi-agent decentralized control of parallel running in HDCE with the distribution of computing resources at the user application level remains a relevant problem.

3   HPCSOMAS Architecture

HPCSOMAS Framework is intended for developing specialized MAS, designed for organizing control of parallel execution of scalable service-oriented applications in HDCE. Such application consists of two main parts – system part, independent from the solving problem class, and problem-oriented one. The system application part consists of the set of components, which are the agents of hierarchical role-based MAS.

There are several possibilities for creating such application. The application based on an autonomously used program (or several programs) is simply designed as a service (or several services) with an appropriate description and web interface. An execution logic of the application is defined by declarative way with using system services. There are two ways for implementation of web-applications based on a package of interrelated programs. The first one is to create a manual description of the subject domain reflecting there interrelations. Then every program is realized as a service. The other way uses for intellectual packages of application programs, which represent a combination of knowledge and processing tools. In this case, there is possibility to use automated conversion of the package knowledge into HPCSOMAS knowledge base. Using an approach, based on knowledge, constitutes the methodological basis of the automation of building web-applications, based on packages of applied programs. We can distinguish three conceptually isolated layers of knowledge: computational, system and production knowledge, over which problems are formulated [20]. Computational knowledge are libraries of sub-programs, provided with specifications and realizing methods of solving problems in the application subject domain under consideration. System knowledge reflect a set of concepts, which are necessary for describing structural peculiarities and characteristics of blocks of mathematical models and research algorithms. Production knowledge allows, depending on the parameters of the model, to select the most appropriate research algorithms as well as numerical values of control parameters of these algorithms. The conceptual model of the subject domain includes the description of the interaction of the following objects: parameters, modules, operations and productions. Developed web-application consolidates objects of the subject domain, which are correspondent to the family of research methods, similar in terms of parameters and operations. Operations are converted into computational services. The HPCSOMAS Framework includes the following components:

- Class library for creating agents, which is used in programmatic implementation of MAS;
- A basic set of configurable system agents, configuration of which is performed according to characteristics of the specific computing resource;
- Agent Based Class Service Wizard (ABCSW) tools for the automation of the process of creating and configuring agents, and developing a service-based application, based on an applied programs package;
- Language tools for subject domain specification, and a set of converters for translating knowledge from one format to a different another;
- Packages of configurable subject-oriented computing services, represented web-applications running in the HPCSOMAS environment.

HPCSOMAS combines the multi-agent approach to organizing the computational process, described in detail in [21, 22] and service-oriented technologies for its implementation. The MAS, created on HPCSOMAS platform, includes three hierarchical levels of agents – user agent (web-interface), server agent-program (agent-manager), system agents and agents of computational applications (computational services). The first level is the frontend part of the platform. The other two levels are the backend part.

The user agent is designed as a thin-client. JavaScript, html language and jQuery library used for its implementation. This agent allows using the HPCSOMAS based application in the browser from computers and mobile devices. The connection is established through AJAX-queries over HTTP messages, and authentication tools with the help of web-sessions protect the transmitted data for user agents. Access keys used for other types of agents.

The agent-manager and computational services are designed as java-servlets based on the REST architectural style. Every agent-manager represent own computational recourse, and is configured accordingly its characteristics. The agent-manager is the access-point to the computational resources for the user-agent and includes operations with the main objects of the system, such as tasks, computational subtasks and resources. These operations are realized as system web-services. Agent-managers are connected peer-to-peer, and share information (about available services, uses, tasks, and current state of resources) over HTTP messages. To distribute tasks between these agents, a tender model is used, where the lots are tasks, and participants are the agent-managers, representatives of computing resources claiming to perform these tasks. Using the Vickrey auction for this model, which is discussed in detail in [23], allows to achieve consistent stable state of participants of the auction upon the end of the bidding. The formed task (subtask) is transferred to the selected agent-manager, which, in turn, sends it to the corresponding computational agent, which independently running task in interaction with the control system of task scheduling of the computational resource.

Agents of computational applications (computational services) represent user programs, transformed into web-services. Computational web-services, designed for performing a single operation (elementary services) can be combined into a composite service with the help of system web-services, describing its execution logic.

HPCSOMAS Framework differs from composite applications performance control systems [9–11] by following functional capabilities:

- The creation of problem-oriented and system agents in the form of services;
- The declarative description of the composite service execution logic;
- The multiagent control of parallel execution of received application composite services in HDCE;
- The distribution of computational resources on the level of a user-application.

4   ABCSW Computational Services Development Subsystem

In order to provide automation of converting application programs of the user into computational services, new Agent Based Class Service Wizard (ABCSW) have been introduced into HPCSOMAS. This tool includes the editor of computational services and agent-manager. ABCSW intended for describing the service schema, and for configuring computational services, in the form of which problem-oriented and system agents are realized with the help of HPCSOMAS Framework (Fig. 1). ABCSW are designed for programmers as well as subject domain experts. The former are provided with code libraries for programmatically implementation of complex computational services, and ABCSW tools help to install these services to computational resources. The latter have access to ready-made realizations of particular scenarios of using HPCSOMAS tools in converting the user's programs into services. Then uses must configure services with the ABCSW editor without compiling the code.

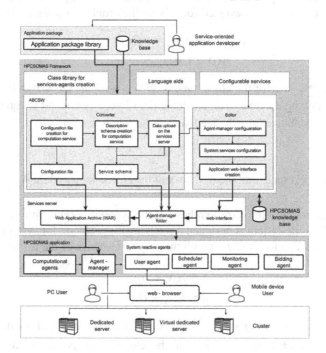

Fig. 1. The structural scheme of developing the service-oriented application in HPCSOMAS environment

5   Service-Oriented PSF Application

The PSF web-application groups services for the parametric synthesis of controller parameters of linear and binary dynamic systems, and constructing stability regions.

The service for the parametric synthesis of the statistical output controller for a continuous linear dynamic system is designed for solving the problem [5]: for the system

$$\dot{x} = Ax + Bu$$
$$y = Cx,$$

(1)

where $x \in R^{n_x}$ – state of the object; $u \in R^{n_u}$ – control; $y \in sR^{n_y}$ – measured output; n_x, n_u and n_y – dimensionalities of vectors x, u and y; A, B, C – matrices of constant coefficients of corresponding dimensionality, figure out whether it is possible to stabilize it by the feedback of the following type

$$u = Ky,$$

(2)

where $K \in R^{n_u \times n_y}$ is such, that the A_c matrix of the closed-loop system $A_c = A + BKC$ is stable. Hereinafter we suppose that interval limitations $\underline{K} \leq K \leq \overline{K}$ are imposed on matrix K (inequalities are considered as componentwise).

The service for the parametric synthesis of the statistical output controller for a linear discrete dynamic system is designed for solving the problem: for the system

$$x(t + 1) = Ax(t) + Bu(t)$$
$$y(t) = Cx(t),$$

(3)

where $x(t) \in R^{n_x}$ – state of the object; $u(t) \in R^{n_u}$ – control vector; $y(t) \in R^{n_y}$ – vector of measured output; n_x, n_u and n_y – dimensionalities of vectors x, u and y, find out whether it is possible to stabilize it by the feedback of the following type

$$u(t) = Ky(t),$$

(4)

where $K \in R^{n_u \times n_y}$ is such, that the A_c matrix of the closed-loop system $A_c = A + BKC$ is stable.

The service for the parametric synthesis of the dynamic output controller for a linear continuous dynamic system is designed for solving the problem: for system (1) find out whether it is possible to stabilize it with the help of feedback of the following type

$$\dot{x} = A_r x_r + B_r y,$$
$$u = C_r x_r + D_r y,$$

where $x_r \in R^{n_{x_r}}$ – controller state, interval limitations are imposed on A_r, B_r, C_r, D_r matrices. With $k \neq 0$ the equation of the closed-loop system is the following: $\dot{x}_c = A_c x_c$, where $x = col(x, x_r)$. Block structure of matrix A_c is shown in [24].

Described above services are based on the parallel algorithm of the directed solution search in the parameter state space, developed by authors [25].

Service for the parametric synthesis of static controller for a binary dynamic system (BDS) is designed to solve the following problem. We consider a linear BDS, where the vector-matrix equation is as follows:

$$x^t = Ax^{t-1} \oplus Bu^{t-1}, \tag{5}$$

where x, u – vector of state and control vector, respectively, $(x \in B^n, u \in B^m, B = \{0,1\}), t \in T = \{1, 2, \ldots, k\}$ – discrete time (measure number), A – $(n \times n)$ binary matrix of state, B – $(n \times m)$ binary input matrix, operations of addition and multiplication are performed by $\mathbf{mod2}$. The problem of static controller synthesis for (5) is in the choice of the control law from the class of inverse linear connections by state as follows

$$u^{t-1} = Px^{t-1},$$

where P – binary matrix of controller parameters of the corresponding order, which provides consistent balance $x = 0$ of the closed-loop system

$$x^t = A_c x^{t-1} = (A \oplus BP)x^{t-1}. \tag{6}$$

Balance position $x = 0$ of autonomous system (6) is considered stable, if for each $x^0 \in B^n$ there is such a moment of time $t \in T$, that trajectory $x(t, x^0)$ for t time steps reaches zero state: $x(t, x^0) = 0$. It is evident, that $x(t, x^0) = 0$ for all following moments of time $t > k$. This service provides the automated specification of the required dynamic property of a closed system on the language of formal logic in the form of a quantified Boolean formula (QBF). The synthesis problem concludes in checking the trueness of QBF (TQBF problem) with the following search of feedback matrix (SAT problem).

Next service group intended for constructing two/three-dimensional (2D/3D) stability region in the space of controller parameters of a closed-loop control system within the given ranges. By varying selected parameters, a numerical grid is constructed. Multivariate calculations are used to determine the stability of the A_c matrix of a closed-loop system in the each point of this grid. Tools for describing the logic of performing services provide the capability to use cycles, logical conditions, and multivariate computations. Computational services based on algorithms providing the natural data parallelism for solving problems referred to above in HDCE.

For example, we find the stabilizing feedback (4) for the system of type (3), matrices of which are defined as follows [26]:

$$A = \begin{pmatrix} 0.7286 & 0.8840 & 0.1568 & 0.3916 & 0.9398 \\ 0.9551 & 0.3472 & 0.4164 & 0.2528 & 0.8328 \\ 0.6564 & 0.0595 & 0.0940 & 0.3544 & 0.4700 \\ 0.7423 & 0.7184 & 0.4499 & 0.7430 & 0.6299 \\ 0.3450 & 0.9582 & 0.8692 & 0.6508 & 0.0582 \end{pmatrix},$$

$$B = \begin{pmatrix} 0.5422 & 0.7869 \\ 0.4557 & 0.6560 \\ 0.8631 & 0.0000 \\ 0.8552 & 0.1312 \\ 0.4723 & 0.4949 \end{pmatrix}, C^T = \begin{pmatrix} 0.0383 & 0.2274 \\ 0.3279 & 0.8995 \\ 0.3137 & 0.2517 \\ 0.4330 & 0.8424 \\ 0.0845 & 0.5082 \end{pmatrix}.$$

with the following limitations for the K feedback matrix:

$$\underline{K} = \begin{pmatrix} -1 & -1 \\ -1 & -1 \end{pmatrix}; \overline{K} = \begin{pmatrix} 1 & 1 \\ 1 & 1 \end{pmatrix}.$$

As a result, we get the following feedback matrix:

$$K = \begin{pmatrix} -0.5 & -0.5 \\ -0.5 & -0.5 \end{pmatrix},$$

which makes the closed-loop system stable. The roots of the characteristic equation of A_c matrix located in the unit circle, and have the following values:

$$\lambda_{1,2} = -0.4751 \pm j0.4397, \lambda_{3,4} = -0.2373 \pm j0.2873 \text{ and } \lambda_5 = -0.5714.$$

The result of constructing 3D region with fixed value of the first parameter K_{11} and variation other parameters in the interval [-1, 1] with the step 0.02 is shown on Fig. 2.

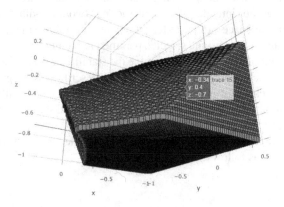

Fig. 2. Stability region in the space of parameters K_{12} (axis x), K_{21} (axis y), K_{22} (axis z)

HPCSOMAS agents (agent-managers) were located on virtual machines of the integrated cluster environment. Computational resources of the supercomputer center [27] of the Matrosov Institute for System Dynamics and Control Theory of SB of RAS were used as HDCE during the experiment.

6 Conclusion

The new automated framework for service-oriented scientific applications and organizing high-performance problem-oriented computations based on multiagent control was developed. The web-application for parametric synthesis of stabilized feedback law for various classes of linear dynamic objects was implemented on the base of this framework. Experimental results [25, 28] show scalability and effectiveness of computations with the help of scientific services, based on the suggested approach.

Acknowledgments. The research was supported by Russian Foundation of Basic Research, projects no. 15-29-07955.

References

1. Somov, Y.I., Oparin, G.A.: Methods and software for computer-aided design of the spacecraft guidance. In: Navigation and Control Systems, MESA, vol. 7, no. 4, CSP, Cambridge, UK, I&S 2016 - Florida, USA, pp. 613–624 (2016)
2. MathWorks: http://www.mathworks.com
3. Aleksandrov, A.G., Isakov, R.V., Mikhailova, L.S.: Structure of the software for computer-aided logical design of automatic control. Autom. Remote Control **66**(4), 664–671 (2005)
4. Aleksandrov, A.G., Mikhailova, L.S., Stepanov, M.F.: GAMMA-3 system and its application. Autom. Remote Control **72**(10), 2023–2030 (2011). doi:10.1134/S0005117911100031
5. Polyak, B.T., Shcherbakov, P.S.: Hard problems in linear control theory: possible approaches to solution. Autom. Remote Control **66**(5), 681–718 (2005). doi:10.1007/s10513-005-0115-0
6. Nemirovskii, A.A.: Several NP-hard problems arising in robust stability analysis. Math. Control Signals Syst. **6**, 99–105 (1993)
7. Kovalchuk, S.V., Smirnov, P.A., Knyazkov, K.V., Zagarskikh, A.S., Boukhanovsky, A.V.: Knowledge-based expressive technologies within cloud computing environments. In: Wen, Z., Li, T. (eds.) Practical Applications of Intelligent Systems. AISC, vol. 279, pp. 1–11. Springer, Heidelberg (2014). doi:10.1007/978-3-642-54927-4_1
8. Sukhoroslov, O., Volkov, S., Afanasiev, A.: Web-based platform for publication and distributed execution of computing applications. In: 14th International Symposium on Parallel and Distributed Computing (ISPDC), pp. 175–184. IEEE (2015)
9. Nasonov, D., Visheratina, A., Butakova, N., Shindyapinaa, N., Melnika, M., Boukhanovskyb, A.: Hybrid evolutionary workflow scheduling algorithm for dynamic heterogeneous distributed computational environment. In: International Joint Conference SOCO 2014-CISIS 2014-ICEUTE 2014, pp. 83–92 (2014)
10. Wolstencroft, K., Haines, R., Fellows, D., et al.: The taverna workflow suite: designing and executing workflows of web services on the desktop, web or in the cloud. Nucleic Acids Res. **41**(Web Server), 557–561 (2013). doi:10.1093/nar/gkt328
11. Deelman, E.: Pegasus in the Cloud: Science Automation through Workflow Technologies. IEEE Internet Comput. **20**(1), 70–76 (2016)

12. Silva, R.F., Deelman, E., Filgueira, R., Vahi, K., Rynge, M., Mayani, R., Mayer, B.: Automating environmental computing applications with scientific workflows. In: Environmental Computing Workshop (ECW 2016) (2016)

13. Silva, R.F., Vicente, R.F., Deelman, E., Pairo-Castineira, E., Overton, I., Atkinson, M.: Using simple PID controllers to prevent and mitigate faults in scientific workflows. In: 11th Workflows in Support of Large-Scale Science (WORKS 2016) (2016)

14. Knyazkov, K.V., Kovalchuk, S.V.: Modeling and simulation framework for development of interactive virtual environments. Procedia Comput. Sci. **29**, 332–342 (2014). Elsevier

15. Kaljaev, A.I., Kaljaev, I.A., Korovin, J.: Metod mul'tiagentnogo dispetchirovanija resursov v geterogennoj oblachnoj srede pri vypolnenii potoka zadach. Herald Comput. Inf. Technol. **11**, 31–40 (2015)

16. Kravari, K., Bassiliades, N.: A survey of agent platforms. J. Artif. Soc. Soc. Simul. **18**(1), 11 (2015)

17. Gorodnichev, M.A., Vaycel, S.A.: Organization of access to supercomputing resources in the HPC community cloud. Comput. Math. Soft. Eng. **3**(4), 85–95 (2014). doi:10.14529/cmse140406

18. Volkov, S., Sukhoroslov, O.A.: Generic web service for running parameter sweep experiments in distributed computing environment. Procedia Comput. Sci. **66**, 477–486 (2015)

19. Krauter, K., Buyya, R., Maheswaran, M.: A taxonomy and survey of grid resource management systems for distributed computing. Soft. Pract. Exper. **32**, 135–164 (2002)

20. Bychkov, I.V., Oparin, G., Tchernykh, A., Feoktistov, A., Bogdanova, V., Gorsky, S.: Conceptual model of problem-oriented heterogeneous distributed computing environment with multi-agent management. Procedia Comput. Sci. **103**, 162–167 (2017)

21. Bychkov, I.V., Oparin, G.A., Feoktistov, A.G., Bogdanova, V.G., Pashinin, A.A.: Service-oriented multiagent control of distributed computations. Autom. Remote Control **76**(11), 2000–2010 (2015)

22. Bychkov, I.V., Oparin, G.A., Feoktistov, A.G., Sidorov, I.A., Bogdanova, V.G., Gorsky, S.A.: Multiagent simulation control of computational systems on the basis of meta-monitoring and imitational. Optoelectron. Instrum. Data Process. **52**(2), 107–112 (2016). doi:10.3103/S8756699016020011

23. Bogdanova, V.G., Bychkov, I.V., Korsukov, A.S., Oparin, G.A., Feoktistov, A.G.: Multiagent approach to controlling distributed computing in a cluster grid system. J. Comput. Syst. Sci. Int. **53**(5), 713–722 (2014). doi:10.1134/S1064230714040030

24. Balandin, D.V., Kogan, M.M.: Synthesis of nonfragile controllers on the basis of linear matrix inequalities. Autom. Remote Control **67**(12), 2002–2009 (2006). doi:10.1134/S0005117906120125

25. Oparin, G., Feoktistov, A., Bogdanova, V., Sidorov, I.: Automation of multi-agent control for complex dynamic systems in heterogeneous computational network. In: AIP Conference Proceedings 1798 (2017). doi:10.1063/1.4972709

26. Bara, G.I., Boutayeb, M.: Static Output feedback stabilization with $\mathcal{H}_\infty$ Performance for linear discrete-time system. IEEE Trans. Autom. Control **50**(2), 250–254 (2005)

27. Irkutsk Supercomputer Center of SB RAS. http://hpc.icc.ru

28. Bychkov, I., Oparin, G., Feoktistov, A., Bogdanova, V., Sidorov, I.: The service-oriented multiagent approach to high-performance scientific computing. In: Dimov, I., Faragó, I., Vulkov, L. (eds.) Numerical Analysis and Its Applications, NAA 2016. LNCS, vol. 10187, pp. 261–268. Springer, Cham (2017). doi:10.1007/978-3-319-57099-0_27

Stopwatch Automata-Based Model for Efficient Schedulability Analysis of Modular Computer Systems

Alevtina Glonina[✉] and Anatoly Bahmurov

Lomonosov Moscow State University, Moscow, Russia
{alevtina,bahmurov}@lvk.cs.msu.su

Abstract. In this paper we propose a stopwatch automata-based model of a modular computer system operation. This model provides an ability to perform schedulability analysis for a wide class of modular computer systems. It is formally proven that the model satisfies a set of correctness requirements. It is also proven that all the traces, generated by the model interpretation, are equivalent for schedulability analysis purposes. The traces equivalence allows to use any trace for analysis and therefore the proposed approach is much more efficient than Model Checking, especially for parallel systems with many simultaneous events. The software implementation of the proposed approach is also presented in the paper.

Keywords: Stopwatch automata · Integrated modular avionics · Simulation · Schedulability analysis

1 Introduction

Nowadays modular approach to computer systems design is replacing the old federated approach. We consider Integrated Modular Avionics (IMA) [1] systems as an example of modular computer systems, but the proposed approach can be also applied for other modular architectures (e.g. [2,3]).

An IMA system consists of standardized hardware modules containing multicore processors connected by a switched network with virtual links. There can be several module types in a system with different processors performance.

A module hardware resources are shared by several applications, called partitions. Every partition is mapped to one of the processing cores. One core can be shared by several partitions. A partition has its own memory space and execution time slots, called windows. A core's scheduling period is divided into windows and each window corresponds to one of the core's partitions.

A partition contains a set of tasks. A task is characterized by priority, period, deadline and worst case execution time (WCET) on every processor type. Every task period an instance of the task (called a job) must be executed. There can be data dependencies between tasks with the same period: current job of receiver task can't be executed until it receives data from corresponding jobs of all senders tasks.

The work is supported by the RFBR grant 17-07-01566.

V. Malyshkin (Ed.): PaCT 2017, LNCS 10421, pp. 289–300, 2017.
DOI: 10.1007/978-3-319-62932-2_28

Every partition has its own task scheduler which controls tasks execution. Schedulers usually work according to dynamic algorithms. The most common algorithm is fixed-priority preemptive scheduling (FPPS) algorithm. Every job must complete within its deadline. If a job's deadline is reached this job can not be executed anymore.

System configuration contains characteristics of hardware modules and partitions, mapping partitions to cores and windows sets for cores. The configuration is called schedulable if all the jobs complete within their deadlines. During system design multiple potential configurations are considered and for each of them schedulability analysis must be performed.

There are many schedulability analysis approaches, but some of them do not consider all modular systems features (e.g. [4]) and others have too high computational complexity (e.g. Model Checking [5]). Another approach is generating system operation trace and then analyzing this trace. Unfortunately, all the existing tools for it have essential drawbacks: some also do not consider all modular systems features (e.g [6]), others support only manual model development (e.g [7]) and almost all do not support any formal proving of model correctness (e.g. [8]). In this work we propose a general model for modular system operation, which can be used for required trace generation and overcomes these drawbacks.

As a modular system consists of standardized components, our model also consists of standardized sub-models. The key idea is to model every component type with a parametric stopwatch automaton with specified interface. The whole model is a parametric Network of Stopwatch Automata (NSA) [9]. System model for a given configuration can be constructed automatically. The formalism of NSA allowed us to prove formally that our model satisfies correctness requirements necessary for using it for schedulability analysis. We also proved that for a given configuration all interpretations of the proposed model are equivalent. This fact allows to use any single model interpretation for schedulability analysis in contrast to Model Checking where all possible interpretations are considered.

The rest of the paper is structured as follows: in Sect. 2 necessary formal definitions are given and our model is presented, in Sect. 3 the model determinism and correctness are proven and in Sect. 4 the software implementation of the proposed approach is described and experimental results are discussed.

2 The Model of Modular System Operation

2.1 Formal Definitions

A system configuration is a tuple $\langle HW, WL, Bind, Sched \rangle$, where

- $HW = \{HW_i\}_{i=1}^N$ — processing cores; $Type : HW \rightarrow \overline{1, N_t}$ — core type ($N_t \in \mathbb{N}$ — number of core types); $Mod : HW \rightarrow \overline{1, N_m}$ — module number for a core ($N_m \in \mathbb{N}$ — number of modules);
- $WL =< Part, G >$ — workload, where:
  - $Part = \{Part_i = \langle T_i, A_i \rangle\}_{i=1}^M$ — partitions, where:

* $T_i = \{T_{ij}\}_{j=1}^{K_i}$ — tasks, each characterized by priority (pr_{ij}), WCETs on different core types $(\overline{C_{ij}} = (C_{ij}^1, ..., C_{ij}^{N_t}))$, period (P_{ij}), deadline (D_{ij});

    * A_i — scheduling algorithm type;

- $G = \langle \cup_{i=1}^M T_i, \{Msg_j\}_{j=1}^H \rangle$ — data flow graph, where Msg_j corresponds to a message and is characterized by sender and receiver tasks and maximum durations of transfer through memory and through network;

- $Bind : Part \to HW$ — partitions binding to cores;

- $Sched = \{\{\langle Start_{ij}, End_{ij} \rangle\}_{j=1}^{N_i^w}\}_{i=1}^M$ — partitions schedule, which is repeated periodically with a period L equal to the least common multiple of all the tasks periods; $N_i^w \in \mathbb{N}$ — number of windows for the i-th partition; $Start_{ij}$, $End_{ij} \in \overline{0, L}$ — start time and end time for j-th window of i-th partition.

Let $CONF$ be a set of all possible system configurations.

For a task T_{ij} a set of jobs $W_{ij} = \{w_{ijk}\}_{k=1}^{L/P_{ij}}$ is defined.

Let $e = \langle Type, Src, t \rangle$ be an event, where $Type \in \{EX, PR, FIN\}$ corresponds to start or continuation of a job execution (EX), job preemption (PR) and finish of a job execution due to its completion or reaching deadline (FIN); $Src \in W_{ij}$ is a source job for the event; $t \in \overline{1, L}$ is a timestamp. Let E be a set of possible events.

A system operation trace is a set of events, therefore it is a subset of E. Let $TR \in 2^E$ be a set of all possible traces.

Design problems for IMA systems are commonly being solved under the following assumptions (e.g. in [8,10], considering industrial avionics systems):

- Every job's execution time is equal to its WCET.
- Every message transfer delay is also equal to its worst case; typical avionics networks (e.g. AFDX) allow to obtain safe estimations for these delays.
- Scheduling algorithms are deterministic (e.g. in ARINC 653 systems [1]).

Under these assumptions system operation is deterministic and corresponds to a worst-case scenario, i.e. only one trace corresponds to a configuration and this trace can be used for schedulability analysis. Therefore the mapping $Q : CONF \to TR$ exists and $\forall conf \in CONF : \exists! Q(conf)$.

Let R_{ijk} be a number of executing intervals for a job w_{ijk}. Then an ordered subtrace for this job is:

- empty, if $R_{ijk} = 0$;
- $\langle EX, w_{ijk}, t_0 \rangle, \langle FIN, w_{ijk}, t_1 \rangle$, if $R_{ijk} = 1$;
- $\langle EX, w_{ijk}, t_0 \rangle, \langle PR, w_{ijk}, t_1 \rangle, ..., \langle EX, w_{ijk}, t_{2R_{ijk}-2} \rangle, \langle FIN, w_{ijk}, t_{2R_{ijk}-1} \rangle$, if $R_{ijk} > 1$.

In these terms the schedulability criterion has the following form:

$\forall w_{ijk}, i \in \overline{1, M}, j \in \overline{1, K_i}, k \in \overline{1, L/P_{ij}}$: $\sum_{r=1}^{R_{ijk}} (t_{2r-1} - t_{2r-2}) = C_{ij}^{Type(Bind(Part_i))}$

In this paper we consider the problem of building the system operation model, the interpretation of which defines the mapping Q. This model is necessary for checking the schedulability criterion for a given configuration.

2.2 Networks of Stopwatch Automata

First of all the mathematical formalism for system operation description must be chosen. It should meet the following requirements:

- ability to model such aspects of system operation as queues, preemption, parallel functioning of different schedulers;
- ability to obtain a time trace of model interpretation;
- ability to formalize and check requirements to models;
- existence of software tools for modeling and verification.

We reviewed several formalisms found in the literature and the formalism of stopwatch automata networks [9,11] was chosen, as it meets all the requirements and has the best program support of modeling and verification. Now we give a brief description of the formalism.

A stopwatch automaton is a finite automaton (denoted graphically by a graph containing a set of nodes or locations and a set of labeled edges) extended with integer variables and clocks. Each variable has a bounded domain and an initial value. A clock is a special real-valued variable, that can be compared with integer variables or with other clocks, reset to zero, stopped and later resumed with the same value. All the clocks are initialized with zero and then increase synchronously (except for the stopped clocks) with the same rate.

An edge represents an action transition and has three labels: a guard label, a synchronization (will be explained later) label and an update label. A transition can be taken when clocks and variables satisfy the guard and synchronization can be performed. During action transitions, synchronizations and updates of clocks and variables are performed. A location has a label called an invariant, which is a predicate over variables and clocks. An automaton may remain in a location as long as the invariant of the location is true. For an automaton an initial location is defined.

In addition to action transitions, represented by automaton edges, there are delay transitions, corresponding to synchronous clock increasing by the same real value. All the clocks (except for the stopped clocks) can be increased by a value of d if their values increased by d satisfy current location invariant. Some locations can be labeled as *committed*. No delay transitions can be performed if an automaton current location is committed.

A network of stopwatch automata (NSA) is a set of several automata, operating synchronously. Communications between the automata are performed by using shared variables and channels.

A channel is a mechanism for automata synchronous communication. Every automaton edge has a synchronization label, which can be either an empty label (for internal transitions) or a synchronization action. There are two complementary types of such actions: *sending* and *receiving* a signal through a channel. And there are two types of channels: binary and broadcast. Two transitions in different automata can synchronize via a binary channel if the guards of both transitions are satisfied, and they have complementary synchronization actions. A transition with binary synchronization action can be performed if and only if the transition in the other automaton with complementary action can be performed. When synchronization is performed the current locations of both automata are changed, i.e. the both transitions are performed simultaneously. $N + 1$ automata can synchronize via a broadcast channel if the transition with sending action is enabled and N transitions with receiving action are enabled.

More formally a stopwatch automaton is a tuple $\langle L, l_0, U, C, V, \overline{v_0}, AU, AS, E, I, P \rangle$, where

- $L, l_0 \in L, U \subseteq L$ — finite set of locations, initial location and set of committed locations;
- C — set of clocks;
- $V, \overline{v_0}$ — set of integer variables and their initial values;
- AU, AS — sets of updating and synchronization actions;
- E — set of edges, $E \subseteq L \times B(C,V) \times AU \times AS \times L$, where $B(C,V)$ is a set of predicates over C and V
- $I : L \to B(C,V)$ associates invariants to locations;
- $P : L \times C \to B(\emptyset, V)$ associates progress conditions to locations and clocks;

Let $A = A_1|...|A_n$ be an NSA, where $A_i = \langle L_i, l_i^0, U_i, C_i, V_i, \overline{v_i^0}, AU_i, AS_i, E_i, I_i, P_i \rangle$. A state of the NSA is a tuple $\langle \overline{l}, \overline{c}, \overline{v} \rangle \in (L_1 \times ... \times L_n) \times \mathbb{R}_{\geq 0}^{|C|} \times \mathbb{Z}^{|V|}$, where $V = \cup V_i$, $C = \cup C_i$. A sequence (may be infinite) of action and delay transitions between states $\langle \overline{l_0}, \overline{c_0}, \overline{v_0} \rangle \to \langle \overline{l_1}, \overline{c_1}, \overline{v_1} \rangle \to ... \to \langle \overline{l_i}, \overline{c_i}, \overline{v_i} \rangle \to ...$ is a run of an NSA. An NSA usually has many (may be infinitely) possible runs.

2.3   General Model of Modular System Operation

To present our model we have to introduce several definitions.

A *parametric stopwatch automata*, or *concrete automata type*, is a tuple $\langle L, l_0, U, C, V, \overline{p}, AU, AS, E, I, P \rangle$, where $\overline{p}$ is a vector of unknown integer-valued parameters. An automaton's shared variables and possible synchronization actions comprise the *automaton interface*. *Base automata type* is a pair of sets $\langle V_b, AS_b \rangle$, where V_b is a set of shared variables and AS_b is a set of synchronization actions. A concrete automata type *implements* a base automata type if $V_b \subseteq V$, $AS_b \subseteq AS$.

A set of base automata types is a *general NSA*. A set of concrete automata types is a *concrete NSA*. A concrete NSA *implements* a general NSA if each base automata type in the general NSA is implemented by one or more concrete automata type in the concrete NSA and logic relations between concrete automata types corresponds to relations (i.e. rules defining, which implementations of base automata types must communicate) between base automata types.

Model time is a value of a special clock, which is never stopped or reset. *Synchronization event* is a tuple $\langle CH, A, t \rangle$, where CH is the channel, A is a set of automata instances, participating in the synchronization, t is the model time of synchronization. *NSA trace* is a set of synchronization events, generated by the network.

We propose to represent the general model of modular system operation as a general NSA.

The following shared variables and channels are used for automata communication in the proposed model:

- variables $\mathtt{is\_ready}_{ij}$, $\mathtt{is\_failed}_{ij}$, $\mathtt{prio}_{ij}$, $\mathtt{deadline}_{ij}$ $i \in \overline{1, M}, j \in \overline{1, K_i}$, each corresponding to a job readiness, reaching its deadline and a task characteristics;
- variables $\mathtt{is\_data\_ready}_h$, $h \in \overline{1, H}$, each corresponding to a message delivery through the hth virtual link;
- channels $\mathtt{wakeup}_i$, $\mathtt{sleep}_i$, $\mathtt{ready}_i$, $\mathtt{finished}_i$ $i \in \overline{1, M}$, each corresponding to a window start and finish, a ready job arrival and its finish; a job finishes either due to its completion or due to reaching its deadline;
- channels $\mathtt{exec}_{ij}$, $\mathtt{preempt}_{ij}$, $i \in \overline{1, M}$, $j \in \overline{1, K_i}$, each corresponding to a job execution start (or resumption) and preemption;
- broadcast channels $\mathtt{send}_{ij}$, $\mathtt{receive}_{ij}$, $i \in \overline{1, M}, j \in \overline{1, K_i}$, each corresponding to receiving data from a sender job and sending data to all receiver jobs.

The general NSA consists of the following base automata types:

1. T base automata type modeling a task. As a task deadline is less or equal to its period, there can be only one active job of a task at a given moment. T is defined by following interface:
   - receiving signals through channels $\mathtt{exec}$ and $\mathtt{preempt}$;
   - sending signals through channels $\mathtt{ready}$, $\mathtt{finished}$, $\mathtt{send}$, $\mathtt{receive}$;
   - changing variables $\mathtt{is\_ready}$, $\mathtt{is\_failed}$, $\mathtt{is\_data\_ready}_h$;
2. TS base automata type modeling a task scheduler for a partition. It is defined by following interface:
   - receiving signals through channels $\mathtt{wakeup}$, $\mathtt{sleep}$, $\mathtt{ready}$, $\mathtt{finished}$;
   - sending signals through channels $\mathtt{exec}_j$, $\mathtt{preempt}_j$; the j-th channel corresponds to the j-th task of the partition;
   - reading variables $\mathtt{is\_ready}_j$, $\mathtt{prio}_j$, $\mathtt{deadline}_j$; the j-th variable corresponds to the j-th task of the partition.
3. CS base automata type modeling a core scheduler (scheduling partitions for a core). It is defined by the following interface:
   - sending signals through channels $\mathtt{wakeup}_i$ and $\mathtt{sleep}_i$; the i-th channel corresponds to the ith partition.
4. L base automata type modeling a virtual link. It is defined by the following interface:
   - receiving signals through a broadcast channel $\mathtt{send}$;
   - sending signals through a broadcast channel $\mathtt{receive}$;
   - changing variable $\mathtt{is\_data\_ready}$.

The structure of the proposed general model of modular system operation is shown on Fig. 1.

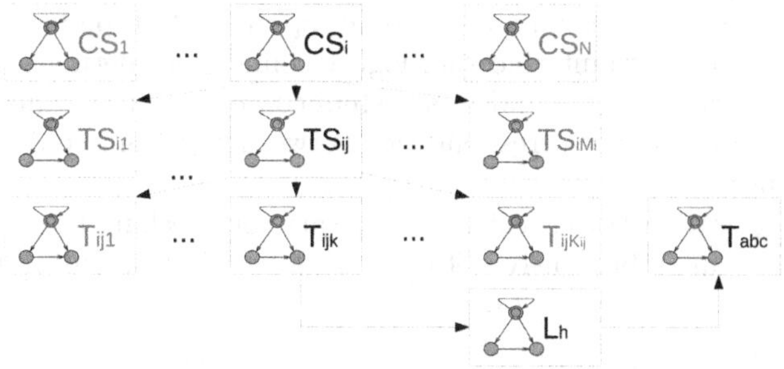

Fig. 1. The structure of the general NSA type modeling modular system operation.

A concrete NSA implementing the proposed general NSA is a parametric model of modular system operation. Our concrete NSA has the following concrete automata types, implementing base automata types: task model, core scheduler model, virtual link model, FPPS scheduler, FPNPS scheduler and EDF scheduler. For a given concrete NSA and a system configuration an NSA instance can be constructed by the Algorithm 1.

Algorithm 1. An NSA instance construction

Data: $conf \in CONF$, concrete NSA
Result: NSA instance modeling system of $conf$ configuration
begin

    for $i \in \overline{1, N}$ **do**

        for $j \in \overline{1, M} : Bind(Part_j) = HW_i$ **do**

            create channels ready_j, $\mathsf{finished}_j$, wakeup_j, sleep_j;

            for $k \in \overline{1, K_i}$ **do**

                create channels exec_{jk}, $\mathsf{preempt}_{jk}$, send_{jk}, $\mathsf{receive}_{jk}$ and variables $\mathsf{is\_ready}_{jk}$, prio_{jk}, $\mathsf{deadline}_{jk}$, $\mathsf{is\_data\_ready}_h$ (each corresponding to a virtual link, where jkth task is a receiver);

                create an automaton implementing T, initialize its interface with channels exec_{jk}, $\mathsf{preempt}_{jk}$, send_{jk}, $\mathsf{receive}_{jk}$, ready_j, $\mathsf{finished}_j$ and variables $\mathsf{is\_ready}_{jk}$, prio_{jk}, $\mathsf{deadline}_{jk}$, $\mathsf{is\_data\_ready}_h$;

            create an automaton implementing TS and corresponding to A_j for jth partition, initialize its interface with channels exec_{jk}, $\mathsf{preempt}_{jk}$, ready_j, $\mathsf{finished}_j$, wakeup_j, sleep_j and variables $\mathsf{is\_ready}_{jk}$, prio_{jk}, $\mathsf{deadline}_{jk}$ $(k \in \overline{1, K_i})$;

        create an automaton implementing CS for ith core and initialize its interface with corresponding channels wakeup_j, sleep_j;

    for $h \in \overline{1, H}$ **do**

        create an automaton implementing L, initialize its interface with corresponding channels $\mathsf{send}_{j_1 k_1}$, $\mathsf{receive}_{j_2 k_2}$ and variable $\mathsf{is\_data\_ready}_h$.

By construction there is an automaton of appropriate type for every system component and automata interfaces for logical connections between components. Automata parameters correspond to a system configuration parameters. Therefore there is unambiguous correspondence between a system configuration and a model instance.

A system operation trace, which is necessary for checking the schedulability criterion, can be unambiguously obtained from the corresponding model trace (i.e. a trace of the NSA instance).

3   Correctness and Determinism

Modular systems specifications contain correctness requirements to system components operation and to the whole system operation. These requirements specify correct events sequences and delays between events of given types. In order to ensure schedulability analysis correctness, our model must satisfy correctness requirements, which are applicable at the chosen abstraction level.

We call a model deterministic if a trace generated by its run is uniquely determined. This determinism is crucial for schedulability analysis of large systems with many simultaneous events, because it allows to use any of the NSA runs for a trace generation in contrast to model-checking where all possible runs are to be considered.

Correctness requirements to system components models (i.e. parametric automata) can be checked automatically by a verifier. For this purpose we chose "observers" approach [12], which is successfully used in practice.

One observer automaton usually corresponds to one requirement. The observer is an automaton, which operates synchronously with a given automaton and does not block any synchronization. The observer has one "bad" location and all incorrect synchronization event sequences or incorrect delays lead the observer to the "bad" location. The reachability of the "bad" location means that an incorrect event sequences can be generated by the given automaton and therefore it does not satisfy the requirement. As the given automaton is parametric and must operate correctly with all possible parameters values, its observer non-deterministically sets each parameter to one of possible values.

We derive correctness requirements to system components from system specifications, construct an observer for each requirement and automatically check with UPPAAL [11] verifier that "bad" locations are unreachable. Such proof was performed for a set of requirements derived from ARINC 653 specification [1] and the set of concrete automata types described in Sect. 2.3.

Let us consider a correctness requirement example and build its observer:

For every partition at any time zero or one job can be executed.

This is the requirement to TS base automata type and all the TS implementations must satisfy this requirement. In terms of synchronization events, a job of task T_{jk} is executed between synchronizations through channels $exec_{jk}$ and $preempt_{jk}$, and through channels $exec_{jk}$ and $finished_j$. It means that any

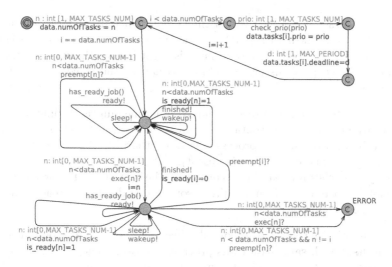

Fig. 2. The observer automaton for the requirement to TS automata

synchronization through $exec_{jk}$ must be followed by a synchronization through $preempt_{jk}$ or $finished_j$. The corresponding observer is shown on Fig. 2.

Satisfaction of the requirements to the whole general model can't be proven automatically because the number of automata of different types in the model is unknown in general. Thus, we have to prove the satisfaction of these requirements manually. This proof implies that all the models instances constructed by the algorithm 1 satisfy these requirements. Our proof is based on the satisfaction of the requirements to components models which are proven automatically.

This is an example of a requirement to the whole model and its proof:

If one task depends on another, then start time for any job of the receiver task is more or equal the completion time for the corresponding job of the sender task plus the upper bound of the message transfer delay.

The satisfaction of the following requirements to components models were proven automatically:

1. Every job sends data to its output virtual links after its completion.
2. A message transfer delay trough a virtual link is equal to its pessimistic upper bound.
3. A job of receiver task can't be executed until it receives data from corresponding jobs of all senders tasks.

The satisfaction of these requirements implies the satisfaction of the given requirement to the whole model.

The model determinism proof is based on the previously proven satisfaction of the correctness requirements.

Suppose by contradiction that two different NSA traces can be generated by the model interpretation for a given configuration. Let the both traces be partially ordered by events time. Thus, a set of events is bound to every time point in every trace. Let t_i be the first time point, which has different events sets for given traces. It means that at least one event is contained in one event set and is absent in the other. Suppose that this event is a synchronization through finished$_j$. As all previous events sets are equal for the traces, there are two alternatives:

1. Some job executes on the processor for WCET time units according to the first trace (where the event is contained). But it means that this job's cumulative time of execution on the processor is more than WCET according to the second trace (where the event is absent).
2. Some job reaches its deadline according to the first trace. But it means that this job is not removed from the processor after its deadline is reached according to the second trace.

Both the alternatives are impossible, because they imply violation of the requirements, which satisfaction was previously proven. Therefore the supposition is impossible. For other events types the proof is similar.

So we proved that the proposed model satisfies correctness requirements and all the traces generated by its interpretation are equal. It was also shown that there is unambiguous correspondence between a system configuration and a generated model instance and between a system trace and a model interpretation trace. Therefore schedulability analysis (checking the criterion specified in Sect. 2.1) performed by using this model is correct.

4   Implementation and Experiments

In order to test the applicability of the proposed approach in practice we implemented it in software. The concrete automata types modeling concrete types of system components were developed and verified using UPPAAL [11] toolset. These concrete automata types are contained in an automata components models library. A user can develop, verify and add to the library own models. As UPPAAL doesn't have commandline interface for NSA interpretation, we developed our own NSA simulation library in C++ and a translator from UPPAAL to C++ automata representation. The library of automata models for components was translated to a library of software models. Models from the library compose the parametric software model of system operation (see Fig. 3).

We compared the proposed approach with Model Checking using the same NSA. The results of the experiments confirm that our approach is much more efficient (see Table 1).

We also integrated the parametric model with an IMA scheduling tool, which searches the optimum IMA configuration among possible configurations [8]. On every iteration the scheduling algorithm chooses a configuration to be checked for schedulability. Then an XML file with the configuration description is generated and passed to the parametric model. After that a model instance is created and run and it trace is passed back to the scheduling tool, which performs schedulability analysis. Unschedulable configurations are discarded by the scheduling

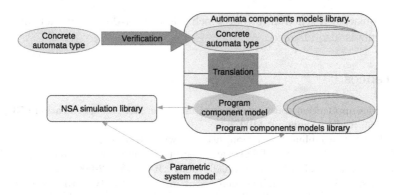

Fig. 3. The scheme of the parametric modular system operation model organization.

Table 1. Execution times for various number of jobs

Number of jobs	10	11	12	13	14	15	16	17	18
Model checking (seconds)	0.57	1.16	2.22	5.05	10.43	23.51	48.13	112.28	215.91
Proposed approach (seconds)	0.027	0.027	0.028	0.030	0.031	0.032	0.033	0.035	0.036

algorithm and schedulable ones are considered as candidate solutions. The experiments showed that a model instance construction and interpretation take about several seconds for configurations of same complexity as configurations of industrial avionics systems (about 11 s for a configuration with 12500 jobs). Thus it was shown that our approach is applicable in practice.

5   Conclusion

We developed a general model of a modular computer system operation based on the NSA formalism. The model can be used for schedulability analysis of such systems configurations. It was proven that our model is deterministic and correct, and therefore the analysis is performed correctly. The model determinism (in terms of jobs start, finish and preemption) makes our approach is significantly more efficient than Model Checking, especially for systems with many multicore processors operating concurrently. The experiments with the model implementation showed the applicability of the proposed approach in practice to real scale systems.

In future work, we plan to extend our components models library with more models of core and task schedulers and models of switched networks components.

Integration with a scheduling tool which allows user-defined models of system components is also planned.

References

1. Avionics application software standard interface: ARINC specification 653. Aeronautical Radio, Annapolis (1997)
2. AUTOSAR. Enabling Innovation. http://www.autosar.org/
3. Obermaisser, R., et al.: DECOS: an integrated time-triggered architecture. Elektrotech. Inftech. **123**(3), 83–95 (2006). doi:10.1007/s00502-006-0323
4. Marinescu, S., et al.: Timing analysis of mixed-criticality hard real-time applications implemented on distributed partitioned architectures. In: Proceedings of 2012 17th IEEE International Conference on Emerging Technologies and Factory Automation (ETFA 2012), Krakow, Poland, pp. 1–4 (2012). doi:10.1109/ETFA. 2012.6489720
5. Macariu, G., Cretu, V.: Timed automata model for component-based real-time systems. In: Proceedings of 2010 17th IEEE International Conference and Workshops on Engineering of Computer Based Systems, Oxford, UK, pp. 121–130 (2010). doi:10.1109/ECBS.2010.20
6. Craveiro, J.P., Silveira, R.O., Rufino, J.: hsSim: an extensible interoperable object-oriented n-level hierarchical scheduling simulator. In: Proceedings of the 3rd International Workshop on Analysis Tools and Methodologies for Embedded and Real-time Systems (WATERS 2012), Pisa, Italy, pp. 9–14 (2012)
7. Khoroshilov, A., et al.: AADL-based toolset for IMA system design and integration. SAE Int. J. Aerosp. **5**(2), 294–299 (2012). doi:10.4271/2012-01-2146
8. Balashov, V.V., Balakhanov, V.A., Kostenko, V.A.: Scheduling of computational tasks in switched network-based IMA systems. In: Proceedings of International Conference on Engineering and Applied Sciences Optimization, Athens, Greece, pp. 1001–1014 (2014)
9. Cassez, F., Larsen, K.: The impressive power of stopwatches. In: Palamidessi, C. (ed.) CONCUR 2000. LNCS, vol. 1877, pp. 138–152. Springer, Heidelberg (2000). doi:10.1007/3-540-44618-4_12
10. Tretyakov, A.: Automation of scheduling for periodic real-time systems (in Russian). Proc. Inst. Syst. Program. **22**, 375–400 (2012). doi:10.1134/S0361768813050046
11. Bengtsson, J., Yi, W.: Timed Automata: Semantics, Algorithms and Tools. In: Desel, J., Reisig, W., Rozenberg, G. (eds.) ACPN 2003. LNCS, vol. 3098, pp. 87–124. Springer, Heidelberg (2004). doi:10.1007/978-3-540-27755-2_3
12. Andre, E.: Observer patterns for real-time systems. In: Proceedings of 2013 18th IEEE International Conference on Engineering of Complex Computer Systems (ICECCS), Singapore, pp. 125–134 (2013). doi:10.1109/ICECCS.2013.26

Parallelizing Inline Data Reduction Operations for Primary Storage Systems

Jeonghyeon Ma$^{(\boxtimes)}$ and Chanik Park

Department of Computer Science and Engineering, POSTECH, Pohang, South Korea
{doitnow0415,cipark}@postech.ac.kr

Abstract. Data reduction operations such as deduplication and compression are widely used to save storage capacity in primary storage system. These operations are compute-intensive. High performance storage devices like SSDs are widely used in most primary storage systems. Therefore, data reduction operations become a performance bottleneck in SSD-based primary storage systems.

In this paper, we propose a parallel data reduction technique on data deduplication and compression utilizing both multi-core CPU and GPU in an integrated manner. First, we introduce bin-based data deduplication, a parallel technique on deduplication, where CPU-based parallelism is mainly applied whereas GPU is utilized as co-processor of CPU. Second, we also propose a parallel technique on compression, where main computation is done by GPU while CPU is responsible only for post-processing. Third, we propose a parallel technique handling both deduplication and compression in an integrated manner, where our technique controls when and how to use GPU. Experimental evaluation shows that our proposed techniques can achieve 15.0%, 88.3%, and 89.7% better throughput than the case where only CPU is applied for deduplication, compression, and integrated data reductions, respectively. Our proposed technique enables easy application of data reduction operations to SSD-based primary storage systems.

Keywords: Primary storage · Inline data reduction scheme · GPU

1 Introduction

Data reduction operations such as data de-duplication and compression are widely used to save storage capacity on primary storage systems. In recent years, however, replacing primary storage systems from HDD-based to SSD-based has exposed the computational overhead of data reduction operations, making it difficult to apply data reduction operations to storage systems. One way to conceal the overhead of data reduction operations is to store all of the data on the storage system and then perform data reduction in the background when the system is idle. However, this generates more write I/O than systems without the data reduction operations. Therefore, it is not applicable to SSD-based storage systems due to write endurance problems. A way to increase the lifetime of SSD-based storage systems is to apply data reduction operations to the critical I/O paths. However, applying them to the critical I/O paths can significantly degrade I/O performance. One way to improve the throughput of data reduction is to take advantage

© Springer International Publishing AG 2017
V. Malyshkin (Ed.): PaCT 2017, LNCS 10421, pp. 301–307, 2017.
DOI: 10.1007/978-3-319-62932-2_29

of GPUs designed to calculate computation-intensive workloads. However, depending on the workload, the performance of the CPU-based parallel data reduction operations may be better than GPU-based techniques.

In this paper, we propose an inline parallel data reduction operations based on multi-core CPU and GPU for primary storage systems. To do this, we design a parallel deduplication and compression method considering multi-core CPU and GPU architecture, and finally we show how to integrate CPU and GPU-based data reduction operations.

2 Background

Data reduction operations such as deduplication and compression are widely used to save storage capacity on primary storage systems. This section describes the basic tasks of data reduction operations and the performance bottlenecks.

Deduplication is performed in four stages: chunking, hashing, indexing, and destaging. Chunking is the process of breaking a data stream into chunks, which is the base unit for checking the redundancy of data. Hashing is the process of calculating the hash value of each chunk. The hash value is used as an identifier for the chunk. Indexing is the process of comparing the hash value of each chunk with the hash values of already stored chunks to determine whether it is a duplicate. If the chunk is found to be unique, a destaging step is performed to store the chunk on the storage device. Of these stages, hashing and indexing are the main performance bottlenecks in deduplication systems. Previous work [1] has also attempted to address these two major performance bottlenecks.

Among the compression algorithms, LZ-based compression algorithms are widely used in main storage systems due to their simplicity and effectiveness [2]. The history buffer and the look-ahead buffer are used to perform LZ compression. If characters in the same order are found in both the history buffer and the look-ahead buffer, the character sequence in the look-ahead buffer is replaced by a pointer to the character sequence in the history buffer. Matching the entire string is a performance bottleneck.

3 Design and Implementation

3.1 Parallel Data Deduplication on Multi-core CPU and GPU

There is no data dependency between chunks when the hash value of the chunk is calculated in the hashing phase. This allows us to easily calculate multiple chunks at once in a natural parallel manner. However, parallelizing the indexing is more complicated than the hashing. This is because the hash table used to determine the chunk's redundancy is globally shared across all computing threads. Therefore, this section describes how indexing is parallelized on the multi-core CPU and GPU, and how it applies to the primary storage system.

(1) How to Parallelize Indexing on the CPU: we divide the hash table into several small hash tables called bin so that multiple computing threads can check the chunks of multiple hash tables at the same time without locking mechanism. This is a technique

that was commonly used in existing DHT-based systems. We call this operation bin-based indexing. In addition, to avoid disk access that significantly degrades performance, hash table entries are kept in memory space only, not disk space. Due to this index management policy, the deduplication module cannot find some duplicate data. However that is not a big deal. Assuming that the storage capacity is 4 TB, the chunk size is 8 KB, and the index size is 32 bytes, including the hash size (SHA1, 20 bytes) and other metadata, the storage system requires 16 GB of memory for the index. That is, if primary storage is the target, it does not require that much memory. In addition, the way to reduce memory consumption is to remove the prefix value of the hash entry. If the prefix value is n bytes, the deduplication system keeps only 20-n bytes for each hash value. If the storage system uses a 2-byte prefix value, we can save 1 GB of memory in this way.

(2) How to Parallelize Indexing on the GPU: parallel processing of GPU indexing needs to take into account the architectural characteristics of the GPU. First, the GPU is connected to the system memory via the PCI interface, and the data used for the calculation must be transferred from the system memory to the GPU device memory. Second, GPU threads in the same workgroup run the same command regardless of branching, even though each thread has its own execution path. Therefore, many branch operations can degrade computational performance. This means we have to design the GPU code in a rather simple way. Third, GPUs have many computing cores and large memory bandwidth. Therefore, we can calculate large amounts of data at a time. This means that allocating data to all computing cores and setting up data layouts is critical to taking full advantage of all GPU resources.

The GPU also performs bin-based indexing just like on a CPU. However, considering the characteristics of the advanced GPU architecture, we organize one bin into a linear table structure rather than a tree structure. This continuous data layout is useful when utilizing the GPU's local memory. This is because copying data from GPU global memory to local memory can be done naturally if the thread accesses the data continuously. It also does not cause multiple branch operations. The GPU can check the redundancy of data by comparing a single hash table. Also, only the hash value persists in GPU memory, and other metadata in the chunk is maintained in system memory. This is because transferring data can be a direct update process. This means that there is no other hash table update overhead on the GPU. Therefore, the result of whether an index is hit or not includes an index number and a hit/miss information pair. The metadata space structure in system memory then uses the results of the GPU.

(3) When to use GPU for indexing: we decide how to apply GPU for indexing. To do this, we compare the CPU and GPU indexing performance. The number of hash table entries used for indexing remains the same on the CPU and GPU for a fair comparison. Preliminary experiments show that CPU performance is 4.16 to 5.45 times better than GPU performance in terms of execution time. For GPU indexing, the execution time is fixed because of the inevitable time at which the GPU kernel starts. This means that even with high-performance GPUs, there is a limit to optimizing indexing on the GPU. Therefore, we decide to use GPU only when CPU utilization is full and there is still some work to do for indexing.

3.2   Parallel Data Compression on Multi-core CPU and GPU

In this section, we focus on the way to parallelize LZ compression schemes that are commonly used in primary storage systems.

(1) How to parallelize compression for CPU: As with hashing operations, there is no data dependency between chunks, so we can run compression independently on each chunk. CPU-based compression algorithms have been well studied previously. Therefore, the compute is parallelized by the CPU by assigning a computing thread that runs the previously studied compression algorithm to each chunk.

(2) How to parallelize compression for GPU: Ozsoy et al. [3] introduced a parallel compression algorithm on the GPU. This algorithm divides the data into several sub-blocks and calculates the compression result in each sub-block and merges it in the CPU. This algorithm has a weakness to apply as a compression algorithm for primary storage systems. This algorithm assumes that the size of the data to be compressed is large enough to take full advantage of the GPU resources. This means that it does not work well for small-sized target data. The size of the chunk is 4 KB. Only a small number of computing cores can be allocated to compute the compression result of 4 KB chunks. Therefore, we design a compression algorithm that computes the chunk compression results at a time. The GPU allocates multiple threads for each chunk. Each stage performs its own LZ compression algorithm with its own history buffer and look-ahead buffer. Adjacent threads inspect overlapping regions by the size of the history buffer. The GPU's compression results are not refined in GPU due to performance issues. Therefore, the CPU must refine the results. It is called as post-processing.

(3) How to use the GPU for compression: we compare the compression performance of the CPU and the GPU to determine when to use the GPU. Experimental results show that GPU performance is 88.3% better than CPU performance in terms of execution time (In Sect. 4). The performance gap is large. Therefore, the GPU performs compression and the CPU is used for refinement.

3.3   Putting It All Together

This section describes how to incorporate two parallel data reduction operations called deduplication and compression. First, we need to determine the order of which operation should be applied. Based on the result of [5], we adopt deduplication-before-compression order for higher data reduction ratio. Second, we add a bin buffer structure to the data deduplication algorithm. The bin buffer is used to temporarily store a hash for each bin before moving each bin to the GPU memory and bin tree. When the buffer is full, the hash is immediately flushed from the buffer to the storage. This creates the appropriate sequential writes for the SSD. Figure 1 shows a workflow that incorporates deduplication and compression operations on the CPU and GPU. GPU indexing is performed if the GPU is available, and CPU indexing is performed if duplicate hashes are not found. For the CPU indexing path, the bin buffer is checked first, because recently updated chunks can reside in the bin buffer and chunks are more likely to find duplicates in the bin buffer due to temporal locality. If there are no duplicates in the bin buffer, check the bin tree to store most of the hash table entries. If we cannot find any duplicate, then the chunk is regarded a unique chunk. Therefore, the chunk becomes the compression target. After compressing the data, the bin buffer is updated because the chunks are unique. If the bin buffer becomes full, the buffer will be flushed to the storage. And then, GPU bin in GPU memory are updated accordingly. Currently, random based replacement policy is applied.

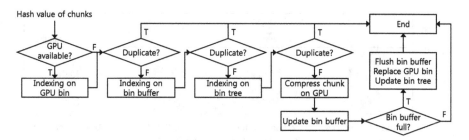

Fig. 1. An integrated workflow of deduplication and compression proposed for data reduction operations

4   Evaluation

This section evaluates the throughput of the parallel data reduction operations on the CPU and GPU. The vdbench is used to generate datasets. Our test machine equipped with Intel i7-3770 k, Radeon HD 7970, and 16 GB main memory. The vdbench [4] is used to generate the dataset. The size of the data stream is about 2 GB. The deduplication and compression ratio are set to 2.0, which is a common ratio for primary storage systems. We compare our schemes with the throughput of Samsung SSD 830. In this section, the Samsung SSD 830 is simply referred to as the SSD.

(1) Parallel data deduplication: the GPU performs indexing of only a small portion of the chunk. The workflow for the integrated CPU and GPU for indexing is the same as in Fig. 1, except for the compression phase. Experimental results show that the GPU-supported data deduplication scheme can improve throughput by 15% over CPU-only data deduplication scheme. In addition, it shows three times the throughput of the SSD.

(2) Parallel data compression: the proposed technique uses the GPU for compression and the CPU for post-processing of compression. Due to the nature of the compression technique, the throughput is high when the compression ratio is high. The CPU-based compression method has lower performance (about 50 K IOPS) than SSD throughput (about 80 K IOPS) when the compression ratio is low, but the GPU-based parallel compression method has the performance of 100 K IOPS even when the compression ratio is low. It always shows higher performance than SSD throughput.

(3) Putting it all together - Parallelizing both data deduplication and compression together: In an environment where CPU and GPU are available, there are several

options for integrating two data reduction operations, deduplication and compression. The first option is to use the GPU in two data reduction operations. The second option is to use the GPU for only one data reduction operation. The last option is that both data reduction operations do not use the GPU at all. The last option may be useful when the performance of the GPU is poor. Figure 2 shows the throughput of these options.

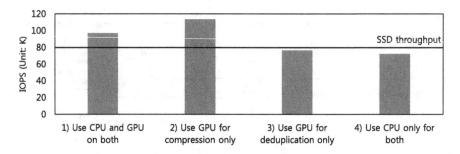

Fig. 2. Throughput comparison of integration methods

Allocating the GPU for compression is the best choice among the integration methods. This is because data compression, which has a high performance gain when using a GPU, monopolizes the GPU. However, because hardware specifications may be different on different platforms, we cannot guarantee that this integration is always right. Therefore, before assigning processors to each data reduction operation, the performance of these integration methods is compared using dummy I/O to determine the best fit for throughput. Therefore, we can ensure the best performance even if the target platform is different.

5 Related Works

There have been lots of previous researches which investigated the way to improve the throughput of data reduction operations.

There exist some researches exploiting parallelism in data deduplication system. Xia W. et al. [6] proposed multicore-based parallel data deduplication approach. However, the problem is that they did not consider the operation of indexing which is known as main bottleneck in data deduplication [1]. Kim et al. [7] proposed GPU-based data deduplication approach for the primary storage. However, they did not consider utilizing CPU that performs better than GPU for indexing operation.

There exist some researches exploiting GPU parallelism for compression operation. Ozsoy et al. [3] introduced the parallel compression algorithms on GPU. However, the compression target data are quite large to utilize GPU resource fully. This feature does not match with primary storage system that conducts compression for 4 KB of several chunks. Moreover, there exist researches introducing CPU parallel algorithms for compression. Shmuel et al. [8] introduce the algorithm for the compression executed using the tree-structured hierarchy. Gonzalo et al. [9] introduce the algorithm dividing data stream into several small subset and allocating each threads to the subset of data. Even they parallelize the compression for CPU, our GPU-based approach is better than at least about 88.3%.

There exists a research analyzing the effect of mixing two data reduction operations, deduplication and compression. Constantinescu et al. [5] analyze the data reduction ratio when deduplication and compression are applied together. However, it focuses only data reduction ratio, not throughput.

6   Conclusion

Throughput is becoming more important as data reduction operations are applied to save space on SSD-based primary storage systems. To solve this problem, we proposed parallel data reduction operations using multi-core CPU and GPU. We also showed how to integrate deduplication and compression technologies on multicore CPUs and GPUs. Applying our parallel approach to deduplication is 3 times better than SSD's throughput. For compression, the throughput of the parallel compression method supported by the GPU is 88.3% better than the average throughput of parallel QuickLZ. Finally, GPU-supported integration shows a performance improvement of 89.7% over parallel data reduction operations using CPU (deduplication ratio 2.0, compression 2.0). This means that our proposed technique enables easy application of data reduction operations to SSD-based primary storage systems.

References

1. Guo, F., Efstathopoulos, P.: Building a high-performance deduplication system: In: USENIX Annual Technical Conference (2011)
2. De Agostino, S.: Lempel-Ziv data compression on parallel and distributed systems. Algorithms **4**, 183–199 (2011)
3. Ozsoy, A., Swany, M., Chauhan, A.: Pipelined parallel LZSS for streaming data compression on GPGPUs. In: Parallel and Distributed Systems, pp. 37–44 (2012)
4. Berryman, A., Calyam, P., Honigford, M., Lai, A.M.: Vdbench: a benchmarking toolkit for thin-client based virtual desktop environments. In: Cloud Computing Technology and Science, pp. 480–487 (2010)
5. Constantinescu, C., Glider, J., Chambliss, D.: Mixing deduplication and compression on active data sets: In: Data Compression Conference, pp. 393–402 (2011)
6. Xia, W., Jiang, H., Feng, D., Tian, L., Fu, M., Wang, Z.: P-dedupe: exploiting parallelism in data deduplication system: In: Networking, Architecture and Storage, pp. 338–347 (2012)
7. Kim, C., Park, K.W., Park, K.H.: GHOST: GPGPU-offloaded high performance storage I/O deduplication for primary storage system. In: Proceedings of the International Workshop on Programming Models and Applications for Multicores and Manycores, pp. 17–26 (2012)
8. Klein, S.T., Wiseman, Y.: Parallel Lempel Ziv coding (extended abstract). In: Amir, A. (ed.) CPM 2001. LNCS, vol. 2089, pp. 18–30. Springer, Heidelberg (2001). doi: 10.1007/3-540-48194-X_2
9. Navarro, G., Raffinot, M.: Practical and flexible pattern matching over Ziv-Lempel compressed text. J. Discrete Algorithms **2**, 347–371 (2004)

Distributed Algorithm of Dynamic Multidimensional Data Mapping on Multidimensional Multicomputer in the LuNA Fragmented Programming System

Victor E. Malyshkin[1,2,3] and Georgy A. Schukin[1,3(✉)]

[1] Institute of Computational Mathematics and Mathematical Geophysics,
Siberian Branch of Russian Academy of Sciences, Novosibirsk, Russia
{malysh,schukin}@ssd.sscc.ru
[2] Novosibirsk National Research University, Novosibirsk, Russia
[3] Novosibirsk State Technical University, Novosibirsk, Russia

Abstract. The distributed algorithm Patch with local communications for dynamic data allocation of a distributed multicomputer in the course of an application LuNA fragmented program execution is presented. The objective of the Patch is to decrease the length and as result the volume of communications while the parallel program is executed. Communications include all the internode interactions for data processing, dynamic data allocation, search and balancing. The Patch takes into account the data dependencies and maximally tries to keep the data locality during all the internode interactions.

Keywords: Dynamic data allocation · Dynamic load balancing · Distributed algorithms with local interactions · Fragmented programming technology

1 Introduction

Supercomputer large-scale numerical modeling is now widely used, especially in science. To achieve good performance and scalability of application parallel programs of numerical modeling the effective resources allocation strategies, control, dynamic load balancing and other means usually should be used. Because of that, the complexity of application parallel programming becomes comparable to the complexity of system parallel programming. To simplify the development of parallel programs of numerical modeling, the LuNA system for automatic construction of parallel programs was developed [1–6]. We consider that in visible future LuNA-like systems should significantly or fully eliminate parallel programming from the process of large-scale numerical models creation.

The LuNA automatically assembles an application parallel numerical program out of pieces (fragments) as data as computations. Each fragment of computations (CF) in the course of a LuNA-program execution defines an independent process. Each CF computes the output data fragments (DF) values from CF's input DFs values. Each DF is the single assignment variable of a LuNA-program and each CF is executed once only. Fragmented structure of a LuNA-program is kept during the LuNA-program execution.

© Springer International Publishing AG 2017
V. Malyshkin (Ed.): PaCT 2017, LNCS 10421, pp. 308–314, 2017.
DOI: 10.1007/978-3-319-62932-2_30

This allows to provide DFs and CFs migration between the nodes of a multicomputer and their execution in parallel.

Efficiency of the LuNA fragmented program execution substantially depends on the quality of distributed resources allocation. Below a distributed Patch algorithm with local interactions is described. Patch is intended for dynamic distributed allocation of distributed resources. It is included into the LuNA system and optimized for allocation of multidimensional data on grid-like communication network.

2   Related Works

There are many domain decomposition methods for data distribution and load balancing. One of them is tiled arrays [7–10], where data are decomposed into orthogonal convex tiles, which are then distributed over nodes of a multicomputer. In [7, 9] hierarchical tiling arrays are used. In [10] arbitrary tiles (i.e. the tiles can have different sizes by any dimension and not necessarily be aligned) are allowed. In [8] user-defined decomposition of arrays into tiles is supported. Restriction of tiled arrays is rectangular shape of tiles, which in some cases can prevent achievement of load balance between tiles. Also, decomposition on tiles is assumed to be static and little information is given about possibility of its dynamic change in a distributed way.

Data decomposition into domains is also widely used in molecular dynamics and particle simulations [11–15]. In [11, 13] two or three dimensional domain (field of particles) is partitioned by placing inner vertices into it, thus creating a mesh of domains, each is initially of rectangular form; one domain is assigned to each PE of a multicomputer. For a purpose of load balancing inner vertices can be moved to change domains size and PEs' workload. Load balancing can proceed in distributed form, i.e. each node communicates only with its neighbors, but sometimes global operations are used. Usually convex shape of domains is required, which may sometimes be a restriction. Also up to 8 PEs (in 3D case) can share a single inner vertex, so a movement of this vertex can require a synchronization between all these PEs.

In [14] recursive orthogonal bisection is used for domain decomposition. For load balancing planes of bisections are recursively moved to change domains size. Because nodes and domains assigned to them are organized in a tree, load balancing requires tree traversal and consequently communications between unfixed number of non-neighboring nodes. Also tree topology may not fit for an actual grid network topology.

In [12] domains consist of Voronoi cells which can be moved between domains to achieve load balance. In [15] usual rectangular cells are used in a similar fashion. To preserve communication pattern between neighboring nodes, some cells are marked as permanent, i.e. can't be moved. Because these algorithms are for molecular dynamics applications and use additional information (such as particles and their movement) for load balancing decisions, they may not be readily applicable for more general cases.

We can conclude that desired data distribution and load balancing algorithm should be distributed, use preferably local communications, and be applicable to a broad range of problems.

3   Distributed Algorithm of Dynamic Data Allocation

Previously, a distributed Rope algorithm for dynamic data allocation was developed [1, 2]. Rope algorithm was designed to support general data structures processing on distributed network. This paper presents new algorithm Patch, which is developed to support mesh data structures processing on distributed grid network, where Rope algorithm has some disadvantages. Further sections present description of Patch algorithm and its comparison with Rope.

3.1   Initial Definitions

Numerical algorithms data are usually multidimensional Cartesian meshes (for example, particles-in-cell or Poisson equation solver meshes). For data decomposition the mesh is divided into parallelepipeds (DFs) that form the grid of DFs. Two DFs, whose planes are adjacent, are called adjacent DFs. Two DFs are called neighbor DFs if the value of one of them is computed by application algorithm with the use of the other DF value.

3.2   Patch Algorithm

Rope algorithm used mapping of multi-dimensional mesh of DFs to one-dimensional numerical domain (for example, with Hilbert space-filling curve) for data distribution [16]. To better exploit multidimensional neighborhood, multi-dimensional numerical domain is required. For this reason, the Patch algorithm maps k-dimensional mesh of DFs to n-dimensional numerical domain. The numerical domain is represented as Cartesian grid of regular cells, each cell has its n-dimensional coordinate. Mapping to n-dimensional domain allows all neighbor DFs be mapped to the same or adjacent cells, thus, neighborhood relation on DFs is better preserved. Mapping is fixed during program execution.

Assuming n-dimensional Cartesian grid network topology is used, the whole numerical domain is decomposed into subdomains ("patches") of cells, one subdomain for each node (Fig. 1, left). Cell coordinates are globally ordered through the nodes by all dimensions. There are no restrictions on a shape of domains, but several constraints should be satisfied:

- No empty subdomains are allowed
- Square of subdomain adjacent planes should be minimal in order to minimize the volume of communications between adjacent subdomains
- Each node's subdomain should be adjacent to subdomain of it's neighbor nodes in a network topology.

These constraints are to enable proper functioning of data allocation algorithm, which is described further.

In LuNA system each DF and each CF are mapped dynamically on the nodes of distributed multicomputer at runtime. Additionally, migration of CFs and DFs also demands dynamically to search where CFs and DFs are located. Thus, data allocation algorithm should provide assignment of any DF on any node and search for any DF from any node. Node, which currently holds a cell to which a DF was mapped, is called a

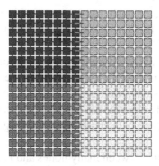 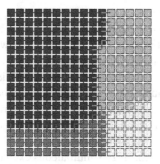

Fig. 1. Distribution of cells on PEs in Patch algorithm, initial (left) and after load balancing (right). Different colors denote different PEs. Adjacency links are shown also. See [16] for dynamic picture.

residence of this DF. If DF's residence is known, the DF can be allocated on the residence node or requested (copied) from it.

To make possible a DF residence determination from any node, in Patch algorithm each cell stores information about current location (node) of all its adjacent cells in a Cartesian grid. The use of this adjacency information makes it possible to find DF's residence from any node for a finite number of steps, using only local communications. First, cell coordinate for required DF is acquired. This coordinate is a constant and can be computed everywhere because DF mapping to cells is fixed. Secondly, if current node doesn't contain a cell with required coordinate, the cell, closest to it in the current node's "patch", is chosen and search can be continued from the neighbor node, adjacent to it in the direction to the required cell. Otherwise, search is over and the residence is found.

Benefits of Patch algorithm are follows:

- Preserves neighborhood of DFs in all dimensions
- Ideally fits for an actual grid network topology
- The worst case DF allocation time is proportional to the diameter of the network topology, which is usually smaller then Rope's algorithm (total number of nodes).

3.3   Dynamic Load Balancing in Patch Algorithm

Diffusion scheme is used in Patch algorithm for dynamic load balancing implementation. All nodes of a multicomputer in a grid topology are divided into overlapping groups of nodes. Each group consists of a central node and all its neighbor nodes in the topology of communicating system. Each cell, assigned to a node, defines a node workload, which is calculated by some formula or criteria, for example, a value of a current total volume of the DFs, on the node mapped to the cell. Total load of a node is a sum of the loads of all cells on the node. Every node keeps the value of its total load and the values of total loads of all its neighbor nodes. Node is considered overloaded, if its total load is greater than average load of all cells of the group, and underloaded otherwise. If a central node of a group is overloaded/underloaded, then some cells should migrate to/from under-loaded/overloaded nodes of the group equalizing their workload (Fig. 1, right). All the CFs, that process migrated DFs, follow the DFs.

Migrated cells are selected in such a way in order to minimize the number of border cells between subdomains, because this number is directly proportional to the volume of communications between the neighbor DFs. Usually, this means selecting connected group of cells lying on subdomains border. Greedy algorithm is used for cells selection. Migration is actually implemented, if the difference between total and average work-loads exceeds a threshold.

To synchronize simultaneous exchange of cells between many nodes and keep cells information consistent, the following transaction mechanism is used. Each transaction comprises transfer of cells from one node to another. If cells are transferred between two nodes, then any other transactions between them are locked. To prevent deadlocks and decide which transaction should be executed next, a random priority is assigned to each transaction. Transaction with max priority is executed first. Because a cell can be adjacent to many nodes' subdomains (not actually participating in transaction), special updates are sent to these nodes to keep adjacency information correct.

Transferring single cells instead of shifting a whole subdomain border allows for more accurate load balancing. Also only two nodes need to be synchronized for each transaction, which allows for more transactions to be done in parallel. Only local communications between neighbor nodes are used during load balancing, which makes the algorithm scalable.

4   Tests

To compare the algorithms, LuNA implementation of Poisson equation solver with an explicit finite-difference scheme on a regular 3D mesh was chosen. Experiments were conducted on cluster with two Intel Xeon E5-2690 processors per PE and Infiniband FDR communication and transport network. GCC 5.3 C++ compiler and MPICH 3.2 MPI library were used.

4.1   Test Results

For testing the regular mesh of 512^3 size was divided into 32^2 three dimensional data fragments. Measurements were done for up to 256 processes (N). Cluster configuration allowed placement up to 4 processes on a single physical node. 1D and 2D grid network topologies were used for testing Rope and Patch algorithms respectively.

Following program execution characteristics were measured: total execution time (ET, seconds), average DF send distance (AvgSD, processes in topology), average summary size of all DFs sent by process (AvgSS, megabytes; also includes DFs move-ment during load balancing), standard deviation from an average computation time per process (AvgCTD, seconds).

Table 1 shows the results when all processes are loaded uniformly and load balancing doesn't need. Showing comparable results of execution time and average migrated data size, Patch, as expected, demonstrates much better average DFs migration distance than Rope due to exploiting of multidimensional DF neighborhood.

Table 1. 512^3 mesh, 32^2 fragments, uniform distribution.

N	1	2	4	8	16	32	64	128	256
ET, Rope	773.8	407.4	217.8	111.9	57.9	31.7	19.7	14.7	17.4
ET, Patch	769.0	409.4	221.3	113.4	68.8	31.7	19.4	12.3	13.6
AvgSD, Rope	0	1	1.5	1.8	2.5	3.24	4.84	6.41	9.66
AvgSD, Patch	0	1	1	1	1	1	1	1	1
AvgSS, Rope	0	20.2	20.2	20.2	15.1	12.6	8.8	6.9	4.7
AvgSS, Patch	0	20.2	20.2	20.2	15.1	12.6	8.8	6.9	4.7

For dynamic load balancing testing initial disbalance was created assigning data and computations on one half of processes only. The goal of balancing was to load all processes as equally as possible. Table 2 shows results of the testing. Patch algorithm generally demonstrates smaller deviation of average computational time, i.e. it managed to load processes better than Rope. Also average DFs migration distance and total size of transferred data are smaller for Patch again because of exploiting more dimensions than Rope. Increased average transferred data sizes and decreased migration distances, comparing with uniform distribution case, are due to data movement during load balancing.

Table 2. 512^3 mesh, 32^2 fragments, non-uniform distribution with dynamic load balancing.

N	2	4	8	16	32	64	128	256
AvgCTD, Rope	53.4	15.99	55.68	47.21	25.18	13.51	7.44	3.93
AvgCTD, Patch	58.11	23.7	56.49	33.33	21.39	12.56	7.01	3.71
AvgSD, Rope	1	1.19	1.42	1.48	1.46	1.80	1.30	1.44
AvgSD, Patch	1	1.03	1.18	1.27	1.31	1.24	1.16	1.04
AvgSS, Rope	7787.6	6246.7	2821.6	1032.2	1192.6	598.0	566.9	392.0
AvgSS, Patch	9917.9	4955.9	2225.3	1226.2	714.6	444.1	193.8	91.8

5 Conclusion

The problematics of data allocation and load balancing automation for implementation of large-scale numerical models for supercomputers are considered. The Patch algorithms for dynamic multidimensional data allocation in LuNA fragmented programming system is proposed. Comparison tests of the algorithm are presented.

References

1. Malyshkin, V.E., Perepelkin, V.A., Schukin, G.A.: Scalable distributed data allocation in LuNA fragmented programming system. J. Supercomput. **73**(2), 726–732 (2017). Springer, US
2. Malyshkin, V.E., Perepelkin, V.A., Schukin, G.A.: Distributed algorithm of data allocation in the fragmented programming system LuNA. In: Malyshkin, V. (ed.) PaCT 2015. LNCS, vol. 9251, pp. 80–85. Springer, Cham (2015). doi:10.1007/978-3-319-21909-7_8
3. Malyshkin, V.E., Perepelkin, V.A.: LuNA fragmented programming system, main functions and peculiarities of run-time subsystem. In: Malyshkin, V. (ed.) PaCT 2011. LNCS, vol. 6873, pp. 53–61. Springer, Heidelberg (2011). doi:10.1007/978-3-642-23178-0_5
4. Malyshkin, V.E., Perepelkin, V.A.: Optimization methods of parallel execution of numerical programs in the LuNA fragmented programming system. J. Supercomput. **61**(1), 235–248 (2012)
5. Malyshkin, V.E., Perepelkin, V.A.: The PIC implementation in LuNA system of fragmented programming. J. Supercomput. **69**(1), 89–97 (2014)
6. Kraeva, M.A., Malyshkin, V.E.: Assembly technology for parallel realization of numerical models on MIMD-multicomputers. J. Future Gener. Comput. Syst. **17**(6), 755–765 (2001)
7. Gonzalez-Escribano, A., Torres, Y., Fresno, J., Llanos, D.R.: An extensible system for multilevel automatic data partition and mapping. J. IEEE Trans. Parallel Distrib. Syst. **25**(5), 1145–1154 (2014). IEEE
8. Chamberlain, B.L., Deitz, S.J., Iten, D., Choi, S.-E.: User-defined distributions and layouts in chapel: philosophy and framework. In: 2nd USENIX Conference on Hot Topics in Parallelism, HotPar 2010, p. 12. USENIX Association, Berkeley (2010)
9. Bikshandi, G., Guo, J., Hoeflinger, D., Almasi, G., Fraguela, B.B., Garzarán, M.J., Padua, D., von Praun, C.: Programming for parallelism and locality with hierarchically tiled arrays. In: 11th ACM SIGPLAN Symposium on Principles and Practice of Parallel Programming, PPoPP 2006, pp. 48–57. ACM, New York (2006)
10. Furtado, P., Baumann, P.: Storage of multidimensional arrays based on arbitrary tiling. In: 15th International Conference on Data Engineering, pp. 480–489. IEEE (1999)
11. Begau, C., Sutmann, G.: Adaptive dynamic load-balancing with irregular domain decomposition for particle simulations. J. Comput. Phys. Commun. **190**, 51–61 (2015). Elsevier B.V.
12. Fattebert, J.-L., Richards, D.F., Glosli, J.N.: Dynamic load balancing algorithm for molecular dynamics based on Voronoi cells domain decompositions. J. Comput. Phys. Commun. **183**(12), 2608–2615 (2012). Elsevier B.V.
13. Deng, Y., Peierls, R.F., Rivera, C.: An adaptive load balancing method for parallel molecular dynamics simulations. J. Comput. Phys. **161**(1), 250–263 (2000). Elsevier B.V.
14. Fleissner, F., Eberhard, P.: Parallel load-balanced simulation for short-range interaction particle methods with hierarchical particle grouping based on orthogonal recursive bisection. Int. J. Numer. Meth. Eng. **74**(4), 531–553 (2008). Wiley, Ltd.
15. Hayashi, R., Horiguchi, S.: Efficiency of dynamic load balancing based on permanent cells for parallel molecular dynamics simulation. In: 14th International Parallel and Distributed Processing Symposium, IPDPS 2000, pp. 85–92. IEEE (2000)
16. Rope and Patch demonstration page. http://ssd.sscc.ru/en/algorithms

Probabilistic Causal Message Ordering

Achour Mostéfaoui[(✉)] and Stéphane Weiss

LS2N, Université de Nantes, 44322 Nantes, France
achour.mostefaoui@univ-nantes.fr

Abstract. Causal broadcast is a classical communication primitive that has been studied for more then three decades and several implementations have been proposed. The implementation of such a primitive has a non negligible cost either in terms of extra information messages have to carry or in time delays needed for the delivery of messages. It has been proved that messages need to carry a control information the size of which is linear with the size of the system. This problem has gained more interest due to new application domains such that collaborative applications are widely used and are becoming massive and social semantic web and linked-data the implementation of which needs causal ordering of messages. This paper proposes a probabilistic but efficient causal broadcast mechanism for large systems with changing membership that uses few integer timestamps.

Keywords: Asynchronous message-passing system · Happened before relation · Logical clock · Message causal ordering · Vector clock

Nowadays, we are facing an increasing number of collaborative applications. The nature of these applications is diverse as they appear as web 2.0 applications such as blogs, wikis or even social networks, as well as applications for mobile devices such as foursquare, yelp, latitude. Moreover, semantic web (web 3.0) and now social semantic web and linked-data (web 4.0) such as DBpedia are gaining more and more interest. The common point to these applications is that they gather the outcome of numerous users in order to provide a service for their users. The more users participate, the more content is created, attracting more users. This virtuous circle tends to create very large scale systems. However, while the content is created by users for users, for many applications, the underlying architecture remains centralized, leading to scalability issues as well as privacy and censorship threats. Stating this observation, several work envision decentralized architectures. The idea behind this concept, is to put the users as nodes of the network, allowing direct communication between users.

Currently, the development of such applications is restricted by several scientific problems. Among them, the problem of data replication has been investigated for many years, and have provided several approaches such as appropriate replicated data structures [11,14] and programming languages [1]. Hence replicated data can be enriched, updated and queried. However, the implementation

© Springer International Publishing AG 2017
V. Malyshkin (Ed.): PaCT 2017, LNCS 10421, pp. 315–326, 2017.
DOI: 10.1007/978-3-319-62932-2_31

of these operations has an underlying requirement: causally ordered communication (causal order for short) [15, 16]. Informally a causal communication primitive imposes some restrictions on the delivery order of sent messages. It can be seen as an extension of the FIFO channel property to a whole communication network. Hence, a causal broadcast communication service imposes that a message is delivered to a some process only if all the messages that have been delivered in its past have been already delivered.

Unfortunately, causal communication has a cost that can be high either in time (message exchanges) or in space (the size of control information carried by messages). This cost becomes unacceptable when we consider a very large scale network of nodes with churn. Moreover, if the set of participating users changes over time, one needs to offer a join/leave decentralized procedure that is theoretically impossible to implement in asynchronous systems [7]. Interestingly, in a real setting, depending on the system we consider, the probability to deliver in a non causal order two messages the sending of which are causally related may be quite low. For example, if the time between the generation of two messages on each peer is bigger than the transit time of a message, most of messages will be received in the causal order without any explicit control or synchronization. This observation is in favor of a probabilistic mechanism.

In this paper, we propose a probabilistic causal broadcast that provides a causal communication with high probability at a low cost for very large systems while allowing continuous joins and leaves. Of course it may happen, in few situations, that causal ordering is not respected. The proposed solution is then evaluated from a theoretical point of view and by simulation.

1   Related Work

The first causal broadcast mechanism was introduced in the ISIS system [2]. The simplest way to implement causal communication consists in piggybacking on each messsage a process want to send the whole set of messages it has delivered prior to this sending. Of course, this is very costly and there is a need to some kind of garbage collector. Otherwise, prior work mainly use either a logical structure (central node, tree, ring, etc.) or are based on the use of timestamps. A timestamp is an integer value that counts events (possibly not all events). A vector clock is a vector of such counters. The first solution based on vector clocks for a broadcast primitive has been proposed is [13] (a solution based on a matrix of counters has been proposed in [12] for point-to-point communication). Vector clocks introduced simultaneously by [6, 9] have been proved to be the smallest data structure that can capture exactly causality [4]. Moreover, vector clocks require to know the exact number of sites involved in the application. As an example, the churn (intempestive join and leave of processes) and the high (and unknown) number of processes make the use of vector clocks unrealistic.

Torres-Rojas and Ahamad presented an approach based on *plausible clocks* [17]. Its aim is to trade the quality of the detected causality (number of false positives and false negative) among events (messages sending events) against timestamp size. When using a vector clock of size 1, a plausible clock boils down to Lamport's clocks [8] then as the timestamp size increases, a more and more accurate causal relation is encoded. Finally, when considering a timestamp size equal to the total number of processes, plausible clocks meet vector clocks. In a vector clock, the entry j of the vector managed by a given process p_i counts the number of messages broadcast by process p_j, to the knowledge of p_i. Indeed due to asynchronism, the different processes do not have the same view of the state of the system at a given time instant. The approach of Torres-Rojas and Ahamad consists in associating several processes to the same entry of each vector clock.

The approach presented in this paper is an extension to the one of Torres-Rojas, namely each entry is associated to several processes and moreover, to each process are associated several entries of the vector clock. To summarize, let us consider the triplet (a, b, c). Where a is the size of the system (number of processes), b the size of the vector and c the number of entries associated with each process. A Lamport clock is $(n, 1, 1)$ where n is the total number of users in the system, a vector clock is $(n, n, 1)$, a plausible clock is $(n, r, 1)$ and the proposed approach is (n, r, k) (r and k being two constants $n \geq r \geq k$).

2   System Model

When we consider the application level, the different users, nodes, processses or whatever we call them share common information by mean of replication to be able to tolerate chrashes and unexpected leaves (each process manages a local copy of part of the whole set of data). The differents processes interact by mean of operations (insert/delete/update a piece of data, make a query, etc.). The frequency and the distribution of operations through time and space depend on the application. At the underlying level, an operation will entail a change in the local state of a process and possibly the sending of messages to inform the other processes as the system is message-passing (no shared memory).

At the abstraction level considered in this paper, a distributed computation is a large set $\varPi$ of n processes/users (n and $\varPi$ are not necessarily known to the different processes). Let p_i and p_j denote any processes in $\varPi$. We assume that processes generate messages at arbitrary rates. Messages are sent to all processes using a broadcast mechanism (broadcast sending primitive). Any process $p_i \in \varPi$ generates three kinds of events. An event e could be a local event, a send event or a delivery event. Local event induce no interaction with other processes and thus will be omitted in the rest of the paper. The events produced by a distributed computation are ordered by Lamport's *happened-before* relation [8].

Definition 1 (Happened-before Relation [8]). *We say that event e_1 happened before event e_2 denoted by $e_1 \rightarrow e_2$ if:*

- *e_1 occurred before e_2 on the same process, or*
- *e_1 is the send event of some message m and e_2 is the delivery event of the same message by some process, or*
- *there exists an event e_3 such that $e_1 \rightarrow e_3$ and $e_3 \rightarrow e_2$ (transitive closure).*

318   A. Mostéfaoui and S. Weiss

Let us note $send(m)$ the sent event of a message m and $del(m)$ the associated delivery event. Note that $del(m) \neq rec(m)$ the receive event. The receive event corresponds to the arrival of a message to the underlying communication level of some process. The delivery of a message corresponds to the arrival of the message at the application level leading to the use of its content. When considering two messages m_1 and m_2, we say that $m_1 \to m_2$ if $send(m_1) \to send(m_2)$.

A distributed computation *respects causal order* if for any pair of messages (m_1, m_2) the following holds $send(m_1) \to send(m_2) \Rightarrow del(m_1) \to del(m_2)$.

To ensure the aforementioned property an arrived message m, event $rec(m)$, at a destination process p_i can be possibly delayed until all the messages sent or delivered by the sending process p_j before the sending of m have been already delivered by the receiving process p_i.

3 Probabilistic Causal Broadcast

Several works in the literature propose to reduce the communication cost to increase scalability by proposing probabilistic solutions. One example is the probabilistic broadcast [5]. While traditional broadcast ensures that each message is delivered exactly once to each recipient when the sender does not crash, a probabilistic broadcast ensures this only with high probability. In addition, this message can be received several times, requiring a mechanism to discard duplicated messages. This paper introduces two definitions: the probabilistic causal ordering mechanism and the probabilistic causal broadcast. Existing mechanisms such as vector clocks are often used to provide a perfect causal ordering mechanism. We believe that their cost is not compatible with large systems, and thus we propose to use a probabilistic causal ordering mechanism that ensures a causal delivery with high probability.

3.1 A Probabilistic Causal Ordering Mechanism

The main idea of the proposed solution is to associate with each process p_i a vector of integer values V_i of size $R < N$ (N being the total number of processes in the system). This vector acts as a logical clock that allows to timestamp a subset of the events generated by this process. This timestamping allows to test whether an event occured before another one. Mainly, the proposed protocol associates a logical date with each send event and this date is attached to the message and is called its timestamp. When a message is received, the receiving process compares its local logical clock with the timestamp carried by the received message. This allows it to know whether there exist messages sent causally before it and that have not yet been delivered. As soon as a message can be delivered it is given to the upper layer application and the local clock is updated to take into account that this message has been delivered.

We denote $V_i[j]$ with $0 \leq j < R$ the j-th entry of the vector clock of the process p_i. A vector clock assigns exactly one entry to each process. With a plausible clock, each process is assigned only one entry, but one entry is assigned to several processes. Finally, our approach proposes to assign several entries to each process, each entry being assigned to several processes. We denote $f(p_i)$ the set of the entries assigned to p_i.

Probabilistic Causal Ordering Delivery Mechanism. The proposed probabilistic causal ordering delivery mechanism is an adaptation of the classical and well-known causal delivery mechanism [2]. When a process p_i wants to broadcast a message m, it executes Algorithm 1 given below. First, process p_i increments all its assigned set of entries $f(p_i)$ in its local vector. Then, a copy of this local vector is attached to the message to be sent. $m.V$ denotes the vector timestamp attached to message m. Finally, m is broadcast to all processes.

Input: m: message to broadcast
$\forall x \in f(p_i), V_i[x] = V_i[x] + 1;$
$m.V = V_i;$
Broadcast(m);
   Algorithm 1: Broadcast of a message m by process p_i

Input: m: message received by p_i from p_j
waitUntil$((\forall x \in f(p_j), V_i[x] \geq m.V[x] - 1) \wedge \forall k \notin f(p_j), V_i[k] \geq m.V[k]);$
$\forall x \in f(p_j), V_i[x] = V_i[x] + 1;$
deliver(m);
   Algorithm 2: Upon reception of message m by process p_i

When a message m broadcast by a process p_j is received by a process p_i, this process executes Algorithm 2. A message m received by a process p_i from a process p_j is queued until it is considered as causally ready, namely all the messages m' sent causally before it $(m' \rightarrow m)$ have been already delivered by process p_i to the application level. Note that:

On the one side, the $f(p_j)$ entries are at least as high as the local vector of process p_j before it generates that message: $\forall x \in f(p_j), V_i[x] \geq m.V[x] - 1$. This means that all the messages sent by process p_j but message m are already known at process p_i. On the other side, the other entries of the vector are at least as high as the local vector of process p_j before it generates that message: $\forall k \notin f(p_j), V_i[k] \geq m.V[k]$. This means that at least all the messages that have been delivered to process p_j before it broadcast message m are known at process p_i before p_i delivers m to its application level.

When the delivery test of a message m holds, process p_i increments the entries of its local vector that belong to the set $f(p_j)$ of entries of its local vector before delivering the message to the application. Hence p_i has recorded the information "m has been delivered to the p_i" in its local vector.

Assume that three processes p_i, p_j and p_k are in the same initial state, meaning they have generated no messages yet, and all values in their vector are set to 0. Also, we consider here that each process has a vector of $R = 4$ entries and each process is assigned $K = 2$ entries. We assume that $f(p_i) = \{0, 1\}$, $f(p_j) = \{1, 2\}$ and $f(p_k) = \{3, 4\}$.

On that state, p_i generates a first message called m. If we assume that $f(p_i) = \{0,1\}$, then after applying the Algorithm 1 at p_i, its vector becomes $[1,1,0,0]$. Notice that it is this vector ($[1,1,0,0]$) that is attached to m. This message is sent to all other processes, here we represent only p_j and p_k. We assume that p_j receives m first: Algorithm 2 is applied and, p_j's vector is updated to $[1,1,0,0]$. Then p_j generates a new message m'. Assuming that $f(p_j) = \{1,2\}$, the generation of m' leads to update p_j's vector to the value $[1,2,1,0]$ which is piggybacked with the message m'. As m' is generated after m, m' is causally dependent of m ($m \to m'$). Although message m' is broadcast and will eventually reach p_i, we only represent its reception by process p_k for the sake of simplicity.

When p_k receives m', the vector p_k is $[0,0,0,0]$ while the vector attached to m' is $[1,2,1,0]$. The delivery of m' is delayed because its delivery condition (see Algorithm 2) is not satisfied ($V_k[1] < (m.V[1] - 1)$ and $V_k[0] < m.V[0]$).

The reception of m turns p_k's vector into $[0,0,1,1]$ which fulfills the condition for delivering m'.

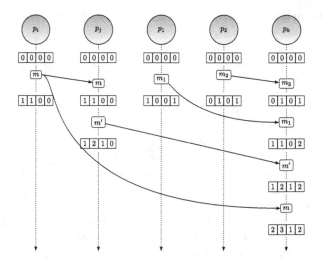

Fig. 1. Example of possible delivery error

This protocol cannot be perfect as it uses control information the size of which is smaller than a vector clock of size N that has been proved to be minimal for ensuring causal delivery of messages. Indeed, it is possible that if process p_k receives some set of messages before receiving m', the vector of p_k could have been updated in such a way that p_k believes that m' is causally ready. This scenario is illustrated by Fig. 1. In addition to the previous processes, we now consider p_1 and p_2 whose assigned entries are respectively $f(p_1) = \{0,3\}$ and $f(p_2) = \{1,3\}$. Processes p_1 and p_2 generate m_1 and m_2, which are received by p_k before m'.

The reception of m_2 and m_1 updates the vector of p_k to the value $[1,1,0,2]$. When m' arrives, p_k evaluates to true the delivery condition of Algorithm 2 and delivers m' while m has not been received yet. This error comes from the fact that all entries of p_i are matched by the combination of the entries of p_1 and p_2: $f(p_i) \subseteq (f(p_1) \cup f(p_2))$. It is interesting to notice that, as each set of entries is assigned to at most one process, the error occurs only if we have at least two concurrent messages, here m_1 and m_2. If only one concurrent message is received, the protocol delivers the message in causal order.

In addition, it is easy to see that the proposed protocol never delays the delivery of an arrived message m' at some process p_i if all the messages m that have been sent causally before m' (i.e. $m \rightarrow m'$) have been already delivered by process p_i. Indeed, between two consecutive sendings of messages, on the same process, the increment is one for some entries and when a message is delivered the corresponding entries are at least incremented by 1 then if the first one is delivered, the second cannot be blocked as the corresponding entries are augmented by at least 1. The same happens for messages sent by different processes as in this case the sender of the second message has necessarily delivered the first message to establish the causal dependency between the two messages. Finally the proofs that all messages are eventually delivered and that a message causally ordered is never delayed are given in [10].

Generation and Distribution of the Sets of Keys. The example given above gives the intuition why the generation of the sets of keys (entries in the vector clock) is the core of the approach, and how it heavily affects the accuracy of the resulting protocol. In this section, we present our approach to assign sets of keys to processes.

Perfect distribution of keys. We assume the existence of a mapping function f used to assign to each process a set of K entries of the vector such that, the function f returns exactly K distinct values between 0 and $R - 1$ (R being the total number of entries, i.e. the size of the vector clock). Moreover, the values are equally distributed among processes. In other words, for all subsets of s values assigned to x processes, there exists no other subset of s values assigned to more than $x + 1$ processes, or less than $x - 1$ processes, for $1 \leq s \leq K$.

Distributed key assignment algorithms. Finding the best solution is a difficult problem. Moreover, such an algorithm would not support dynamicity. Indeed, the addition or the removal of one process would lead to re-assign all new entries to all or part of the processes. Therefore, to select the K values, we propose to use a random algorithm. Each process generates randomly a value called set_{id} chosen between 1 and C_K^R. An algorithm that ensures all peers have different identities, all the generated sets are distinct, and the intersection between all the sets is at most $K - 1$ is proposed in [10].

3.2   Detecting Delivery Errors

Most of the applications that require causal ordering of messages assume that no message is wrongly delivered. The proposed approach that aims to scale, may deliver a message before a message that causally precedes it. This would lead to inconsistencies, however, we make the assumption that a recovery procedure does exist (e.g., anti-antropy). This procedure may be costly, and we must determine when it is required. A simple solution could be to run it at an arbitrarily chosen period of time. However, this period impacts the system performance: on the one hand, if it is underestimated, we will waste time by running unnecessary costly recovery procedure. On the other hand, the over estimation of this period will cause heavy changes on the application and will hurt the usability of the system and the recovery from this inconsistent global state. Two mechanisms to detect delivery errors and to improve the accuracy of this protocol are proposed in [10].

4   Theoretical Error Analysis

In this section, we evaluate the error rate of our probabilistic causal ordering mechanism depending on the estimation of the system load and the different parameters of our approach (N, R and K).

Let us first say that the higher R the better is the resulting protocol (better in the sense less messages that violate causal ordering). At the extreme, if $R = 1$ we have a linear clock similar to Lamport's clock. At the other extreme, $R = N$ we get the perfect and optimal (no causal order violation and delivery at the earliest) solution to causal ordering. Concerning K, it is easy to see that the situation is less clear. Indeed, if K is too small; at the extreme $K = 1$ we get the plausible clocks of Torres and Ahamad, and we are sure that if two process are assigned the same entry they will interfere at each message sending. At the other extreme $K = R$ the protocol boils down to the use of a Lamport's clock merging all processes within one single entry. Hence the intuition of the proposed approach is that there is some value of K to determine that lies between 1 and R that is optimal. The aim of this theoretical analysis and the simulations presented in the coming sections is to determine the best value for K and to evaluate the impact of K on the error rate.

First, we need to compute the probability that a message m is delivered before a message m' that precedes it causally ($m' \rightarrow m$). As explained in Sect. 3.1 if such messages m and m' are received by some process p_i with m received first, then m can be delivered before m' with a probability that we note P_{nc} the probability that a message m bypasses a message m' sent causally before it. Depending on the system we consider, this probability may be quite low. Therefore, we have to consider this probability to dimension precisely the size of the vector and the number of entries each process chooses. A necessary condition, though not sufficient, to wrongly delivered a message is that this message is received after a preceding message, and the entries of the delayed messages have been all matched by concurrent messages. Let us note P_{error} the probability that all the K entries of the missing message are covered by a set of concurrent messages (see the example given in Sect. 3.1). Consequently, the probability P of wrongly delivering a message is bounded by the probability that a delayed message has its entry matched concurrently $P \leq P_{nc} * P_{error}$.

The probability that a message is replaced by a set combination of previous messages is computed following the same scheme as the false positive error of a bloom filter [3]. The probability that one entry is incremented is $1/R$. So the probability that it is not incremented is: $1 - 1/R$ and that it is not set by X messages is $(1 - 1/R)^{k*X}$. Then the probability that one entry is incremented by X messages is $1 - (1 - 1/R)^{k*X}$. Finally the probability of an error delivery is $(1 - (1 - 1/R)^{k*X})^k$.

We need to find the value K that minimizes the probability of an error. We can easily show that $(1 - (1 - 1/R)^{k*X})^K$ is minimal when $K_{min} = ln(2) * \frac{R}{X}$.

4.1 Experiments

In this section, we detail the model used to run our simulation. In the first part, we show that the estimation of the optimal value of K is sound. Then, in a second part, the experiments are run using this optimal value. We, therefore, show the behavior of the mechanism based on the size of the vector. In a third part, we show the accuracy of the estimation of the probability of an error occurrence.

Methodology: In order to evaluate our proposal, we have developed a simple event-based simulator. Each process generates messages according to a Poisson distribution of parameter λ. Each message has its own propagation time d described as a random value which follows a Gaussian distribution $N(\mu, \sigma^2)$ law. Each process receives a message whose propagation time is according to a $N(d, \sigma_m^2)$. In the average, each node generates a message each second. The message propagation time d follows a normal distribution law $N(100, 20)$ and the skew between a message reception on all nodes follows also a normal distribution law $N(d, 20)$.

Detecting Delivery Errors. One of the challenges to evaluate the proposed approach is to measure the error rate. When a message is said to be "causally ready" by our mechanism, we need to verify that it is really causally ready, therefore, in our simulator we also need to implement a perfect causal broadcast. This additional mechanism should have the lowest cost possible as it limits the simulator scalability. To detect an error, we must know all the messages the sending of which happened-before a given message. A simple solution would be to attached a set of messages to each sent message. Obviously, this would limit drastically the scalability of the simulator. Therefore, we use a mechanism based on vector clocks. Unfortunately, a vector clock cannot capture wrongly delivered messages.

Indeed, when a non-causally ready message arrives, the causal ordering mechanism delays it, while in our case, it may be delivered to the application. To deal with this case, we update the local vector clock by taking the maximum of the local vectors and the wrongly delivered messages. Therefore, missing messages will be dropped by the perfect causal delivery mechanism. Detecting precisely if the missing messages are causally ready is costly, instead we propose two metrics. The first one, ϵ_{min}, simply assumes that all missing messages are delivered in a causal order, while the second one ϵ_{max} assume that all of them are delivered in a non-causal order. Finally, we have two bounds on the error rate: the lower bound ϵ_{min} and the upper bound ϵ_{max}.

Choosing the Optimal Number of Keys. The first step in validating our approach is to verify that we choose the best value for the parameter K. Therefore, we ran several experiments by changing only the value of K and then compare the value that minimizes it with the theoretical optimal value. Figure 2 shows the error rate for respectively 500, 1000, 1500 and 2000 peers. In this experiment, the average number of messages received by a process is constant (200 msgs/s) and, the number of exchanged messages is more than a hundred million.

As our simulation considers an average message propagation time of 100 ms, the average number of messages that are received concurrently is 20. We use 100-entry vectors, hence, the optimal number of keys is theoretically $ln(2)*100/20 \approx 3.5$ and the experimental results show that the value for K that minimizes the error rate for this configuration is 4.

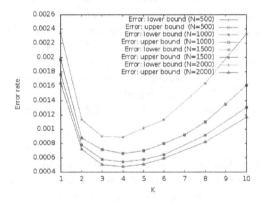

Fig. 2. Number of errors for different value of K (Theoretical best value is 3.5)

Impact of the Different Parameters. For the different simulations, we assumed a system composed on $n = 1000$ processes each managing a vector of 4 entries ($K = 4$) and generates a message every 5 s in the average ($\lambda = 5000$). As a real system may behave differently from the estimation, we are now interested in the impact on the error rate when we vary only one parameter. We studied respectively the effect on the error rate of the message generation rate λ, the total number of process in the system, and for different constant number of message per process. The different results are given in [10]. The different simulations show that indeed, it is not λ and n by themselves that directly impact the error rate but the "concurrency". We mean by concurrency the mean number of messages that are broadcast during the transit time of some message (latency of the network).

5   Conclusion

In this paper we presented a new approach that allows to heavily reduce the cost of causal broadcast communication primitive. This reduction of the cost leads to a small rate of errors. The errors being the cases when a message is delivered while there are causally related messages that need to be delivered that are not yet delivered. We have shown that the approach is theoretically sound. The main parameter K optimizes the protocol and may vary from 1 and R the two extreme already existing cases. The second contribution of the paper is an alert mechanism that allows to check the bad cases. In case there is no alert, we are sure there is no error.

Acknowledgments. This work has been partially supported by the Franco-German DFG-ANR Project DISCMAT (40300781) devoted to connections between mathematics and distributed computing, and the French ANR project O'Browser (ANR-16-CE25-0005-01) devoted to decentralized computing on networks of browsers.

References

1. Alvaro, P., Conway, N., Hellerstein, J.M., Marczak, W.R.: Consistency analysis in bloom: a CALM and collected approach. In: Proceedings of the CIDR 2011, pp. 249–260 (2011)
2. Birman, K.P., Joseph, T.A.: Reliable communication in the presence of failures. ACM Trans. Comput. Syst. **5**, 47–76 (1987)
3. Bloom, B.H.: Space/time trade-offs in hash coding with allowable errors. Commun. ACM **13**(7), 422–426 (1970)
4. Charron-Bost, B.: Concerning the size of logical clocks in distributed systems. Inf. Process. Lett. **39**, 11–16 (1991)
5. Eugster, P.T., Guerraoui, R., Handurukande, S.B., Kouznetsov, P., Kermarrec, A.M.: Lightweight probabilistic broadcast. In: DSN, pp. 443–452 (2001)
6. Fidge, C.: Logical time in distributed computing systems. Computer **24**(8), 28–33 (1991)
7. Fischer, M.J., Lynch, N.A., Paterson, M.: Impossibility of distributed consensus with one faulty process. J. ACM **32**(2), 374–382 (1985)
8. Lamport, L.: Time, clocks, and the ordering of events in a distributed system. Commun. ACM **21**(7), 558–565 (1978)
9. Mattern, F.: Virtual time and global states of distributed systems. In: Proceedings of the International Workshop on Parallel and Distributed Algorithms, pp. 215–226 (1989)
10. Mostefaoui, A., Weiss, S.: A Probabilistic Causal Message Ordering Mechanism. Research report, Université de Nantes (2017). Open archive HAL ref. hal-01527110
11. Oster, G., Urso, P., Molli, P., Imine, A.: Data consistency for p2p collaborative editing. In: CSCW, pp. 259–268 (2006)
12. Raynal, M., Schiper, A., Toueg, S.: The causal ordering abstraction and a simple way to implement it. Inf. Process. Lett. **39**(6), 343–350 (1991)
13. Schiper, A., Eggli, J., Sandoz, A.: A new algorithm to implement causal ordering. Distrib. Algorithms **392**, 219–232 (1989)

14. Shapiro, M., Preguiça, N., Baquero, C., Zawirski, M.: Conflict-free replicated data types. In: Défago, X., Petit, F., Villain, V. (eds.) SSS 2011. LNCS, vol. 6976, pp. 386–400. Springer, Heidelberg (2011). doi:10.1007/978-3-642-24550-3_29

15. Sun, C., Jia, X., Zhang, Y., Yang, Y., Chen, D.: Achieving convergence, causality preservation, and intention preservation in real-time cooperative editing systems. ACM Trans. Comput.-Hum. Interact. 5(1), 63–108 (1998)

16. Terry, D., Demers, A., Petersen, K., Spreitzer, M., Theimer, M., Welch, B.: Session guarantees for weakly consistent replicated data. In: Proceedings of the PDIS, pp. 140–149 (1994)

17. Torres-Rojas, F.J., Ahamad, M.: Plausible clocks: constant size logical clocks for distributed systems. Distrib. Comput. 12(4), 179–195 (1999)

An Experimental Study of Workflow Scheduling Algorithms for Heterogeneous Systems

Alexey Nazarenko and Oleg Sukhoroslov$^{(\boxtimes)}$

Institute for Information Transmission Problems of the Russian Academy of Sciences,
Moscow, Russia
nazar@phystech.edu, sukhoroslov@iitp.ru

Abstract. The paper studies the efficiency of nine state-of-the-art algorithms for scheduling of workflow applications in heterogeneous computing systems (HCS). The comparison of algorithms is performed on the base of discrete-event simulation for a wide range of workflow and system configurations. The developed open source simulation framework based on SimGrid toolkit allowed us to perform a large number of experiments in a reasonable amount of time and to ensure reproducible results. The accuracy of the used network model helped to reveal drawbacks of simpler models commonly used for studying scheduling algorithms.

Keywords: Distributed computing · Heterogeneous systems · Scheduling · Workflow · Simulation

1   Introduction

Heterogeneous computing systems (HCSs) composed of different computational units or standalone resources, which can be local or geographically distributed, are widely used nowadays for executing parallel applications. Workflows [13] is an important class of such applications that consist of many tasks with logical or data dependencies which can be modeled as directed acyclic graphs (DAGs).

The efficiency of executing workflows in HCS critically depends on the methods used to schedule the workflow tasks, i.e. decide when and which resource must execute the tasks of the workflow. The main objective is to minimize the overall completion time or makespan subject to possible additional constraints such as meeting a deadline or using a fixed budget. In comparison to homogeneous systems, the task scheduling problem in HCS is more complicated because of the different execution rates of individual resources and different communication rates of links between these resources.

The DAG scheduling problem has been shown to be NP-complete [9], even for the homogeneous case. This makes it practically impossible to obtain the optimal schedule even for the simplest formulations of practical interest. Therefore the research effort in this field has been mainly to obtain low complexity heuristics that produce good schedules. Since the late 1990s and until now, a multitude of workflow scheduling algorithms [18] based on different heuristics

© Springer International Publishing AG 2017
V. Malyshkin (Ed.): PaCT 2017, LNCS 10421, pp. 327–341, 2017.
DOI: 10.1007/978-3-319-62932-2_32

and metaheuristics have been proposed. While each scheduling algorithm is generally compared by its authors with the best known ones, such comparisons are ad-hoc, use different methodologies and assumptions, and can not be easily reproduced by other researchers. Thus there is a lack of comprehensive and reproducible comparative studies of such algorithms for a wide range of application and system configurations.

Simulation, involving computer modeling of the process of application execution in HCS, is widely used in scheduling algorithm research. In comparison to the full-scale experiments on real systems, simulation allows to perform a statistically significant number of experiments in a reasonable amount of time while ensuring the reproducibility and having moderate hardware resource requirements. While individual researchers often rely on different simulators and underlying models, the comparative study necessitates the use of a common simulation framework with accurate and validated models.

In this paper we attempt to address the aforementioned issues by comparing the performance of nine state-of-the-art algorithms for scheduling of workflows in HCS using the developed simulation framework. The proposed framework provides tools for implementation of scheduling algorithms, generation of synthetic systems and workflows, execution of simulation experiments and analysis of results. It leverages SimGrid toolkit for the discrete-event simulation of parallel applications in distributed environments. In contrast to WorkflowSim [7], a previously published toolkit for simulating workflows, the developed framework includes more algorithms and relies on a thoroughly validated network model implemented in SimGrid. The obtained experimental results, while confirming the advantage of HEFT [15] and Lookahead [5] algorithms, also provide a strong evidence against the widely used simple network models that disregard network topology and bandwidth allocation.

The paper is structured as follows. Section 2 provides a brief description of the studied algorithms. Section 3 introduces the used system and application models along with the developed simulation framework. Section 4 presents and discusses the results of simulation experiments. Section 5 concludes and discusses future work.

2   Workflow Scheduling Algorithms

For our experiments, we have selected a number of scheduling algorithms. To assess the possible advantages of static scheduling, we have employed a well-known Heterogeneous Earliest Finish Time (HEFT) [15] algorithm and a few related list scheduling heuristics, using the simplest dynamic scheduling algorithm (OLB) as a reference point. We have also added a few more advanced dynamic algorithms filling the gap between whole workflow analysis of static methods and per-task planning of OLB.

In this section we provide a brief overview of each studied algorithm. To describe the algorithms, we will introduce a few common notations:

- Individual workflow tasks — T_a;
- Data transfer size between task T_a and its child T_b — c_{ab};
- Individual computing resources — R_i;
- Estimated execution time of a task on a particular resource — $EET(T_a, R_i)$. It must be noted that obtaining the EET is a separate and non-trivial problem. In this study, we compute EET as the ratio of known computational complexity of the task to the resource performance;
- Estimated communication time between tasks — $ECOMT(T_a, T_b, R_i, R_j)$. Yet again, it may be difficult to estimate this value on practice. In this study, we use a simple yet widely applied Hockney's model:

$$ECOMT(T_a, T_b, R_i, R_j) = \frac{c_{ab}}{\text{bandwidth between } R_i \text{ and } R_j} + \text{network latency}$$

- Estimated task start time — $EST(T_a, R_i)$. At any given moment, non-zero values of EST show that either the resource R_i is currently busy or that the task T_a is not ready to run yet.
- Estimated task completion time on a particular — $ECT(T_a, R_i)$. This value can be computed as

$$ECT(T_a, R_i) = EST(T_a, R_i) + \max_{T_b \in \text{parents of } T_a} (ECOMT(c_{ab})) + EET(T_a, R_i)$$

2.1 Static Algorithms

Static algorithms schedule all workflow tasks before the actual execution of the workflow. These algorithms inherently rely on some performance models to estimate the task completion and communication times.

It is worth noting that, while this approach does not take into account the dynamic nature of real HCSs or any inaccuracies in used models, it could be used in a dynamic setting by recomputing the schedule during the workflow execution if some of the previous assumptions are violated.

Heterogeneous Earliest Finish Time (HEFT): Probably the most cited workflow scheduling algorithm [15]. The tasks are sorted in descending order of their rank computed as

$$rank(T_a) = \overline{EET}(T_a) + \max_{T_b \in \text{children of } T_a} \left(\overline{ECOMT}(c_{ab}) + rank(T_b) \right),$$

and then each task is scheduled to a resource with minimum ECT. Note the important feature of the rank function — it defines a valid topological order for the tasks. All tasks are scheduled after their parents, so we can compute the communication time estimate.

Heterogeneous Critical Parent Trees (HCPT): This algorithm implements an alternative ordering approach [10]. For each task we compute two criteria. The first criterion is the average earliest start time

$$AEST(T_a) = \max_{T_b \in \text{parents of } T_a} \left(\overline{ECOMT}(c_{ab}) + \overline{EET}(T_b) + AEST(T_b) \right).$$

This criterion is calculated by a forward traversal of the workflow graph. $AEST$ value for the entry task is defined to be zero. The second criterion is the average latest start time

$$ALST(T_a) = \min_{T_b \in \text{children of } T_a} \left(ALST(T_b) - \overline{ECOMT}(c_{ab}) \right) - \overline{EET}(T_a).$$

This criterion is calculated by a reverse traversal of the workflow graph. $ALST$ value for the exit task is set to be equal to its AEST value. Tasks with equal values of $AEST$ and $ALST$ are considered to be critical tasks. Then we perform a guided topological sort that prioritizes critical tasks. On a scheduling stage, tasks are considered in resulting order. Each task is scheduled to a resource that minimizes the ECT.

Lookahead (LA): This algorithm can be considered as an extension of the HEFT algorithm [5]. It uses the same task ranking approach, however on a scheduling step the best resource is selected not by the ECT of a current task, but by an estimated total makespan of the workflow if all remaining tasks are scheduled using HEFT. Hence a particular task can get a less performant resource if this reduces the total makespan, making the approach less greedy. The algorithm tends to produce efficient schedules, but has a high computational complexity.

Predict Earliest Finish Time (PEFT): This algorithm attempts to achieve the benefits of the Lookahead algorithm while keeping the computational complexity low [1]. To do this it precomputes the values of Optimistic Cost Table (OCT) for each task-resource pair as follows:

$$OCT(T_a, R_i) = \max_{T_b \in \text{children of } T_a} \min_{R_j} (OCT(T_b, R_j) + EET(T_b, R_j)$$
$$+ \overline{ECOMT}(c_{ab}, R_i, R_j)).$$

The idea of this criterion is to estimate the remaining workflow execution time disregarding the resource availability. The tasks are scheduled in decreasing order of the mean OCT value across all resources. A task is assigned to a resource that minimizes the sum of ECT and OCT.

2.2 Dynamic Algorithms

Dynamic algorithms schedule workflow tasks incrementally during the workflow execution. On each scheduling cycle, the algorithm considers a subset of tasks currently available for scheduling. In most cases, the available tasks include all unscheduled tasks ready for execution, i.e. tasks whose parent tasks are already completed. While the simplest algorithms blindly schedule available tasks to idle resources, more sophisticated algorithms try to estimate task completion and communication times using performance models and consider all resources just like the static algorithms.

Opportunistic Load Balancing (OLB): A simple algorithm that dynamically assigns available tasks to resources currently being idle [3]. The order in which the tasks and resources are considered during the scheduling cycle is undefined in a general case. In this paper we order tasks by their name and resources by their performance to ensure reproducible results. The advantages of this algorithm include its simplicity and adaptability. It also avoids the need to estimate task completion times for each resource which is often non-trivial. Although it may produce significantly suboptimal schedules, OLB is among the most used scheduling algorithms in modern HCSs.

Minimum Completion Time (MCT): An algorithm that assigns each available task to a resource that is expected to finish the task the earliest. The tasks can be scheduled in an undefined order, however in this paper we use a predefined order for reproducible results. Unlike OLB, this algorithm considers all available resources regardless of whether they are idle. To achieve this, the algorithm maintains an estimate of the earliest start time for each resource $EST(R_i)$. This approach allows to combine the strengths of static and dynamic scheduling: computing the ECT for each task implicitly takes into account the local workflow structure (as ECT includes communication times from task's parents) while reducing the schedule degradation when the used estimates are inaccurate.

Min-Min: A dynamic batch-mode scheduling algorithm [8,12]. The resource selection procedure is the same as in the MCT algorithm. However, it resolves the task ordering problem by scheduling all available tasks in one batch using the following heuristic. The ECT matrix is computed for each task-resource pair, and for each task the minimum ECT value and the corresponding best resource are determined. Then the task with a minimum best ECT is selected (hence the name of this algorithm) and is assigned to its best resource. The scheduled task is then removed from the batch and the whole process is repeated until the batch is fully scheduled.

Max-Min: A dynamic batch-mode scheduling algorithm [8,12]. Max-Min is structurally identical to the Min-Min heuristic, the only difference being the task ordering criterion — the tasks with a maximum best ECT are scheduled first. The intuition behind this heuristic is to schedule the long-running tasks as earlier as possible.

Sufferage: A dynamic batch-mode scheduling algorithm introduced in [12]. Again this algorithm is structurally identical to the Min-Min heuristic, the only difference being the task ordering criterion — the tasks with a maximum difference between ECT values on best and second-best resources are scheduled first. The intuition behind this heuristic is to give preference to tasks which completion time could suffer the most if they are not scheduled now.

3 Simulation Framework

To study the workflow scheduling algorithms in this paper we use simulation. This approach, involving computer modeling of the process of application execution in a distributed computing system, is widely used for such kind of research.

In comparison with the full-scale experiments on real systems, simulation allows to significantly reduce the time needed to run an experiment and to ensure the reproducibility of produced results, while having moderate requirements to the used hardware resources. This allowed us to perform a statistically significant number of experiments for a wide range of workflow and system configurations in a reasonable amount of time. However, when using simulation it is important to ensure the accuracy, i.e. minimal deviation from the results of real-world experiments, and the scalability, i.e. the ability to conduct large-scale experiments, of the used simulation model.

The simulation model used in this paper is implemented on the base of SimGrid[1] [6], a simulation toolkit for studying the behavior of large-scale distributed systems. The toolkit provides the required fundamental abstractions for the discrete-event simulation of parallel applications in distributed environments. The choice of SimGrid was motivated by the maturity of the toolkit, the soundness and high level of verification of embedded models, and the active support of developers. An important factor is also the versatility of the toolkit that allows one to simulate grids, cloud infrastructures, peer-to-peer systems and MPI applications.

Many studies also used WorkflowSim [7], an open source toolkit for simulating scientific workflows based on CloudSim simulator. We avoided the use of WorkflowSim as it has been shown that CloudSim among other simulators has flaws in its network model [17].

3.1 System and Application Models

The heterogeneous computing system is modeled as a set of hosts and network links between them as depicted on Fig. 1. Each host is characterized by its performance expressed in FLOPS. In this study, it is assumed that each host has a single processor core, which resources are evenly distributed among the tasks running on the host. The execution of any task is considered nonpreemptive. Network links are characterized by their bandwidth and latency.

While the simulation is widely used for studying scheduling algorithms, the researchers often neglect the accuracy of the used models, especially network ones. In particular, in many papers authors assume a contention-free network model in which a network host can simultaneously send to or receive data from as many hosts as possible without experiencing any performance degradation. However, this model is not representative of real-world networks. In this study, we use the bounded multiport model provided by SimGrid. In this model, a host can communicate with several other hosts simultaneously, but each communication flow is limited by the bandwidth of the traversed route, and communications using a common network link have to share bandwidth. This scheme corresponds well to the behavior of TCP connections on a LAN. The validity of this network model has been demonstrated in [16].

---

[1] http://simgrid.gforge.inria.fr/.

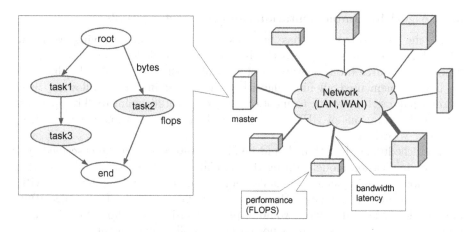

Fig. 1. Workflow and heterogeneous computing system models.

SimGrid supports simulation of various network topologies including hierarchies and combinations of autonomous systems with different internal routing strategies. We consider systems with a simple topology where each host is connected to a central backbone via a dedicated link as depicted on Fig. 1, and a route between any two hosts contains the two respective links. The backbone, which can correspond to the LAN switch or the WAN, doesn't impose additional latency or bandwidth constraints in this model. Therefore the rate of communication between any pair of hosts is determined only by characteristics of the corresponding pair of links.

The workflow application is modeled as a directed acyclic graph (DAG), whose vertices correspond to individual tasks and directed edges represent the data dependencies between tasks as depicted on Fig. 1. Each vertex is characterized by its size, i.e. the amount of computations in flops associated with the corresponding task. Similarly, each edge is characterized by the amount of communication in bytes between the corresponding pair of tasks. The size of task input data equals to the sum of sizes of incoming edges.

The two special tasks with zero size are introduced in order to model the staging of workflow input and output data. The *root* task passes through its outgoing edges the input data to the initial tasks, i.e. those that do not depend on other workflow tasks. The *end* task receives through its incoming edges the output data from the final tasks, i.e. those that do not pass their data to other workflow tasks.

The *root* and *end* tasks are executed on a dedicated host called *master*, which does not participate in computations. This host corresponds to the machine, which stores the input data and where the output data should be placed after the application execution. In practice, this host often performs submission and management of the workflow.

3.2 Algorithm Implementations and Supporting Tools

While the SimGrid toolkit has been used previously for studying workflow scheduling algorithms [2,11], to the best of our knowledge there are no published open source implementations of such algorithms for SimGrid. Therefore we have implemented all nine algorithms described in Sect. 2 following their original papers. It is worth to mention that the WorkflowSim toolkit[2] includes only implementations of HEFT and simple dynamic algorithms (FCFS, MCT, MinMin, MaxMin).

The scheduling algorithm should be able to interact with the described system and application models during the simulation. In the case of SimGrid, this interaction can be implemented via its application programming interfaces (API) for C programming language. However, the implementation of each algorithm in C using low-level interfaces would require a considerable effort. SimGrid also provides a Java API, but it does not cover all toolkit functionality.

To simplify the implementation of scheduling algorithms for our experiments we have developed a *pysimgrid* library. This library implements a thin wrapper around the native SimGrid API and provides a convenient interface for development of scheduling algorithms in Python language. The library also includes auxiliary tools for generation of synthetic systems and workflows, batch execution of simulation experiments and analysis of simulation results. The use of ubiquitous high-level programming language with a wide range of third-party libraries helped to significantly accelerate the development of the simulation framework. The code of *pysimgrid* library including the implementations of studied algorithms is publicly available on the GitHub[3].

4 Experimental Results

In this section, we present the results of simulation experiments that compare the performance of studied algorithms for a wide range of workflow and system configurations using the described simulation framework.

We use the *makespan*, i.e. the measured total run time of a workflow in a given system according to a schedule produced by an algorithm, as the basis for comparison of algorithm performance. For each simulated pair system-application we run all algorithms and then normalize their makespans by the makespan achieved by the OLB algorithm. The use of OLB as a baseline is motivated by its wide use in modern HCSs. Finally, to reduce the variance, we compute the mean of normalized makespans across all simulations and report these values in tables for each experiment.

4.1 Experiments with Real Workflows

The first series of experiments uses a fixed set of workflows while varying the system characteristics. The used workflows are based on real world scientific applications [4]:

---

[2] https://github.com/WorkflowSim/.
[3] https://github.com/alexmnazarenko/pysimgrid.

- **CyberShake:** characterize earthquake hazards in a region (SCEC);
- **Epigenomics:** automates various genome sequencing operations (USC Epigenome Center);
- **LIGO Inspiral:** analyse and filter the time-frequency data from the Laser Interferometer Gravitational Wave Observatory experiment (LIGO);
- **Montage:** stitch together multiple images of the sky to create large-scale custom mosaics (NASA/IPAC).

The simulated systems have 5, 10 or 20 hosts with performance varying in a range of 1 to 4 GFlops. The network links have identical characteristics selected to be close to the Gigabit Ethernet network (bandwidth: 100 MBytes/sec, latency: 100 us). For each host count 100 distinct systems are randomly generated.

The mean normalized makespans achieved by the studied algorithms are presented in the Table 1. The results for Montage, LIGO and Epigenomics workflows are similar — OLB performs the worst, dynamic heuristics follow and the best results are produced by static algorithms. Among the static algorithms, Lookahead performs the best, closely followed by HEFT. The maximum speedup achieved in comparison to OLB varies among the workflows due to the different amount of inherent parallelism.

Table 1. Experiment 1, mean normalized makespan

Hosts count	OLB	MCT	MinMin	MaxMin	Sufferage	HCPT	HEFT	Lookahead	PEFT
CyberShake, 100 tasks									
5	1.0000	1.0031	1.0160	0.9995	1.0125	1.1000	1.0525	1.0424	1.0607
10	1.0000	0.9950	0.9920	0.9817	0.9855	1.1151	1.0828	1.0599	1.1020
20	1.0000	0.9833	0.9623	0.9833	0.9537	1.1111	1.1027	1.0497	1.1130
Epigenomics, 100 tasks									
5	1.0000	0.9679	0.9692	0.9684	0.9652	0.9710	0.9349	0.9052	0.9669
10	1.0000	0.9002	0.9229	0.9107	0.9092	0.9338	0.8482	0.8088	0.9155
20	1.0000	0.7706	0.8071	0.7725	0.7836	0.8349	0.6908	0.6619	0.8305
LIGO Inspiral, 100 tasks									
5	1.0000	0.9820	0.9813	0.9819	0.9801	1.0895	0.9619	0.9428	1.0043
10	1.0000	0.9038	0.9106	0.9100	0.9094	1.1099	0.8619	0.8320	0.9487
20	1.0000	0.7900	0.7914	0.8245	0.8153	0.9045	0.6899	0.6699	0.8132
Montage, 100 tasks									
5	1.0000	0.9777	0.9779	0.9765	0.9780	1.0296	0.9752	0.9674	0.9791
10	1.0000	0.9579	0.9596	0.9549	0.9579	1.0071	0.9573	0.9435	0.9632
20	1.0000	0.9124	0.9159	0.9080	0.9135	0.9243	0.9163	0.8912	0.9200

CyberShake workflow, however, is different. All dynamic algorithms show similar results and outperform the static algorithms. Investigation shows that this workflow has two distinguishing properties — high parallelism and high communication-to-computation ratio (CCR). This could lead to a network contention resulting in a significant mismatch between the simple model used in the static algorithms for estimation of $ECOMT$ and the accurately modeled network in the simulator. To check this hypothesis, we obtained the predicted makespan values from the internal state of the static algorithms. Each predicted makespan is normalized by the simulated OLB makespan so that the values are compatible with the previous table. The results are presented in the Table 2.

Table 2. Experiment 1, mean normalized makespan predicted by the static algorithms

Hosts count	OLB	HCPT	HEFT	Lookahead	PEFT
CyberShake, 100 tasks					
5	1.0000	0.5594	0.5074	0.5035	0.5269
10	1.0000	0.4069	0.3789	0.3706	0.4101
20	1.0000	0.3003	0.2958	0.2922	0.3284
Epigenomics, 100 tasks					
5	1.0000	0.9671	0.9311	0.9017	0.9631
10	1.0000	0.9261	0.8405	0.8023	0.9079
20	1.0000	0.8197	0.6740	0.6471	0.8157
LIGO Inspiral, 100 tasks					
5	1.0000	1.0883	0.9608	0.9418	1.0031
10	1.0000	1.1076	0.8602	0.8305	0.9468
20	1.0000	0.9006	0.6865	0.6666	0.8096
Montage, 100 tasks					
5	1.0000	1.0223	0.9683	0.9603	0.9720
10	1.0000	0.9972	0.9478	0.9338	0.9537
20	1.0000	0.9107	0.9023	0.8772	0.9070

As we can see, the static algorithms expect to achieve drastically different values of makespan than the ones produced by the simulation. It means that for the CyberShake workflow, ignoring the network contention effect caused by competing data transfers produces more than 200% error in the makespan estimation. This result emphasizes the importance of accurate network simulation for studying and benchmarking of task scheduling algorithms for HCSs.

To strengthen our conclusion, we repeated the simulation for the same set of workflows and systems using the virtually infinitely fast network. The obtained results are presented in the Table 3. In this setup, we get consistent results across all workflows, with the simplest OLB algorithm producing the worst schedules and the most advanced Lookahead algorithm producing the best ones.

4.2 Experiments with Synthetic Workflows

To investigate the observed effects related to the workflow inherent parallelism and CCR ratio, we have conducted another series of experiments with randomly

Table 3. Experiment 1, mean normalized makespan with infinitely fast network

Host count	OLB	MCT	MinMin	MaxMin	Sufferage	HCPT	HEFT	Lookahead	PEFT
CyberShake, 100 tasks									
5	1.0000	0.9853	0.9861	0.9554	0.9640	1.0605	0.9584	0.9468	0.9890
10	1.0000	0.9636	0.9668	0.9032	0.9247	0.9754	0.9065	0.8828	0.9812
20	1.0000	0.8982	0.9512	0.8145	0.8810	0.8912	0.8001	0.7865	0.9248
Epigenomics, 100 tasks									
5	1.0000	0.9758	0.9760	0.9707	0.9676	0.9647	0.9330	0.9036	0.9533
10	1.0000	0.8977	0.9174	0.9081	0.9085	0.9325	0.8451	0.8071	0.8816
20	1.0000	0.7553	0.8009	0.7643	0.7718	0.8350	0.6920	0.6629	0.8045
LIGO Inspiral, 100 tasks									
5	1.0000	0.9787	0.9792	0.9785	0.9763	1.0879	0.9605	0.9398	0.9960
10	1.0000	0.9043	0.9063	0.9113	0.9118	1.1133	0.8680	0.8371	0.9487
20	1.0000	0.7979	0.8025	0.8331	0.8288	0.9136	0.6953	0.6739	0.8426
Montage, 100 tasks									
5	1.0000	0.9770	0.9771	0.9752	0.9769	1.0318	0.9766	0.9684	0.9797
10	1.0000	0.9582	0.9596	0.9544	0.9574	1.0082	0.9605	0.9467	0.9681
20	1.0000	0.9097	0.9120	0.9036	0.9094	0.9375	0.9207	0.8942	0.9240

generated synthetic workflows. These experiments are run on a single system consisting of 10 hosts with performance ranging from 1 to 4 Gflops. Network links have the same bandwidth and latency as previously.

The synthetic workflows are generated using the *daggen* utility, which implements the layered workflow generation algorithm specifically designed for studying workflow scheduling algorithms [14]. Two parameters of generated workflows are varied: the average number of tasks per layer (graph width) and the CCR ratio. The CCR is expressed as the ratio of the total size of task inputs in MBs to its computational cost in Gflops. The CCR value was fixed across all workflow tasks. 100 random workflows with 100 tasks were generated for each combination of graph width and CCR.

The simulation results are presented in the Table 4. It appears that the results of static algorithms become less and less consistent both with the increase of the graph width and of the CCR. To quantify the error of their internal models, we have collected the makespans predicted by the algorithms using the same approach as in the first experiment in the Table 5.

As we can see, for CCR=1000 the makespan prediction error gets as high as 2500%! Although the total predicted makespan is not important by itself, this result also shows that the algorithm's internal model completely diverges from the simulation for all static algorithms. Interestingly enough, the simpler algorithms such as HEFT and HCPT are still able to produce decent schedules while the more advanced Lookahead and PEFT algorithms end up worse than the simplest dynamic OLB due to the overoptimized schedule built using an inaccurate performance model.

Finally, we repeated the simulation for the same synthetic workflows and the system using an infinitely fast network. The results are presented in the Table 6.

Table 4. Experiment 2, mean normalized makespan

CCR	OLB	MCT	MinMin	MaxMin	Sufferage	HCPT	HEFT	Lookahead	PEFT
Workflows with 100 tasks, 10 tasks per layer									
1	1.0000	0.9954	1.0149	0.9979	1.0186	0.9308	0.8157	0.8117	0.9085
10	1.0000	1.0245	1.0367	1.0503	1.0837	0.8452	0.8004	0.7761	0.9223
100	1.0000	0.4934	0.4882	0.5111	0.5282	0.5206	0.5969	0.5929	0.7692
500	1.0000	0.2540	0.2538	0.2670	0.2643	0.2489	0.2732	0.2810	0.3590
1000	1.0000	0.2193	0.2191	0.2262	0.2295	0.2367	0.2496	0.1999	0.3387
Workflows with 100 tasks, 20 tasks per layer									
1	1.0000	1.0238	1.0996	1.0563	1.0614	0.9411	0.8295	0.8265	0.9242
10	1.0000	1.1279	1.1837	1.1215	1.1653	0.9057	0.8830	0.8637	1.0594
100	1.0000	0.7148	0.7268	0.7787	0.7801	0.7381	0.7748	0.7424	1.0609
500	1.0000	0.5284	0.5151	0.5740	0.5731	0.4622	0.4755	0.6458	0.6637
1000	1.0000	0.4466	0.4456	0.4961	0.4948	0.4057	0.4083	0.6205	0.4991
Workflows with 100 tasks, 40 tasks per layer									
1	1.0000	1.0589	1.0821	1.0468	1.0816	0.8613	0.8160	0.8226	0.9341
10	1.0000	1.1852	1.1984	1.2057	1.2076	0.9464	0.9303	0.9648	1.2927
100	1.0000	1.0190	0.9934	1.0648	1.0802	0.9357	0.9463	0.8855	1.7834
500	1.0000	0.8159	0.8169	0.8734	0.8897	0.7725	0.7926	1.3575	1.4882
1000	1.0000	0.7890	0.7460	0.8370	0.8425	0.7590	0.7499	1.3137	0.9202

Table 5. Experiment 2, mean normalized makespan predicted by the static algorithms

CCR	OLB	HCPT	HEFT	Lookahead	PEFT
Workflows with 100 tasks, 10 tasks per layer					
1	1.0000	0.9096	0.7932	0.7831	0.8803
10	1.0000	0.6186	0.5375	0.5270	0.5983
100	1.0000	0.1462	0.1557	0.1392	0.1817
1000	1.0000	0.0279	0.0293	0.0262	0.0397
Workflows with 100 tasks, 20 tasks per layer					
1	1.0000	0.9148	0.8001	0.7922	0.8905
10	1.0000	0.6260	0.5584	0.5493	0.6186
100	1.0000	0.1430	0.1381	0.1248	0.1513
1000	1.0000	0.0655	0.0646	0.0527	0.0554
Workflows with 100 tasks, 40 tasks per layer					
1	1.0000	0.8307	0.7818	0.7730	0.8893
10	1.0000	0.6167	0.5756	0.5638	0.6485
100	1.0000	0.1377	0.1293	0.1204	0.1494
1000	1.0000	0.0629	0.0625	0.0545	0.0547

Table 6. Experiment 2, mean simulated makespan with infinitely fast network

CCR	OLB	MCT	MinMin	MaxMin	Sufferage	HCPT	HEFT	Lookahead	PEFT
Workflows with 100 tasks, 10 tasks per layer									
1	1.0000	1.0243	1.0357	1.0211	1.0192	0.9793	0.8536	0.8438	0.9540
10	1.0000	0.9951	1.0429	1.0006	1.0313	0.9646	0.8386	0.8298	0.9305
100	1.0000	1.0098	1.0276	1.0105	1.0293	0.9808	0.8559	0.8450	0.9537
500	1.0000	1.0202	1.0362	1.0125	1.0518	0.9869	0.8717	0.8561	0.9663
1000	1.0000	1.0199	1.0254	1.0079	1.0480	0.9803	0.8670	0.8525	0.9578
Workflows with 100 tasks, 20 tasks per layer									
1	1.0000	1.0958	1.1173	1.0803	1.0815	0.9780	0.8494	0.8408	0.9566
10	1.0000	1.1062	1.1463	1.1060	1.1209	0.9675	0.8498	0.8419	0.9500
100	1.0000	1.0953	1.1398	1.0778	1.0947	0.9774	0.8505	0.8426	0.9596
500	1.0000	1.1043	1.1275	1.0812	1.0925	0.9645	0.8429	0.8343	0.9446
1000	1.0000	1.0961	1.1149	1.0736	1.0985	0.9732	0.8476	0.8380	0.9523
Workflows with 100 tasks, 40 tasks per layer									
1	1.0000	1.1137	1.1054	1.1177	1.1367	0.8942	0.8378	0.8278	0.9575
10	1.0000	1.1400	1.1354	1.1095	1.1353	0.9130	0.8431	0.8339	0.9687
100	1.0000	1.1427	1.1331	1.1210	1.1411	0.9149	0.8519	0.8421	0.9766
500	1.0000	1.1404	1.1375	1.1137	1.1300	0.9098	0.8450	0.8342	0.9668
1000	1.0000	1.1455	1.1390	1.1057	1.1233	0.9042	0.8407	0.8299	0.9709

As anticipated, we are back to the expected results — the static algorithms are outperforming the dynamic ones with Lookahead performing the best.

It is interesting to note that in these experiments dynamic heuristics are often outperformed by the OLB algorithm. The analysis of simulation traces shows that these heuristics do save computation and communication time for individual tasks, but still increase the makespan. The main cause is that dynamic heuristics disregard the task dependencies. The tasks can accumulate on the most performant hosts, delaying the execution of the next layer of the workflow. This effect is further amplified by overloading the network links of those most performant hosts.

5   Conclusion and Future Work

In this study, we have reviewed and compared the performance of nine state-of-the-art algorithms for scheduling of workflows in heterogeneous computing systems. The comparison was performed by running a large number of simulation experiments for a wide range of workflow and system configurations.

For this purpose, we have developed an open source simulation framework that provides tools for implementation of scheduling algorithms, generation of synthetic systems and workflows, execution of simulation experiments and analysis of results. The developed framework leverages SimGrid toolkit for the discrete-event simulation of parallel applications in distributed environments. The accuracy of the used network model allowed us to ensure realistic simulations and reveal drawbacks of simpler models commonly used for studying scheduling algorithms.

The presented experimental results provide a strong evidence against the widely used experimental approach based on linear performance models that disregard network topology and bandwidth allocation. The schedules produced by static algorithms clearly demonstrate that even for the modestly parallel workloads with sufficiently large data items the effect of competing data transfers may lead to the drastic underestimation of the communication time and the makespan degradation. However, when the parallelism and/or communication-to-computation ratio are low enough, the static algorithms, even based on simple models, can significantly outperform the opportunistic load-balancing approach widely used in practice.

Future work will include further development of the presented simulation framework, implementation of other known algorithms, i.e. based on metaheuristics, carrying out additional experiments and validation on real systems. We also plan to incorporate more accurate network models into static algorithms to improve their performance for highly parallel and data-intensive workflows.

Acknowledgments. This work is supported by the Russian Science Foundation (project No. 16-11-10352).

References

1. Arabnejad, H., Barbosa, J.G.: List scheduling algorithm for heterogeneous systems by an optimistic cost table. IEEE Trans. Parallel Distrib. Syst. **25**(3), 682–694 (2014)
2. Arabnejad, H., Barbosa, J.G., Prodan, R.: Low-time complexity budget-deadline constrained workflow scheduling on heterogeneous resources. Future Gener. Comput. Syst. **55**, 29–40 (2016)
3. Armstrong, R., Hensgen, D., Kidd, T.: The relative performance of various mapping algorithms is independent of sizable variances in run-time predictions. In: Proceedings of 1998 Seventh Heterogeneous Computing Workshop, HCW 1998, pp. 79–87. IEEE (1998)
4. Bharathi, S., Chervenak, A., Deelman, E., Mehta, G., Su, M.H., Vahi, K.: Characterization of scientific workflows. In: 2008 Third Workshop on Workflows in Support of Large-Scale Science, pp. 1–10, November 2008
5. Bittencourt, L.F., Sakellariou, R., Madeira, E.R.M.: Dag scheduling using a look ahead variant of the heterogeneous earliest finish time algorithm. In: 2010 18th Euromicro Conference on Parallel, Distributed and Network-Based Processing, pp. 27–34, February 2010
6. Casanova, H., Giersch, A., Legrand, A., Quinson, M., Suter, F.: Versatile, scalable, and accurate simulation of distributed applications and platforms. J. Parallel Distrib. Comput. **74**(10), 2899–2917 (2014)
7. Chen, W., Deelman, E.: Workflowsim: a toolkit for simulating scientific workflows in distributed environments. In: 2012 IEEE 8th International Conference on E-science (e-science), pp. 1–8. IEEE (2012)

8. Freund, R.F., Gherrity, M., Ambrosius, S., Campbell, M., Halderman, M., Hensgen, D., Keith, E., Kidd, T., Kussow, M., Lima, J.D., et al.: Scheduling resources in multi-user, heterogeneous, computing environments with smartnet. In: Proceedings 1998 Seventh Heterogeneous Computing Workshop, (HCW 1998), pp. 184–199. IEEE (1998)
9. Graham, R.L., Lawler, E.L., Lenstra, J.K., Kan, A.R.: Optimization and approximation in deterministic sequencing and scheduling: a survey. Ann. Discret. Math. **5**, 287–326 (1979)
10. Hagras, T., Janecek, J.: A simple scheduling heuristic for heterogeneous computing environments. In: Proceedings of Second International Symposium on Parallel and Distributed Computing, pp. 104–110, October 2003
11. Hunold, S., Rauber, T., Suter, F.: Scheduling dynamic workflows onto clusters of clusters using postponing. In: 8th IEEE International Symposium on Cluster Computing and the Grid, CCGRID 2008, pp. 669–674. IEEE (2008)
12. Maheswaran, M., Ali, S., Siegal, H.J., Hensgen, D., Freund, R.F.: Dynamic matching and scheduling of a class of independent tasks onto heterogeneous computing systems. In: Proceedings of the Eighth Heterogeneous Computing Workshop, (HCW 1999), pp. 30–44. IEEE (1999)
13. Taylor, I.J., Deelman, E., Gannon, D.B., Shields, M.: Workflows for e-Science: Scientific Workflows for Grids. Springer Publishing Company Incorporated, London (2014)
14. Tobita, T., Kasahara, H.: A standard task graph set for fair evaluation of multiprocessor scheduling algorithms. J. Sched. **5**(5), 379–394 (2002)
15. Topcuoglu, H., Hariri, S., Wu, M.Y.: Performance-effective and low-complexity task scheduling for heterogeneous computing. IEEE Trans. Parallel Distrib. Syst. **13**(3), 260–274 (2002)
16. Velho, P., Legrand, A.: Accuracy study and improvement of network simulation in the simgrid framework. In: Proceedings of the 2nd International Conference on Simulation Tools and Techniques, p. 13. ICST (Institute for Computer Sciences, Social-Informatics and Telecommunications Engineering) (2009)
17. Velho, P., Schnorr, L.M., Casanova, H., Legrand, A.: On the validity of flow-level TCP network models for grid and cloud simulations. ACM Trans. Model. Comput. Simul. (TOMACS) **23**(4), 23 (2013)
18. Yu, J., Buyya, R., Ramamohanarao, K.: Workflow scheduling algorithms for grid computing. In: Xhafa, F., Abraham, A. (eds.) Metaheuristics for Scheduling in Distributed Computing Environments. SCI, vol. 146, pp. 173–214. Springer, Heidelberg (2008)

PGAS Approach to Implement Mapreduce Framework Based on UPC Language

Shomanov Aday$^{(\boxtimes)}$, Akhmed-Zaki Darkhan, and Mansurova Madina

Al-Farabi Kazakh National University, Almaty, Kazakhstan
adai.shomanov@gmail.com, {darhan_a,mansurova01}@mail.ru

Abstract. Over the years from its introduction Mapreduce technology proved to be very effective parallel programming technique to process large volumes of data. One of the most prevalent implementations of Mapreduce is Hadoop framework and Google proprietary Mapreduce system.

Out of other notable implementations one should mention recent PGAS (partitioned global address space) – based X10, UPC (Unified Parallel C) versions. These implementations present a new viewpoint when Mapreduce application developers can benefit from using global address space model while writing data parallel tasks. In this paper we introduce a novel UPC implementation of Mapreduce technology based on idea of using purely UPC based implementation of shared hashmap data structure as an intermediate key/value store. Shared hashmap is used in to perform exchange of key/values between parallel UPC threads during shuffle phase of Mapreduce framework. The framework also allows to express data parallel applications using simple sequential code.

Additionally, we present a heuristic approach based on genetic algorithm that could efficiently perform load balancing optimization to distribute key/values among threads such that we minimize data movement operations and evenly distribute computational workload.

Results of evaluation of Mapreduce on UPC framework based on WordCount benchmark application are presented and compared to Apache Hadoop implementation.

Keywords: UPC · PGAS · Mapreduce

1 Introduction

Large-scale data processing nowadays is widely used in many domains of science and industry. There is a large number of sophisticated tools and algorithmic solutions that allow to achieve high efficiency in handling and processing enormous amount of data. Main driving forces of modern big data development are powerful Mapreduce - based frameworks. The idea of Mapreduce was first presented in paper [1] by Google researchers Jeffrey Dean and Sanjay Ghemawat in 2004. In general, the main idea behind Mapreduce is to divide processing of the big data set between concurrently running map and reduce processes such that each process performs processing of smaller data chunk. The processing work in Mapreduce is done in several steps:

© Springer International Publishing AG 2017
V. Malyshkin (Ed.): PaCT 2017, LNCS 10421, pp. 342–350, 2017.
DOI: 10.1007/978-3-319-62932-2_33

- **Init phase.** Specify map and reduce functions, provide input and output directory paths and etc.
- **Map phase.** Each mapper scans the input chunk of data and emits key/value pairs based on user provided map function.
- **Shuffle phase.** Distribute key/value pairs among reducers in a way that each reducer operates on list of key/value pairs with some assigned to that reducer unique key.
- **Reduce phase.** Each reducer performs operations on assigned key based on user provided reduce function.

The main complexity in efficiently implementing Mapreduce lies in developing scalable and optimized code for shuffle phase. To achieve these goals it is required to distribute key/value pairs with minimized network latencies. In distributed environment due to necessity of data movements between processes that belong to different nodes, network latencies can be very high and significantly degrade overall performance.

To overcome that we need to consider efficient tools that will allow to perform sufficiently transparent and optimized remote data access operations. For that purpose in our current work we will use UPC programming language. UPC programming language [2] belongs to a family of PGAS languages. PGAS (Partitioned Global Address Space) is a parallel programming model in which memory address space is divided into two non-overlapping logical areas: private and shared. Private space is local to every thread and can be accessed only by its own thread. Shared space has a more complex structure where each thread has an access to shared memory and each memory element has additionally affinity to the owner thread. The benefit of PGAS model is that each thread has a transparent view of shared memory layout hence locality can be preserved where it is needed to optimize data distribution for specific purposes of the application. UPC language provides a set of operations with shared memory such as: pointers arithmetic, write and read functions, memory allocation and de-allocation functions and other. UPC uses specifically designed GasNet communication system that enables high-performance one-sided communications in order to implement remote data access operations on shared memory.

2   Related Work

The implementation of PGAS-based Mapreduce model requires careful consideration and solving of many problems associated with the organization of the computational process, the process of data exchange between computing nodes, distribution and load balancing between concurrent map and reduce processes. In the article [3], authors describe Mapreduce framework, implemented on the UPC language. The approach described in this article applies collective functions for data exchange in shuffle phase. Map and reduce functions in that approach operate on the local storage of each node, and for that reason the authors were forced to change the implementation of collective UPC functions to make them work with local memory space of each thread. In our implementation we used different approach based on shared hashmap data structure to perform key/value exchange. Hashmap instances reside in shared address space and each instance has an affinity to a single thread. Accordingly, every thread has an access to hashmap instance of any other thread. In different paper [4] authors presented a similar approach where they applied X-10 library implementation of hashmap data structure to store locally in each thread intermediate key/value pairs and then merge all the values to one thread. X-10 enabled Mapreduce merging procedure is poorly scalable since all data is moved to a single place and therefore such an approach possesses inherent limitations associated with processing and storage capabilities of a single node. In our approach we keep one instance of shared hashmap per thread such that each thread works on local portion of its own shared hashmap and other threads when needed could perform remote operations on that thread-local instance of shared hashmap. Hence, processing is not limited by resources of a single node and only requires efficient data exchange after finishing map phase. Additionally, this way we can control locality of operations on each instance of hashmap and as a result later on can optimize key distribution among threads for reduce stage. Shared hashmap allows to efficiently extract and write key/value pairs in average O(1) time complexity. Consequently, based on features of hashmap data structure we attempted to reduce overhead associated with searching and extracting keys.

3   Main Part

3.1   Mapreduce on UPC Framework

Presented in the paper Mapreduce on UPC framework aims to bring together programmability benefits associated with UPC model with advanced processing power of Mapreduce technique. Implications of such architectural solution is that it is become very convenient to be able to express complex Mapreduce logic in a more concise form of UPC - Mapreduce by using global memory abstraction.

In order to implement Mapreduce in UPC we first wrote code for shared hashmap data structure based on shared memory operations such as upc_memput, upc_memcpy, upc_alloc, upc_memget and other. Operations on shared hashmap are controlled by our API functions such as shared_hashmap_put, shared_hashmap_get, shared_hashmap_resize, shared_hashmap_remove.

To store key/value elements we created globally addressable array of shared hashmap instances with default blocking factor of one in shared address space. Such layout of shared array corresponds to one-to-one mapping of threads and hashmap array entries. Consequently, each hashmap is designed to store key/value elements that are local to the thread executing map functions (see Fig. 1).

Map and reduce functions are specified by the application developer and are passed as parameters to init_mapreduce function that launches and controls the entire processing cycle of Mapreduce execution.

Each thread is assigned a number of map tasks. Each map task operates on exactly one input file. Therefore, in order to avoid imbalance, before map phase runtime distributes files among threads in such a way that each thread has approximately the same proportion of input files.

After all map functions are finished their execution, the shuffle phase take place. The shuffle phase is divided into 2 main stages:

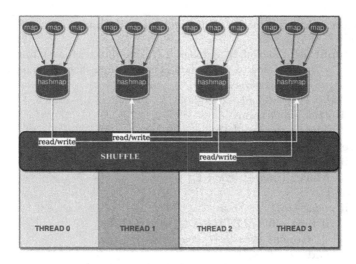

Fig. 1. UPC on Mapreduce map and shuffle design.

1. Data movement optimization and load balancing step.
2. Distribution of key/values among reducers.

In the process of load balancing we are using integer indexing of keys. We have to assign each key unique integer identifier. It is turn out that this operation is very expensive to perform since we need to traverse all hashmap entries in every thread by using only processing power of a single thread.

This thread is responsible for fetching remote hashmap entries and checking if that entry (key) already has been assigned identifier or not. If identifier already has been assigned to that entry (key) then we can skip it, otherwise it is required to update *id* field of that hashmap entry by remote write operation. Fetching and updating remote values by fine-grained operations incur a lot of communication and software overhead that should be avoided or substituted by coarse-grained bulk operations.

Hashmap element consists of the following fields: integer *id*, shared [] char * *key*, integer *in_use*, shared [] shared_vector * *data*. Since all field values are located in shared memory they can only be accessed by shared pointers. Shared pointers orders of magnitude slower than ordinary private pointers and therefore amount of accesses to shared memory area by shared pointers should be minimized.

Therefore, in order to minimize fine-grained access operations we developed more scalable and efficient in terms of running time method to assign each key unique integer identifier. We store keys in a shared array of string entries. A new method works by merging local to each thread keys stored in a shared array into a single shared array that has an affinity to thread number 0. The goal was to minimize number of copy operations. In a new method this number is equal to $O(\log n)$ compared to $O(n)$ operations in a previous implementation. There n represents number of threads.

```
 1  function Merge (int left, int right,int turn,shared string *
    keys)
       Input: left index of array range left, right index of array range
              right, side to which data is copied turn,shared pointer to
              keys array keys
 2     mid = left + (right − left)/2 ;
 3     if left < right then
 4      |  Merge (left,mid,0);
 5     end
 6     if mid + 1 < right then
 7      |  Merge (mid + 1,right,1);
 8     end
 9     if right − left ≥ 1 then
10        if turn = 0 ∧ MYTHREAD = left then
11         |  Concat (keys[left],keys[right]);
12        end
13        else if turn = 1 ∧ MYTHREAD = right then
14         |  Concat (keys[right],keys[left]);
15        end
16     end
17     Barrier;
```

Listing 1. Procedure for coarse-grained merge of key arrays

Merge procedure uses divide and conquer method that works according to Listing 1.

3.2 Data Movement and Load Balancing Optimization

For load balancing and data movement optimization we employ heuristic approach based on genetic algorithm [5]. Genetic algorithms are used in many problems in domain of combinatorial and multi-objective optimization. The problem with many instances of combinatorial optimization tasks is that they belong to NP class of problems. Therefore they cannot be solved by means of polynomial time algorithms and only hope to find a feasible solution for sufficiently large dimensions is to apply different heuristic approaches.

The following set of equations describes the problem:

$$min \sum_{i=0}^{threads-1} \sum_{j=1}^{keys} x_{ij} \times cost_{ij} \tag{1}$$

$$x_{ij} \in \{0, 1\} \tag{2}$$

$$min\left(\max_{i,j=0..threads-1} \left| load_i - load_j \right| \right) \tag{3}$$

$$load_i = \sum_{t=0}^{threads-1} \sum_{j=1}^{keys} x_{ij} \times size_{tj} \qquad (4)$$

The optimization problem we have stated above is a modification of "Generalized assignment problem" which is known to be NP-hard. Genetic algorithms for solving GAP has been presented in different sources before, e.g. in [6, 7].

In order to find cost of assigning key j to thread i we construct cost matrix in which each entry $cost_{ij}$ is corresponding cost value of moving key j to thread i. Quantitatively, cost represents number of elements of some particular key that needs to be moved to some other thread. Formula (3) defines load balancing function. Load balancing function is calculated as minimum value over maximum difference of loads assigned to different pairs of threads. We need to perform distribution of key/values among threads with aim to optimize both functionals defined in formulas (1) and (3). Formula (2) defines the domain of x_{ij} variable to be consisting of two integer values of either 0 or 1. For thread i and key j the value of $x_{ij} = 0$ when thread i is not assigned to process key j and $x_{ij} = 1$ otherwise. Load value for each thread i is defined in formula (4). Genetic algorithm works according to following procedure:

```
1  function LoadBalance (int n, chromosone p, int m)
      Input   : Initial population p, Max number of generations n,
                Population size m
      Output: shared array sol
2     i = 1;
3     np ← ∅ ;
4     while i ≤ n ∨ stopping criteria is not met do
5         ComputeFitness (p);
6         for j = 1 to m do
7             p1 ←TournamentSelection(p);
8             p2 ←TournamentSelection(p);
9             child ←Crossover (p1,p2);
10            Mutation (child);
11            Enque (child,np);
12        end
13        p ← np;
14        np ← ∅;
15        i = i + 1;
16    end
17    sol ←SelectBestFitnessSolution (p);
18    return sol ;
```

Listing 2. Genetic algorithm for load balancing of keys among reducers

In order to be able to adapt genetic algorithm to solve our problem we first need to identify how to represent solution in the language of genetic algorithm. Solution (chromosone) is represented by vector, where i-th entry contains number of the thread that is assigned to process i-th key. Population is defined as set of all solutions and can be selected and correspondingly adjusted depending on specific needs and limitations of the task. Fitness value is an objective function that can be calculated for each particular solution. The task of genetic algorithm is to find specific solution with best fitness value. Fitness function in our problem is represented by combination of functionals described in (1) and (3).

Then, after genetic algorithm generates a solution, runtime can proceed to perform shuffle procedure.

3.3   Shuffle Phase

To perform shuffle procedure we need to appropriately distribute key/values among reducers such that each reducer can then schedule to perform reduce function calls on input elements with same key. In our program we have implemented shuffle procedure as follows:

- To store key/value elements on reduce side we created a new array of shared hashmap data structures with default layout in shared address space
- Each hashmap of the old array on each thread is traversed in parallel and according to the thread-keys mappings, obtained by solving optimization problem, elements are copied to threads that are assigned to process current element (key).
- After key/value distribution completes, each thread is ready to run reduce functions

Reduce stage is organized such that on each thread shared hashmap is traversed and each hashmap entry of <key, set of values> is assigned as input to a single reduce function. After completing their execution each reduce function writes final result to a single resulting file.

3.4   WordCount Implementation

For experimental evaluation of our Mapreduce framework we have chosen WordCount benchmark application. WordCount program computes number of occurrences of each word in a set of documents. This problem is a standard application for evaluating Mapreduce-based frameworks. The main idea behind implementing WordCount on Mapreduce is to divide processing such that each mapper emits for every word a pair of <word, 1> and each reducer then add all entries in the list of 1's that has been assigned to it and emits as final result pair of <word, overall_count>. In code listings 3 and 4 below our map and reduce function implementations for WordCount application are presented. The code for map and reduce functions must be written in C language with possible use of UPC-related functions for shared memory operations.

```
void * map (string filename)
{
  char * file_data;
  file_data = read_file_contents (filename);
  Vector tokens;
  vector_init(&tokens);
Tokenize (file_data,&tokens);
for (int i = 0;i<tokens.size;i++)
{
  collect (vector_get (&tokens,i),1);
}
free(file_data);
}
```

Listing 3. Implementation of map function for WordCount application

```
void reduce (string key,shared [] vector_sh
*values)
{
  int i;
  int cnt = 0;
  for (i = 0;i<values->size;i++)
  {
    int v = vector_get_shared_copy (values,i);
    cnt+=v;
  }
  reduce_collect (key,cnt);
}
```

Listing 4. Implementation of reduce function for WordCount application

4 Experimental Results

In this section we present results of evaluation of UPC on Mapreduce framework based on Google cloud platform architecture. The setup consisted of one instance of n1-highmem-8 (8 vCPUs, 52 GB memory). In our experiments we used the following software:

- Berkeley UPC runtime version 2.24.0
- Apache Hadoop version 2.7.3
- The Berkeley UPC-to-C translator, version 2.24.0

WordCount application has been tested for different input sizes ranging from 50 to 200 megabytes. Based on results of running WordCount on Apache Hadoop and UPC on Mapreduce (see Fig. 2) we can conclude that Mapreduce on UPC shows better performance on all inputs besides smallest 50 Mb input in which both frameworks show the same performance.

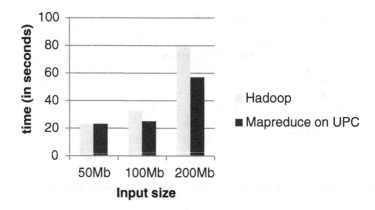

Fig. 2. Hadoop and Mapreduce on UPC running time for different input sizes

5   Conclusion

The paper presented UPC on Mapreduce framework that allows to users to implement data parallel applications by expressing them in the form of map and reduce functions. By analyzing results of evaluation of Mapreduce on UPC framework we observed better performance results compared to Hadoop, but algorithm have some scalability issues in case of small number of threads performing WordCount task.

References

1. Dean, J., Ghemawat, S.: MapReduce: simplified data processing on large clusters. In: Sixth Symposium on Operating System Design and Implementation (OSDI2004), p. 10. USENIX Association, San Francisco (2004)
2. Carlson, W.W., Draper, J.M., Culler, D.E., Yelick, K., Brooks, E., Warren, K.: Introduction to UPC and language specification. Technical report, IDA Center for Computing Sciences (1999)
3. Teijeiro, C., Taboada, G.L., Tourino, J., Doallo, R.: Design and implementation of Mapreduce using the PGAS programming model with UPC. In: 17th International Conference on Parallel and Distributed Systems (ICPADS 2011), pp. 196–203. IEEE Computer Society, Washington (2011). doi:10.1109/ICPADS.2011.162
4. Dong, H., Zhou, S., Grove, D.: X10-enabled MapReduce. In: 4th Conference on Partitioned Global Address Space Programming Model (PGAS 2010), pp. 1–6. ACM, New York (2010). doi:10.1145/2020373.2020382
5. Man, K.F., Tang, K.S., Kwong, S.: Genetic algorithms: Concepts and applications. IEEE Trans. Industr. Electron. **43**(5), 519–534 (1996). doi:10.1109/41.538609
6. Chu, P.C., Beasley, J.E.: A genetic algorithm for the generalised assignment problem. Comput. Oper. Res. **24**(1), 17–23 (1997). doi:10.1016/S0305-0548(96)00032-9
7. Liu, Y.Y., Wang, S.: A scalable parallel genetic algorithm for the generalized Assignment Problem. Parallel Comput. **46**, 98–119 (2015). doi:10.1016/j.parco.2014.04.008

Islands-of-Cores Approach for Harnessing SMP/NUMA Architectures in Heterogeneous Stencil Computations

Lukasz Szustak[1(✉)], Roman Wyrzykowski[1], and Ondřej Jakl[2]

[1] Czestochowa University of Technology,
Dabrowskiego 69, 42-201 Czestochowa, Poland
{lszustak,roman}@icis.pcz.pl
[2] Institute of Geonics of the Czech Academy of Sciences,
Studentská 1768, 708 00 Ostrava-Poruba, Czech Republic
ondrej.jakl@ugn.cas.cz

Abstract. SMP/NUMA systems are powerful HPC platforms which could be applied for a wide range of real-life applications. These systems provide large capacity of shared memory, and allow using the shared-variable programming model to take advantages of shared memory for inter-process communications and synchronizations. However, as data can be physically dispersed over many nodes, the access to various data items may require significantly different times. In this paper, we face the challenge of harnessing the heterogeneous nature of SMP/NUMA communications for a complex scientific application which implements the Multidimensional Positive Definite Advection Transport Algorithm (MPDATA), consisting of a set of heterogeneous stencil computations.

When using our method of MPDATA workload distribution, which was successfully applied for small-scale shared memory systems with several CPUs and/or accelerators, significant performance losses are noticeable for larger SMP/NUMA systems, such as SGI UV 2000 server used in this work. To overcome this shortcoming, we propose a new islands-of-cores approach. It exposes a correlation between computation and communication for heterogeneous stencils, and enables an efficient management of trade-off between computation and communication costs in accordance with the features of SMP/NUMA systems. In consequence, when using the maximum configuration with 112 cores of 14 Intel Xeon E5-4627v2 3.3 GHz processors, the proposed approach accelerates the previous method more then 10 times, achieving about 390 Gflop/s, or approximately 30% of the theoretical peak performance.

1 Introduction

In the last years, it appears evident [7, 22] that emerging computing platforms will combine multi- and manycore architectures. In particular, this trend is noticeable in an environment of large-scale computations (High Performance Computing, HPC) where supercomputers are built with massively parallel components [24], such as multicore processors and manycores accelerators. The most

© Springer International Publishing AG 2017
V. Malyshkin (Ed.): PaCT 2017, LNCS 10421, pp. 351–364, 2017.
DOI: 10.1007/978-3-319-62932-2_34

common solutions for such systems are based on the cluster architectures, that
are delivered by many vendors. More than 80% of supercomputers in the TOP500
list for November 2016 refer to these systems (http://top500.org).

However, despite the considerable popularity of clusters, other powerful computing platforms are also perceptible in HPC environments. Among them are systems based on the SMP/NUMA (symmetric multiprocessor/non-uniform memory access) architectures [4], which are usually built around high-performance
networks as distributed shared memory (DSM) systems. DSM is a form of memory architecture where physically separated memories can be addressed as one
logically shared address space. These systems provide extremely large capacity of shared memory, and are able to achieve high levels of memory throughput performance. At the same time, because data can be physically dispersed
over many nodes, the access time for different data items may well be different
which explains the term non-uniform data access. In SMP/NUMA architectures,
the parallelism can be successfully expressed with the OpenMP library, or the
MPI standard - a common solution for clusters. Also, the mixture of MPI and
OpenMP is possible. However, is worthwhile to mention that OpenMP is capable
of itself to fully utilize such systems without demanding more complex message
passing operations [5,23] required by MPI.

One of leading vendors of these systems is SGI, that has been delivering
SMP/NUMA architectures for more than 20 years. Its newest SMP/NUMA product series, SGI UV [11] is based on Intel multicore processors and the high-speed
NUMAlink system interconnect, offering up to thousands of cores in a single
system which shares large main memory capacity. An example of using the SGI
UV 2000 server for accelerating a complex real-world application, MapReduce
is presented in [2], where a topology-aware placement algorithm is proposed to
speed up the data shuffling phase of MapReduce. The first generation of SGI
UV platforms is applied in [3] to parallelize the Generalized Conjugate Residual
(GCR) elliptic solver with preconditioner, using a mixture of MPI and OpenMP.
In order to place properly all MPI processes and OpenMP threads on the underlying hardware, a specialized scheduler was developed to take into account the
network topology. Apart from numerical applications, the SGI UV 2000 systems
are also reported to be efficiently used in other areas, such as computation on
graphs [25] and combinatorial optimization problems [1].

In this paper, we face the challenge of efficient utilization of SMP/NUMA systems in practice, for a rather complex scientific application. The application we
study implements the Multidimensional Positive Definite Advection Transport
Algorithm (MPDATA) [13,14], which consists of a set of heterogeneous stencils.
Besides the GCR solver, MPDATA is the second major part of the dynamic
core of the EULAG (Eulerian/semi-Lagrangian) geophysical model [15]. It is an
established numerical model developed for simulating thermo-fluid flows across
a wide range of scales and physical scenario. In particular, it can be used in
numerical weather prediction, simulation of urban flows, turbulences, and ocean
currents [9,16,17].

In our previous works [18–20], we successfully developed a new version of MPDATA, dedicated to small-scale shared memory systems with several processors and/or accelerators. In particular, we proposed a new (3+1)D decomposition of MPDATA computations that allows us to significantly reduce the main memory traffic. This implementation provides a much better usage of capabilities of novel CPUs and Intel Xeon Phi coprocessors. However, although the proposed new strategy of workload distributions gives a gain at the desired performance level not only for Intel Xeon processors, but also for the first generation of Intel Xeon Phi accelerators ([19]), significant performance losses are noticeable for larger SMP/NUMA systems, such as SGI UV 2000 server used in this work.

In this paper, to overcome this shortcoming and to improve the efficiency of the MPDATA application, we propose an islands-of-cores approach dedicated to heterogeneous stencils such as those of MPDATA. This approach reveals a correlation between computation and communication for heterogeneous stencil computations, and enables a better management of the balance between computation and communication costs in accordance with the features of SMP/NUMA systems such as the SGI UV 2000 server. The proposed approach is based on the analysis of two scenarios for the parallel execution of a set of heterogeneous stencils. While the first scenario performs less computations but requires more data transfers, the second one allows us to replace the implicit data traffic between nodes by extra computations, and overcome the non-uniform memory constraints. In consequence, when using the maximum number $P = 14$ of processors with 112 cores totally, the proposed approach accelerates the pure (3+1)D decomposition more then 10 times, achieving approximately 30% of the theoretical peak performance of the system.

To our best knowledge, there exists no investigations of the correlation between computation and communication for heterogeneous stencils computations which consist of a set of stencils with different patterns. The closest approaches were proposed in papers [6,26]. Similarly to our study, these works consider the code transformation using the overlapped tiling technique. It enables removing the synchronization and enhancing the data locality at the cost of redundant computations. However, these works take into account only the homogeneous stencil computations, with a single pattern only. Opposite to our study, these approaches are addressed to small computing platforms with one or two processors.

2   SMP/NUMA Architecture: SGI UV 2000 Server

The parallel computer architecture that we are interested in this paper has all its processing elements interconnected to a shared main memory. One of the most prominent manufacturers of shared-memory systems is SGI. The latest SGI UV ("UltraViolet") product line is delivered since 2009. In all the experiments described in this paper we employ a machine of the second UltraViolet generation known as UV 2 [12], launched in 2012. For a single system, its cache-coherent shared memory can be extended up to 64 TB, and accessed from up to

2048 Intel CPU cores, thanks to the high-speed NUMAlink 6 proprietary inter-connect with a point-to-point bandwidth of 6.7 GB/s per direction; doubled with respect to NumaLink 5. This allows putting hundreds of NUMA nodes together to behave as a single multicore system.

The target SGI UV 2000 server was acquired by the IT4Innovations National Supercomputing Center in Ostrava [8] to support applications with extraor-dinary large memory requirements. It consists of one "individual rack unit" (IRU) that features 3328 GB of RAM and 112 cores in total, distributed across 14 NUMA nodes in 7 compute/memory modules called blades, connected to each other via a backplane (one blade position in this IRU enclosure is empty). Each NUMA node is based on the 8-core Intel Xeon E5-4627v2 3.3 GHz proces-sor with roughly 236 GB RAM. IRU has ports that are brought out to external NUMAlink 6 connectors. This UV 2000 server shares some infrastructure with the Salomon supercomputer of the IT4Innovations center, in June 2015 placed #40 on the TOP500 list (http://top500.org).

3 Parallelization of MPDATA for Shared-Memory Model

3.1 Introduction to MPDATA Application

The MPDATA application implements a general approach for integrating the conservation laws of geophysical fluids on micro-to-planetary scales [10,13]. The MPDATA algorithm enables solving advection problems, and offers sev-eral options to model a wide range of complex geophysical flows. The MPDATA computations correspond to the group of iterative, forward-in-time algorithms. This application is used typically for long running simulations, such as the numerical weather prediction, that require execution of several thousand time steps for a given size of domain. Moreover, since the accuracy of computation plays a key role for MPDATA, these simulations usually are performed using the double-precision floating-point format. The application allows solving 1-, 2- or 3-dimensional problems. In this paper, we consider the last case, when the MPDATA algorithm is defined on 3D grids with i, j, and k dimensions.

Every MPDATA time step performs the same computations, which consist of the set of 17 stages [19,20]. The MPDATA stages represent the heterogeneous stencils codes which update grid elements according to different patterns. All the stages are dependent on each other: outcomes of prior stages are usually input data for the subsequent computations. A single MPDATA time step loads five 3D input arrays from the main memory, and saves one output 3D array that is necessary for the next steps. In the original version of code, a lot of intermediate results (3D arrays) are also transferred to/from the main memory. In consequence, a significant data traffic to the main memory is generated, which mostly limits the attainable performance on novel architectures.

3.2 (3+1)D Decomposition

In our previous works [18–20], we proposed a new strategy of workload dis-tribution for the MPDATA application. This strategy contributes to ease the memory and communication bounds, and to better exploit computation resources of shared-memory systems including CPUs and the first generation of Intel Xeon Phi accelerators. The main challenge of these works was to minimize data transfers between the main memory and the cache hierarchy. To improve the overall performance, we reorganized computation inside each time step of MPDATA.

The main aim of the new computational flow for the MPDATA application is to eliminate accesses to the main memory associated with all the intermediate computations. This idea implies that all the intermediate outcomes of computations have to be kept in cache only - without transferring them to the main memory. As a result, for each MPDATA time step, the main memory traffic will be generated only by transfers required by input/output data (arrays). To reach this goal, we proposed the (3+1)D decomposition of MPDATA computation [19,20] that is based on a combination of loop fusion and loop tiling optimization techniques.

The implementation of the (3+1)D decomposition requires to partition the MPDATA domain (grid) onto a set of sub-domains of size that enables to kept all the necessary intermediate data in the cache memory. The consecutive sub-domains are processed sequentially, one by one, while every sub-domain is processed in parallel by available computing resources. Every sub-domain is responsible for computing all the MPDATA stages that perform computations on chunks (or blocks) of the corresponding arrays, and returns an adequate part of the output array.

The proposed (3+1)D decomposition allows us to significantly reduce the main memory traffic, where the real profit depends on the size of domains, as well as computational characteristic of a given computing platform. For example, using a single Intel Xeon CPU E5-2660v2 processor, the volume of the main memory traffic is reduced from 133 GB to 30 GB, and computations are accelerated about 2.8 times for domains of the size $256 \times 256 \times 64$, and 50 time steps. In this research, the `likwid-perfctr` tool [21] is used for the performance analysis of developed codes. However, although the proposed (3+1)D decomposition gives a gain at the desired performance level not only for Intel Xeon processors, but also for the first generation of Intel Xeon Phi accelerators (see [19]), significant performance losses are noticeable for large shared-memory architectures, such as SMP/NUMA systems.

Table 1 presents the comparison of execution times of MPDATA obtained for the SGI UV 2000 server introduced in Sect. 2, for different versions of code. The performance results are generated for the various number of processors, benchmarking both versions: original and after (3+1)D decomposition. It should be noted that in order to get the optimal performance for this server, it is necessary for each thread of execution to allocate memory *closest* to a core on which it is executed. This is achieved by initializing memory using the technique known [22] as the first-touch policy with parallel initialization.

The obtained performance results reveals the performance limitations for the proposed (3+1)D decomposition. Particularly, the performance gain at the

Table 1. Execution times of 50 MPDATA time steps and grid of size 1024 × 512 × 64 obtained for the original parallel version of code and after the (3+1)D decomposition, using the SGI UV 2000 server

#CPUs	1	2	3	4	5	6	7	8	9	10	11	12	13	14
Original	30.4	44.5	58.2	61.5	64.3	70.1	71.6	73.7	75.4	77.6	78.4	78.2	80.6	82.2
Original[a]	30.4	15.4	10.5	7.9	6.6	5.6	5.0	4.3	4.0	3.6	3.3	3.1	3.0	2.8
(3+1)D[a]	9.0	8.2	7.4	8.0	7.1	7.2	7.3	7.7	9.1	9.5	10.2	10.1	10.3	10.4

[a]The first-touch policy with parallel initialization is used

desired level is achieved only for a single processor (3.37× faster than the original version), while significant performance losses take place for the MPDATA executions with a higher number of processors. It should be also underlined here that the original version returns even better execution times than the (3+1)D decomposition, for all the benchmarks with the number of processors greater than 4. To overcome this shortcoming and to improve the efficiency of the MPDATA application, we propose the islands-of-cores approach, dedicated to heterogeneous stencils such as those of MPDATA.

4   Islands-of-Cores Approach for MPDATA

4.1   Trade-off Between Computation and Communication for Heterogeneous Stencils

To eliminate the revealed performance losses, the analysis of computational flow for heterogeneous stencils has to be considered. Figure 1(a) presents an example of forward-in-time computations with a set of heterogeneous stencils, when every time step consists of three stages. Here each stage corresponds to execution of an 1D stencil. Figure 1(b) and (c) show two scenarios for parallelization of this example using two processors.

The first scenario (Fig. 1b) reveals an implicit data traffic between processors in a shared-memory system because of data dependencies. This data traffic takes place on borders of sub-domains distributed between processors. For example, the output element $C[d]$ computed by CPU_B within the 3rd stage depends on the element $B[c]$ that is computed by CPU_A as a result of stage 2. However, $B[c]$ depends on the element $A[d]$ which is returned by CPU_B in the 1st stage. In consequence, the implicit transfers of two elements take place between the processors CPU_A and CPU_B of a shared-memory system. Furthermore, three synchronization points have to be added in order to ensure the correctness of parallel computations.

The second scenario (Fig. 1c) demonstrates how to avoid exchanging data between processors at the cost of extra computations. To compute the output element $C[d]$, the processor CPU_B has to provide the element $B[c]$ computed in the 2nd stage by CPU_A in the first scenario. Instead of transferring this element, let CPU_B compute the required element $B[c]$ once more. However, this element depends on the element $A[c]$ from the Stage 1, which is returned by CPU_A in the

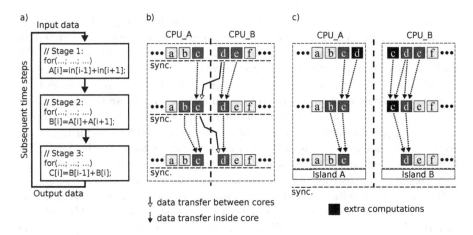

Fig. 1. Idea of Islands-of-cores approach: (a) computations corresponding to three exemplary heterogeneous stencils; (b) parallelization with transfers of data between CPUs; (c) parallelization without transfers of data and synchronization points between CPUs, at the cost of extra computations

first scenario. Again, let both processors compute the element $A[c]$ twice, rather than transferring it from CPU_A to CPU_B. In consequence, CPU_B computes two elements more, independently of CPU_A. This strategy can be also applied for CPU_B that requires the element $A[d]$ computed by CPU_A in the first scenario. As a result, something like independent islands, both processors are enabled to perform computations independently of each other within every time step, at the total cost of computing three extra elements.

As shown in Fig. 1, both scenarios enable performing parallel computations. The first scenario performs less computations but requires more data traffic, while the second one allows us to replace the implicit data traffic between processors by replicating some computations. In fact, both solutions should be considered, but the key point is how they fit to the architecture of a computing system. It is expected that the second scenario will be able to get a higher performance in the case of powerful computing resources with relatively less efficient interconnects. On the contrary, the first scenario is more suitable for systems with more efficient networks that connect less powerful computing resources.

Taking into account the architecture of the SGI UV 2000 server, the second scenario seems to fit perfectly to processors connected each other by NUMAlink. At the same time, the first scenario should be well suited inside each processor, where a more efficient, internal memory hierarchy is used to implement the data traffic between available cores.

4.2   Implementation: From Islands-of-Cores to Work-Teams

The (3+1)D decomposition moves the data traffic from the main memory to the cache hierarchy. In consequence, a lot of intra- and inter-cache communications between cores/processors is generated. This approach corresponds to the first scenario (Fig. 1b). When using a single processor, this traffic is restricted to the cache hierarchy of this CPU. However, in the case of the whole server, the required data are implicitly transferred between caches of neighbor processors through the NUMAlink interconnect [12]. As Table 1 shows, it is particularly significant when more than two processors cooperate to execute the application.

To face this issue, we propose to adopt the islands-of-cores approach to the MPDATA application, which has a significantly more complex computing structure than the example shown in Fig. 1. In this work, we focus on the MPDATA algorithm defined on a 3D grid, where every time step consists of 17 stages with heterogeneous stencils that depend on each other in all three dimensions.

Based on the conclusion formulated in the end of the previous subsection, the abstraction of islands-of-cores is applied across P processors of an SMP/NUMA platform. In consequence, the MPDATA domain is partitioned into P parts that are mapped onto P islands. Following the islands-of-cores approach, each processor is now an island of cores, and these islands perform the following phases:

1. All islands share all input data for each MPDATA time step, utilizing the first-touch policy with parallel initialization.
2. Every island processes the part of MPDATA assigned to it according to the (3+1)D decomposition.
3. Each island performs independent computations within every time step, at the cost of extra computations (see Fig. 1).
4. All islands return common outcomes to the main memory, after each time step.
5. All islands synchronize their works after each time step, in order to ensure correctness of input data for subsequent time steps.

Since every island consists of the same number of cores, the MPDATA domain is decomposed into equals parts, where the number of parts is equal to the number of processors used in computations. Each part is further partitioned into the set of sub-domains according to the proposed (3+1)D decomposition. While different sub-domains are executed sequentially, each of sub-domain is processed in parallel by utilizing a work team of cores which belong to every island. Each work team of cores performs computations corresponding to all the 17 stages of MPDATA, including computing extra elements instead of transferring them from other teams. As a result, every work team is able to perform computations for each MPDATA time step independently of other teams.

To adapt the proposed islands-of-cores approach to the MPDATA application, an efficient method of mapping parts of MPDATA onto processors has to be developed. It is expected to obtain too large communication overheads when the MPDATA domain is partitioned in all three dimensions. The reason is that data layouts of all the MPDATA arrays allow performing required transfers of the continuous areas of memory only in the first and second dimensions. As a result, only 1D and 2D variants of partitioning the MPDATA domain should be taken into account. When evaluating the proposed approach, the 1D partitioning is considered as a starting point in this paper, while investigating more complex 2D variants will be among the main goals of our future works.

As data transfers take place only between neighbour parts of the MPDATA domain, to reduce the communication paths through the network topology, all the neighbour parts should be assigned to the adjacent processors that are closely connected each other within the interconnect. It can be achieved by controlling the OpenMP Thread Affinity interface that allows us to bind threads to physical processing units.

The total amount of extra elements which have to be computed redundantly depends on the problem size, number of islands, and shape of partitioning, as well as data dependencies between all the MPDATA stages. Table 2 presents an example how the total number of extra elements increases with the number of work teams, in comparison with the original version. We compare results for two variants of mapping the MPDATA domain onto 1D grids of processors - across either the first (A) or second (B) dimension of the MPDATA grid. It can be concluded that the first variant gives fewer extra elements, for any number of islands.

Table 2. The total amount of extra elements in percentage in comparison with the original version, obtained for mapping the MPDATA grid onto 1D grids of processors using variants A and B, for different number of islands and the domain of size $1024 \times 512 \times 64$

# of islands	1	2	3	4	5	6	7	8	9	10	11	12	13	14
Variant A [%]	0.00	0.25	0.49	0.74	0.99	1.24	1.48	1.73	1.98	2.22	2.47	2.72	2.96	3.21
Variant B [%]	0.00	0.49	0.99	1.48	1.98	2.47	2.96	3.46	3.95	4.45	4.94	5.43	5.93	6.42

5 Performance Results

This section outlines the performance results obtained for the new implementation of MPDATA developed using the approach described in the previous sections. The new strategies proposed for the workload distribution and data parallelism require also to develop a proprietary scheduler with the affinity-aware placement of threads/cores. To achieve this goal, the OpenMP application programming interface is used only for creating threads and controlling their affinity policy, while all parallel computations are managed by our scheduler which supports the proposed approach.

All the benchmarks are compiled using the Intel compiler icpc v.17.0.1 with the compilation flags: `-O3 -xavx -fp-model precise -fp-model source`, and executed using the SGI UV 2000 server equipped with 14 CPUs. All performance results are obtained for the double-precision floating-point format, the grid of size $1024 \times 512 \times 64$, and 50 times steps. Such a relatively small number of time steps is sufficient to provide the performance evaluation because of homogeneity of all time steps.

In this work, we test the 1D mapping of parts of the MPDATA domain onto a grid of processors. Two variants of experiments are performed, which correspond to distributing the MPDATA domain across either its first or second dimension. Only the results for the first variant are presented in the rest of the paper as it gives better results for all the benchmarks. This is a consequence of a smaller number of extra elements provided by this variant (Table 2).

Table 3 and Fig. 2 present the execution times achieved for the proposed islands-of-cores approach in comparison with the original version and the pure (3+1)D decomposition. Also, we show the partial S_{pr} and overall S_{ov} speedups which define the performance gains of the proposed approach against the pure (3+1)D decomposition and original version, respectively.

The main conclusion is that the proposed islands-of-cores approach, which combines the (3+1)D decomposition and the second scenario of parallelizing stencil computations, allows us to improve radically the efficiency of the MPDATA computations in comparison with the pure (3+1)D decomposition. As expected, despite the extra computations (Table 2), MPDATA is now executed faster for all values of P. What should be underlined here, the usage of the islands-of-core approach together with the (3+1)D decomposition permits preserving the high efficiency of such a decomposition.

Table 3. Execution times for the original version, pure (3+1)D decomposition, and the proposed islands-of-cores approach, as well as partial S_{pr} and overall S_{ov} speedups of the proposed approach

#CPUs	1	2	3	4	5	6	7	8	9	10	11	12	13	14
Execution times														
Original	30.40	15.40	10.50	7.87	6.55	5.61	4.95	4.27	4.01	3.58	3.31	3.14	2.95	2.81
(3+1)D	9.00	8.20	7.38	7.98	7.06	7.22	7.26	7.69	9.11	9.48	10.20	10.10	10.30	10.40
Islands of cores	9.00	5.62	4.17	2.93	2.34	1.97	1.72	1.49	1.36	1.25	1.12	1.06	1.05	1.01
Speedups														
S_{pr}	1.00	1.46	1.77	2.72	3.02	3.66	4.22	5.16	6.70	7.58	9.11	9.53	9.81	10.30
S_{ov}	3.38	2.74	2.52	2.69	2.80	2.85	2.88	2.87	2.95	2.86	2.96	2.96	2.81	2.78

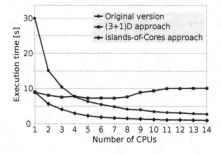 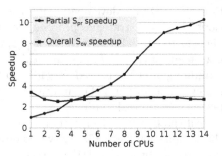

Fig. 2. Performance results for the different number P of processors: (a) comparison of the execution time for the different versions of MPDATA; (b) partial and overall speedups of the islands-of-cores approach

In fact, only for configurations with one 8-core processor and two processors (16 cores totally), the pure (3+1)D decomposition is able to shorten radically the execution time in comparison with the original version, while already for $P = 4$ this decomposition gives a worse performance. For larger values of P, it is the original version that overpowers the pure (3+1)D decomposition. The disadvantage of utilizing only the (3+1)D decomposition increases with the growing number of processors, achieving the ratio of about 3.7 for $P = 14$ (112 cores totally). The reason for such a disappointing behavior of the pure (3+1)D decomposition is large overheads of data transfers between NUMA nodes (processors) when data should be extracted from the deep memory hierarchy of each node before performing transfers, while in the original version these data are simply located in the main memory.

On the contrary, the acceleration of the proposed island-of-core approach against the original version is kept on a similar level, independently of the number of processors, with $S_{ov} = 2.74$ and $S_{ov} = 2.78$ for $P = 2$ and $P = 14$, respectively. As shown in Fig. 2, the performance gain of the combined approach against the pure (3+1)D decomposition increases with the growing number of processors that together perform computations. Finally, when using the maximum number $P = 14$ of processors, the proposed approach accelerates the (3+1)D decomposition more then 10 times.

Table 4 presents the sustained performance (in Gflop/s) for the islands-of-cores approach, as well as the utilization rate in comparison with the theoretical peak performance of the server. As shown in this table, approximately 30% of the theoretical peak is achieved when using less than 12 processors, while it decreases up to the level of 26% for larger values of P. In this benchmark, the maximum sustained performance of about 390 Gflop/s is obtained for $P = 14$, which corresponds to about 77% of the linear scaling. For smaller values of P, the parallel efficiency decreases from 96.6% for $P = 4$ to 80.7% for $P = 12$.

Table 4. Sustained performance [Gflop/s] obtained for the islands-of-cores approach when using the SGI UV 2000 server, as well as utilization rate [%], and parallel efficiency expressed as percentage of linear scaling

Number of processors												
1	2	3	4	5	6	7	8	9	10	11	12	14
Theoretical performance												
105.6	211.2	316.8	422.4	528.0	633.6	739.2	844.8	950.4	1056.0	1161.6	1267.2	1478.4
Sustained performance												
42.7	68.5	92.5	131.9	165.5	197.0	226.1	261.4	287.0	325.9	349.8	370.3	390.1
Utilization rate [%]												
40.4	32.4	29.2	31.2	31.3	31.1	30.5	30.9	30.2	30.8	30.1	29.2	26.3
Parallel efficiency: % of linear scaling												
100.0	98.7	96.5	96.6	92.8	90.3	87.7	89.0	84.2	84.9	83.5	80.7	77.3

6   Conclusions and Future Work

Accelerating memory access by arranging data and computations in an appropriate way is vital for achieving the high application performance on modern computing architectures. Applications with a poor data locality reduce the effectiveness of the memory hierarchy, causing long stall times waiting for data accesses. A purposeful management of data locality plays the primary role for enabling applications to run on different architectures efficiently. The above statement refers in particular to SMP/NUMA systems, which are characterized by heterogeneous network structures. In consequence, since data can be physically dispersed over many nodes, the access to various data items may require significantly different times. This favours accesses to the local memory as fastest.

The present paper faces this challenge for heterogeneous stencil computations, where MPDATA is an important example of such scientific codes. For this purpose, the new islands-of-cores approach is proposed aiming at increasing the efficiency of stencil computations on SMP/NUMA platforms, by improving the data locality. This approach exposes a correlation between computation and communication for heterogeneous stencils, enabling a better management of the trade-off between computation and communication costs in accordance with the features of SMP/NUMA systems, such as the SGI UV 2000 server used in this work. To overcome the non-uniform memory access constraints, the proposed approach combines the previously developed $(3+1)$D decomposition and the scenario of parallelizing stencil computations when the implicit data traffic between nodes is replaced by extra computations. As a result, the resulting parallel code scales well with increasing the number of processors, and radically better than both the original version and pure $(3+1)$D decomposition.

In particular, for the MPDATA grid of size $1024 \times 512 \times 64$, approximately 30% of the theoretical peak is achieved when using less than 12 processors, while it decreases up to the level of 26% for large configurations. In this benchmark, the maximum sustained performance of about 390 Gflop/s is obtained for the maximum configuration with 112 cores of 14 Intel Xeon E5-4627v2 3.3 GHz processors. It corresponds to about 77% of the linear scaling. For smaller values of P, the parallel efficiency decreases from 96.6% for $P = 4$ to 80.7% for $P = 12$.

The achieved results justify further research on improving the efficiency of heterogeneous stencil computations on modern architectures. In particular, the proposed islands-of-cores approach can be applied to optimize computations within every multicore CPU (or manycore accelerator). At the opposite edge of the scale, we plan to study the usage of MPI for extending the scalability of our approach for much large system configurations. This requires to build performance models and methods for modeling and management of the correlation between computation and communication costs, to study its impact on the sustained performance. The optimal trade-off between computations and communications inside and between processors should be determined on this basis.

Acknowledgments. This work was supported by the National Science Centre (Poland) under grant UMO-2015/17/D/ST6/04059, as well as partially supported by the Ministry of Education, Youth and Sports of Czech Republic from the project "IT4Innovations National Supercomputing Center LM2015070", and by EU under the COST Program Action IC1305 "Network for Sustainable Ultrascale Computing (NESUS)" and its Czech supporting project LD15105 "Ultrascale Computing in Geosciences".

References

1. Cao, X., et al.: Accelerating data shuffling in MapReduce framework with a scale-up NUMA computing architecture. In: Proceedings of the 24th High Performance Computing Symposium, HPC 2016. International Society for Computer Simulation (2016)
2. Castro, M., Francesquini, E., Nguélé, T.M., Méhaut, J.F.: Analysis of computing and energy performance of multicore, NUMA, and manycore platforms for an irregular application. In: Proceedings of the 3rd Workshop on Irregular Applications: Architectures and Algorithms. ACM (2013)
3. Ciznicki, M., Kulczewski, M., Kopta, P., Kurowski, K.: Methods to load balance a GCR pressure solver using a stencil framework on multi-and many-core architectures. Sci. Program. (2015)
4. Culler, D., Pal Singh, J., Gupta, A.: Parallel Computer Architecture: A Hardware/Software Approach. Morgan Kaufmann Publishers Inc., San Francisco (1999)
5. Czarnul, P.: Benchmarking performance of a hybrid Xeon/Xeon Phi system for parallel computation of similarity measures between large vectors. Int. J. Parallel Program. 1–17 (2017)
6. Guo, J., Bikshandi, G., Fraguela, B.B., Padua, D.: Writing productive stencil codes with overlapped tiling. Concurr. Comput. Pract. Exp. **21**(1), 25–39 (2009)
7. Hager, G., Treibig, J., Habich, J., Wellein, G.: Exploring performance and power properties of modern multi-core chips via simple machine models. Concurr. Comput. Pract. Exp. **28**(22), 189–210 (2016)
8. National Supercomputing Center IT4Innovations (2017). http://www.it4i.cz
9. Kumar, S., Bhattacharyya, R., Joshi, B., Smolarkiewicz, P.: On the role of repetitive magnetic reconnections in evolution of magnetic flux ropes in solar corona. Astrophys. J. **830**(2), 80 (2016)
10. Lastovetsky, A., Szustak, L., Wyrzykowski, R.: Model-based optimization of EULAG kernel on Intel Xeon Phi through load imbalancing. IEEE Trans. Parallel Distrib. Syst. **28**(3), 787–797 (2017)
11. SGI Products: Servers SGI UV (2015). https://www.sgi.com/products/servers/uv/
12. SGI UV 2000 System User Guide. Document Number 007–5832-002 (2013)
13. Smolarkiewicz, P.: Multidimensional positive definite advection transport algorithm: an overview. Int. J. Numer. Methods Fluids **50**(10), 1123–1144 (2006)
14. Smolarkiewicz, P., Margolin, L.: MPDATA: a finite-difference solver for geophysical flows. J. Comput. Phys. **140**(2), 459–480 (1998)
15. Smolarkiewicz, P.K., Charbonneau, P.: EULAG, a computational model for multiscale flows: an MHD extension. J. Comput. Phys. **236**, 608–623 (2013)
16. Smolarkiewicz, P.K., Szmelter, J., Xiao, F.: Simulation of all-scale atmospheric dynamics on unstructured meshes. J. Comput. Phys. **322**(C), 267–287 (2016)

L. Szustak et al.

17. Strugarek, A., Beaudoin, P., Brun, A., Charbonneau, P., Mathis, S., Smolarkiewicz, P.: Modeling turbulent stellar convection zones: sub-grid scales effects. Adv. Space Res. **58**(8), 1538–1553 (2016)
18. Szustak, L., Rojek, K., Gepner, P.: Using Intel Xeon Phi coprocessor to accelerate computations in MPDATA algorithm. In: Wyrzykowski, R., Dongarra, J., Karczewski, K., Waśniewski, J. (eds.) PPAM 2013. LNCS, vol. 8384, pp. 582–592. Springer, Heidelberg (2014). doi:10.1007/978-3-642-55224-3_54
19. Szustak, L., Rojek, K., Olas, T., Kuczynski, L., Halbiniak, K., Gepner, P.: Adaptation of MPDATA heterogeneous stencil computation to Intel Xeon Phi coprocessor. Sci. Program. (2015). doi:10.1155/2015/642705
20. Szustak, L., Rojek, K., Wyrzykowski, R., Gepner, P.: Toward efficient distribution of MPDATA stencil computation on Intel MIC architecture. In: Proceedings of the 1st International Workshop on High-Performance Stencil Computations, HiStencils 2014, pp. 51–56 (2014)
21. Treibig, J., Hager, G., Wellein, G.: LIKWID: a lightweight performance-oriented tool suite for x86 multicore environments. In: Proceedings of the First International Workshop on Parallel Software Tools and Tool Infrastructures, PSTI 2010, San Diego, CA (2010)
22. Unat, D., et al.: Programming abstractions for data locality. (2014). http://web.eecs.umich.edu/akamil/papers/padal14report.pdf
23. Utrera, G., Gil, M., Martorell, X.: In search of the best MPI-OpenMP distribution for optimum Intel-MIC cluster performance. In: 2015 International Conference on High Performance Computing and Simulation (HPCS), pp. 429–435. IEEE (2015)
24. Xue, W., et al.: Ultra-scalable CPU-MIC acceleration of mesoscale atmospheric modeling on Tianhe-2. IEEE Trans. Comput. **64**(8), 2382–2393 (2015)
25. Yasui, Y., Fujisawa, K., Goh, E.L., Baron, J., Sugiura, A., Uchiyama, T.: NUMA-aware scalable graph traversal on SGI UV systems. In: Proceedings of the ACM Workshop on High Performance Graph Processing, pp. 19–26. ACM (2016)
26. Zhou, X., Giacalone, J.P., Garzarán, M.J., Kuhn, R.H., Ni, Y., Padua, D.: Hierarchical overlapped tiling. In: Proceedings of the Tenth International Symposium on Code Generation and Optimization, pp. 207–218. ACM (2012)

The Algorithm of Control Program Generation for Optimization of LuNA Program Execution

Anastasia A. Tkacheva[1,2(✉)]

[1] Institute of Computational Mathematics and Mathematical Geophysics,
Siberian Branch of Russian Academy of Sciences, Novosibirsk, Russia
tkacheva@ssd.sscc.ru
[2] Novosibirsk State University, Novosibirsk, Russia

Abstract. LuNA fragmented programming system is a high-level declarative system of parallel programming. Such systems have the problem of achieving on appropriate program execution performance in comparison with MPI. The reasons are a high degree of parallel program execution non-determinism and execution overhead. The paper presents an algorithm of control program generation for LuNA programs. That is a step towards automatic improvement of LuNA program execution performance. Performance tests presented show effectiveness of the proposed approach.

Keywords: High performance computing · Fragmented programming technology · Fragmented programming system LuNA · Parallel program generation

1 Introduction

Implementation of large-scale numerical models on supercomputers is difficult and to achieve good performance the programmer has to have knowledge of parallel programming. For example, to program the particle-in-cell method [2] it requires providing dynamic load balance, virtual layers and so on. The LuNA system [1] is being developed. Its main aim is to simplify the parallel programming process for the case of large-scale numerical models. An application program is represented in a cross-platform form with explicit parallelism. The such form increases parallel program code reuse and portability, but requires complex execution algorithms in parallel programming system. It is for these reasons that there is the lack of LuNA program execution efficiency in comparison with the similar implementation using MPI. On the other hand, the programmer does not have to define resources distribution. Most dynamic properties are provided automatically in LuNA system.

In LuNA program can be divided into parts (subroutines). For some of those parts the most decisions on resources distribution can be made statically at compiling stage. In the paper the authors are studying ways to optimize LuNA program execution performance. One of them is to create and use framework for

© Springer International Publishing AG 2017
V. Malyshkin (Ed.): PaCT 2017, LNCS 10421, pp. 365–371, 2017.
DOI: 10.1007/978-3-319-62932-2_35

implementation of these parts. The LuNAFW framework is being developed. In LuNAFW the application execution is based on the model of event-driven type. For using it control program has to be developed. Inside control program most decisions on resources distribution and operations order were partially made, and they are formulated using event handlers and LuNAFW API functions for distributed or shared memory environment. The efficiency of suggested approach is presented in [4]. The manual development of control program for LuNA program is a separate time-consuming task and it does not respond to the LuNA system development objectives. Therefore, an algorithm of control program generation is developed and proposed in the paper.

2   Related Works

There is the lack of efficiency of parallel program execution in most high-level parallel programming systems in comparison with MPI for the large-scale task on supercomputers. The main reason is that runtime-system can not make good decisions on resources distribution and organize parallel computation without knowledge of the application problem. To improve performance these systems usually narrow down object domain or include annotations in the language that help the run-time system to make more appropriate decisions.

For example, PaRSEC [5] system was developed for DPLASMA [6] library contained linear algebra subroutines for dense matrices. In spite of small object domain the system includes a way to define priorities of operations execution. The priorities are also presented in SMP Superscalar [7]. Charm++ [8] system has annotation language Charisma [9] to show the run-time system which functions are related to communications. To improve performance functional language Haskell [10] uses a coordinate language Eden [11]. The engine to create skeleton for object subdomain are also provided.

3   LuNA Fragmented Programming System

LuNA (Language for Numerical Algorithms) is a language and a parallel programming system [1] intended for implementation of large-scale numerical models on supercomputers. It is being developed in the Institute of Computation Mathematics and Mathematical Geophysics of the Siberian Branch of Russian Academy Of Sciences.

In LuNA application program is represented in a single-assignment coarse-grained explicitly parallel language LuNA as a bipartite graph of *data fragments* (DF) and *computational fragments* (CF). DFs are basically blocks of data (submatrixes, array slices, etc.). CFs are applications of pure functions on DFs. A CF has a set of input DFs and a set of output DFs. Values of output DFs are computed by the CF from the values of input DFs. Such representation is called *fragmented algorithm*(FA).

LuNA program consists of the FA description in LuNA language and a dynamic load library with a set of conventional sequential procedures. CFs are implemented as calls to these procedures with input and output DFs. Execution of all the CFs is done in accordance with partial order, that is imposed on the set of CFs by the information dependencies, forms the FA execution.

A FA is executed by the LuNA run-time system. Fragmented structure of the FA is kept in run-time, allowing the run-time system to dynamically assign CFs and DFs to different computing nodes, execute CFs in parallel (if possible), balance computational workload by redistributing CFs and DFs and so on.

The run-time system makes most decisions on FA execution dynamically. That is the reason of significant execution overheads.

The overall overhead may be divided into the following types:

- Overhead to organize computations inside a node.
- Overhead to organize computations among different nodes.

The previous work [12] presents ways to decrease overhead inside a node by optimizing checks for CFs being ready. This is applied for loop execution with using Petri nets or by monolithization. It is especially important for small-grained FAs when CFs computational time is too short. The experiments in distributed computing environment show that the benefit of using those ways is limited in comparison with overhead to organize computations among different nodes. So the next aim is to optimize overhead related to distributed computing.

For optimization an approach of framework developing for an object sub-domain was chosen. The LuNAFW is such a framework based on a model of event-driven type. To use it for LuNA program it is required to develop a control program where most decisions on resources distribution and CFs execution order are partially made. They represented using event handlers and framework basic API function. The efficiency of the approach is presented in [4].

Since the main aim of LuNA system development is to automate the process of parallel programs development, the manual control program development is not appropriate. Thus, the algorithm of control program generations was developed.

4   The Algorithm of Control Program Generation for LuNA Program

In LuNA program each DF and CF has to be identified by unique (program-wide) identifiers. The identifier has atomic or indexed form. In LuNAFW framework CFs are distributed among different nodes using the same resources distribution strategy as in LuNA system [13].

The input of suggested algorithm is FA. It includes the elements of following types:

- CF description. It contains:
  - CF identifier.
  - The set of input DFs' identifiers.
  - The set of output DFs' identifiers.

- Name of pure function on C/C++ language.
- The set of DFs' identifiers should be destroyed after CF execution is finished (optional parameter).
- The set of CF descriptions using loop construction. It contains:
  - The name of cycle counter.
  - The value of lower boundary (must be integer).
  - The value of upper boundary (must be integer).
  - List of CF descriptions or the set of CF descriptions using loop construction.
- The set of output DFs of FA.

The requirements for input FA is that all information dependences should be able to be analyzed in compiling stage. In case of the identifier has indexed form the index must be integer constant, the name of cycle counter or expression of type the name of cycle counter plus/minus integer constant.

The output of the algorithm is the generated control program represented as C++ class. LuNAFW framework can execute control program in distributed or shared memory environment.

The LuNAFW program execution is based on model of event-driven type. The following handlers have to be defined:

- *onInit()* - the beginning event handler is used for initialization.
- *onComputed(df_id)* - the handler is called after each DF is computed.
- *onReceived (df_id)* - the handler is called after each DF is received from other nodes.
- *onCfFinished (cf_id)* - the handler is called after each CF is finished execution.

Inside handlers the following functions (actions) supported by LuNAFW framework API can be used:

- *startCF (CF disctiption)* - the action to start CF execution.
- *checkCF (CF disctiption)* - if all input DFs are available the action *startCF* is called.
- *destroyDF(df_id)* - the action to destroy DF with identifier df_id.
- *sendDF (df_id, rank)* - the action to send DF with identifier *df_id* to node *rank*.
- *exit* - the action to stop of execution.
- *int getRank(identifier)* - the action returns node where CF is distributed.

The algorithm of control program generation can be divided into two stages:

1. Converter is to convert FA from data-flow-based to event-driven computation model.
2. Generator is to generate control program from Converter output taking into account resources distribution strategy.

The Converter output includes:

- *Init* is the list of descriptions of CFs which have no input DFs.

- B is the list of descriptions of CFs which have no output DFs.
- Out is the list of output DFs of FA.
- Dictionary $GarbageCollection$:
  - Key is CF identifier.
  - $Value$ is the list of DF identifiers which should be destroyed after CF execution is finished. If DF or CF identifier has indexed form then the boundary use for each index is also defined.
- Dictionary DAG:
  - Key is DF identifier.
  - $Value$ the list of CF descriptions for which key is the input DF identifier. If DF or CF identifier has indexed form then the boundary use for each index is also defined.

In Coverter the identifier is the key of the dictionary. So the functioning of the algorithm is evident in case when identifier has atomic form. In case of the identifier has indexed form, then transformation to common indexed form is needed. For these reasons each index is substituted for new name which is chosen depended on the order of the indices in indexed form. If index is the expression of type the name of cycle counter plus/minus integer constant, then the old name is substituted by new name minus/plus integer constant, and new name boundary use is defined as old name boundary use plus/minus integer constant.

The output of Converter is the Generator input. To generate handler $onInit$ the list $Init$ is used. For each CF from it checks if CF is distributed on the node then the action $startCF$ is called. To generate handler $onCfFinished\ (cf\_id)$ the dictionary $GarbageCollection$ and the list B are used. The following conditions are checked:

- If $GarbageCollection$ has the key cf_id then action $destroyDF$ with the corresponding value is called.
- If the list B contains cf_id, it is kept track of in the exit algorithm.

To generate handlers $onComputed(df\_id)$ and $onReceived(df\_id)$ is used the dictionary DAG. If DAG has the key df_id then:

- In both cases: The value with key df_id is viewed and if a CF from it is distributed on this node, then action $checkCF$ is called. If the input DFs of FA contain df_id it is kept track of in the exit algorithm.
- In case of handler $onComputed$: If CF is distributed on different node action $sendDF$ is called. If DF needs to execute many CFs optimization to send once is supported.

Control program is considered to be finished if all CFs from list B were executed and all DFs from the set of output DFs of FA were computed. In that case action $exit$ is called.

The order of CF execution in generated control program does not contradict information dependence described in input FA.

5   Perfomance Tests

To investigate the efficiency of the proposed algorithm an explicit finite differ-ence method (FDM) for 3D Poisson equation solution [3] was chosen as a test application.

The experiments were conducted on MVS-10P cluster of Joint Supercom-puter Center of RAS (each cluster node has two Xeon E5-2690 processors with 64 GB RAM (16 cores per node); nodes are connected by Infiniband FDR net-work). GCC 5.2.0 compiler and MPICH 3.1.4 communication library were used.

The three versions of parallel program were tested: MPI, LuNA and LuNAFW. The LuNAFW version is automatically generated using the suggested algorithm of control program generation. One MPI process per a core is used. The LuNA version was tested with one thread per MPI process. The goal of the test is to evaluate weak scalability, when the problem size increases with increas-ing a number of processes. In ideal case the computation time is the same, but in reality communication overhead make effect and time is growing.

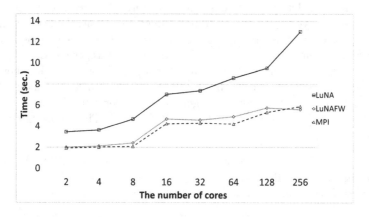

Fig. 1. Weak scalability: computation time (in sec.) dependency on number of cores.

In Fig. 1 computation times are shown for the case of a fragment size of $100 \times 200 \times 200$ per core and for 20 iteration of FDM. The LuNAFW implemen-tation is more efficient than LuNA version and allows achieving good perfor-mance of parallel program in comparison with MPI. The average benefit is 40%.

6   Conclusion

The ways to optimize LuNA programs are studied. The way to develop LuNAFW framework based on the model of event-driven type is chosen. To automate con-trol program development for LuNAFW the algorithm is developed and consid-ered. Performance of evaluation is presented. It showed the efficiency of suggested approach.

Acknowledgments. The author would like to thank his supervisor Dr. Victor E. Malyshkin for his professional guidance and Vladislav A. Perepelkin for his constructive suggestions during the development of this research work.

References

1. Malyshkin, V.E., Perepelkin, V.A.: LuNA fragmented programming system, main functions and peculiarities of run-time subsystem. In: Malyshkin, V. (ed.) PaCT 2011. LNCS, vol. 6873, pp. 53–61. Springer, Heidelberg (2011). doi:10.1007/978-3-642-23178-0_5
2. Kraeva, M.A., Malyshkin, V.E.: Assembly technology for parallel realization of numerical models on MIMD-multicomputers. J. Future Gener. Comput. Syst. **17**(6), 755–765 (2001)
3. Kireev, S.E., Malyshkin, V.E.: Fragmentation of numerical algorithms for parallel subroutines library. J. Supercomput. **57**(2), 161–171 (2011)
4. Akhmed-Zaki, D.Z., Lebedev, D.V., Perepelkin, V.A.: Implementation of a three dimensional three-phase fluid flow (oilwatergas) numerical model in LuNA fragmented programming system. J. Supercomput. **73**(2), 624–630 (2017)
5. Bosilca, G., Bouteiller, A., et al.: DAGuE: a generic distributed DAG engine for high performance computing. In Proceedings of IPDPS 2011 Workshops, pp. 1151–1158 (2011)
6. Bosilca, G., Bouteiller, A., et al.: Flexible development of dense linear algebra algorithms on massively parallel architectures with DPLASMA. In: Proceedings of IPDPS 2011 Workshops, pp. 1432–1441 (2011)
7. Perez, J.M., Badia, R.M., Labarta, J.: A flexible and portable programming model for SMP and multi-cores. Technical report 03/2007, Barcelona Supercomputing Center (2007)
8. Kale, L.V., Krishnan, S.: CHARM++: a portable concurrent object oriented system based on C++. In: Proceedings of OOPSLA 1993, pp. 91–108. ACM, New York (1993)
9. Huang, C., Laxmikant, V.K.: Charisma: orchestrating migratable parallel objects. In: Proceedings of the 16th International Symposium on High Performance Distributed Computing (HPDC), pp. 75–84 (2007)
10. Coutts, D., Loeh, A.: Deterministic parallel programming with Haskell. Comput. Sci. Eng. **14**(6), 36–43 (2012)
11. Loogen, R., Ortega-Malln, Y., Pea-Mar, R.: Parallel functional programming in Eden. J. Funct. Program. **15**(3), 431–475 (2005)
12. Malyshkin, V.E., Perepelkin, V.A., Tkacheva, A.A.: Control flow usage to improve performance of fragmented programs execution. In: Malyshkin, V. (ed.) PaCT 2015. LNCS, vol. 9251, pp. 86–90. Springer, Cham (2015). doi:10.1007/978-3-319-21909-7_9
13. Malyshkin, V.E., Perepelkin, V.A., Schukin, G.A.: Scalable distributed data allocation in LuNA fragmented programming system. J. Supercomput. **73**(2), 726–732 (2017)

Cyclic Anticipation Scheduling in Grid VOs with Stakeholders Preferences

Victor Toporkov[1(✉)], Dmitry Yemelyanov[1], Anna Toporkova[2],
and Petr Potekhin[1]

[1] National Research University "MPEI",
ul. Krasnokazarmennaya, 14, Moscow 111250, Russia
{ToporkovVV, YemelyanovDM, PotekhinPA}@mpei.ru
[2] National Research University Higher School of Economics,
ul. Myasnitskaya, 20, Moscow 101000, Russia
atoporkova@hse.ru

Abstract. In this work, a job-flow scheduling approach for Grid virtual organizations (VOs) is proposed and studied. Users' and resource providers' preferences, VOs internal policies, resources geographical distribution along with local private utilization impose specific requirements for efficient scheduling according to different, usually contradictive, criteria. With increasing resources utilization level the available resources set and corresponding decision space are reduced. This further complicates the problem of efficient scheduling. In order to improve overall scheduling efficiency, we propose an anticipation scheduling approach based on a cyclic scheduling scheme. It generates a near optimal but infeasible scheduling solution and includes a special replication procedure for efficient and feasible resources allocation. Anticipation scheduling is compared with the general cycle scheduling scheme and conservative backfilling using such criteria as average jobs' response time (start and finish times) as well as users' and VO economic criteria (execution time and cost).

Keywords: Scheduling · Grid · Resources · Utilization · Heuristic · Job batch · Virtual organization · Cycle scheduling scheme · Anticipation · Replication

1 Introduction and Related Works

In Grids with non-dedicated resources the computational nodes are usually partly utilized by local high-priority jobs coming from resource owners. Thus, the resources available for use are represented with a set time intervals (slots) during which the individual computational nodes are capable to execute parts of independent users' parallel jobs. These slots generally have different start and finish times and a performance difference. The presence of a set of slots impedes the problem of resources allocation necessary to execute the job flow from VOs users. Resource fragmentation also results in a decrease of the total computing environment utilization level [1, 2].

Application level scheduling [3] is based on the available resources utilization and, as a rule, does not imply any global resource sharing or allocation policy. Job flow

V. Malyshkin (Ed.): PaCT 2017, LNCS 10421, pp. 372–383, 2017.
DOI: 10.1007/978-3-319-62932-2_36

scheduling in VOs [4, 5] suppose uniform rules of resource sharing and consumption, in particular based on economic models [2, 6]. This approach allows improving the job-flow level scheduling and resource distribution efficiency. VO policy may offer optimized scheduling to satisfy both users' and VO common preferences. The VO scheduling problems may be formulated as follows: to optimize users' criteria or utility function for selected jobs [6, 7], to keep resource overall load balance [8, 9], to have job run in strict order or maintain job priorities [10], to optimize overall scheduling performance by some custom criteria [11, 12], etc.

Users' preferences and VO common preferences (owners' and administrators' combined) may conflict with each other. Users are likely to be interested in the fastest possible running time for their jobs with least possible costs whereas VO preferences are usually directed to balancing of available resources load or node owners' profit boosting. Thus, VO policies in general should respect all members to function properly and the most important aspect of rules suggested by VO is their fairness. A number of works understand fairness as it is defined in the theory of cooperative games [7], such as fair job flow distribution [9], fair quotas [13, 14], fair user jobs prioritization [10], and non-monetary distribution [15]. In many studies VO stakeholders' preferences are usually ensured only partially: either owners are competing for jobs optimizing only users' criteria [6, 16], or the main purpose is the efficient resources utilization not considering users' preferences [17]. Sometimes multi-agent economic models are established [3, 18]. Usually they do not allow optimizing the whole job flow processing.

The goal of the current study is to design a general scheduling approach which will be able to find a tradeoff between VO stakeholders' contradictory preferences based on the cyclic scheduling scheme (CSS). CSS [19] has fair resource share in a sense that every VO stakeholder has mechanisms to influence scheduling results providing own preferences. The downside of a majority centralized metascheduling approaches is that they lose their efficiency and optimization features in distributed environments with a limited resources supply. For example in [2], a traditional backfilling algorithm provided better scheduling outcome when compared to different optimization approaches in resource domain with a minimal performance configuration. The general root cause is that in fact the same scarce set of resources (being efficient or not) have to be used for a job flow execution or otherwise some jobs might hang in the queue. Under such conditions, user jobs priority and ordering greatly influence the scheduling results. At the same time, application-level brokers are still able to ensure user preferences and optimize the job's performance under free-market mechanisms.

Main contribution of this paper is a CSS-based job-flow scheduling approach which retains efficiency even in distributed computing environments with limited resources. A special replication procedure is proposed and studied to ensure a feasible scheduling solution.

The rest of the paper is organized as follows. Section 2 presents a general CSS fair scheduling concept. The proposed heuristic-based scheduling technique is presented in Sect. 3. Section 4 contains experiment setup and results for the proposed scheduling approach. Finally, Sect. 5 summarizes the paper.

2   Cyclic Alternative-Based Scheduling

Scheduling of a job flow using CSS is performed in time cycles known as scheduling intervals, by job batches [19]. The actual scheduling procedure consists of two main steps. The first step involves a search for alternative scenarios of each job execution, or simply alternatives [20]. During the second step the dynamic programming methods [19] are used to choose an optimal alternatives' combination. One alternative is selected for each job with respect to the given VO and user criteria. An example for a user scheduling criterion may be an overall job running time, an overall running cost, etc. This criterion describes user's preferences for that specific job execution and expresses a type of an additional optimization to perform when searching for alternatives. Alongside with time (T) and cost (C) properties each job execution alternative has a user utility (U) value: user evaluation against the scheduling criterion. A common VO optimization problem may be stated as either minimization or maximization of one of the properties, having other fixed or limited, or involve Pareto-optimal strategy search involving both kinds of properties [4, 19, 21, 22].

We consider the following relative approach to represent the user utility U: each alternative gets its utility in relation to the "best" and the "worst" optimization criterion values user could expect according to the job's priority. Accordingly $U \in [0\%; 100\%]$ and the more some alternative corresponds to user's preferences (the smaller the difference from the "best" alternative) the smaller is the U value.

For a fair scheduling model the second step VO optimization problem could be in form of: $C \rightarrow$ max, lim U (maximize total job flow execution cost, while respecting user's preferences to some extent); $U \rightarrow$ min, lim T (meet user's best interests, while ensuring some acceptable job flow execution time) and so on [19, 21].

The launch of any job requires a co-allocation of a specified number of slots, as well as in the classic backfilling variation. A single slot is a time span that can be assigned to run a part of a parallel job. The target is to scan a list of available slots and to select a window of parallel slots with a "length" of the required resource reservation time. The user job requirements are arranged into a resource request containing a resource reservation time, characteristics of computational nodes (clock speed, RAM volume, disk space, operating system etc.), limitation on the selected window maximum cost.

ALP, AMP and AEP window search algorithms were discussed in [20]. The job batch scheduling performs consecutive allocation of a multiple nonintersecting in terms of slots alternatives for each job. Otherwise irresolvable collisions for resources may occur if different jobs will share the same time-slots. Sequential alternatives search and resources reservation procedures help to prevent such scenario. However in an extreme case when resources are limited or over utilized only at most one alternative execution could be reserved for each job. In this case alternatives-based scheduling result will be no different from First Fit resources allocation procedure [2]. First Fit resource selection algorithms [23] assign any job to the first set of slots matching the resource request conditions without any optimization.

3   Cyclic Anticipation Scheduling

In order to address the scheduling optimization problem the following heuristic job batch scheduling scheme is proposed. It consists of three main steps.

First, a set of all possible execution alternatives is found for each job not considering time slots intersections and without any resources reservation. The resulting intersecting alternatives found for each job reflect a full range of different job execution possibilities which user may expect on the current scheduling interval.

Second, CSS scheduling procedure [19, 21] is performed to select alternatives combination (one alternative for each job of the batch) optimal according to VO policy. The resulting alternatives combination most likely corresponds to an infeasible scheduling solution as possible time slots intersection will cause collisions on resources allocation stage. The main idea of this step is that obtained infeasible solution will provide some heuristic insights on how each job should be handled during the scheduling. For example, if time-biased or cost-biased execution is preferred, how it should correspond to user criterion and VO administration policy and so on.

Third, a feasible resources allocation is performed by replicating alternatives selected in step 2. The base for this replication step is an Algorithm searching for Extreme Performance (AEP) described in details in [20]. In the current step AEP helps to find and reserve feasible execution alternatives most similar to those selected in the near-optimal infeasible solution. After these three steps are performed the resulting solution is both feasible and efficient as it reflects scheduling pattern obtained from a near-optimal reference solution from step 2.

We used AEP modification to allocate a diverse set of execution alternatives for each job. Originally AEP scans through a whole list of available time slots and retrieves one alternative execution satisfying user resource request and optimal according to the user custom criterion. During this scan, we saved all intermediate AEP search results to a dedicated list of possible alternatives.

For the replication purpose a new Execution Similarity criterion was introduced which helps AEP to find a window with a minimum distance to a reference alternative. Generally, we define a distance between two different alternatives (windows) as a relative difference or error between their significant criteria values. For example if reference alternative has C_{ref} total cost, and some candidate alternative cost is C_{can}, then the relative cost error E_C is calculated as $E_C = \frac{|C_{ref} - C_{can}|}{C_{ref}}$. If one needs to consider several criteria the distance D between two alternatives may be calculated as a linear sum of criteria errors: $D_l = E_C + E_T + .. + E_U$, or as a geometric distance in a parameters space: $D_g = \sqrt{E_C^2 + E_T^2 + ..E_U^2}$.

AEP with Execution Similarity scans through the whole list of available time slots and checks every feasible slots combination. The main difference from the original AEP is that instead of searching for a window with a maximum single criterion value, we retrieve window with a minimum distance D_g or D_l to a reference execution alternative. Generally, this distance can reflect job execution preferences in terms of multiple criteria such as job execution cost, runtime, start time, finish time, etc.

For a feasible job batch resources allocation AEP consequentially allocates for each job a single execution window with a minimum distance to a reference corresponding alternative from an infeasible solution. Time slots allocated for some job are reserved and excluded from the slot list when AEP search algorithm is performed for the following jobs of the batch. Thus this procedure prevents any conflicts for resources and provides scheduling solution which in some sense reflects near-optimal reference solution.

4 Simulation Study

An experiment was prepared as follows using a custom distributed environment simulator [2, 19, 21].

Simulation environment was configured with the following features. The resource pool includes 80 heterogeneous computational nodes. A specific cost of a node is an exponential function of its performance value (base cost) with an added variable margin distributed normally as ±0.6 of a base cost. The scheduling interval length is 800 time quanta. The initial resource load with owner jobs is distributed hyper-geometrically resulting in 5% to 10% time quanta excluded in total.

Jobs number in a batch is 125. Nodes quantity needed for a job is a whole number distributed evenly on [2; 6]. Node reservation time is a whole number distributed evenly on [100; 500]. Job budget varies in the way that some of jobs can pay as much as 160% of base cost whereas some may require a discount. Every request contains a specification of a custom user criterion which is one of the following: job execution runtime or overall execution cost.

4.1 Replication Scheduling Accuracy

The first experiment is dedicated to a replication scheduling accuracy study. For this matter we conducted and collected data from more than 1000 independent job batch scheduling simulations. First, the general CSS was performed in each experiment for the following job-flow execution cost maximization problem $C \to \max$, $\lim U_a = 10\%$. U_a stands for the average user utility for one job, i.e. $\lim U_a = 10\%$ means that at average resulting deviation from the best possible outcome for each user did not exceed 10%. Next, linear and geometric replication algorithms were executed to replicate CSS solution using linear D_l and geometric D_g distance criteria. In the current experiment we used job execution cost error E_c and processor time usage error E_t to calculate distances D_l and D_g.

In order to evaluate the resulting difference in scheduling outcomes, we additionally performed CSS algorithm for $C \to \max$, $\lim U_a = 0\%$ (ensuring users' individual preferences only) and $C \to \max$, $\lim U_a = 100\%$ (ensuring VO preference, i.e. maximizing overall cost without taking into account users' criteria) problems. These additional problems reflect extreme boundaries for scheduling results, which can be used to evaluate a relative replication error. Table 1 contains scheduling results for all these three problems and two replication algorithms.

The results indicate that both linear and geometric replication algorithms provided average scheduling parameters very close to the reference solution (indicated as bold in Table 1), and especially close against job execution cost and processor time usage, i.e.

Table 1. CSS replication average scheduling results

Job execution characteristic	$C \to$ max, lim $Ua = 0\%$	$C \to$ max, lim $Ua = 10\%$	Linear replication	Geometric replication	$C \to$ max, lim $Ua = 100\%$
Cost	1283	*1349*	1353	1353	1475
Processor Time	191.6	*191.2*	190.6	190.5	202.3
Finish time	367.1	*353.8*	356.2	356.4	358.5
$U_a\%$	0	*9.9*	17.6	17.8	65

integral characteristics which were used for a replication distance calculation. For example, borderline problems $C \to$ max, lim $U_a = 100\%$ and $C \to$ max, lim $U_a = 0\%$ provided average job execution cost (main job-flow optimization criterion) values 1283 and 1475 correspondingly. Reference intermediate solution provided 1349. And both replication algorithms ensured average job execution cost 1353 with only 2% deviation from reference solution against [1283; 1475] interval of possible scheduling outcomes. However individual user's preferences were considered to a lesser extent as both replication algorithms provided average user utility U_a almost twice as much as the reference problem.

4.2 Anticipation Scheduling Simulation

The second experiment series consider anticipation scheduling efficiency. During each experiment a VO domain and a job batch were generated and the following scheduling schemes were simulated and studied. First, a general CSS solved the optimization problems $T \to$ min, lim U with different limits $U_a \in \{0\%, 1\%, 4\%, 10\%, 16\%, 32\%, 100\%\}$. Second, a near-optimal but infeasible reference solution REF was obtained for the same problems. Third, a replication procedure CSS_{rep} was performed based on CSS solution to demonstrate a replication process accuracy. For the heuristic anticipation scheduling ANT the same replication procedure was performed based on REF solution. We used a geometric distance as a replication criterion. Finally two independent job batch scheduling procedures were performed to find scheduling solutions most suitable for VO users ($USER_{opt}$) and VO administrators (VO_{opt}). $USER_{opt}$ was obtained by using only user criteria to allocate resources for jobs without taking into account VO preferences. VO_{opt} was obtained by using one VO optimization criterion ($T \to$ min) for each job scheduling without taking into account user preferences.

1000 single scheduling experiments were conducted. Average number of alternatives found for a job in CSS was 2.6. This result shows that while for relatively small jobs usually a few alternative executions have been found, large jobs usually had at most one possible execution option (remember that according to the simulation settings, the difference between jobs execution time could be up to 15 times). At the same time REF algorithm at average considered more than 100 alternative executions for each job. CSS failed to find any alternative executions for at least for one job of the batch in 209 experiments; ANT - in 155 experiments. These results show that simulation settings at the same time provided quite a diverse job batch and a limited set of resources not allowing executing all the jobs during every experiment.

Figure 1 shows average job execution time (VO criterion) in $T \rightarrow$ min, lim U optimization problem. Different limits $U_a \in \{0\%, 1\%, 4\%, 10\%, 16\%, 32\%, 100\%\}$ specify to what extent user preferences were taken into account. Two horizontal lines USER$_{opt}$ and VO$_{opt}$ represent practical T values when only user or VO administration criteria are optimized correspondingly.

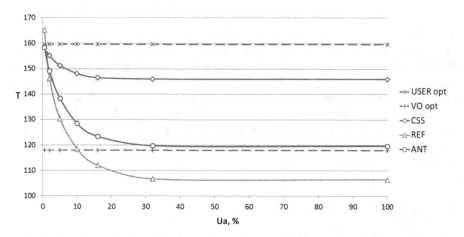

Fig. 1. Average job execution time in $T \rightarrow$ min, lim U problem

First thing that catches the eye in Fig. 1 is that REF for $U > 10\%$ provides job execution time value better (smaller) than those of VO$_{opt}$. However such behavior is expected as REF generates an infeasible solution and may use time-slots from more suitable (according to VO preferences) resources several times for different jobs. Otherwise ANT provided better VO criterion value than CSS for all $U > 0\%$. The relative advantage reaches 20% when $U > 20\%$ is considered. ANT algorithm graph gradually changes from USER$_{opt}$ value at $U = 0\%$ to almost VO$_{opt}$ value at $U = 100\%$ just with changing average user utility limit. Thereby ANT represents a general scheduling approach allowing balancing between VO stakeholder's criteria according to specified scenario, including VO or user criteria optimization.

A similar pattern can be observed in Fig. 2 where $C \rightarrow$ max, lim U scheduling problem is presented. However, in this case ANT advantage over CSS amounts to 10% against VO criterion.

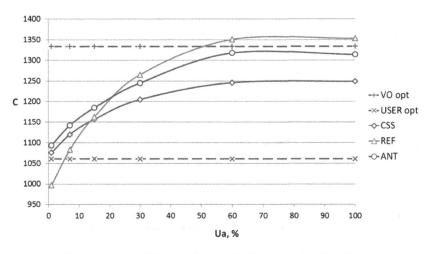

Fig. 2. Average job execution cost in $C \rightarrow$ max, $\lim U$ problem

4.3 Anticipation and Backfilling Scheduling Comparison

The third experiment setup reiterates work [2] and is intended to compare anticipation scheduling procedure with a traditional backfilling algorithm. The main criteria for comparison include average jobs' response time (or start and finish times) as well as users' and VO economic criteria (such as execution time and cost).

We used the following three algorithms for the comparison: **CSS** – the general cycle scheduling scheme; **ANT** – the anticipation scheduling procedure; **BF** – the conservative backfilling algorithm.

In a single experiment CSS and ANT solved $C \rightarrow$ max, $\lim U_a = 10\%$ problem. Execution cost ($C \rightarrow$ min) and processor time ($T \rightarrow$ min) criteria were uniformly distributed between 75 user jobs generated in each experiment.

Important addition was introduced for ANT scheduling. In contrast with experiment series in Subsects. 4.1 and 4.2, job replication geometric distance D_g was calculated as $D_g = \sqrt{E_c + E_T + E_s}$, where additional element E_s stands for a job start time error. As a reference start time value for each job we used start time obtained for a particular job by a prior backfilling scheduling. Thus, when searching for a job execution window we used infeasible solution for time and cost reference values, and a feasible backfilling solution as a reference for an attainable start time values complying with a queue priority.

To observe the behavior of the main scheduling parameters we conducted experiments with a different number N of computing nodes available during the scheduling: $N \in \{20, 25, 30, 40\}$.

Average job's start and finish times are presented in Figs. 3 and 4. As can be seen in Figs. 3 and 4, backfilling provided better start and finish times for a job-flow execution compared to CSS and this result is consistent with [2]. In the current problem setup backfilling was able to finish the job flow execution almost twice earlier then CSS. At the same time anticipation algorithm during each experiment solved $C \rightarrow$

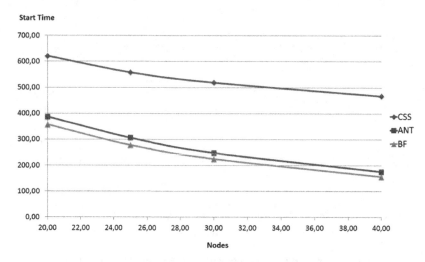

Fig. 3. Average jobs' start time in $C \to$ max, lim U problem

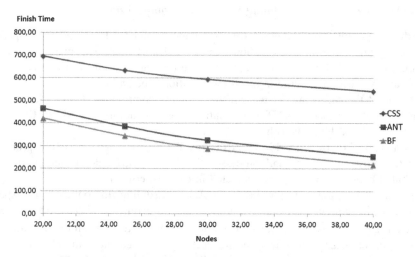

Fig. 4. Average jobs' finish time in $C \to$ max, lim U problem

max, lim U problem and provided jobs' start and finish times only 10% behind the backfilling scheduling outcome.

The details of anticipation scheduling can be examined in Figs. 5 and 6.

Figure 5 shows average job execution time provided by backfilling (BF) and anticipation algorithm (ANT). Additionally ANT T and ANT C represent average execution times obtained by anticipation scheduling for jobs with time minimization and cost minimization criteria correspondingly. As it can be observed, ANT and BF generally provided comparable execution times, which is not a direct optimization criterion for either of them. At the same time ANT applied completely different scheduling policies for jobs with different private scheduling criteria. So that ANT T

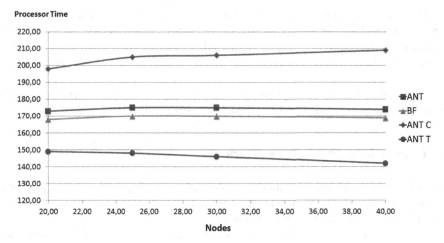

Fig. 5. Average jobs' execution time in $C \to \max$, $\lim U$ problem

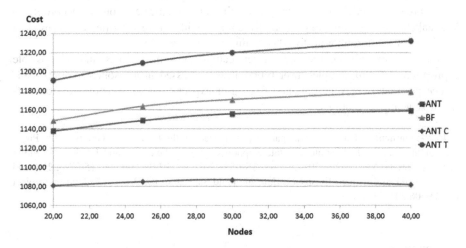

Fig. 6. Average jobs' execution cost in $C \to \max$, $\lim U$ problem

jobs used 25%–33% less processor time then ANT C jobs and 15% less compared to BF solution.

A similar pattern can be observed in Fig. 6, where average jobs' execution cost is presented. ANT and BF provided comparable general job-flow execution cost value. However ANT was able to consider user preferences and shared resources so that ANT C jobs execution cost was 10–15% less then ANT T jobs and 6–9% less compared to backfilling.

Summarizing the results, ANT is able to provide a general scheduling outcome similar to backfilling (with at most 10% error on job's start and finish times), and at the same time considers users' and VO preferences by efficiently solving $C \to \max$, $\lim U$ problem.

Thereby the available resources are distributed between user jobs according to the predefined scheduling requirements (see Figs. 5 and 6). They include individual jobs execution preferences (for example, certain job's execution cost minimization) and a common job-flow scheduling policy (total job-flow execution cost maximization in our example).

Even better VO job-flow optimization results may be obtained when users' preferences are fully matched with VO scheduling criterion [19, 21].

5   Conclusions and Future Work

In this paper, we study the problem of fair job batch scheduling with a relatively limited resources supply. The main problem arise is a scarce set of job execution alternatives which eliminates scheduling optimization efficiency.

We study a heuristic scheduling scheme which generates a near-optimal but infeasible reference solution and then replicates it to allocate a feasible accessible solution. The obtained results show that in computing environments with a limited set of resources the anticipation algorithm is still able to allocate resources according to VO stakeholders' preferences, generally comply with queue priorities and provide a job-flow completion time up to 10% behind backfilling solution.

Future work will be focused on replication algorithm studies and its possible application to fulfill complex user preferences expressed in a resource request. Reference parameters may be obtained from user expectations or transformed from different scheduling solutions.

Acknowledgments. This work was partially supported by the Council on Grants of the President of the Russian Federation for State Support of Young Scientists and Leading Scientific Schools (grants YPhD-2297.2017.9 and SS-6577.2016.9), RFBR (grants 15-07-02259 and 15-07-03401), and by the Ministry on Education and Science of the Russian Federation (project no. 2.9606.2017/BCh).

References

1. Dimitriadou, S.K., Karatza, H.D.: Job scheduling in a distributed system using backfilling with inaccurate runtime computations. In: Proceedings of 2010 International Conference on Complex, Intelligent and Software Intensive Systems, pp. 329–336 (2010)
2. Toporkov, V., Toporkova, A., Tselishchev, A., Yemelyanov, D., Potekhin, P.: Heuristic strategies for preference-based scheduling in virtual organizations of utility grids. J. Ambient Intell. Humanized Comput. **6**(6), 733–740 (2015)
3. Buyya, R., Abramson, D., Giddy, J.: Economic models for resource management and scheduling in grid computing. J. Concurr. Comput. **14**(5), 1507–1542 (2002)
4. Kurowski, K., Nabrzyski, J., Oleksiak, A., Weglarz, J.: Multicriteria aspects of grid resource management. In: Nabrzyski, J., Schopf, J.M., Weglarz, J. (eds.) Grid Resource Management. State of the Art and Future Trends, pp. 271–293. Kluwer Academic Publishers, Boston (2003)
5. Rodero, I., Villegas, D., Bobroff, N., Liu, Y., Fong, L., Sadjadi, S.M.: Enabling interoperability among grid meta-schedulers. J. Grid Comput. **11**(2), 311–336 (2013)

6. Ernemann, C., Hamscher, V., Yahyapour, R.: Economic scheduling in grid computing. In: Feitelson, D.G., Rudolph, L., Schwiegelshohn, U. (eds.) JSSPP 2002. LNCS, vol. 2537, pp. 128–152. Springer, Heidelberg (2002). doi:10.1007/3-540-36180-4_8

7. Rzadca, K., Trystram, D., Wierzbicki, A.: Fair game-theoretic resource management in dedicated grids. In: IEEE International Symposium on Cluster Computing and the Grid (CCGRID 2007), pp. 343–350. IEEE Computer Society, Rio De Janeiro (2007)

8. Vasile, M., Pop, F., Tutueanu, R., Cristea, V., Kolodziej, J.: Resource-aware hybrid scheduling algorithm in heterogeneous distributed computing. J. Future Gener. Comput. Syst. **51**, 61–71 (2015)

9. Penmatsa, S., Chronopoulos, A.T.: Cost minimization in utility computing systems. Concurr. Comput. Pract. Exp. **16**(1), 287–307 (2014). Wiley

10. Mutz, A., Wolski, R., Brevik, J.: Eliciting honest value information in a batch-queue environment. In: 8th IEEE/ACM International Conference on Grid Computing, New York, USA, pp. 291–297 (2007)

11. Blanco, H., Guirado, F., Lérida, J.L., Albornoz, V.M.: MIP model scheduling for multi-clusters. In: Caragiannis, I., et al. (eds.) Euro-Par 2012. LNCS, vol. 7640, pp. 196–206. Springer, Heidelberg (2013). doi:10.1007/978-3-642-36949-0_22

12. Takefusa, A., Nakada, H., Kudoh, T., Tanaka, Y.: An advance reservation-based co-allocation algorithm for distributed computers and network bandwidth on QoS-guaranteed grids. In: Frachtenberg, E., Schwiegelshohn, U. (eds.) JSSPP 2010. LNCS, vol. 6253, pp. 16–34. Springer, Heidelberg (2010). doi:10.1007/978-3-642-16505-4_2

13. Carroll, T., Grosu, D.: Divisible load scheduling: an approach using coalitional games. In: Proceedings of the Sixth International Symposium on Parallel and Distributed Computing, ISPDC 2007, p. 36 (2007)

14. Kim, K., Buyya, R.: Fair resource sharing in hierarchical virtual organizations for global grids. In: Proceedings of the 8th IEEE/ACM International Conference on Grid Computing, pp. 50–57. IEEE Computer Society, Austin (2007)

15. Skowron, P., Rzadca, K.: Non-monetary fair scheduling cooperative game theory approach. In: Proceedings of the Twenty-Fifth Annual ACM Symposium on Parallelism in Algorithms and Architectures, pp. 288–297. ACM, New York (2013)

16. Dalheimer, M., Pfreundt, F., Merz, P.: Agent-based grid scheduling with Calana. In: Proceedings of Parallel Processing and Applied Mathematics, 6th International Conference, pp. 741–750 (2006)

17. Jackson, D., Snell, Q., Clement, M.: Core algorithms of the Maui scheduler. In: Feitelson, D. G., Rudolph, L. (eds.) JSSPP 2001. LNCS, vol. 2221, pp. 87–102. Springer, Heidelberg (2001). doi:10.1007/3-540-45540-X_6

18. Thain, T., Livny, M.: Distributed computing in practice: the condor experience. Concurr. Comput. Pract. Exp. **17**, 323–356 (2005)

19. Toporkov, V., Toporkova, A., Tselishchev, A., Yemelyanov, D., Potekhin, P.: Metascheduling and heuristic co-allocation strategies in distributed computing. Comput. Inform. **34**(1), 45–76 (2015)

20. Toporkov, V., Toporkova, A., Tselishchev, A., Yemelyanov, D.: Slot selection algorithms in distributed computing. J. Supercomput. **69**(1), 53–60 (2014)

21. Toporkov, V., Yemelyanov, D., Bobchenkov, A., Potekhin, P.: Fair resource allocation and metascheduling in grid with VO stakeholders preferences. In: Proceedings of the 45th International Conference on Parallel Processing Workshops, pp. 375–384. IEEE (2016)

22. Farahabady, M.H., Lee, Y.C., Zomaya, A.Y.: Pareto-optimal cloud bursting. IEEE Trans. Parallel Distrib. Syst. **25**, 2670–2682 (2014)

23. Cafaro, M., Mirto, M., Aloisio, G.: Preference-based matchmaking of grid resources with CP-nets. J. Grid Comput. **11**(2), 211–237 (2013)

Parallel Computing Applications

Comparison of Auction Methods
for Job Scheduling with Absolute Priorities

Anton Baranov$^{(\boxtimes)}$, Pavel Telegin, and Artem Tikhomirov

Joint Supercomputer Center of the Russian Academy of Sciences,
Branch of Federal State Institution "Scientific Research Institute
for System Analysis of the Russian Academy of Sciences", Moscow, Russia
antbar@mail.ru, ptelegin@jscc.ru, tema4277@rambler.ru

Abstract. The model of geographically distributed computing system with absolute priorities of jobs is described in the paper. Authors designed the decentralized scheduling algorithm using the auction methods. Two auction methods were researched and compared: the first-price sealed-bid auction and the English auction. The paper includes results of experimental comparison of researched auction methods.

1 Introduction

To improve the performance and reliability of computations individual supercomputer computing facilities (CF) are often integrated into geographically distributed systems (GDS). In this kind of systems absolute priorities can be used for industrial problems. These priorities can be combined with well-known auction methods of scheduling computational jobs (or simply jobs). The main advantages of the auction methods and scheduling algorithms are ease of organization and high speed of operation. There are known several auction models, including the English auction and the first-price sealed-bid auction. The goal of this paper is to compare efficiency of these two auction models for scheduling jobs in GDS with absolute priorities.

2 GDS Architecture

Grid technologies are often used [1] for integration of CFs into a GDS. CFs as are typically computer clusters [2], consisting of individual computational nodes, combined with high-speed communication networks. It is important to note that GDS can be a heterogeneous system (CFs in one GDS may vary by number of nodes and their performance), and a separate CF can be heterogeneous as well (e.g., it may contain different generations of computational nodes).

A single CF in GDS runs under a local resource management system (LRMS), like common systems: PBS, SLURM, Moab, or domestic batch system SUPPZ. The main functions of LRMS are: scheduling computational jobs, start and monitoring their execution on the computational resources of individual CFs.

© Springer International Publishing AG 2017
V. Malyshkin (Ed.): PaCT 2017, LNCS 10421, pp. 387–395, 2017.
DOI: 10.1007/978-3-319-62932-2_37

The basic unit of information processing in GDS is a job, which is defined as a set containing input data, program and job passport. Job passport is a special object that describes the resource request: number of processors (cores), memory and disk space, ordered execution time and some others.

In classification [3] every computational job can be of one of the following types: rigid, moldable and evolving. Job with a rigid resource request requires only one resource request for execution. Moldable job requires several resource requests for execution; scheduling system selects and satisfies one of the requests just once immediately before job start. Evolving job allows changes in the list and amount of resources used during execution of job.

Resource request for a scalable job is described by the vector of values: number of computational nodes required for the job, estimation of running time, the weight coefficient – priority that allows to specify the scheduler which queries are the most preferred options for the job.

The general scheme of processing jobs flow is the following. Jobs are independently entered to any of the CFs in GDS, then they are placed into a queue of Global Resource Management System (GRMS). GRMS determines a target CF for each job. GRMS organizes delivery of input data and job to the target CF, resulting in the job to be placed into local LRMS queue of the target CF.

It is known that in order to improve the resilience and scalability the GRMS should be based on a decentralized scheme [3,4]. This means the absence of a single control center, which operates on a dedicated CF and makes decisions on jobs scheduling. Decentralized management is based on the joint coordinated work of a team of peer dispatchers, which run locally on all GDS CFs.

Each manager schedules jobs flow in accordance with the decentralized scheduling algorithm. This algorithm should be based on the principle of levelling the GDS CF load [5]. Consistency of managers decisions on jobs allocation is achieved by their interaction through a uniform information system [6,7], which is responsible for maintenance of the of the same global jobs queue [8] for all dispatchers.

3   Decentralized Scheduling Algorithms Using the Auction Method

Earlier studies [9–11] show that economic methods can be applied for scheduling jobs in a distributed computing system. Each computing job is considered to be a subject of trade, i.e. the goods. Every dispatcher at different time points may operate as Customer or Seller. Customers compete for the right to process the jobs gaining maximum benefit for themselves. Here benefit means minimum idle time of the computing resources. Sellers, in turn, are interested in obtaining results of their jobs execution as soon as possible.

Two basic economic models used in a distributed system are: commodities markets and auctions [11]. The commodities market model [8] assumes availability of a large number of the same type of goods (computing jobs), while many Customers are ready to buy (process) it. In this case, the price for the job can be established on the base of statistics of completed sales. For the future sales, the established price will be used to all other jobs. The model of the commodity market can be applied particularly in a system with homogeneous computing resources.

The auction model is efficient when the product is unique or limited in quantity, or when the number of Customers (who are ready to participate in the auction) is unknown [12]. Note that the price of the goods can not be determined initially. Every participant who wants to buy goods, sets a bid and sends it to the auctioneer [13]. The auctioneer is another possible role for the dispatcher in addition to the Customer and Seller. The main task of the auctioneer is to accept bids from auction participants. After the end of the auction, the auctioneer ranks the bids offered by the participants and determines the best. The participant who offered the best bid is considered the winner of the auction and receives the goods. Advantages of the auction model are the ease of implementation and the possibility of usage in the decentralized scheme of dispatchers interaction. The disadvantage limiting use of the auction model in some areas (like balancing the workload of CDN-servers), is relatively long time for making a job assignment decision.

There are known several models of the auction [14–18] including first price sealed-bid auction, English auction, Vickrey auction, Double auction, and combined auctions.

Note that in most cases for auction scheduling methods, a GDS model with equal or relative priorities and fixed resource request for assignments is used. For GDS model with absolute priorities let us consider a scheme for planning scalable jobs using two auction models: the first price sealed-bid auction [19] and the English auction.

First-price sealed-bid auction is the most widely used model for scheduling computer resources. In this model, all participants of the auction (dispatchers) bid on job, they do not see the bids of opponents and cannot change their own. The winner is the participant who offered the highest rate. English auction is a multi-round open ascending price auction, which starts with setting a minimum price. The participants are aware of the bids made by the others, and bet only if their rate exceeds all earlier bids.

The difficulty in using this auction model is determination of rate which the dispatcher can offer, since there must be a way of rate increase.

While scheduling scalable jobs for more than one resource request indicating their preference, the dispatcher may increase the rate by offering to run a scalable job according to the most preferred resource request.

4   Bid Problem

Determination of rate that the dispatcher can offer for the job is the key issue of the auction methods. In [20] to determine bid the authors suggested to use the heuristic "compatibility" coefficient of job characteristics and target CF, and considered different ways of its definition. In [11], the authors studied the GDS model with absolute priorities, where time of initial data transfer sometimes is comparable to or even much greater than the execution time of job. For this model the following compatibility coefficient P_{send} is suggested.

$$P_{send} = A \cdot \frac{V}{C} \qquad (1)$$

where
   V is volume of the original job data,
   C – communication bandwidth between the CF, which enqueued the job and the target CF,
   A is a weight coefficient.
   Bid of dispatcher P_{total} for a job is defined as

$$P_{total} = \frac{1}{P_{send} + 1} \qquad (2)$$

The rate of job is determined by the formula (2) and takes into account communications heterogeneity of GDS, but does not take into account the computational heterogeneity, i.e., different number of computational nodes in different CFs inside the GDS. So, it is suggested to form the rate with several components: price for computational resources, price for interrupts and price for transfer of input data.

Price for Computational Resources. P_{work} can be defined as follows:

$$P_{work} = B \cdot N \qquad (3)$$

where
   N is number of modules used for job on CF
   B is weight coefficient. It is reasonable to assume that for CF with large number of modules probability to get greater number of modules is higher, and the price
   P_{work} takes into account computational heterogeneity of GDS.

Price for Interrupts. P_{speed} is defined as follows:

$$P_{speed} = D \cdot G \qquad (4)$$

where
   G is the maximum priority of displaced jobs,
   D is a weight coefficient that allows you to take into account the workload of computational resources. Price for interrupts takes into consideration the assumption that the high-priority job should displace from execution (interrupt) job with minimum priority.

The Final Bid. P_{total} of dispatcher is defined as follows:

$$P_{total} = \frac{P_{work}}{P_{send} + P_{speed} + 1} \tag{5}$$

Auction is won by dispatcher, which offered the maximum bid. To win, the dispatcher must offer to as much computing power as possible, the time for transfer of initial data should be minimal, while proposed computational resources should be either free or occupied by the least priority jobs.

5 Scheduling Algorithm

Scheduling algorithm developed by the authors enables the use of both auction models: English and first-price sealed-bid ones. The algorithm is executed independently by every dispatcher on each CF in a GDS, while the dispatcher executing the algorithm is regarded as a potential job performer.

Step 1. Dispatcher interacts with LRMS and generates a list of running jobs on the CF. Information on priority and the number of occupied nodes is placed in the list for each job.

Step 2. Dispatcher scans the global queue for high-priority jobs and determines whether these jobs can displace running jobs contained in the list from step 1. If there is no such kind of job in the global queue, then the algorithm stops. Otherwise, dispatcher generates the list of high-priority jobs from the global queue, for which it is ready to participate in the auction, and proceeds to step 3.

Step 3. Dispatcher auctions high-priority jobs selected in step 2, and becomes the auctioneer and the auction participant at the same time. If selected in step 2 jobs have already been put on the auction by another dispatcher, the current dispatcher becomes only participant in the auction for these jobs.

Step 4. Each participant in accordance with (5) determines the rate for each job and sends it to the auctioneer.

Step 5. The auction lasts for the time set for the auction. After this time, the auctioneer proceeds to step 6, and participants to step 7.

Step 6. Determination of winner. The auctioneer takes bids from all the participants. Then the auctioneer ranks the bids aby their value. The auctioneer appoints the winner: the dispatcher, which offered the maximum rate for the job. The rate for the job is P_{total} value, which determined according to (5). If the more than one participant offered maximum, the winner is the dispatcher, which offered it first.

Step 7. If the dispatcher has won the auction, it organizes preparation of the job initial data and puts the job into the queue of LRMS.

Steps 1–7 are repeated until all the jobs in the global queue are distributed on CFs.

Sequence of steps described above can be applied to the both auction models. In the case of a first-price sealed-bid auction steps 1–5 are passed once, then the job is selected according to the bids on step 6. In the case of an English auction, the auctioneer sets a minimum bid on step 3, then it starts to accept bids from other dispatchers. Steps 1–5 for each job can be repeated many times during the time of the auction, and the participants can change their bids, increasing job price.

Dispatcher bids are determined by the formula (5). It is assumed that the dispatcher bid can change when the amount of available computational resources changes, i.e., changes the price P_{work} according to (3). At the end of the auction the dispatcher with the highest rate gets the job.

6   Experimental Comparison of Two Auction Models

To test technique above the authors created model of GDS, see Fig. 1. The model contains two CFs with different performance: the first CF contains 8 nodes, and the second one contains 13 nodes. Global job queue is placed in a special distributed information system.

Experiments were performed on supercomputer MVS-10P in JSCC RAS. We used MPI-programs from NAS Parallel Benchmarks (NPB) for the test. The global queue received steady flow of $M = 400$ jobs. Test jobs were selected pseudo-randomly. Number of resource requests was set by gamma distribution with parameters $\alpha = 3$ $\beta = 1$ (the average number of resource requests for one job was equal to 2). For each resource request number of required CFs for job was generated accordingly gamma distribution with parameters 2 and 3. Jobs entered the global queue with time intervals t, exponentially distributed with intensity value $\lambda = 350$.

The following indicators were selected for evaluating efficiency of scheduling algorithms:

1. Average processing time of high priority jobs T.
2. Number of jobs in the global queue for each priority level.

   In addition, two more characteristics were investigated:
3. Number of jobs, which execution of was interrupted by a priority job.
4. Held auctions, i.e. proportion of jobs for which two or more dispatchers competed.

The goal of the experiment was to compare the English auction model and the first-price sealed-bid auction model for scheduling jobs in GDS with absolute priorities. During experiment it was established that the first-price sealed-bid auction can not guarantee allocation of maximum resources for the job. Job execution time with first-price sealed-bid model is greater for most priority levels. Distribution of processing time for both models depending on priority levels is presented in Fig. 1.

On the one hand, when jobs does not get the maximum resources, it is possible to run a greater number of jobs simultaneously compared to the case when all jobs run using the maximum resources. On the other hand, increase in execution

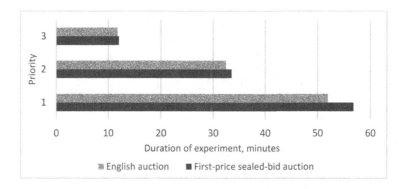

Fig. 1. Average processing time of jobs depending on priority level

time of job results the increase in interrupted lower priority jobs. Figure 2 demonstrates proportion of interrupted jobs for both auction methods. Interrupting a job is associated with the following time-consuming:

1. Time spent on preparation to the job interrupt.
2. Time spent on reinitialization of the computational resources for the interrupted job.

The displaced job does not return to the global queue and remains in the local queue of CF. Impossibility to reallocate interrupted job leads to long wait in the local queue of the CF, this results the increase in average job processing time.

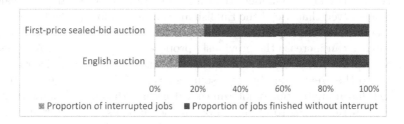

Fig. 2. Proportion of jobs, which execution was interrupted

Increase in run time for almost every job, as well as increase in number of interrupted jobs results decrease in intensity of handling jobs from the global queue. Increasing intensity of coming jobs will increase number of interrupts for both types of auctions; in this case advantage of the English auction becomes even more significant.

Figure 3 demonstrates that the English auction model gives smaller average number of jobs in the queue. Each of five charts on Fig. 3 corresponds to one of five priority levels, the upper chart corresponds to priority level 3 (top priority), the bottom chart - to priority level 1 (bottom priority).

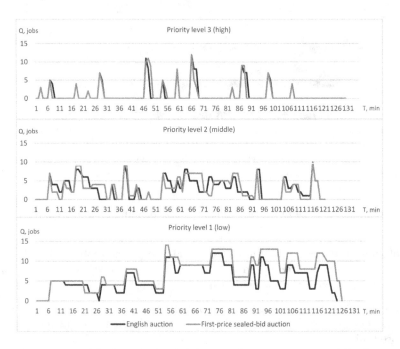

Fig. 3. Comparison of the length the global queue for two auction models: Q is number of jobs in the queue, T is duration of the experiment

7   Conclusions

Comparison of two auction models: the English auction and the first-price sealed-bid auction showed that use of the English auction model for scheduling scalable jobs in GDS with absolute priorities is more efficient. Efficiency is greater because the dispatcher can increase the previously proposed rate for the job. As a result of the experiment, it was found that raising the rate allows to minimize run time of job due to the greater number of resources assigned for job execution. It was discovered that minimization of job run time decreases the number of interrupted jobs, which in turn leads to reduction of time losses caused by interrupts, and, as a consequence, to reduction of the average job processing time.

References

1. Foster, I.: The physiology of the grid: an open grid services architecture for distributed systems integration. Comput. Netw. Int. J. Comput. Telecommun. Netw. **40**(1), 5–17 (2002)
2. Kovalenko, V.N., Koryagin, D.A.: Organization of Grid resources Keldysh Institute of Applied Mathematics RAS, no. 63, p. 25 (2004)
3. Khoroshevskii, V.G.: Virtualization architecture of distributed computing systems in Student, Scientist, Teacher. p. 69. Avtograf, Novosibirsk (2015)
4. Korneev, V.V., Monakhov, O.G.: About allocation of tasks in computer systems with programmable structure in Architecture of computer systems with programmable structure, pp. 3–17. Sobolev Institute of mathematics, Novosibirsk (1982)

5. Hamscher, V., Schwiegelshohn, U., Streit, A., Yahyapour, R.: Evaluation of job-scheduling strategies for grid computing. In: Buyya, R., Baker, M. (eds.) GRID 2000. LNCS, vol. 1971, pp. 191–202. Springer, Heidelberg (2000). doi:10.1007/3-540-44444-0_18
6. Kovalenko, V.N., Kovalenko, E.I., Shorin, O.N.: Development of grid job dispatcher based on lookahead scheduling. Keldysh Institute of Applied Mathematics RAS, Moscow (2005)
7. Bobchenkov, A.V.: Development of models and management practices in virtual organizations distributed computing systems (in Russian). MPEI, Moscow (2011)
8. Buyya, R., Abramson, D., Giddy, J., Stockinger, H.: Economic models for resource allocation and scheduling in grid computing. Concurrency Comput. Pract. Exp. 14, 1507–1542 (2002). doi:10.1002/cpe.690
9. Nabrzyski, J., Schopf, J.M., Weglarz, J.: Grid Resource Management. State of the Art and Future Trends (2003). doi:10.1007/978-1-4615-0509-9
10. Wolski, R., Plank, J.S., Bryan, T., Brevik, J.: G-commerce: market formulations controlling resource allocation on the computational grid. In: Proceedings of the 15th IEEE International Parallel and Distributed Processing Symposium, April 2000. doi:10.1109/IPDPS.2001.924985
11. Vazhkudai, S., von Laszewski, G.: A greedy grid - the grid economic engine directive. In: Proceedings of the 15th IEEE International Parallel and Distributed Processing Symposium, April 2000. doi:10.1109/IPDPS.2001.925170
12. Chen, C., Maheswaran, M., Toulouse, M.: Supporting coallocation in an auctioning-based resource allocator for grid systems. In: Proceedings of the 11th IEEE Heterogeneous Computing Workshop, April 2001. doi:10.1109/IPDPS.2002.1015666
13. Hurwicz, L.: The design of resource allocation mechanisms. In: Arrow, K., Hurwicz, L. (eds.) Studies in Resource Allocation Processes, pp. 3–38. Cambridge University Press, Cambridge (1977). doi:10.1017/CBO9780511752940.002
14. Gomoluch, J., Schroeder, M.: Market-based resource allocation for grid computing: a model and simulation. In: Proceedings of the 1st International Workshop on Middleware for Grid Computing, pp. 211–218, June 2003
15. Grosu, D., Das, A.: Auction-based resource allocation protocols in grids. In: Proceedings of the 16th IASTED International Conference on Parallel and Distributed Computing and Systems, pp. 20–27 (2004)
16. Wolski, R., Plank, J.S., Brevik, J., Bryan, T.: Analyzing market-based resource allocation strategies for the computational grid. Int. J. High Perform. Comput. Appl. 15(3), 258–281 (2001). doi:10.1177/109434200101500305
17. Vohra, R.V.: Combinatorial auctions. In: Handbook of Game Theory with Economic Applications, vol. 4, pp. 455–476 (2015). doi:10.1016/B978-0-444-53766-9.00008-2
18. Kale, L.V., Kumar, S., Potnuru, M., DeSouza, J., Bandhakavi, S.: Faucets: efficient resource allocation on the computational grid. In: Proceedings of the International Conference on Parallel Processing (ICPP 2004), pp. 396–405 (2004). doi:10.1109/ICPP.2004.1327948
19. Baranov, A.V., Tikhomirov, A.I.: Use closed bid auction in a territorially distributed computing system with absolute priorities. In: Proceedings of the NSCF 2016, Pereslavl-Zalessky (2016)
20. Toporkov, V.V., Emelyanov, D.M., Potehin, P.A.: Job batch generation and scheduling in distributed computing environments. Bulletin of the South Ural State University. Series: Computational Mathematics and Software Engineering, no 2. pp. 21–24 (2015)

Parallel Algorithm for Solving Constrained Global Optimization Problems

Konstantin Barkalov[✉] and Ilya Lebedev

Lobachevsky State University of Nizhny Novgorod, Nizhny Novgorod, Russia
{barkalov,lebedev}@vmk.unn.ru

Abstract. This work considers a parallel algorithm for solving multiextremal problems with non-convex constraints. The distinctive feature of this algorithm, which does not use penalty functions, is the separate consideration of each problem constraint. The search process can be conducted by reducing the original multidimensional problem to a number of related one-dimensional problems and solving this set of problems in parallel. An experimental assessment of parallel algorithm efficiency was conducted by finding the numeric solution to several hundred randomly generated multidimensional multiextremal problems with non-convex constraints.

Keywords: Global optimization · Constrained problems · Non-convex constraints · Dimension reduction · Parallel algorithms

1 Introduction

This work considers parallel methods for solving global optimization problems with non-convex constraints. The objective function and constraints are assumed to satisfy the Lipschitz condition with a priori unknown Lipschitz constants. The analytical form of the problem's functions may also be unknown, i.e. they can be set by an algorithm computing their values at various points within the search domain (so-called "black-box" functions). Moreover, it is supposed that even a single computation of the function value can be time-consuming, as in the applied problems it requires performing numerical simulation. These assumptions are typical for many approaches to building parallel algorithms for unconstrained global optimization [1–4].

At the same time, it is common for applied constrained optimization problems to be in a situation where violating one constraint results in all other functions returning indeterminate values. An example includes optimal control problems, described through systems of ordinary differential equations with a certain matrix A on the right side [5, 6]. It is only possible to calculate the optimality criteria for these problems if the matrix A is a Hurwitz matrix, i.e. every eigenvalue of A has strictly negative real part. Otherwise the value of the criteria is indeterminate.

This partial computability of the functions in constrained optimization problems substantially complicates the application of the well-known penalty function method (in some cases making it completely impossible). Thanks to its simplicity, this method is one of the most popular approaches to solving problems with constraints. However, calculating the penalty function requires first calculating the values of all of the

© Springer International Publishing AG 2017
V. Malyshkin (Ed.): PaCT 2017, LNCS 10421, pp. 396–404, 2017.
DOI: 10.1007/978-3-319-62932-2_38

problem's functions at the given point, which is impossible if they are partially indeterminate.

In this work, the authors consider an approach to minimizing multiextremal functions under non-convex constraints, developed in [7–9] and called the index method. The approach is based on separate considering every constraint in the problem and is not related to using penalty functions. According to the index method, each iteration (*a trial*) at a respective point in the search domain includes a sequential check of the problem constraints at that point. As soon as the first constraint violation is found, the trial is interrupted and the method proceeds with the next iteration; no other problem functions are calculated at that point. This allows problems to be solved in which function values may not be determined for the entire search domain. Under this approach, solving multidimensional problems is reduced (using Peano-type space-filling curves) to solving equivalent one-dimensional problems.

It should be noted that standard approaches to algorithm parallelization are not quite applicable to global optimization. For example, the rules for selecting another iteration point are quite simple and do not require parallelization (as overheads associated with organizing parallel computations will nullify any possible acceleration). Some acceleration can be achieved by parallelizing the computation of function values describing the object to be optimized; however, this approach is specific to each individual problem being solved.

The following approach looks more promising. The algorithm can be modified to run several trials in parallel. This approach provides the efficiency (as parallelization is applied to the most computation-intensive part of the problem solving process) and generality (in that it applies to a wide range of global optimization algorithms). The approach, described in [10] for unconstrained optimization, was used in this work for parallelizing constrained optimization algorithms.

The main part of the paper has the following structure. Section 2 states the constrained optimization problem, reviews the index method and an approach to reducing dimensionality by using Peano curves. Section 3 presents a parallel implementation of the index method using a set of space-filling curves. Section 4 presents the results of numerical experiments. Section 5 concludes the paper.

2 Problem Statement

Let us consider the N-dimensional optimization problem

$$\min\{\varphi(y): y \in D, \ g_i(y) \leq 0, \ 1 \leq i \leq m\} \tag{1}$$

$$D = \{y \in R^N: a_j \leq y_j \leq b_j, \ 1 \leq j \leq N\}. \tag{2}$$

The objective function $\varphi(y)$ (henceforth denoted by $g_{m+1}(y)$) and the left-hand sides $g_i(y)$, $1 \leq i \leq m$, of the constraints satisfy Lipschitz condition

$$|g_i(y_1) - g_i(y_2)| \le L_i\|y_1 - y_2\|, \ 1 \le i \le m+1,$$

with a priory unknown constants L_i, $1 \le i \le m+1$, and may be multiextremal. It is assumed that functions $g_i(y)$ are defined and computable only at the points $y \in D$ satisfying the conditions

$$g_k(y) \le 0, \ 1 \le k < i. \tag{3}$$

Employing the continuous single-valued Peano curve $y(x)$ mapping the unit interval $[0,1]$ on the x-axis onto the N-dimensional domain (2) it is possible to find the minimum in (1) by solving the one-dimensional problem

$$\varphi(y(x^*)) = \min\{\varphi(y(x)): x \in [0, 1], \ g_i(y(x)) \le 0, \ 1 \le i \le m\}.$$

Algorithms for numerical construction of Peano curve approximation (*evolvent*) are given in [11]. Due to (3) the functions $g_i(y(x))$ are defined and computable in the domains

$$Q_1 = [0, 1], \ Q_{i+1} = \{x \in Q_i : g_i(y(x)) \le 0\}, \ 1 \le i \le m.$$

These conditions allows to introduce a classification of the points $x \in [0, 1]$ according to the number $v(x)$ of the constraints computed at this point. The index $v(x)$ can also be defined by the conditions

$$g_i(y(x)) \le 0, \ 1 \le i < v, \ g_v(y(x)) > 0,$$

where the last inequality is inessential if $v = m+1$.

The considered dimensionality reduction scheme juxtaposes to a multidimensional problem with lipschitzian functions a one-dimensional problem, where the corresponding functions satisfy uniform Hölder condition (see [11]), i.e.,

$$g_i(y(x')) - g_i(y(x'')) \le H_i|x' - x''|^{\frac{1}{N}}, \ x',x'' \in [0, 1], \ 1 \le j \le m+1.$$

Here N is the dimensionality of the initial multidimensional problem and the coefficients H_i are related with Lipschitz constant L_i of the initial problem as $H_i \le 2L_i\sqrt{N+3}$.

Thus, a trial at a point $x^k \in [0, 1]$ executed at the k-th iteration of the algorithm will consist in the following sequence of operations.

- To determine the image $y^k = y(x^k)$ in accordance with the mapping $y(x)$.
- To compute the values $g_1(y^k), \ldots, g_v(y^k)$, where the index $v \le m$ is determined by the conditions

$$g_i(y^k) \le 0, \ 1 \le i < v, \ g_v(y^k) > 0, \ v \le m.$$

The occurrence of the first violation of the constraint terminates the trial. In the case, when the point y^k is a feasible one, i.e. when $y(x^k) \in Q_{m+1}$, the trial includes the computation of the values of all functions of the problems and the index is accepted to be $v = m + 1$. The pair of values

$$v = v(x^k), \quad z^k = g_v(y(x^k))$$

is a *result of the trial*.

The scheme of the serial index algorithm is as follows. The first trial is executed at an arbitrary internal point $x_1 \in (0, 1)$. The selection of the point $x^{k+1}, k \geq 1$, of any next trial is carried out by the following steps.

Step 1. Renumber the points $x^1, \dots, x^k$ of the preceding trials by the lower indices in increasing order of the coordinate values, i.e.

$$0 = x_0 < x_1 < \cdots < x_k < x_{k+1} = 1,$$

and juxtapose to them the values $z_i = g_v(y(x_i))$, $v = v(x_i)$, $1 \leq i \leq k$, computed at these points. The points $x_0 = 0$ and $x_{k+1} = 1$ are introduced additionally and the values z_0 and z_{k+1} are not defined.

Step 2. For each interval (x_{i-1}, x_i), $1 \leq i \leq k+1$, compute the *characteristics* $R(i)$ using some formulae.

Step 3. Find the interval (x_{t-1}, x_t) with the maximal characteristic

$$R(t) = \max\{R(i): 1 \leq i \leq k+1\}.$$

Step 4. Execute the next trial in the inner point of the interval (x_{t-1}, x_t), i.e. $x^{k+1} \in (x_{t-1}, x_t)$.

Step 5. Check termination condition $|x_t - x_{t-1}|^{\frac{1}{N}} \leq \epsilon$, where t is the number of interval with the maximal characteristic and $\epsilon > 0$ is the predefined accuracy.

Detailed description of this algorithm and the corresponding theory of convergence are presented in [7–9].

3   Parallel Index Algorithm with the Set of Evolvents

The reduction of the multidimensional problems to the one-dimensional ones using evolvents has such important properties as the continuity and preservation of boundedness of function divided differences. However, a partial loss of information on the nearness of the points in the multidimensional space takes place since a point $x \in [0, 1]$ has only the left and the right neighbors while the corresponding point $y(x) \in R^N$ has the neighbors in $2N$ directions. As a result, when using the mappings like Peano curve the images y', y'', which are close to each other in the N-dimensional space can correspond to the preimages x', x'', which can be far away from each other in the interval $[0, 1]$. This property results in the excess computations since several limit points x', x'' of the trial sequence generated by the index method in the interval $[0, 1]$ can correspond to a single limit point y in the N-dimensional space.

One of the possible ways to overcome this disadvantage consists in using the multiple mapping $Y^S(x) = \{y^1(x), \dots, y^S(x)\}$ instead of single evolvent $y(x)$.

To construct the set $Y^S(x)$ different approaches can be used. For example, in [7] a scheme was implemented, according to which each evolvent $y^i(x)$ from $Y^S(x)$ is constructed as a result of shifting the original evolvent $y^0(x)$ along the main diagonal of the hypercube D. The set of Peano curves thus constructed allows one to obtain y', y'' from D for any close multidimensional images, which differ only in one coordinate, close preimages x', x'' from the interval $[0,1]$ for the evolvent $y^s(x)$, $1 \leq s \leq S$.

Using the multiple mapping allows solving initial problem (1) by parallel solving the problems

$$\min\{\varphi(y^s(x)): x \in [0,1], \; g_i(y^s(x)) \leq 0, \; 1 \leq i \leq m\}, \; 1 \leq s \leq S.$$

on a set of intervals [0,1] by the index method. Each one-dimensional problem is solved on a separate processor. The trial results at the point x^k obtained for the problem being solved by particular processor are interpreted as the results of the trials in the rest problems (in the corresponding points $x^{k_1}, \ldots, x^{k_S}$). In this approach, a trial at the point $x^k \in [0,1]$ executed in the framework of the s-th problem, consists in the following sequence of operations.

1. Determine the image $y^k = y^s(x^k)$ for the evolvent $y^s(x)$.
2. Inform the rest of processors about the start of the trial execution at the point y^k (*the blocking* of the point y^k).
3. Compute the values $g_1(y^k), \ldots, g_v(y^k)$, where the index $v \leq m$ is determined by the conditions

$$g_i(y^k) \leq 0, \; 1 \leq i < v, \; g_v(y^k) > 0, \; v \leq m.$$

The occurrence of the first violation of any constraint terminates the trial at the point y^k. In the case when y^k is a feasible one, i.e., when $y^s(x^k) \in Q_{m+1}$, the trial includes the computation of all problem functions. In this situation, the index is set to $v = m+1$. The triplet

$$y^s(x^k), \; v = v(x^k), \; z^k = g_v(y^s(x^k))$$

is the result of the trial at the point x^k.

4. Determine the preimages $x^{k_s} \in [0,1]$, $1 \leq s \leq S$, of the point y^k and interpret the trial executed at the point $y^k \in D$ as the execution of the trials in the S points $x^{k_1}, \ldots, x^{k_S}$ with the same results

$$v(x^{k_1}) = \cdots = v(x^{k_S}) = v(x^k),$$

$$g_v(y^1(x^{k_1})) = \cdots = g_v(y^S(x^{k_S})) = z^k.$$

5. Inform the rest of processors about the trial results at the point y^k.

The decision rules for the proposed parallel algorithm, in general, are the same as the rules of the sequential algorithm (except the method of the trial execution). Each processor has its own copy of the software realizing the computations of the problem functions and the decision rule of the index algorithm. For the organization of the interactions among the processors, the queues are created on each processor, where the processors store the information on the executed iterations in the form of the tuples: the processor number s, the trial point x^{k_s}, the index $v(x^{k_s})$, and the value $g_v(y^s(x^{k_s}))$. Moreover, the index of the blocked point is assumed to be equal to -1; the function value at this point is undefined.

The proposed parallelization scheme was implemented with the use of MPI technology. Main features of implementation consist in the following. A separate MPI-process is created for each of S one-dimensional problems being solved, usually, one process per one processor employed. Each process can use p threads, usually one thread per an accessible core.

At every iteration of the method, the process with the index s, $0 \leq s < S$ performs p trials in parallel at the points x^{s+iS}, $0 \leq i < p$. At that, each process stores all Sp points, and an attribute indicating whether this point is blocked by another process or not is stored for each point. Let us remind that the point is blocked if the process starts the execution of a trial at this point.

At every iteration of the algorithm, operating within the s-th process, determines the coordinates of p «its own» trial points. Then, the interchange of the coordinates of images of the trial points y^{s+iS}, $0 \leq i < p$, $0 \leq s < S$ is performed (from each process to each one). After that, the preimages x^{q+iS}, $0 \leq q < S$, $q \neq s$ of the points received by the s-th process from the neighbor ones are determined with the use of the evolvent $y^s(x)$. The points blocked within the s-th process will correspond to the preimages obtained. Then, each process performs the trials at the non-blocked points, the computations are performed in parallel using OpenMP. The results of the executed trials (the index of the point, the computed values of the problem functions, and the attribute of unblocking of this point) are transferred to all rest processes. All the points are added to the search information database, and the transition to the next iteration is performed.

4   Results of Numerical Experiments

A well-known approach to the investigation and comparing of the multiextremal optimization algorithms is based on testing these methods by solving a set of problems, chosen randomly from some specially designed class.

GKLS generator for the functions of arbitrary dimensionality with known properties (the number of local minima, the size of their domains of attraction, the global minimizer, etc.) has been proposed in [12]. Four GKLS classes of differentiable test functions of the dimensions $N = 4$ and 5, have been used. For each dimension, both *Hard* and *Simple* classes have been considered. The difficulty of a class was increased either by decreasing the radius of the attraction region of the global minimizer, or by decreasing the distance from the global minimizer y^* to the domain boundaries. Application of the generator for studying some optimization algorithms has been described in [13–15].

In this study we will use GKLS generator to produce the constrained problems. The scheme that allows to form the constrained global optimization problems is proposed in [16]. In the previous investigations, the index method has been confirmed experimentally to be not inferior to well-known analogues. The comparing of the method to well known DIRECT one [1] in solving the unconstrained optimization problems has been performed in [17, 18]. In the present study, an experimental investigation of the speedup, which is obtained by the use of the index method in combination with the two-level parallelization scheme from [10].

The experiments have been carried out by solving a series of 100 problems with two constraints and the objective functions from the *Simple* and *Hard* GKLS classes with the dimensionalities $N = 4$, $N = 5$. The number of the used cluster nodes S and, correspondingly, the number of evolvents as well as the number of cores p employed at each node have been varied. The problem was considered to be solved, if the algorithm generated trial point y^k in δ-vicinity of the global minimum, i.e., $\|y^k - y^*\| \leq \delta$. The size of the vicinity was selected as $\delta = 0.03\|b - a\|$, where a and b are borders of the search domain. For the purpose of simulation of the computational complexity inherent to applied problems of optimization, calculation of the problem functions in all performed experiments was made more complex by additional calculations without changing the type of function and arrangement of its minima (series summation of 80 thousand elements).

The average time and number of iterations, which were required to solve the problems of the series at various parallelization parameters are reflected in Tables 1 and 2. Here *Node/core* are the numbers of employed nodes and cores per a node, correspondingly.

Table 1. Average time

Node/core	N = 4		N = 5	
	Simple	Hard	Simple	Hard
1/1	220.5	334.8	1223.6	1386.6
1/16	31.3	49.1	211.8	547.2
2/1	158.4	260.0	1052.9	1458.1
2/16	22.1	35.9	227.5	603.0
4/1	127.7	286.4	951.3	1362.2
4/16	20.9	45.0	206.0	925.7
8/1	99.3	141.8	700.1	897.3
8/16	31.0	77.7	264.6	374.0

The results demonstrate the presence of the speedup when using the common memory at a node (performing several trials within a problem in parallel) as well as the distributed memory (parallel solving of several subproblems at different nodes). At that, the highest time speedup was 10 (when using 64 cored on 4 cluster nodes), the highest iteration speedup was 95 (when using 128 cores on 8 cluster nodes). The difference in the speedups in time and in the number of iterations can be explained by the effect of

Table 2. Average number of iterations

Node/core	N = 4		N = 5	
	Simple	Hard	Simple	Hard
1/1	58320	84546	266943	287102
1/16	4297	6601	22655	56754
2/1	34791	52126	188465	241369
2/16	2029	3239	16689	40763
4/1	22223	47771	135734	180489
4/16	1281	2483	9241	35024
8/1	13844	18933	77748	94563
8/16	608	1473	5820	23033

the overheads of the data transmission between the processes. Note that when solving the applied optimization problems, the computing of the problem function values even in one point is a computation costly operation. The data transfer overheads will not affect the total computational costs predominately in this case, and the time speedup will not differ from the iteration one so strongly.

5 Conclusions

The parallel index method for solving constrained global optimization problems considered in the present work:

- allows solving the initial problem directly, without the use of the penalty functions (thus, the issues of selection the penalty coefficient and of solving a series of unconstrained problems with different penalty coefficients are eliminated);
- allows solving the problems, which the values of the problem function are not defined everywhere (for example, the objective function values are undefined out of the feasible domains of the problem constraints);
- allows using the two-level parallelization scheme with the shared and distributed memory proposed earlier for the unconstrained optimization methods.

The parallel algorithm has demonstrated speedup with respect to the number of processors/cores employed. This was confirmed by the results of the numerical solving of several hundred test problems using 128 cores of UNN computer cluster. The direction of further research is the generalization of the considered parallel algorithm for solving multicriteria problems.

Acknowledgements. The study was supported by the Russian Science Foundation, project No 16-11-10150.

References

1. Jones, D.R.: The direct global optimization algorithm. In: Floudas, C.A., Pardalos, P.M. (eds.) The Encyclopedia of Optimization, 2nd edn., pp. 725–735. Springer, Heidelberg (2009). doi:10.1007/978-0-387-74759-0_128
2. Evtushenko, Y., Malkova, V.U., Stanevichyus, A.A.: Parallel global optimization of functions of several variables. Comput. Math. Math. Phys. **49**(2), 246–260 (2009)
3. Paulavicius, R., Zilinskas, J., Grothey, A.: Parallel branch and bound for global optimization with combination of Lipschitz bounds. Optim. Methods Softw. **26**(3), 487–498 (2011)
4. Evtushenko, Y., Posypkin, M.: A deterministic approach to global box-constrained optimization. Optim. Lett. **7**(4), 819–829 (2013)
5. Balandin, D.V., Kogan, M.M.: Optimal linear-quadratic control: from matrix equations to linear matrix inequalities. Autom. Remote Control **72**(11), 2276–2284 (2011)
6. Balandin, D.V., Kogan, M.M.: Pareto-optimal generalized H_2-control and vibration isolation problems. Autom. Remote Control **8**, 76–90 (2017). [in Russian]
7. Strongin, R.G., Sergeyev, Y.D.: Global Optimization with Non-convex Constraints: Sequential and parallel algorithms. Springer, New York (2000). doi:10.1007/978-1-4615-4677-1
8. Sergeyev, Y.D., Famularo, D., Pugliese, P.: Index branch-and-bound algorithm for Lipschitz univariate global optimization with multiextremal constraints. J. Glob. Optim. **21**(3), 317–341 (2001)
9. Barkalov, K.A., Strongin, R.G.: A global optimization technique with an adaptive order of checking for constraints. Comput. Math. Math. Phys. **42**(9), 1289–1300 (2002)
10. Gergel, V., Sidorov, S.: A two-level parallel global search algorithm for solution of computationally intensive multiextremal optimization problems. In: Malyshkin, V. (ed.) PaCT 2015. LNCS, vol. 9251, pp. 505–515. Springer, Cham (2015). doi:10.1007/978-3-319-21909-7_49
11. Sergeyev, Y.D., Strongin, R.G., Lera, D.: Introduction to Global Optimization Exploiting Space-Filling Curves. Springer, New York (2013). doi:10.1007/978-1-4614-8042-6
12. Gaviano, M., Kvasov, D.E., Lera, D., Sergeyev, Y.: Software for generation of classes of test functions with known local and global minima for global optimization. ACM Trans. Math. Softw. **29**(4), 469–480 (2003)
13. Sergeyev, Y.D., Kvasov, D.E.: Global search based on efficient diagonal partitions and a set of Lipschitz constants. SIAM J. Optim. **16**(3), 910–937 (2006)
14. Paulavicius, R., Sergeyev, Y., Kvasov, D., Zilinskas, J.: Globally-biased DISIMPL algorithm for expensive global optimization. J. Glob. Optim. **59**(2–3), 54–567 (2014)
15. Sergeyev, Y.D., Kvasov, D.E.: A deterministic global optimization using smooth diagonal auxiliary functions. Commun. Nonlinear Sci. Numer. Simul. **21**(1–3), 99–111 (2015)
16. Gergel, V.: An approach for generating test problems of constrained global optimization. In: Proceedings of Learning and Intelligent Optimization Conference (to appear)
17. Barkalov, K., Gergel, V., Lebedev, I.: Use of Xeon Phi coprocessor for solving global optimization problems. In: Malyshkin, V. (ed.) PaCT 2015. LNCS, vol. 9251, pp. 307–318. Springer, Cham (2015). doi:10.1007/978-3-319-21909-7_31
18. Barkalov, K., Gergel, V.: Parallel global optimization on GPU. J. Glob. Optim. **66**(1), 3–20 (2016)

Parallelizing Metaheuristics for Optimal Design of Multiproduct Batch Plants on GPU

Andrey Borisenko[1($\boxtimes$)] and Sergei Gorlatch[2]

[1] Tambov State Technical University, Tambov, Russia
borisenko@mail.gaps.tstu.ru
[2] University of Muenster, Muenster, Germany
gorlatch@uni-muenster.de

Abstract. We propose a metaheuristics-based approach to the optimal design of multi-product batch plants, with a particular application example of chemical-engineering systems. Our hybrid approach combines two metaheuristics: Ant Colony Optimization (ACO) and Simulated Annealing (SA). We develop a sequential implementation of the proposed method and we parallelize it on Graphics Processing Units (GPU) using the CUDA programming environment. We experimentally demonstrate that the results of our hybrid metaheuristic approach (ACO+SA) are very near to the global optimal solutions, but they are produced much faster than using the deterministic Branch-and-Bound approach.

Keywords: Hybrid metaheuristics · Ant Colony Optimization · Simulated Annealing · GPU computing · CUDA · Parallel metaheuristics · Combinatorial optimization · Multiproduct batch plant design

1 Motivation and Related Work

A *heuristic* for an optimization problem is an algorithm that explores not all possible states of the problem, but rather the most likely ones. Purely heuristics-based solutions may be inconsistent, therefore, *metaheuristics* are used that usually perform better than simple heuristics [14]. A metaheuristic is a generic algorithmic template that can find high-quality solutions of optimization problems [4] exploiting a trade-off of local search and global exploration. Metaheuristics find good-quality solutions for optimization problems in a reasonable amount of time, but there is no guarantee that the optimal solution is always reached [26].

In this paper, we consider a challenging area of optimisation – optimal design of multiproduct batch plants, e.g., in the chemical industry for producing pharmaceuticals, polymers, food etc. There has been an active research on efficiently solving such and similar problems. The classical n-queens problem was addressed using the Ant Colony Optimization (ACO) [17,24] and its combination with a Genetic Algorithm (GA) [2]. Paper [21] solves process engineering problems by the Differential Evolution (DE) algorithm and demonstrates its advantages over the exact optimization by Branch-and-Bound (B&B) and using a GA. In [13],

V. Malyshkin (Ed.): PaCT 2017, LNCS 10421, pp. 405–417, 2017.
DOI: 10.1007/978-3-319-62932-2_39

a particle swarm algorithm and a GA are exploited for multiproduct batch plant design. Paper [10] develops a multiobjective GA which demonstrates high flexibility and adaptability for various engineering problems. The problem of the optimal design of batch plants with imprecise demands on product amounts is addressed in [3] by integrating an analytic hierarchy process strategy for the analysis of the GA Pareto-optimal solutions. Paper [19] uses ACO and SA to solve a stochastic facility layout problem in which product demands are normally distributed random variables.

In order to reduce the run time of metaheuristics-based approaches, their implementation on different parallel architectures has been studied. In particular, Graphics Processing Units (GPU) are widely used by employing the CUDA platform [20]. GPU were used for solving the classical TSP problem by simulated annealing [27] and by an ant system [8]. The problem of scheduling transit stop inspection and maintenance was studied by using Harmony Search and ACO [16], with alternative implementations on CPU and GPU.

Our contribution in this paper is two-fold: (1) we develop a novel, hybrid approach which combines two metaheuristics – Ant Colony Optimization (ACO) [11] and Simulated Annealing (SA) [18], and (2) we implement it on a CPU-GPU system using CUDA and we show that it is preferable to the Branch-and-Bound approach used in our previous work [7]. Section 2 describes the mathematical problem formulation, Sect. 3 – the methodology of our hybrid ACO+SA approach, and Sect. 4 – its parallelization. Section 5 reports our experimental results, and Sect. 6 concludes the paper.

2   Problem Formulation

Our application use case is optimizing a *Chemical-Engineering System* (CES) – a set of equipment (tanks, filters, dryers etc.) which manufacture some products. A CES consists of a sequence of I processing stages; i-th stage is equipped with equipment units from a finite set X_i, with J_i being the number of equipment units variants in X_i. All equipment unit variants of a CES are described as $X_i = \{x_{i,j}\}, i = \overline{1, I}, j = \overline{1, J_i}$, where $x_{i,j}$ is the main size j (working volume, working surface) of the unit suitable for processing stage i. A CES variant $\Omega_e, e = \overline{1, E}$ (where $E = \prod_{i=1}^{I} J_i$ is the number of all possible variants) is an ordered set of available equipment unit variants. The goal is finding the optimal number of units at processing stages and their sizes while the input data are: demand for each product of assortment, production horizon, available equipment set, etc. Each variant Ω_e of a system must be in an operable condition (*compatibility constraint*), i.e., it must satisfy the condition of a joint action for its processing stages expressed by function S: $S(\Omega_e) = 0$ if the compatibility constraint is satisfied. An operable variant of a CES must also satisfy a *processing time constraint*: $T(\Omega_e) \leq T_{max}$, where T_{max} is the total available time (horizon).

Thus, designing an optimal CES is formulated as follows [5,6]: find a variant $\Omega^* \in \Omega_e, e = \overline{1, E}$ of a CES, that minimizes the objective function – equipment costs $Cost(\Omega_e)$, and both compatibility and processing time constraint are satisfied:

$$\Omega^* = argmin\ Cost(\Omega_e), e = \overline{1, E} \tag{1}$$

$$\Omega_e = \{x_{1,j_1}, x_{2,j_2}, \ldots, x_{I,j_I} | j_i = \overline{1, J_i}, i = \overline{1, I}\}, e = \overline{1, E} \tag{2}$$

$$x_{i,j} \in X_i,\ i = \overline{1, I}, j = \overline{1, J_i} \tag{3}$$

$$S(\Omega_e) = 0, e = \overline{1, E} \tag{4}$$

$$T(\Omega_e) \leq T_{max}, e = \overline{1, E} \tag{5}$$

The search space can be represented as a tree of height I (Fig. 1). Each tree level corresponds to one processing stage of the CES, each edge corresponds to a selected equipment variant taken from the set of possible variants X_i at stage i. Each node $n_{i,k}$ at the tree layer $N_i = \{n_{i,1}, n_{i,2}, \ldots, n_{i,k}\},\ i = \overline{1, I}, k = \overline{1, K_i}, K_i = \prod_{l=1}^{i}(J_l)$ corresponds to a variant of equipment units for stages 1 to i.

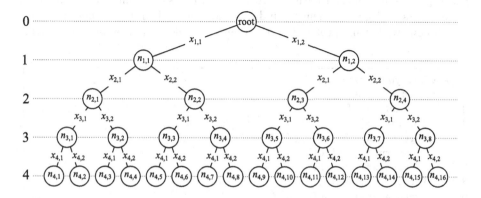

Fig. 1. The search tree for a CES with 4 stages.

Figure 1 shows an example CES consisting of 4 stages ($I = 4$), where each stage can be equipped with 2 devices ($J_1 = J_2 = J_3 = J_4 = 2$), i.e., the number of all possible system variants is $2^4 = 16$.

3   Hybrid Metaheuristic Approach

Our approach to the optimal design of multi-product batch plants is based on two metaheuristics: Simulated Annealing (SA) and Ant Colony Optimization (ACO). SA is widely used for solving optimization problems [18,26]; its key advantage is escaping from local optima by allowing hill-climbing moves to find a global optimum. SA can deal with arbitrary systems and objective functions; it often finds an optimal solution and generally finds a good-quality solution.

Before searching for a solution using SA, we need a feasible initial solution. For classical optimization problems, e.g., Traveling Salesman Problem (TSP), it is possible to use a random initial solution. However, for our problem described in Sect. 2, random initialization is unacceptable, because the compatibility (4) and processing time (5) constraints must be satisfied. Our search for a feasible initial solution is a *Constraint Satisfaction Problem* (CSP) [23] without cost optimization, which consists in finding an operable variant of a CES, satisfying (4) and (5). For solving it, we use the Ant Colony Optimization (ACO) metaheuristic [15] that provides good-quality results in many applications, including CSP [24].

3.1   Ant Colony Optimization (ACO)

The Ant Colony Optimization (ACO) metaheuristic can be viewed as a multi-agent system in which agents (ants) interact with each other in order to reach a global goal [26]. It is inspired by the behaviour of ant colonies: while walking from food source to the nest and vice versa, ants deposit a chemical substance called *pheromone* on their path. Pheromone is used as a communication medium among ants and guides them to find the shortest path from the nest to food: ants follow, with some probability, the pheromone deposited by previous ants.

```
1  AntColonyOptimization(){
2   isFound = false; /* repeat while solution not found */
3   while(!isFound){
4     Initialize(); /* initialize pheromone value */
5     foreach(ant in swarm){/* for each ant in colony */
6      ConstructSolution(); }
7     if(isFound) return; /* if solution is found, then end */
8     PheromoneUpdate(); /* update pheromone */
9     EvaporatePheramone(); }
```

Listing 1. The pseudocode of ACO algorithm.

Listing 1 shows the pseudocode of the ACO algorithm for our problem. The number M of ants is the algorithm parameter which determines the trade-off between the number of iterations and the breadth of the search per iteration: the larger the number of ants per iteration, the fewer iterations are needed [25]. All ants behave in a similar way: each ant moves from the top of the tree in Fig. 1 to the bottom. Once an ant selected a node $r = n_{i,j}$ at level i, it can pick the next child node $s = n_{i+1,j}$. The tour of an ant ends at the last tree level I; each path corresponds to a potential solution of the problem. The ant transition from node r to s is probabilistically biased by two values: pheromone trail τ_{rs} and heuristic information η_{rs}, as follows: $p_{rs} = \tau_{rs}^{\alpha} \cdot \eta_{rs}^{\beta} / \sum_{k \in C_r} (\tau_{rk}^{\alpha} \cdot \eta_{rk}^{\beta})$, where C_r is the set of child nodes for r [12,24,26]. The factors α and β influence the pheromone value and heuristic value respectively. These parameters control the relative importance of the pheromone trails and the heuristic information.

Our approach to calculating heuristic information is based on the fact that a CES with bigger units is usually more expensive, but it has a bigger batch size of products and so produces faster than a CES with smaller units. This is favourable for satisfying the time constraint (5). Therefore, we make a unit which satisfies the compatibility constraint (4) for the beginning part of the CES and a larger basic size more preferable than a unit with the unsatisfied compatibility constraint and smaller basic size. We use the following rule for the pheromone update (line 8): $\tau_{rs} = \tau_{rs} + Q/\sum_{m=1}^{M} L_m$, where Q is some constant and L_m is the tour length of the m-th ant, M is the swarm size. The smaller is the value of L_m the larger is the value added to the previous pheromone value. We use L_m as a fitness value that indicates how close is a given solution to achieving the required goals. Listing 2 shows our approach to computing the fitness value L[m] (line 2). Function NumberS() (lines 4–8) counts the number of stages of the beginning part of the CES, composed of devices for stages 1 to i (lines 6–7), for which (4) is satisfied. We add 1 to NumberS() if constraint (5) is satisfied (line 2). Therefore, the minimal fitness value is 0 (no constraint is satisfied), the maximal value is $I + 1$ (all constraint are satisfied, the problem is solved). We use the maximal fitness value as constant Q: $Q = I + 1$. With time, the concentration of pheromone decreases due to evaporation. The evaporation (Listing 1 line 9) is performed at a constant rate after the completion of each iteration. It allows the ant colony to avoid an unlimited increase of the pheromone value and to forget poor choices made previously [25]. We implement this as follows: $\tau_{rs} = \tau_{rs} \cdot \rho$, where $\rho \in [0, 1]$ is the trail persistence parameter.

```
1   ...
2      L[m] = NumberS(W[m]) + (T(W[m]) <= Tmax ? 1 : 0);
3   ...
4   int NumberS(W){
5     count = 0;
6     for (i = 1; i <= I; i++){ /* check constraint (4) */
7       if(PartS(W, i) == 0) count++; }
8     return count; }
```

Listing 2. Pseudocode of the fitness value computing for ACO.

3.2   Simulated Annealing (SA)

The basic idea of SA is to use random search which accepts not only changes that improve the objective function, but also some changes that are not ideal, in order to escape local minima. A parameter t called *temperature* governs the search behaviour.

Listing 3 shows a pseudocode of our SA version that performs two loops: the inner loop (line 4) to search for a neighbouring solution, and the outer loop (line 3) to decrease the temperature in order to reduce the probability of accepting the non-improving neighbouring solutions in the inner loop. W is a vector of length I, each element W[i] specifying the device variant at each

```
1   SimulatedAnnealing(){
2     t = Tinit; W = Winit;/* initialize temperature and guess */
3     while(t > Tfinal) {/* loop until t don't reaches Tfinal */
4       for(l = 0; l < Lmax; l++) { /* repeat Lmax times */
5         Wcand = Perturb(W); /* construct neighbour solution */
6         /* check compatibility and processing time constraints */
7         if (S( Wcand ) == 0 && T( Wcand ) <= Tmax ){
8           deltaCost = Cost(Wcand) - Cost(W);
9           if (deltaCost < 0){ /* if new solution is better */
10            W = Wcand;} /* accept the new solution */
11          else {
12            r = rand(0, 1); /* generate a random number */
13            p = exp (-deltaCost / t); /* calculate probability */
14            if (p > r) {
15              W = Wcand; }}}} /* accept the new solution */
16      t = sigma * t; }} /* decrease the temperature value */
17  Perturb(W){
18    stage = (int) rand(1, I); /* select random stage */
19    W[stage] = (int) rand(1, J[stage]);/* select random unit */
20    return W; }
```

Listing 3. The pseudocode of SA algorithm.

stage of the problem solution (1)–(5). At each iteration of the inner loop, we generate a new candidate solution Wcand in the neighbourhood of the current feasible solution (line 5) using our procedure Perturb() (lines 17–20): we select a random stage (line 18) in the feasible solution, for which we select a random unit (line 19) from the equipment set accessible for this stage. Thus at each iteration we change only one unit in the feasible solution at a stage. We avoid getting trapped in a local optimum by randomly generating neighbours and accepting a solution that worsens the value of the objective function with certain probability [22] which depends on the change of the objective function $\Delta\mathcal{E}$ and parameter t: the acceptance probability p decreases over time as t decreases. Consequently, SA first performs a wide investigation of the solution space and then restricts the solution space gradually, converging to the best solution. We initialize W with an initial feasible solution Winit obtained as the result of ACO, and the temperature t with initial value Tinit (line 2). We choose Tinit as a difference between the cost of the most expensive and the cheapest CES variant, as recommended in [1]. The transition probability p (line 13) is determined by $p = \exp(-\Delta\mathcal{E}/(k_{\mathrm{B}} \cdot t))$, where k_{B} is the Boltzmann's constant, $\mathcal{E}$ is the change of the energy level [28]. We use $k_{\mathrm{B}} = 1$ and $\gamma = 1$ [28]. Thus, the probability becomes p = exp (-deltaCost / t) (line 13).

A finite-time implementation of SA is obtained by generating a sequence of homogeneous Markov chains of finite length L_{max} which depends on the size of the problem [1]. The iterations at a given value of t repeat Lmax times (line 4). We compute Lmax as the total size of equipment set, i.e., $L_{max} = \sum_{i=1}^{I} J_i,$

where J_i is the number of equipment units variants for stage i. By permuting the feasible solution, we select at each iteration a random unit (line 19) in one random stage (line 18). Temperature t is decreased at the end of each iteration using a cooling schedule defined by an initial temperature Tinit, a rule for reducing t, and a final temperature Tfinal which is fixed at a small value chosen as the smallest possible difference in cost between two neighboring solutions; in our case, we use for Tfinal the price of the cheapest unit. We use (line 16) the fast cooling rule $t = \sigma \cdot t$ [26], where $0.8 \leq \sigma \leq 0.99$ as recommended in [1].

4   Parallelization for GPU

Figure 2 illustrates our parallel implementation of the hybrid (ACO+SA) approach described in Sect. 3 on a system comprising a CPU and a GPU.

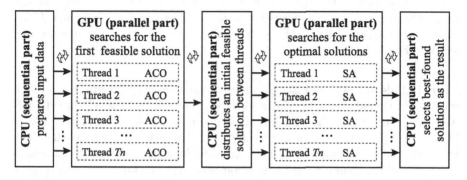

Fig. 2. The hybrid algorithm structure.

Application code consists of a sequential code (*host* code for CPU) that invokes parallel execution of hundreds or thousands of threads on the *device* (GPU), where all threads execute the same *kernel* code. The implementation consists of the following five steps (from left to right in Fig. 2):

1. CPU reads the input data (number of CES stages I, number of accessible equipment set J_i, production horizon T_{max} etc.) from a file, initializes the metaheuristics' parameters for ACO and SA, sends this data to GPU, and starts on the GPU the kernel function for ACO.
2. The ACO kernel on the GPU searches for the first feasible solution – the initial CES-variant, as described is Sect. 3.1. We use the Multiple Ant Colonies approach [9]: all colonies work as threads in parallel to solve the problem independently. If some thread finds a solution then all threads terminate. With an increasing number of threads, the probability of finding a solution increases, and therefore the search time is typically reduced.
3. CPU receives the obtained solution, distributes it between threads as an initial solution for SA, and starts the SA kernel function on the GPU.

412   A. Borisenko and S. Gorlatch

4. The SA kernel on the GPU searches in each thread for the optimal solution
   with the initial solution found by ACO, i.e., we do not try to reduce the
   time of one iteration, but rather increase the number of iterations executed
   simultaneously. Each thread executes an independent instance of SA, thus,
   the chance of the algorithm to converge to the global optimum increases, even
   if all instances use the same initial solution. A larger number of threads does
   not reduce the run time of the algorithm, but rather increases the probability
   that some thread eventually finds a nearly optimal solution.
5. CPU receives the SA solutions obtained by the GPU threads and chooses the
   best among them – this is the final solution of our problem.

Host Code. The host starts its work by loading the input data from a file.
The number of threads is a program launch parameter taken as a command-
line argument. The host sends data to the GPU and starts the kernel `ACO()`
that implements the ACO-algorithm. A CUDA kernel launch is asynchronous,
i.e. it returns control to the CPU immediately after starting the kernel. Using
`cudaDeviceSynchronize()`, the CPU waits until the GPU terminates and
receives the results from it.

Kernel Code for ACO. Listing 4 shows our parallel implementation of ACO,
where each thread simulates the work of one ant colony. For all threads, initially, all edges are assigned small random pheromone values from interval $[0, 1]$
(lines 4–5). The global flag `isFound` and the local iteration counter `iterCounter`
are used to control threads. The flag is changed by a thread using `atomicAdd()`
if this thread has found a feasible solution (line 21). The local iteration counter
is used by each thread as a nonstop operation protection: if ants in this thread
cannot find the solution after `maxIterNumber` iterations (which is possible for
stochastic algorithms) then the thread terminates. After initialization, each ant
`m` in swarm `M` generates a path (lines 9–17). Here, `Want` is a local two-dimensional
array of length `M`, each element of which is a vector of length `I` specifying the
device variant at each stage of the solution.

   We do not discuss the kernel code of SA – it largely follows Listing 3.

5 Experimental Results

Our experiments are conducted on a heterogeneous system comprising: (1) a
CPU: Intel Xeon E5-1620 v2, 4 cores with Hyper-Threading, 3.7 GHz with 16 GB
RAM, and (2) a GPU: NVIDIA Tesla K20c with altogether 2496 CUDA cores
and 5 GB of global memory. We use Ubuntu 16.04.2, NVIDIA Driver version
367.57, CUDA version 8.0 and GNU C++ Compiler version 5.4.0.

   As our test case, we evaluate the design of a CES consisting of 16 processing
stages with 2 to 12 variants of devices at every stage (total 2^{16} to 12^{16} CES
variants). In our previous work [6,7], we used the Branch-and-Bound (B&B)
algorithm to find the global optimal solution. Here we solve the same problem

```
1   __global__ void ACO(){ /* obtaining thread identifier */
2     threadID = blockDim.x * blockIdx.x + threadIdx.x;
3     if(threadID < numThreads){ /* pheromone initialization */
4      for (i = 1; i <= I; i++) {
5       for (j = 1; j <= J[i]; j++){tau[i][j] = curand(0, 1);}}
6      iterCounter = 0; /* while solution is not found */
7      while (isFound == 0 && iterCounter < maxIterNumber)){
8       /* generate path for each ant m in swarm M */
9       for(m = 1; m <= M && isFound == 0; m++){sum = 0.0;
10       for (i = 1; i <= I - 1; i++){
11        for (j = 1; j <= J[i]; j++){
12         eta[i][j] = (S(Want[m], i + 1) ? 1:0) + X[i][j];
13         sum += pow(tau[i][j], alpha) * pow(eta[i][j], beta);}
14        r = curand(0, 1); sump = 0.0;
15        for (j = 1; j <= J[i]; j++){
16         p = pow(tau[i][j],alpha) * pow(eta[i][j],beta) / sum;
17         sump += p; if(sump > r) {Want[m][i] = j; break; }}}}
18       /* calculate new pheromone values */
19       for(m = 1; m <= M && isFound == 0; m++){
20        L[m] = NumberS(Want[m],I) + (T(Want[m]) <= Tmax ? 1:0);
21        if (L[m] == Q) {atomicAdd(isFound, 1); bestAntId = m;}
22        for (i = 1; i <= I; i++) {
23         for (j = 1; j <= J[i]; j++) {dtau[i][j] = 0.0; }}
24        for (i = 1; i <= I; i++){
25         idx = Want[m][i]; dtau[i][idx] += Q / L[m]; }
26       /* pheromone update and evaporation */
27       for (i = 1; i <= I; i++){
28        for (j = 1; j <= J[i]; j++){
29         tau[i][j] = tau[i][j] * rho + dtau[i][j]; }}}
30        iterCounter++; }
31       /* save feasible solution and its thread identifier */
32       if(bestAntId != -1) {Wfirst[threadID] = Want[bestAntId];
33        threadIdx[threadID] = threadID; }}}
```

Listing 4. The kernel pseudocode for ACO.

on the same test system using our hybrid metaheuristic approach (ACO+SA), and we compare the results with the solution obtained by B&B. Since both SA and ACO are probability-based algorithms, their results will be different if run multiple times on the same instance of a problem; therefore, we run each instance for 100 times and we take the average of the measured values.

Figure 3 shows how the run time of the (ACO+SA) parallel program depends on the number of threads. We run our CUDA-based implementation with the number of threads from 100 to 2500 with step 100, for the CES example of 16 processing stages with 10 variants of units. We observe that the run time is decreasing with the increasing number of threads. While on 100 threads, the ACO takes 91% of the total run time, the portion of ACO decreases to only about

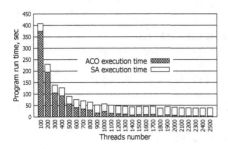

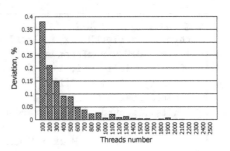

Fig. 3. Run time of (ACO+SA) depending on threads number.

Fig. 4. Deviation of the found solution from global optimum.

10% on more than 2000 threads. This is because, with more threads, ACO finds a solution faster with a higher probability, whereas using more threads for SA can improve the quality of solution, but not the speed.

Figure 4 shows the deviation of the observed objective value (CES cost) by our hybrid algorithm, calculated as $|observed - expected|/expected \cdot 100\%$. As the expected value we use the global optimal value obtained by the B&B algorithm. On 100 threads, the deviation is about 0.38% and then it decreases to almost 0% for more than 2000 threads, so we achieve an almost optimal solution.

Figure 5 shows the deviation (vertical axis has a logarithmic scale) of solutions obtained by our hybrid (ACO+SA) algorithm (sequential on CPU, parallel on GPU using up to 2500 threads) from the global optimum obtained by sequential B&B. We observe that our parallel hybrid algorithm produces a nearly global optimal solution (for problem size 2^{16}–9^{16} the deviation is 0%, and for problem size 10^{16}–12^{16} the deviation is less than 0.01%). The deviation obtained by our sequential algorithm is less than 1% for small size problem 2^{16}, but it increases to about 14% for the problem size of 12^{16}. This is because the sequential implementation performs only a single run of the SA: for small problem sizes, the probability of finding a good solution is higher than for larger problem sizes.

Figure 6 shows the program run time (vertical axis has a logarithmic scale) of our hybrid approach vs. B&B depending on the problem size. The program

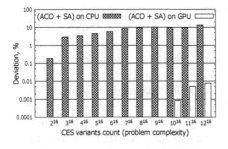

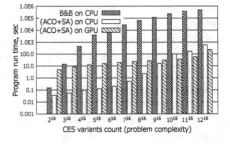

Fig. 5. Deviation (logarithmic scale) for different problem sizes.

Fig. 6. Run time (logarithmic scale) depending on problem sizes.

run time of B&B increases exponentially: while for problem size 2^{16} the run time is less than 1 s, for size 12^{16} the run time of B&B becomes prohibitively long at about 134 h. The run time of the sequential implementation of our hybrid algorithm is less than 0.1 s for the smallest problem, and it increases to about 580 sec for our maximal problem size. The run time of the parallel implementation smoothly increases from 5 to 227 sec. The parallel implementation is slower than the sequential implementation for smaller problem sizes (2^{16} to 9^{16}), but for larger problem sizes (10^{16} to 12^{16}) it is faster than the sequential version by about 2.6–2.8 times. At first glance, the speedup of 2.7 times compared to the sequential case is small, but the quality of the solutions obtained by the parallel implementation is significantly higher: the deviation from the global optimum for the parallel implementation is less than 0.01% against more than 10% deviation for the sequential version. This good quality of solutions is achieved by the independent runs of SA: for a larger number of threads the probability that one of the threads finds a nearly optimal solution is higher, because running a parallel algorithm on 2500 threads is equivalent to the launch of sequential algorithm 2500 times and the choice of the best among the found solutions. The parameters of metaheuristic algorithms influence both the run time and the quality of solutions. Empirically we have found that $\alpha = 0.4$ and $\beta = 0.6$ with the colony size $M = 100$ are good values for our application: they were selected after numerous experiments. In our experiments, we use $\rho = 0.9$ and the cooling rule constant $\sigma = 0.9$, also chosen empirically for our problem.

6   Conclusion

Our contribution is the novel hybrid (ACO+SA) metaheuristic approach to solving the optimization problem for multiproduct batch plants design and its parallel implementation on a CPU-GPU platform. We have found out that increasing the number of threads accelerates finding the solution with ACO and increases the reliability and quality of the solutions obtained by SA. We compare our results with the global optimal solution obtained by the B&B method. Our experiments confirm that our parallel hybrid approach obtains good-quality solutions which are very near to the global optimal values obtained by a deterministic algorithm like B&B, but our approach finds the solution much faster.

Acknowledgement. This work was supported by the DAAD (German Academic Exchange Service) and by the Ministry of Education and Science of the Russian Federation under the "Mikhail Lomonosov II"-Programme, as well as by the German Research Agency (DFG) in the framework of the Cluster of Excellence CiM at the University of Muenster. We also thank the Nvidia Corp. for the donated hardware used in our experiments.

References

1. Aarts, E., Korst, J., Michiels, W.: Simulated annealing. In: Search Methodologies, pp. 265–285. Springer Science & Business Media, Heidelberg (2014)
2. Agarwal, K., Sinha, A., Hima Bindu, M.: A novel hybrid approach to N-Queen problem. In: Wyld, D., Zizka, J., Nagamalai, D. (eds.) Advances in Computer Science, Engineering & Applications. AISC, vol. 166, pp. 519–527. Springer, Heidelberg (2012). doi:10.1007/978-3-642-30157-5_52
3. Aguilar-Lasserre, A.A., Bautista, M.A.B., Ponsich, A., Huerta, M.A.G.: An AHP-based decision-making tool for the solution of multiproduct batch plant design problem under imprecise demand. Comput. Oper. Res. **36**(3), 711–736 (2009)
4. Birattari, M.: Tuning Metaheuristics: A Machine Learning Perspective. Springer, Heidelberg (2009)
5. Borisenko, A.B., Karpushkin, S.V.: Hierarchy of processing equipment configuration design problems for multiproduct chemical plants. J. Comput. Syst. Sci. Int. **53**(3), 410–419 (2014)
6. Borisenko, A., Haidl, M., Gorlatch, S.: A GPU parallelization of branch-and-bound for multiproduct batch plants optimization. J. Supercomput. **73**(2), 639–651 (2017)
7. Borisenko, A., Kegel, P., Gorlatch, S.: Optimal design of multi-product batch plants using a parallel branch-and-bound method. In: Malyshkin, V. (ed.) PaCT 2011. LNCS, vol. 6873, pp. 417–430. Springer, Heidelberg (2011). doi:10.1007/978-3-642-23178-0_36
8. Dawson, L., Stewart, I.: Improving ant colony optimization performance on the GPU using CUDA. In: 2013 IEEE Congress on Evolutionary Computation, pp. 1901–1908. IEEE, June 2013
9. Delévacq, A., Delisle, P., Gravel, M., Krajecki, M.: Parallel ant colony optimization on graphics processing units. J. Parallel Distrib. Comput. **73**(1), 52–61 (2013)
10. Dietz, A., Azzaro-Pantel, C., Pibouleau, L., Domenech, S.: Strategies for multiobjective genetic algorithm development: Application to optimal batch plant design in process systems engineering. Comput. Ind. Eng. **54**(3), 539–569 (2008)
11. Dorigo, M., Blum, C.: Ant colony optimization theory: a survey. Theoret. Comput. Sci. **344**(2–3), 243–278 (2005)
12. Dorigo, M., Stützle, T.: Ant colony optimization: overview and recent advances. In: Gendreau, M., Potvin, J.-Y. (eds.) Handbook of Metaheuristics. International Series in Operations Research & Management Science, vol. 146, pp. 227–263. Springer, New York (2010). doi:10.1007/978-1-4419-1665-5_8
13. El Hamzaoui, Y., Bassam, A., Abatal, M., Rodríguez, J.A., Duarte-Villaseñor, M.A., Escobedo, L., Puga, S.A.: Flexibility in biopharmaceutical manufacturing using particle swarm algorithms and genetic algorithms. In: Schütze, O., Trujillo, L., Legrand, P., Maldonado, Y. (eds.) NEO 2015. SCI, vol. 663, pp. 149–171. Springer, Cham (2017). doi:10.1007/978-3-319-44003-3_7
14. Gandomi, A.H., Yang, X.S., Talatahari, S., Alavi, A.H.: Metaheuristic algorithms in modeling and optimization. In: Metaheuristic Applications in Structures and Infrastructures, pp. 1–24. Elsevier BV (2013)
15. Gonzalez-Pardo, A., Camacho, D.: A new CSP graph-based representation for ant colony optimization. In: 2013 IEEE Congress on Evolutionary Computation, pp. 689–696. Institute of Electrical and Electronics Engineers (IEEE), June 2013
16. Kallioras, N.A., Kepaptsoglou, K., Lagaros, N.D.: Transit stop inspection and maintenance scheduling: a GPU accelerated metaheuristics approach. Transp. Res. Part C Emerg. Technol. **55**, 246–260 (2015)

17. Khan, S., Bilal, M., Sharif, M., Sajid, M., Baig, R.: Solution of n-queen problem using ACO. In: 2009 IEEE 13th International Multitopic Conference, pp. 1–5. Institute of Electrical and Electronics Engineers (IEEE), December 2009
18. Kirkpatrick, S., Gelatt, C.D., Vecchi, M.P., et al.: Optimization by simulated annealing. Science **220**(4598), 671–680 (1983)
19. Lee, T.S., Moslemipour, G., Ting, T.O., Rilling, D.: A novel hybrid ACO/SA approach to solve stochastic dynamic facility layout problem (SDFLP). In: Huang, D.-S., Gupta, P., Zhang, X., Premaratne, P. (eds.) ICIC 2012. CCIS, vol. 304, pp. 100–108. Springer, Heidelberg (2012). doi:10.1007/978-3-642-31837-5_15
20. NVIDIA Corporation: CUDA C programming guide 8.0, September 2016. http://docs.nvidia.com/cuda/pdf/CUDA_C_Programming_Guide.pdf
21. Ponsich, A., Coello, C.C.: Differential evolution performances for the solution of mixed-integer constrained process engineering problems. Appl. Soft Comput. **11**(1), 399–409 (2011)
22. Pourvaziri, H., Azimi, P.: A tuned-parameter hybrid algorithm for dynamic facility layout problem with budget constraint using GA and SAA. J. Optim. Ind. Eng. **7**(15), 65–75 (2014)
23. Rossi, F., Van Beek, P., Walsh, T.: Handbook of Constraint Programming. Elsevier, Amsterdam (2006)
24. Solnon, C.: Ant Colony Optimization and Constraint Programming. Wiley Inc., Hoboken (2010)
25. Stützle, T., López-Ibánez, M., Pellegrini, P., Maur, M., de Oca, M.M., Birattari, M., Dorigo, M.: Parameter adaptation in ant colony optimization. In: Hamadi, Y., Monfroy, E., Saubion, F. (eds.) Autonomous Search, pp. 191–215. Springer, Heidelberg (2011). doi:10.1007/978-3-642-21434-9_8
26. Valadi, J., Siarry, P.: Applications of Metaheuristics in Process Engineering. Springer Science & Business Media, Heidelberg (2014)
27. Wei, K.C., Wu, C.C., Yu, H.L.: Mapping the simulated annealing algorithm onto CUDA GPUs. In: 2015 10th International Conference on Intelligent Systems and Knowledge Engineering (ISKE), pp. 1–8, November 2015
28. Yang, X.S.: Nature-Inspired Metaheuristic Algorithms. Luniver Press, Bristol (2010)

The Optimization of Traffic Management for Cloud Application and Services in the Virtual Data Center

Irina Bolodurina and Denis Parfenov[✉]

Orenburg State University, Orenburg, Russia
{prmat,fdot_it}@mail.osu.ru

Abstract. Nowadays one of the problems of optimization is the control of the traffic in cloud applications and services in the network environment of virtual data center. Taking into account the multitier architecture of modern data centers, we need to pay a special attention to this task. The advantage of modern infrastructure virtualization is the possibility to use software-defined networks and software-defined data storages. However, the existing optimization of algorithmic solutions does not take into account the specific features of the heterogeneous network traffic routing with multiple application types. The task of optimizing traffic distribution for cloud applications and services can be solved by using software-defined infrastructure of virtual data centers. We have developed a simulation model for the traffic in software-defined networks segments of virtual data centers involved in processing user requests to cloud application and services within a network environment. Our model enables to implement the traffic management algorithm of cloud applications and optimize the access to storage systems through the effective use of data transmission channels. During the experimental studies, we have found that the use of our algorithm enables to decrease the response time of cloud applications and services and, therefore, increase the productivity of user requests processing and reduce the number of refusals.

Keywords: Software-Defined Network · Virtual Data Center · Cloud computing · Traffic · Simulation model · Software-defined infrastructure

1 Introduction

Nowadays, we see a steady growth in the use of cloud computing in modern business. This enables to reduce the cost of IT infrastructure owning and operation; however, there are some issues related to the management of data centers. At present, the solutions for virtual infrastructure are dynamically developing. Thus, the container technology has been lately used for placing cloud applications and services within virtual data centers. Container technologies are mostly based on Docker. Besides, modern data centers rather

© Springer International Publishing AG 2017
V. Malyshkin (Ed.): PaCT 2017, LNCS 10421, pp. 418–426, 2017.
DOI: 10.1007/978-3-319-62932-2_40

use virtual infrastructures instead of physical infrastructures especially based on software-defined components: networks, data storages [1, 2], etc. This changes the mechanisms launch and placement management of applications and services. Thus, it is important to develop effective scheduling and resource distributing methods for cloud systems that optimize the response time in user requests.

2   The Multilevel Model of the Software-Defined Infrastructure

We developed the multilevel model of the software-defined infrastructure of the virtual data center (VDC), which supports containerization method of cloud applications and services.

The first level is the hardware component of any data center, which includes computing nodes (Nodes), storages systems (Storages) and physical network objects (NetObj). Let us introduce it as a set of solutions: PhysLayer = {Nodes, Storages, NetObj}.

The next level represents the software-defined layer. This layer consists of the same number of objects as the first level but the main difference is that all the infrastructure elements are dynamic, easily transformed and adjusted within the limits of the physical database network environment. The second level can be presented as the following set of connections: SDLayer = {SDNodes, SDStorages, SDNetwork}, where SDNodes – software-defined computing nodes; SDStorages – software-defined storages (SDS); SDNetwork – software-defined network (SDN).

Above the layer of the software-defined infrastructure, there is a level of the specific objects virtualization. The main objects are: virtual computing nodes (VirtNodes), virtual data storages and other elements of the software-defined network. Virtual objects are used for work in the cloud platform and consolidated in next set of VirtLayer = {VirtNodes, VirtStorages, VirtNetwork}. In the software-defined infrastructure, computing nodes and data storages are more often presented as virtual machines (VM) that discharge the set of given functions.

To control such multi-layer infrastructure, a separate orchestration layer is needed (the forth level). Its include functions of orchestration main types of the virtualization objects OrchLayer = {ONodes, OStorages, ONetwork}.

The next level (service level) ServiceLayer = {$Service_1$, ..., $Service_n$}. Its represents the services used in the working process of the cloud platform or by cloud applications distributed there. For example, DBMS, Hadoop, Nginx and others. All the multitude of ServiceLayer cloud services that work in the virtual data center infrastructure can be divided into two disjoint subsets ServVM ∪ ServDocker = ServiceLayer. The first set (ServVM) involves services that use virtualization based on other machines. In the second set (ServDocker), there are services based on containers under Docker control.

The top level includes cloud applications that are exploited by users for flexible scalability providing AppLayer = {App_1, ..., App_m}. Like at the previous level, cloud applications App_i can be placed in containers and form the AppVM set. Or they can form the AppDocker set using containerization. At the same time AppLayer = AppVM ∪ AppDocker.

Thus, the set of objects of the software-defined infrastructure can be divided into two groups by the methods of placing. Virtual objects that use a container placing method can be referred to the first group. Let us describe them in this way: Docker = {Serv-Docker, AppDocker}. In the second group, there are services and applications that use virtual machines as a placing platform: VM = {ServVM, AppVM}.

3   Research Methods

Generally, the software-defined infrastructure of a virtual data center has several heterogeneous applications and services. We can assume that the network of a virtual data center encompasses at least three types of application traffic: web-applications, case-applications, and video services. To generate user requests in the simulation model, we apply weight coefficients k_1, k_2, k_3 for each traffic type. Each coefficient allows us to classify requests into types and affects the following set of parameters: running time, routes, priority in the process queue, request intensity, and the distribution law for each type of traffic.

Presented as a multi-channel queuing system, the simulation model of the software-defined infrastructure of the virtual data center includes a user request source (I), a queue (Qs) and a scheduler (S) who manages application hosting and its launch (App). Besides, it contains computing cluster (Srv) and systems of data center storage (Stg). The queuing system is represented in Fig. 1.

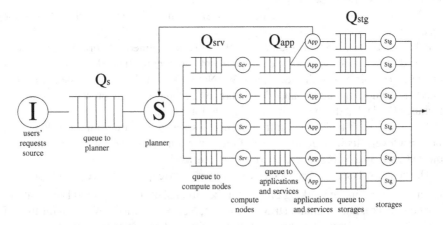

Fig. 1. A queuing system scheme of the software-defined infrastructure of the VDC

A queuing system model is stochastic. For its operation, it is necessary to make a user request flow to cloud applications and services with account for the distribution laws and request intensities for each type of cloud applications and services.

To optimize application distribution in the cloud environment of a virtual data center, it is necessary to determine the traffic distribution laws for each application type and distribute the traffic into access objects. For this purpose, it is necessary to set a certain route and make the control law for it within the time interval $T = [t_1, t_2]$.

The dynamic of traffic in cloud applications and services of the software-defined infrastructure of a virtual data center can be described by the following discrete system:

$$x_{i,j}(t + \Delta t) = x_{i,j}(t) - \sum_{k=1}^{K} \sum_{l=1}^{N} s_{i,j}(t) u_{i,l}^{j,k}(t) + \sum_{m=1}^{N} s_{m,i}(t) u_{m,l}^{j}(t) + y_{i,j}(t) \tag{1}$$

where N is the number of virtual nodes within the network; K is the number of application types within the network; $s_{i,j}(t)$ is the capacity of the channels between i-th computing node and j-th storage system $(i \neq j)$; $y_{i,j}(t) = \lambda_{i,j}(t) \Delta t$ i is the traffic volume (the number of user requests) at the moment t on the virtual node i-th, intended for transferring to the storage system j-th; $\lambda_{i,j}(t)$ is the intensity of incoming load, which is defined as the total intensity of the user request flow connecting to the virtual node i-th and using the storage system j-th; $u_{i,l}^{j,k}(t)$ is a part of the channel transmission capacity in a certain segment of the software-defined network (i, l) at the moment t for the user request flow to the application of type k, working with the data storage system j-th.

To exclude the possibility of overloading the objects of the virtual data center due to the limited queue buffers on compute nodes, as well as to use data transmission channel capacity efficiently, a number of restrictions are introduced for network parameters of cloud applications in software-defined networks (SDN).

The restrictions related to channel capacity limits can be written as follows:

$$0 \leq u_{i,l}^{j}(t) \leq u_{i,l}^{j(\max)} \leq 1; \sum_{l=1}^{N} u_{i,l}^{j}(t) \leq \varepsilon_{i,l}^{j,k} \leq 1 \tag{2}$$

where $u_{i,l}^{j(\max)}$ $u_{i,l}^{j(\max)}$ is the limit of channel capacity available for the computing node i-th in the SDN segment l-th for traffic transfer to the storage j-th; $\varepsilon_{i,l}^{j,k}$ is the part of the channel capacity for the compute node i-th in the SDN segment l-th for the transmission of user requests to the application of type k to the storage system j-th.

Let us consider the system performance as a criterion of optimality, gained through a fixed period T = [t₁, t₂], which is formalized within the model as an objective function in the following form:

$$\sum_{t=0}^{t-1} \sum_{k=1}^{K} \sum_{i=1}^{N} \sum_{j=1}^{N} s_{i,j}(t) u_{i,l}^{j,k}(t) \rightarrow \max \tag{3}$$

To solve the optimization task, we use an iterative method that allows us to explore the dynamics of the system at the interval T = [t₁, t₂] and control channel capacity for a certain type of applications in a software-defined network.

4 Algorithm of Adaptive Routing

Based on the constructed models, we have developed an optimization algorithm of adaptive routing and balancing of the application and services flows in a heterogeneous cloud platform, which is located in the data center. The algorithm aims to ensure the efficient management of the application and services' flows under dynamic changes in the load on the communication channels used to deploy data center software-defined networks by reducing the design complexity for optimal route schemes. Generalized algorithm is as follows:

Application of the proposed algorithm for optimizing the adaptive routing of data flow balancing has allowed us to reduce the complexity of calculating the optimum route to the value $O(kN)$, where k is the number of completed transitions to alternative routes, and N is the number of objects in the software-defined network of data center. Thus, the algorithm is designed to speed up the search and selection of optimal routes for application and service data flows arranged in a heterogeneous cloud platform under dynamic load changes on the communication channels.

5   Data Assignment Algorithm for Cloud Applications

The data assignment algorithm for cloud applications provides heuristic analysis of application requests and traffic classification on data types at the performance time. The flexibility of the algorithm is due to virtualization of data storage. This makes it possible to dynamically change the physical location of the application within the cloud system for providing uninterrupted access to services.

The suggested solution is transparent to the client and scales cloud applications into multiple virtual storage devices. This provides a reduction in the application response time, and also improves the fault tolerance of the whole system.

The data assignment algorithm for cloud applications is based on a cloud resource model describing its structure and links between virtual storage devices, machines and cloud applications. The model uses multi-agent approach in data storage. Agents get information about the system state. Analyzing the maps the cloud control system makes decisions on reconfiguring or migrating virtual storage devices as well as data redistributing between nodes.

For a user request service several resources can be used with different access parameters. In this case the cloud control system has to optimize read time. Our data assignment algorithm forms internal assignment rules and changes them according to resource demands. Such approach allows to balance dynamically the resource load.

6   Experimental Results

To assess the efficiency of the developed algorithm for optimizing the adaptive routing of applications and services data flow balancing in a heterogeneous data center cloud platform, we have conducted a pilot study. We have chosen the Openstack cloud system as a basic platform. For comparison, we have applied algorithms used in the OpenFlow version 1.4 for route control of the software-defined network of the data center in the experiment. We have created a virtual data center prototype with basic nodes and software modules for the developed algorithms that redistribute data and applications flows for the pilot study. To verify the developed algorithm of optimal routing and traffic balancing under conditions of dynamic changes channels in the software-defined network of the data center, we have deployed several experimental networks consisting of from 25 to 400 objects. All generated requests have been reproduced consequently at two pilot sites: with the traditional routing technology (platform 1, NW) and with the technology of the software-defined networks (platform 2, SDN). This restriction is caused by the need to compare the results with traditional network infrastructure incapable of dynamic reconfiguration. Two tests were carried out on platform 2. In the first case, we have used the model OpenFlow version 1.4 routing algorithms, in the second case (NEW SDN), we have applied the developed routing optimization algorithm. Experiment time was one hour. We have chosen the response time of applications and

---

1. Divide all channels in network in two subsets, *ST* and *SR*.

2. Generate optimal routes for data flow of the particular class of applications.

3. Determine the points of entry in two subset, *ST* and *SR*.

4. Search for all the alternative routes at minimum cost.

5. Calculate alternative routes for dynamic load change in the network channel.

6. Form the full list of alternative paths in network.

7. Define whether there were load changes.

If "Yes" then **GO TO** Step **8**, else **GO TO** Step **7**.

8. Define whether you need to change the route for the current data flow.

If "Yes" then **GO TO** Step **9**, else **GO TO** Step **13**.

9. Calculate metrics connection.

10. Define list of network objects placed higher in the hierarchy, which performance has decreased. If network objects found then **GO TO** Step **11**, else **GO TO** Step **12**.

11. Define the new minimum length route for every networks object.

12. Design the new optimal route tree.

13. Transfer the current data flows to new routes. Reshape the list of alternative routes. **GO TO** Step **7**.

Fig. 2. Generalized plan of optimization algorithm of adaptive routing and balancing of the application and services flows in the virtual data center

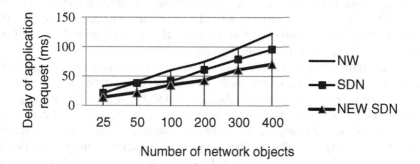

Fig. 3. A schedule of dependence of the response time of applications and services in a heterogeneous cloud platform from the quantities of network objects in the data center

services that work in a cloud platform as a basic metrics to assess the efficiency of the proposed solutions. The results of the experiment are provided in Fig. 2 (Fig. 3).

7 Discussion

Traditional approaches to route traffic based on load-balancing are reactive. They use simple classical algorithms for distributed computing tasks First Fit or Best Fit. Such algorithms as [3–5] First Come First Served Scan, Most Processors First Served Scan, and Shortest Job First Scan are popular too. Their main disadvantage is poor utilization of a computer system due to a large number of windows in the task launch schedule and problem with "hanging up" when their service is postponed indefinitely due to tasks of higher priority [9]. The solution proposed by D. Lifka from Argonne National Laboratory is usually applied as an alternative method of load distribution between nodes. It is based on the aggressive variant of Backfill algorithm [3, 5] and has two conflicting goals – a more efficient use of computing resources by filling the empty windows schedule and prevention of problems with "hanging up" due to redundancy mechanism. Further, various modifications have been created by B. Lawson and E. Smyrni, S. Srinivasan [6, 7] and etc. The main drawback of these algorithms is the time lag during calculation, which is not acceptable for critical services at the time of failure.

In addition to the traditional reactive fault-tolerant technology, such as replication and redundancy to ensure reliability of networked storage cloud platforms, a group of scientists from Nankai University proposed an approach based on the Markov model, which provides secure storage of data without excessive redundancy [8]. However, a significant drawback of this model is the lack of classification and analysis of the types and sources of data to be placed in their consumption. Nevertheless, the model demonstrates a proactive approach that gives certain advantages to achieve the desired resiliency of cloud storage.

Reliability and availability of applications and services play an important role in the assessment of its cloud platform performance. A major shortcoming of existing software reliability solutions in the data center infrastructure is the use of traditional data flow routing methods. In this work, we offer to use the software-defined network technology to adjust the network to the current load of the applications and services that are hosted in a cloud platform before they start using pre-computed and installation routes of transmission. The principles of a software-defined network first emerged in research laboratories at Stanford and Berkeley, and are currently being developed by the Open Network Foundation consortium, GENI project.

The algorithms for routing data flows in a software-defined network in case of track selection published in scientific sources do not take into account the need to ensure the QoS parameters for the previously installed and routed data flows. We are going to do it within a framework of the developed methods of adaptive network communications routing. The existing QoS algorithms to provide a software-defined network are also quite inefficient. The paper [10] describes an approach to dynamic routing of multimedia flows transmission that provide a guaranteed maximum delay via the LARAC algorithm. However, the authors consider only the cases of single delays on each network connection and do not take into account the minimum guaranteed bandwidth. A similar approach is described in the paper [11]; the authors pose and solve the optimization problem for the transfer of multimedia traffic without losses on alternative routes, leaving the shortcuts for common data.

The researchers from Stanford have offered an algorithm for adaptive control of QoS Shortest Span First, which enables to calculate the optimal priorities for each flow mathematically, to minimize crosstalk influence of flows on delay, to manage priorities dynamically depending on the current situation, and to lay the flow of data transmission through specific port queues [12].

We are going to formulate optimization problems for laying routes with QoS constraints and load balancing within a framework of adaptive routing methods of network communications cloud services and applications developed in this research. In their solution, we may use heuristics similar to the Shortest Span First algorithm.

The analysis of scientific sources on the topic of the study has shown that:

(a) so far, there are no effective algorithmic solutions for planning virtual machines, cloud services, application-oriented accounting topology of the computer system, and communication tasks schemes;
(b) the existing solutions for managing distributed scientific computing on multi-cloud platforms plan computing tasks without subsequent adjustment of network to their communication schemes and use traditional routing methods;
(c) the existing methods of data flow routing can be enhanced by taking into account the QoS requirements and distributed nature of a heterogeneous cloud platform.

This demonstrates the novelty of the solutions offered by the project.

8 Conclusions

The experimental studies found that the application of the developed algorithm allows reducing the response time of cloud applications and services, and as a consequence, to improve the performance of processing user requests and to reduce the number of failures. As shown in our research, the algorithm for optimizing the adaptive routing of data flow balancing based on collected information about alternative routes has enabled to reduce the response time for applications and service of a heterogeneous cloud platform with a dynamically changing load on channels by 40% compared to traditional networks, and by 25% compared to the model algorithms of the Protocol version 1.4 of OpenFlow. Thus, the algorithm is efficient in designing optimal routes and traffic balancing in SDN of the virtual data center in case of dynamic changes of load on communication channels.

Acknowledgements. The research has been supported by the Russian Foundation of Basic Research (grants 16-37-60086 mol_a_dk, 16-07-01004 a), and the President of the Russian Federation within the grant for state support of young Russian scientists (MK-1624.2017.9).

References

1. Bolodurina, I., Parfenov, D.: Development and research of models of organization storages based on the software-defined infrastructure. In: 39th International Conference on Telecommunications and Signal Processing, pp. 1–6. IEEE Press, Vienna (2016). doi: 10.1109/TSP.2016.7760818
2. Parfenov, D., Bolodurina, I., Shukhman, A.: Approach to the effective controlling cloud computing resources in data centers for providing multimedia services. In: International Siberian Conference on Control and Communications, pp. 1–6. IEEE Press, Omsk (2015). doi:10.1109/SIBCON.2015.7147170
3. Garey, M., Graham, R.: Bounds for multiprocessor scheduling with resource constraints. SIAM J. Comput. **4**(2), 187–200 (1975)
4. Arndt, O., Freisleben, B., Kielmann, T., Thilo, F.: A comparative study of online scheduling algorithms for networks of workstations. Cluster Comput. **3**(2), 95–112 (2000)
5. Feitelson, D., Weil, A.: Utilization and predictability in scheduling the IBM SP2 with backfilling. In: Parallel Processing Symposium, pp. 45–52 (1998)
6. Lawson, B.G., Smirni, E.: Multiple-queue backfilling scheduling with priorities and reservations for parallel systems. In: Feitelson, D.G., Rudolph, L., Schwiegelshohn, U. (eds.) JSSPP 2002. LNCS, vol. 2537, pp. 72–87. Springer, Heidelberg (2002). doi: 10.1007/3-540-36180-4_5
7. Srinivasan, S., Kettimuthu, R., Subramani, V., Sadayappan, P.: Selective reservation strategies for backfill job scheduling. In: Feitelson, D.G., Rudolph, L., Schwiegelshohn, U. (eds.) JSSPP 2002. LNCS, vol. 2537, pp. 55–71. Springer, Heidelberg (2002). doi: 10.1007/3-540-36180-4_4
8. Li, J., Li, M., Wang, G., Liu, X., Li, Z., Tang, H.: Global reliability evaluation for cloud storage systems with proactive fault tolerance. In: Wang, G., Zomaya, A., Perez, G.M., Li, K. (eds.) ICA3PP 2015. LNCS, vol. 9531, pp. 189–203. Springer, Cham (2015). doi: 10.1007/978-3-319-27140-8_14
9. Rahme, J., Xu, H.: Reliability-based software rejuvenation scheduling for cloud-based systems. In: The 27th International Conference on Software Engineering and Knowledge Engineering, pp. 1–6 (2015)
10. Lin, T.: Enabling SDN applications on software-defined infrastructure. In: Network Operations and Management Symposium (NOMS), pp. 1–7. IEEE Press (2011)
11. Ibanez, G., Naous, J., Rojas, E., Rivera, D., Schuymer, T.: A small data center network of ARP-path bridges made of openflow switches. In: The 36th IEEE Conference on Local Computer Networks (LCN), pp. 15–23. IEEE Press (2011)
12. Tavakoli, A., Casado, M., Koponen, T., Shenker, S.: Applying NOX to the datacenter. In: 8th ACM Workshop on Hot Topics in Networks (HotNets-VIII). IEEE Press, New York (2009)

Distributed Data Fusion
for the Internet of Things

Rustem Dautov[1](✉) and Salvatore Distefano[1,2]

[1] Higher Institute of Information Technology and Information Systems (ITIS),
Kazan Federal University (KFU), Kazan, Russia
{rdautov,s_distefano}@it.kfu.ru, sdistefano@unime.it
[2] University of Messina, Messina, Italy

Abstract. The ubiquitous Internet of Things is underpinned by the recent advancements in the wireless networking technology, which enabled connecting previously scattered devices into the global network. IoT engineers, however, are required to handle current limitations and find the right balance between data transferring range, throughput, and power consumption of wireless IoT devices. As a result, existing IoT systems, based on collecting data from a distributed network of edge devices, are limited by the amount of data they are able to transfer over the network. This means that some sort of data fusion mechanism has to be introduced, which would be responsible for filtering raw data before sending them further to a next node through the network. As a potential way of implementing such a mechanism, this paper proposes utilising Complex Event Processing and introduces a hierarchical distributed architecture for enabling data fusion at various levels.

Keywords: Data fusion · Complex Event Processing · Distributed architecture · Internet of Things · Edge computing · Cloud computing

1 Introduction

The development of the Internet of Things (IoT) and ubiquitous penetration of 'smart' devices in almost every aspect of people's everyday life have been supported by the rapid progress in the networking area and – more specifically – wireless technologies. Wireless communication enabled connecting previously disconnected embedded devices into the global network, facilitating device discovery, querying and interaction. Examples of such wireless networking technologies, actively used in the context of complex distributed IoT systems, include LPWAN, Bluetooth Low Energy, ZigBee, Wi-Fi, etc. These technologies differ in their data transferring range, throughput, and power consumption. These three aspects are typically seen as the key factors when choosing a particular wireless technology to a be applied to a scenario at hand. There also exists a dependency between these three factors. Usually, the larger the data transferring range, the lower the throughput, and vice versa. Increasing any of the two – either the range or the throughput – typically leads to an increased power consumption.

© Springer International Publishing AG 2017
V. Malyshkin (Ed.): PaCT 2017, LNCS 10421, pp. 427–432, 2017.
DOI: 10.1007/978-3-319-62932-2_41

As a result, these dependencies introduce certain constraints to the amount of transferred data – that is, in order to be sent to a remote device at a distance ranging between hundreds and thousands meters, data packets have to be considerably small. Given the extreme amounts of generated data, a potentially promising solution is to perform data filtering/aggregation – i.e. *data fusion* – as close to the source of data as possible, thus minimising the amount of 'noisy' data being sent over throughput-limited wireless links. However, this is not always possible due to the lack of computing resources on embedded systems and constrained devices. Therefore, a solution, able to meet local resource restrictions, while reducing the overhead by keeping computation as close to data sources as possible, is required. To this end, this paper introduces Complex Event Processing (CEP) as a potential way of implementing sensor data fusion in distributed IoT systems, aiming to leverage local processing capabilities wherever possible, or off-load tasks to Edge/Cloud Computing otherwise – thereby paving the way for a multi-layered, hierarchical data fusion approach, aiming at reducing network latencies and amounts of transferred data.

2   Background and Related Work

A common solution adopted for data management and processing is to off-load related tasks to remote servers, typically located in data centers and cloud platforms, which collect, store and process data, thus promoting a convergence between the IoT and the cloud paradigms [2]. Indeed, the IoT has primarily adopted a 'vertical' offloading paradigm, in which raw sensor data are collected by edge devices and transferred over the network to a central processing location (e.g. an IoT-Cloud platform) via several network links, such as network gateways and routers. This inevitably implies that raw sensor data are sent out immediately upon generation, thus putting strict dependency on the underlying network bandwidth. Likely, this requirement appears not affordable or unsustainable due to restrictions of wireless networks and related (3G/4G) providers, which impacts on the overall processing latency. This way, (resource-constrained) edge devices are not expected to perform data processing themselves, but rather to push collected data through the network topology, albeit they may already have enough processing, storage and computing resources on-board, which can be exploited in filtering, pre-processing, local analytics, sensor fusion and similar activities on nodes with great benefits in terms of performance and network overhead.

Indeed, network devices as well as communication and processing units, such as Mobile Edge Cloud (MEC) servers, are usually quite powerful and therefore can support this computation. This pattern has been recently proposed in the context of the Edge/Fog computing paradigm, aiming at pushing intelligence towards the edge [4]. In Edge/Fog computing, network devices, switches, routers, servers, or even 'cloudlet' machines, are widely used to support computational tasks, incoming from IoT edge nodes. This complements the traditional offloading to the cloud and overcomes bandwidth constraints, while mitigating latency and delays. Therefore, a solution able to minimise costs and latency, while taking into account edge devices' resource constraints, is possible by exploiting and properly combining local resources with those provided by the network (i.e. Edge computing) and cloud nodes.

Despite the increasing processing capabilities, edge devices are typically not yet equipped with full-featured hard disks to store large data sets. In this light, it naturally follows that edge devices are more suited for in-memory Stream Processing – i.e. data processing, which does not write data to local mass storage, but rather keeps all the computation in memory, thus potentially achieving better performance. Complex Event Processing (CEP) – one of the existing approaches to Stream Processing – goes beyond simple data querying and aims to detect complex event patterns, themselves consisting of simpler atomic events, within a data stream. Accordingly, from CEP's point of view, constantly arriving tuples can be seen as notifications of events happening in the external world – e.g. a fire alarm signal, social status update, a stock exchange fluctuation, etc. The focus of this perspective is on detecting occurrences of particular patterns of lower-level events that represent higher-level events. A standing query fetches results, if and only if a corresponding pattern of lower-level events is detected. For example, a common CEP task is to detect situation patterns, where one atomic event happened after another in time. To achieve this functionality, CEP systems rely on event timestamps; they extend continuous query languages with sequential operators, which allow specifying the chronological order of events.

CEP's capabilities to enable data fusion over incoming streams have been utilised in IoT scenarios, for example to enable run-time monitoring and data fusion [1,3,5,6]. There are two main aspects, however, which seem to be not addressed in the existing works. First, existing approaches tend to implement CEP only at the highest level of cloud computing, thus neglecting the possibility of introducing intermediate data fusion at the levels of networking and edge devices. Second, there is little evidence of the bi-directional communication – that is, existing approaches only focus on data collection, and do not consider coordination of lower-level devices by modifying data fusion policies in a top-down manner.

3 Proposed Solution

In the presence of increasing processing capabilities of edge devices, it is natural to think on how to exploit this untapped potential to support IoT data fusion tasks. To this end, given different types of devices and their location within an IoT network topology, as well as the CEP technology, this paper proposes a hierarchical multi-level architecture for data fusion in IoT systems. According to the proposed approach, data should be firstly processed on-board (i.e. locally on IoT nodes) whenever possible, or pushed to network devices and services, and, finally, to the cloud in a hierarchical manner. The proposed architecture includes three conceptual levels, which are also aligned with geographical areas, from which sensor data are collected.

Local area data fusion (LADF) is supposed to take place on edge devices, which collect data, coming from embedded sensors. The amount of data is relatively small, and data fusion can be performed on-board.

Wide area data fusion (WADF) refers to performing more intensive data analytics, pushed from a wider network of edge devices. Following the principles of Edge Computing, WADF is performed on communication and processing units.

Global area data fusion (GADF) refers to the highest level of data fusion, which provides a global view on the whole managed system of edge devices and networking nodes. This involves processing of large amounts of data, and therefore is expected to be implemented in a data center or a cloud platform.

As a result, the described high-level conceptual data fusion architecture is realised through deploying and running instances of CEP middleware on devices, constituting multi-level IoT systems. The CEP middleware implements main two functions – namely, (i) the actual data fusion via event correlation across multiple data sources, and (ii) communication with devices, located at lower levels of the IoT network topology, and orchestration of the data processing tasks distributed across lower-level nodes. The reference architecture in Fig. 1 depicts this conceptual separation of concerns – all three levels are equipped with dedicated CEP engine instances (i.e. Local Area CEP – LACEP, Wide Area CEP – WACEP, and Global Area CEP – GACEP), whereas the upper two levels also include coordination components, responsible for bi-directional communication between lower- and higher-level nodes, while coordinating and managing all the offloading requests incoming from these bottom layer nodes. It is worth remarking that the presence of one of the two upper layers is optional, since a LACEP Engine can directly interact with a WACEP Coordinator, or a LACEP Coordinator can assign requests only to WACEP and LACEP engines.

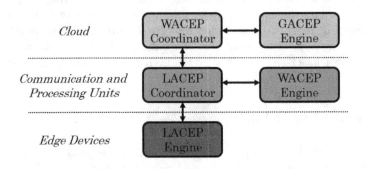

Fig. 1. Reference architecture of the multi-level hierarchical CEP data fusion.

4 Towards a Proof of Concept

As a first step towards validating the proposed hierarchical data fusion approach, an initial proof-of-concept prototype was implemented. The prototype utilises Drools Fusion[1] as the underlying CEP middleware. Being just one of the pluggable modules of the larger modular platform Drools, Fusion is a light-weight open-source CEP implementation, which combines CEP expressiveness (i.e. temporal reasoning over events and sliding windows of interest), relatively low resource requirements, and well-maintained client libraries. As a result, the Fusion middleware was deployed at three levels of the hierarchical architecture (Fig. 2).

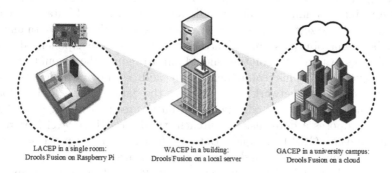

LACEP in a single room: WACEP in a building: GACEP in a university campus:
Drools Fusion on Raspberry Pi   Drools Fusion on a local server   Drools Fusion on a cloud

Fig. 2. Prototype implementation of the multi-level hierarchical data fusion in a university campus.

The low-level smart objects, equipped with multiple sensing devices (i.e. temperature, humidity, light sensors and motion detectors) are represented by Raspberry Pi boards. These boards are responsible for collecting relevant raw sensor data from a local area (i.e. a room) and perform CEP – that is, data fusion takes place as close to the source of data as possible, such that only filtered/aggregated values are transferred to an upper-level processing node. The goal of this LACEP is to detect if someone is still indoors, or if electrical appliances (e.g. heaters, A/C, lights) are still on, and to send a corresponding alert.

The middle-level communication/processing units are represented by a server, responsible for collecting data from multiple Raspberry Pi boards and perform data fusion over a larger area of interest. This way, WACEP provides a more extensive perspective on the managed area (i.e. a university building). By being notified of rooms, in which someone is still working, the system can identify empty floors in the building and activate the alarm on those floors. Same way, the system can detect a potentially dangerous situation (e.g. fire or burglary) and report on this.

The highest-level processing location is represented by the Amazon Web Services[2] cloud, which is responsible for GACEP – i.e. collecting data from all the managed buildings and performing data fusion over the whole university campus area. This way, the Cloud platform is able to provide a global real-time view on

---

[1] http://drools.jboss.org/drools-fusion.html
[2] https://aws.amazon.com/

the campus, including, for example, an emergency situation which is spreading and already involves several buildings.

5   Conclusions

The number of connected devices has already exceeded 10 billions, and is expected to reach hundreds of billions in the next 4–5 years. Equipped with processing, storage, networking, sensing and actuation capabilities, such devices are able to intercommunicate and handle mega-/gigabytes of data, and collectively approach the zettabyte frontier on a daily basis. In this light, data management is becoming of utmost importance – i.e. to support the exponential growth of IoT devices, a potential solution should aim to reduce the amount of redundant data exchanged. This requirement can be addressed by pushing intelligence to the edge of an IoT network – i.e. closer to the data source – and exploiting resources and capabilities of edge nodes for data (pre-)processing. Accordingly, this paper proposes a data fusion approach based on a three-level hierarchical architecture using CEP techniques. Data are first sensed and processed locally on-board (i.e. sensor fusion) and then, if required, further processed by higher-level (wide and/or global) data fusion engines deployed on a server and/or a cloud platform. This way, a scalable architecture is achieved, in which local computational resources, if insufficient, are extended by the higher levels. The main idea and architecture of the proposed solution are described in this paper, while further implementation and experimentation on a real use case are ongoing.

References

1. Brunelli, D., Gallo, G., Benini, L.: Sensormind: virtual sensing and complex event detection for internet of things. In: Gloria, A. (ed.) ApplePies 2016. LNEE, vol. 409, pp. 75–83. Springer, Cham (2017). doi:10.1007/978-3-319-47913-2_10
2. Díaz, M., Martín, C., Rubio, B.: State-of-the-art, challenges, and open issues in the integration of Internet of things and cloud computing. J. Netw. Comput. Appl. **67**, 99–117 (2016)
3. Fonseca, J., Ferraz, C., Gama, K.: A policy-based coordination architecture for distributed complex event processing in the internet of things: doctoral symposium. In: Proceedings of the 10th ACM International Conference on Distributed and Event-Based Systems, pp. 418–421. ACM (2016)
4. Garcia Lopez, P., Montresor, A., Epema, D., Datta, A., Higashino, T., Iamnitchi, A., Barcellos, M., Felber, P., Riviere, E.: Edge-centric computing: vision and challenges. SIGCOMM Comput. Commun. Rev. **45**(5), 37–42 (2015)
5. Guo, Q., Huang, J.: A complex event processing based approach of multi-sensor data fusion in IoT sensing systems. In: 2015 4th International Conference on Computer Science and Network Technology (ICCSNT), vol. 1, pp. 548–551. IEEE (2015)
6. Wang, Y., Cao, K.: A proactive complex event processing method for large-scale transportation internet of things. Int. J. Distrib. Sens. Netw. **10**(3), 159052 (2014)

Scalable Computations of GeRa Code on the Base of Software Platform INMOST

Igor Konshin[1,2(✉)] and Ivan Kapyrin[1,2]

[1] Institute of Numerical Mathematics of the Russian Academy of Sciences,
Moscow 119333, Russia
igor.konshin@gmail.com
[2] Nuclear Safety Institute of the Russian Academy of Sciences,
Moscow 115191, Russia

Abstract. The hydrogeological modeling code GeRa is based on INMOST software platform, which operates with distributed mesh data and allows to assemble and solve the system of linear equations. The set of groundwater flow models with filtration, transport, and chemical processes are considered. The comparison of parallel efficiency for different linear solvers in the INMOST framework is performed. The analysis of scalability of GeRa code on different computer platforms from multicore laptop to Lomonosov supercomputer is presented.

Keywords: Numerical modelling · Software platform · Distributed meshes · Subsurface flow and transport

1 Introduction

At present the problem of safe radioactive waste (RW) disposal is of great interest for the countries utilizing nuclear energy and radionuclides in their national economy. Along with relatively successful practice of low level waste disposal in surface repositories the advances in creation of national high-level waste (HLW) disposals are moderate. No country except Finland issued a license for the construction of such an object while the national programs for the creation of HLW deep geological disposals had been conducted for several decades. The reason for that is the complexity of safety assessment problem stemming from extremely large time and space scales, large variety of coupled processes and uncertainties.

Hydrogeological modeling codes able to model the groundwater flow and transport processes are the basis for disposal safety assessment. By order of the State Atomic Energy Corporation ROSATOM Nuclear Safety Institute and Institute of Numerical Mathematics of the Russian Academy of Sciences develop the GeRa numerical code designed for the solution of a broad class of surface and deep geological RW disposals safety assessment problems. GeRa features the application of 3D unstructured adaptive grids, initially established means of parallelization and integral modeling approach. The latter means that the code shall allow to solve the problem as a whole, starting from geological model

© Springer International Publishing AG 2017
V. Malyshkin (Ed.): PaCT 2017, LNCS 10421, pp. 433–445, 2017.
DOI: 10.1007/978-3-319-62932-2_42

generation and ending with doses for the population calculations with the proper uncertainty analysis. At present the following major processes can be modeled in GeRa:

- groundwater flow in confined, unconfined and unsaturated conditions;
- transport in uniform and dual-porosity media (advection, dispersion, diffusion);
- equilibrium chemical reactions either governed by sorption isotherms or with real chemical calculations;
- radioactive decay chains;
- heat generation caused by radioactive decay;
- density and temperature driven convection.

The discretizations of GeRa are based on finite volume (FV) method. Besides the conventional two-point (TPFA) and multi-point flux approximation (MPFA) schemes [1] a nonlinear monotone FV method [2,3] is applied for the diffusion operator approximation. The advection operator may be discretized either using TVD-schemes with limiters or an upwind first order accurate scheme. The discretizations are aimed at use on polyhedral conformal grids. Two grid generators were implemented in GeRa [4]. The first one is the generator of triangular-prismatic grids, the second is a hexahedral grid generator based on octree structures with the ability of cell cutting. The geochemical module iPhreeqc [5] is used for chemical reactions calculation (see [6] for an example).

The INMOST [7,8] software platform is used in GeRa to support the distributed mesh and data storage and operations as well as assembling and solution of linear systems. In this work we analyze for the first time the parallel efficiency of GeRa on several computer architectures from multicore laptop to clusters and supercomputers.

The article is organized as follows. In Sect. 1 a brief overview of INMOST platform based GeRa code is given; in Sect. 2 the test problems are defined; the available linear solvers are described in Sect. 3; the results of numerical experiments are presented in Sect. 4; while the conclusions are given in Sect. 5.

2   Model Problems Description

A set of model problems common for hydrogeological modeling was chosen for numerical experiments. Different physical and chemical processes are taken into account and meshes of different sizes are generated in these tests. In the following when solving problems on a series of refined grids the following notation is used: the letter is the first letter in the name of the test ("g" – "geos", "c" – "chemistry", "t" – "transport") and the following it digits denote the number of mesh cells measured in thousands. For example model "t5740" denotes the "transport" model with 5740 thousand cells in the computational grid.

The model "f262" is a steady groundwater flow problem in a rectangular domain $[0;1] \times [0;1] \times [0;0.1]$. A regular rectangular grid containing $128 \times 128 \times 16$ cells (approximately 262 thousand cells) were used.

The "geos" set contain the groundwater flow models that are solved in a real-life domain with heterogeneous parameters. Three geological layers are present in the model. The top and bottom layers are aquitards (hydraulic conductivity $K = 0.001$ m/day), the middle layer is an aquifer with hydraulic conductivity $K = 1$ m/day. The coarsest grid contains approximately 28 thousand cells, the major part of these are triangular prisms, but also there are 102 tetrahedra and 108 pyramids caused by the top layer pinch-out. A stationary saturated groundwater flow problem is solved using the MPFA scheme. The series contains "g28", "g185", "g402", "g1425" and "g5740" models.

In the "chemistry" set of tests a reactive transport advection problem is solved. A full description of the problem is available in [6]. Five wells with balanced rates are working in a uniform layer $200 \times 200 \times 10$ meters in size: four of them being injection wells located in the corners of the domain; the fifth being a production well in the middle. The chemical calculations are done using iPhreeqc [5]. Hexahedral octree-based grids with local refinement to well screens are used. The series contains five tests: "c18", "c60", "c254", "c700", "c1120".

In the "transport" set of tests one dimensional advective transport along the X-axis is modeled in a rectangular domain $[0; 1000] \times [0; 100] \times [0; 1]$. The discretization is done using the TVD scheme which implies local optimization problem solution on each mesh cell in the process of concentration gradient cellwise reconstruction. Triangular prismatic meshes are used. Five models with different grid sized are in the set: "t20", "t70", "t377", "t1215", "t5740".

3 Linear System Solvers Available in INMOST

Except for distributed mesh operations, INMOST software platform provides a user the interface to collect the coefficient matrix and the right-hand side of the discretization linear system and than to solve it. The main feature of this interface is the handling to the matrix row/column indices i and j by its global value just as for the dense one. It gives an opportunity for the problem discretization to simplify the collection of the coefficient matrix.

INMOST provides a common interface to linear solvers: both set of inner solvers and the third party (PETSc, Trilinos, SuperLU, Ani3D) ones. The most of inner solvers are based on advanced second order incomplete triangular factorization ILU2(τ) or in other words two-threshold ILU2(τ_1, τ_2) factorization [9]. In our experiments we used the theoretically approved values $\tau_1 = \tau$ and $\tau_2 = \tau^2$. It should be noted that the partial case $\tau_1 = \tau_2 = \tau$ is reduced to the conventional one-threshold ILU(τ) factorization.

The preconditioner parallelization is based on either Additive Schwarz AS(q) or BIILU(q) scheme with the overlap size parameter q. The geometrical interpretation of this parameter is the number of layers in the subdomains overlap.

In the present paper we consider two inner linear solvers InnerILU2 and BIILU2 ones based on ILU2(τ) with AS(q) and BIILU(q) preconditioning, respectively, as well as the conventional linear solver from PETSc package [10] based on structural factorization ILU(k) and AS(q) preconditioning. All the considered linear solvers are accelerated by BiCGStab iterations.

436   I. Konshin and I. Kapyrin

The BIILU2 linear solver is of the special care in our team: the first time it was presented [11] namely at the PaCT conference in 1999. It was a symmetric version with the second order incomplete Cholesky factorization $IC2(\tau)$ and block incomplete inverse Cholesky $BIIC(q)$ as a parallel scheme with Preconditioned Conjugate Gradient (PCG) iterations. Next, at PaCT-2009 [12] the version with post filtration of triangular factors were presented.

4 Numerical Experiments

4.1 The Parallel Computer Platforms Available

In the present paper we performed the comparative analysis of parallel run properties for the developed GeRa code. The following parallel computer platforms were used:

- quad-core laptop Intel i7-4810MQ (2.80 GHz) with 16 GB RAM under Ubuntu 16.04.1 using compiler gcc v.5.4.0 and mpicc for MPICH v.3.2.
- INM RAS cluster [13] consisting of nodes with two six-core Intel Xeon X5650 (2.67 GHz) and 24 GB RAM per node under SUSE Linux Enterprise Server 11 SP1 (x86_64) using compiler Intel C v.4.0.1 and Intel MPI v.5.0.3.
- "test" and "regular4" partitions of "Lomonosov" supercomputer [14] located in the Moscow State University consisting of nodes with quad-core Intel Xeon E5-2697 v3 (2.60 GHz) and 12 GB RAM per node.

When performing numerical experiments, we analyzed the discretization stage time $T = T(p)$ obtained on one of the mentioned platform using p cores. The relative speedup $S = T(1)/T(p)$ and the resulting computational efficiency $E = S/p$ were calculated as well. The PETSc linear solver BiCGStab+AS(q)+ILU(k) with the parameters $q = 1$ and $k = 1$ was used as a default one.

4.2 Numerical Experiment on a Multicore Laptop

Table 1 presents the computation time $T(p)$ using p cores for some described above models. Numerical experiments performed on quad-core laptop showed a fairly good monotonic acceleration of computation time (up to about 3-fold) with increasing number of used cores to 4. This may allow to a GeRa user to accelerate the calculations carried out even on the local PC without exploiting an external computing cluster or in the case when the cluster is unavailable.

The data for calculation for "t70" model on 8 threads are missing in the table due to RAM restrictions. It should be noted that the use of hyper-threading technology with 8 threads allows to actually reduce the computation time for the models considered.

Table 1. Computation times (in sec.) for some models on quad-core laptop.

p	"c60"	"d224"	"g1425"	"t70"
1	320.485	4.339460	59.388	11.406
2	242.605	3.217543	35.801	7.762
4	150.606	2.927468	30.258	4.455
8	119.705	2.586953	28.005	—

4.3 Preliminary Experiments on INM RAS Cluster

In Table 2 we compare the performance of different linear solvers of INMOST software platform when solving "f262" model using $p = 64$ cores of INM RAS cluster. Besides the already mentioned default linear solver PETSc (with parameters $q = 1$, $k = 1$) we used the inner INMOST solvers InnerILU2 (with parameters $q = 1$, $\tau = 0.005$) and solver BIILU2 (with parameters $q = 1$, $\tau = 0.03$). The default parameter settings were used in all linear solvers.

The analysis of the results in Table 2 shows the most efficient use of the linear solver BIILU2, which gives a reason for a more detailed study of its properties.

Table 2. Computation times for model "f262" on 64 cores of INM RAS cluster for different linear solvers.

	T_{discr}
InnerILU2 ($q = 1$, $\tau = 0.005$)	2.749
PETSc ($q = 1$, $k = 1$)	2.230
BIILU2 ($q = 1$, $\tau = 0.03$)	1.371

In the next experiment we analyze the influence of the parameters choice to the BIILU2 solver performance behavior. The experiment was performed on INM RAS cluster using 64 cores. The results of parameters tuning are presented on Fig. 1. The tuning of threshold parameter τ (for $q = 2$) is presented on Fig. 1a. One can observe the very smooth behavior of solution time in wide range of the parameter τ from $5 \cdot 10^{-3}$ to 10^{-6}. Next, on Fig. 1b the tuning of overlap parameter q ones again demonstrates the smooth behavior of solution time depending on q. The most important conclusion is the crucial importance of the overlap usage ($q > 0$) as well as very stable behavior up to overlap $q = 5$. The optimal values for the considered model "t1215" for $p = 64$ are $q = 3$ and $\tau = 0.001$, which are more strict ones than the usually exploited default set $q = 1$ and $\tau = 0.03$. It means that the considered model "t1215" is a more difficult to solve among the other ones.

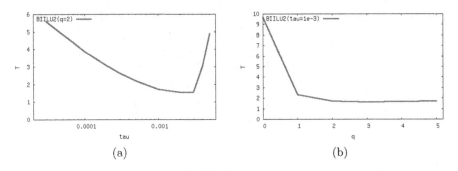

Fig. 1. Tuning of (a) threshold parameter τ (for $q = 2$) and (b) overlap parameter q (for $\tau = 0.001$) of the BIILU2 linear solver for "t1215" model on INM RAS cluster.

4.4   Solution of Chemical Models

In the present subsection we consider the numerical results obtained on different computer platforms: INM RAS cluster and the "Lomonosov" supercomputer specified in Sect. 4.1. Both platform have about the same scalar performance, but communication rate of "Lomonosov" is appreciably higher to provide a possibility of efficient parallel computations on several thousand cores.

The number of used cores on both platforms varied from 1 to 128. Let us consider the solution of some of the model series on these two computer platforms in more detail.

Table 3 presents the speedup with respect to the serial run for the set "c" models on the INM RAS cluster using the solver PETSc($q = 3, k = 3$). The table above shows that with increasing number of cores there is a significant acceleration in computation time, and in most cases with an increase in the dimension of the problem the obtained speedup is growing. The latter can be seen most clearly in Fig. 2a. This effect is associated with a decrease in the portion of communications regarding that of calculations when the size of the local subproblem is increasing.

The maximum speedup obtained on 128 cores is 56.17, which means about 50% of parallel efficiency reached.

Table 4 presents the speedup with respect to the serial run for the set "c" models on "Lomonosov" supercomputer using the same solver PETSc($q = 3, k = 3$). The table above shows the acceleration in computation time is even better than for INM RAS (see Fig. 2b). This fact is in agreement with the above remarks on communication rate. The maximum speedup for "Lomonosov" supercomputer obtained on 128 cores is 80.28, which means more than 60% of parallel efficiency reached.

It should be noted the growth in speedup for the model "c18" on 128 cores, which means the acceleration even for about 150 unknowns per core. Effective functioning with such a small dimension subproblems on one core means high efficiency parallel implementation of INMOST platform as well as the GeRa code itself.

Table 3. Speedups for the set "c" models on INM RAS cluster with PETSc($q = 3, k = 3$) for $p = 1, ..., 128$ cores.

p	"c18"	"c60"	"c254"	"c700"	"c1120"
1	1.00	1.00	1.00	1.00	1.00
2	1.84	1.86	1.82	1.82	1.88
4	3.36	3.66	3.54	3.64	3.69
8	5.08	5.81	6.13	6.25	6.41
16	8.39	8.17	9.89	9.62	10.45
32	13.35	14.05	17.59	18.04	17.75
64	20.29	23.98	30.75	29.91	31.64
128	20.15	38.85	51.18	53.13	56.17

Table 4. Speedups for the set "c" models on "Lomonosov" supercomputer with PETSc($q = 3, k = 3$) for $p = 1, ..., 128$ cores.

p	"c18"	"c60"	"c254"	"c700"	"c1120"
1	1.00	1.00	1.00	1.00	1.00
2	1.54	1.52	1.66	1.79	1.78
4	3.14	2.87	2.93	3.27	3.28
8	4.80	4.59	4.59	5.27	5.57
16	8.75	8.79	8.69	10.19	10.84
32	16.59	17.63	17.04	20.40	21.54
64	27.82	32.48	35.60	40.19	42.84
128	30.65	56.20	63.57	76.35	80.28

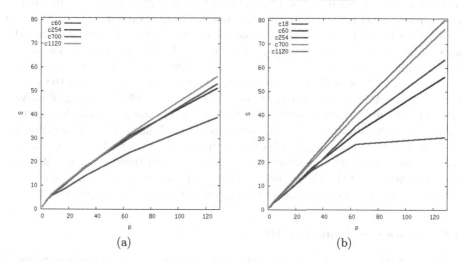

(a)   (b)

Fig. 2. Speedups for the set "c" models on both (a) INM RAS cluster and (b) "Lomonosov" supercomputer with the PETSc($q = 3, k = 3$) linear solver.

4.5   Solution of "geos" Models

As shown in Table 1 the "geos" problem of dimension more than 1 million of cells can be solved on a regular laptop. In the present section we continue to analyze the parallel performance for this set of models.

Table 5 presents the speedups with respect to serial run for some models from set "g" both on INM RAS cluster and "Lomonosov" supercomputer, respectively. The default PETSc($q = 3, k = 3$) solver was used for $p = 1, ..., 128$ cores.

Table 5. Speedups for two models of the set "g" on both INM RAS cluster and "Lomonosov" supercomputer by PETSc($q = 3, k = 3$) solver for $p = 1, ..., 128$ cores.

p	INM RAS cluster			"Lomonosov"	
	"g402"	"g1425"	"g5740"	"g402"	"g1425"
1	1.00	1.00	—	1.00	1.00
2	1.85	1.97	—	1.90	2.09
4	3.18	3.49	—	3.14	3.60
8	4.64	4.81	—	4.59	3.49
16	4.50	4.28	1.00	9.06	9.78
32	10.84	8.97	1.64	16.68	19.12
64	13.63	21.03	3.16	28.46	36.07
128	16.47	22.81	5.86	48.56	62.32

From the above data, it can be seen that for "g1425" model the monotonous increase of the speedup can be obtained up to 128 cores, besides the maximal speedups for INM RAS cluster and "Lomonosov" supercomputer are 22.81 and 62.32, respectively (see also Fig. 3). From the analysis of the above data it should also be noted that for the largest model "g5740" RAM limit on "Lomonosov" supercomputer does not allow to obtain the problem solution, as well as when using from 1 to 8 cores of INM RAS cluster. However, when using from 16 to 128 cores it is possible to obtain a solution, and even with sufficiently high relative speedup 5.86 for 128 cores with respect to run on 16 cores. On one hand, the latter shows the ability to solve the problem of over 5 million of computational cells, and of the other hand, it indicates the existence of problems for which the resources of personal computer are insufficient and there is a necessity for a parallel version of the GeRa code. The latter is not only due to increasing of computation efficiency, but namely the opportunities to solve the problem itself.

4.6   Solution of Transport Models

In dealing with transport models it is required to solve the problem for groundwater flow and than the respective transport problem. In GeRa two separate default

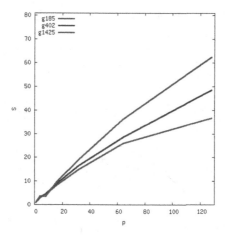

Fig. 3. Speedups for the medium size models from "g" set on "Lomonosov" supercomputer with the PETSc($q = 3, k = 3$) linear solver.

set of linear solver parameters used for these two problems: PETSc($q = 3, k = 3$) for flow problem and PETSc($q = 1$, $k = 1$) for the transport one.

When solving "t" set of models using the solver PETSc with the above default parameters, there was no convergence for the flow problem. By this reason, it was necessary to "enhance" the PETSc parameter for the flow equation up to values ($q = 7, k = 7$).

The comparative numerical results for two models from the set "t" on INM RAS cluster and "Lomonosov" supercomputer are shown in Table 6. As expected, the obtained results on "Lomonosov" supercomputer were much more scalable (as can be seen in Fig. 4).

Table 6. Speedups for two models of the set "t" on both INM RAS cluster and "Lomonosov" supercomputer by PETSc($q = 7, k = 7$) solver for $p = 1, ..., 128$ cores.

	INM RAS cluster		"Lomonosov"	
p	"t70"	"t377"	"t70"	"t377"
1	1.00	1.00	1.00	1.00
2	1.67	1.67	1.69	1.68
4	3.24	3.08	3.30	3.09
8	5.06	5.25	5.20	5.39
16	7.03	8.05	8.77	9.49
32	10.68	12.94	15.30	18.09
64	12.92	18.68	22.80	27.15
128	12.19	24.89	30.10	44.92

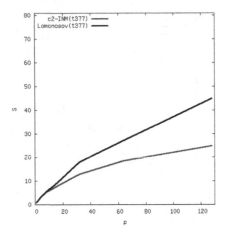

Fig. 4. Speedups for the model "t377" on INM RAS cluster and "Lomonosov" super-computer with the PETSc($q = 7, k = 7$) linear solver.

At the next step of the linear solvers properties study we used the BIILU2 solver with default set of parameters $q = 5$ and $\tau = 0.003$. In this case all the problems at once have been successfully solved.

The calculation results are shown in Tables 7, 8 and Fig. 5 for INM RAS cluster and "Lomonosov" supercomputer, respectively. As expected, on "Lomonosov" supercomputer, it was able to achieve more significant speedup. When using 128 cores for the model "t1215" it was attained the value of 78.62, that means the parallel efficiency more than 60%. It should be noted, that the time reduction for the smallest model "t20" were observed until the use of 64 cores, which is about 300 computational cells per one core.

Table 7. Speedups for the set "t" models on INM RAS cluster by BIILU2($q = 5, \tau = 0.003$) for $p = 1, ..., 128$ cores.

p	"t20"	"t70"	"t377"	"t1215"
1	1.00	1.00	1.00	1.00
2	1.98	1.84	1.81	1.79
4	3.74	3.46	3.33	3.08
8	6.98	6.77	5.34	5.20
16	4.43	7.82	9.09	6.47
32	6.65	12.62	14.64	17.26
64	5.00	15.13	26.98	30.06
128	1.05	10.12	40.50	54.64

Table 8. Speedups for the set "t" models on "Lomonosov" supercomputer by BIILU2($q = 5, \tau = 0.003$) for $p = 1, ..., 128$ cores.

p	"t20"	"t70"	"t377"	"t1215"
1	1.00	1.00	1.00	1.00
2	1.90	1.79	1.75	1.80
4	3.69	3.31	3.29	3.23
8	7.01	6.65	5.75	5.23
16	13.02	12.91	11.64	10.18
32	22.65	24.35	24.17	24.11
64	27.35	42.70	45.41	45.78
128	20.13	50.21	76.63	78.62

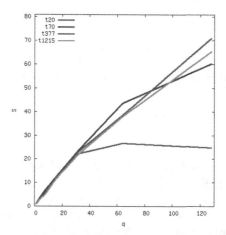

Fig. 5. Speedups for the set "t" models on "Lomonosov" supercomputer for the BIILU2($q = 3, \tau = 0.001$) linear solver.

4.7 The Results for the Largest Models

As a final illustration of the achieved parallel efficiency for the model problems, Fig. 6 shows the plots of the speedups of numerical experiments on "Lomonosov" supercomputer for the largest size models using a variety of INMOST linear solvers. From these plots, one can observe a sufficiently high parallel efficiency, which in most of the considered runs amounted more than 50 %.

To conclude this section, it should be noted the high reliability and efficiency of developed parallel linear solver BIILU2 from INMOST software platform. With BIILU2 it was able to solve all without an exception linear system for the considered models as well as for the above examples to get solutions in less time than a linear solver PETSc.

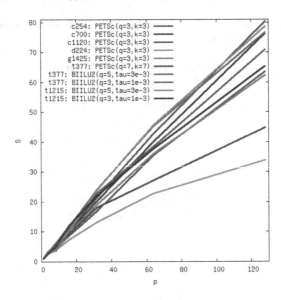

Fig. 6. Speedups for the large size models from all sets on "Lomonosov" supercomputer.

5   Conclusions

- A speedup up to 3 for GeRa code on regular quad-core laptop can be achieved for sufficiently large size models.
- The nonoverlapping block Jacobi preconditioning can be used for the most simple models only, the more complicated problems require the use of overlapping preconditioning such as AS(q) or BIILU(q).
- A conventional AS(q)+ILU(k) preconditioning from PETSc is unable to obtain the solution for the most hard-to-solve linear systems, the usage of advanced BIILU2(q, τ) preconditioning is required.
- The communication rate of the computer is of importance; Having about the same scalar performance as INM RAS cluster, the Lomonosov supercomputer takes less time for parallel runs.
- The speedup 80 on 128 cores can be achieved of Lomonosov supercomputer for sufficiently large size models.
- A speedup can be obtained even for sufficiently small size models with up to 150 unknowns per each of 128 cores.
- Ranging the models types by degree of scalability, it should be noted that the worse scalable physical process is diffusion, the transport is the sufficiently good one, while the chemical processes are the best scalable due the large amount of independent work in each mesh cell.
- Both the distributed mesh operations and linear solvers from INMOST demonstrate hight reliability and parallel efficiency in the framework of groundwater flow modelling by GeRa code.

Acknowledgements. The authors express their gratitude to V. Kramarenko for his permanent assistance with assembling the GeRa code, for installing GeRa on different computer platforms, and for generating the model sets from the GeRa verification tests.

References

1. Aavatsmark, I., Barkve, T., Boe, O., Mannseth, T.: Discretization on unstructured grids for inhomogeneous, anizotropic media. Part I: derivation of the methods. SIAM J. Sci. Comput. **19**(5), 1700–1716 (1998)
2. Danilov, A., Vassilevski, Y.: A monotone nonlinear finite volume method for diffusion equations on conformal polyhedral meshes. Russ. J. Numer. Anal. Math. Model. **24**(3), 207–227 (2009)
3. Kapyrin, I., Nikitin, K., Terekhov, K., Vassilevski, Y.: Nonlinear monotone FV schemes for radionuclide geomigration and multiphase flow models. In: Fuhrmann, J., Ohlberger, M., Rohde, C. (eds.) Finite Volumes for Complex Applications VII-Elliptic, Parabolic and Hyperbolic Problems. Springer Proceedings in Mathematics & Statistics, vol. 78, pp. 655–663. Springer, Berlin (2014)
4. Plenkin, A.V., Chernyshenko, A.Y., Chugunov, V.N., Kapyrin, I.V.: Adaptive unstructured mesh generation methods for hydrogeological problems. Vychisl. Metody Program. **16**(4), 518–533 (2015)
5. Charlton, S.R., Parkhurst, D.L.: Modules based on the geochemical model PHREEQC for use in scripting and programming languages. Comput. Geosci. **37**(10), 1653–1663 (2011)
6. Boldyrev, K.A., Kapyrin, I.V., Konstantinova, L.I., Zakharova, E.V.: Simulation of strontium sorption onto rocks at high concentrations of sodium nitrate in the solution. Radiochemistry **58**(3), 243–251 (2016)
7. INMOST: a toolkit for distributed mathematical modelling. URL: http://www.inmost.org
8. Vassilevski, Y.V., Konshin, I.N., Kopytov, G.V., Terekhov, K.M.: INMOST - a software platform and a graphical environment for development of parallel numerical models on general meshes, 144 pp. Moscow State University Publication, Moscow (2013) (in Russian)
9. Kaporin, I.E.: High quality preconditioning of a general symmetric positive definite matrix based on its $U^T U + U^T R + R^T U$-decomposition. Numer. Lin. Alg. Applic. **5**(6), 483–509 (1998)
10. PETSc - Portable, Extensible Toolkit for Scientific Computation. https://www.mcs.anl.gov/petsc
11. Kaporin, I.E., Konshin, I.N.: Parallel solution of large sparse SPD linear systems based on overlapping domain decomposition. In: Malyshkin, V. (ed.) PaCT 1999. LNCS, vol. 1662, pp. 436–446. Springer, Heidelberg (1999). doi:10.1007/3-540-48387-X_45
12. Kaporin, I.E., Konshin, I.N.: Load balancing of parallel block overlapped incomplete cholesky preconditioning. In: Malyshkin, V. (ed.) PaCT 2009. LNCS, vol. 5698, pp. 304–315. Springer, Heidelberg (2009). doi:10.1007/978-3-642-03275-2_30
13. INM RAS cluster. http://cluster2.inm.ras.ru (in Russian)
14. "Lomonosov" supercomputer. https://parallel.ru/cluster/lomonosov.html (in Russian)

Parallel Computing for Time-Consuming Multicriterial Optimization Problems

Victor Gergel$^{(\boxtimes)}$ and Evgeny Kozinov

Lobachevsky State University of Nizhny Novgorod, Nizhny Novgorod, Russia
gergel@unn.ru, evgeny.kozinov@itmm.unn.ru

Abstract. In the present paper, an efficient method for parallel solving the time-consuming multicriterial optimization problems, where the optimality criteria can be multiextremal, and the computation of the criteria values can require a large amount of computations, is proposed. The proposed scheme of parallel computations allows obtaining several efficient decisions of a multicriterial problem. During performing the computations, the maximum use of the search information is provided. The results of the numerical experiments have demonstrated such an approach to allow reducing the computational costs of solving the multicriterial optimization problems essentially – several tens and hundred times.

Keywords: Decision making · Multicriterial optimization · Parallel computing · Dimensionality reduction · Criteria convolution · Algorithm of global search · Computation complexity

1 Introduction

The statements of the multicriterial optimization (MCO) problems belong to the most general models of the decision making problems. A general state of the art in the field of multicriterial optimization is presented in the monographs [2–4, 19], the reviews of the scientific and practical results are given in [1, 5–8, 20].

At the same time, the MCO problems are the most complicated ones. As a rule, the solving of the MCO problems is reduced to finding some compromised (efficient) decisions, where the best values obtained with respect to particular partial criteria are coordinated with each other.

The necessity to find several efficient decisions increases the computational complexity of solving the MCO problems. In such conditions, finding even a single efficient solution requires a large amount of computations whereas the finding of several efficient decisions (or of the whole Pareto set) becomes a problem of high computation complexity. Addressing this problem becomes possible by using the huge computational capabilities of high-performance systems. In addition, full utilization of the search information obtained in the course of computations is necessary for efficient computations.

Further structure of the paper is as follows. In Sect. 2, the multicriterial optimization problem statement is given. In Sect. 3, a scheme of parallel computations for the simultaneous solving of a set of multicriterial global optimization problems is

© Springer International Publishing AG 2017
V. Malyshkin (Ed.): PaCT 2017, LNCS 10421, pp. 446–458, 2017.
DOI: 10.1007/978-3-319-62932-2_43

proposed. In Sect. 4, a parallel algorithm of multicriterial global search is presented. Section 5 describes the results of numerical experiments. In Conclusion, the obtained results are discussed and main directions of further investigations are outlined.

2 Multicriterial Optimization Problem Statement

The problem of multicriterial (or multi-objective) optimization (MCO) can be defined as follows:

$$f(y) = (f_1(y), f_2(y), \ldots, f_s(y)) \to min, y \in D, \tag{1}$$

where $y = (y_1, y_2, \ldots, y_N)$ is a vector of varied parameters, N is the dimensionality of the problem being solved, and D is the search domain being an N-dimensional hyperparallelepiped

$$D = \{y \in R^N : a_i \le y_i \le b_i, 1 \le i \le N\}$$

with given boundary vectors a and b. Without a loss in generality, the partial criteria values in the problem (1) are supposed to be non-negative, and the decrease of these ones corresponds to the increase of the efficiency of the considered solutions $y \in D$. It is also supposed that the partial criteria $f_i(y)$, $1 \le i \le s$ satisfy the Lipschitz condition

$$|f_i(y') - f_i(y'')| \le L_i ||y' - y''||, y', y'' \in D, 1 \le i \le s \tag{2}$$

where L_i, $1 \le i \le s$ are the Lipschitz constants for the corresponding partial criteria $f_i(y)$, $1 \le i \le s$, and $||\cdot||$ denotes the norm in the R^N space.

As a solution of the MCO problem, any efficient decision (*a partial solution*) is considered. In the general case, it is required to find the whole set of Pareto-optimal solutions $PD(f, D)$ (*a full solution of a MCO problem*).

In the present work, the MCO problems are considered in the context of the most complicated problems of decision making, where the partial criteria $f_i(y)$, $1 \le i \le s$ can be multiextremal, and the obtaining of the criteria values at the points of the search domain $y \in D$ can require a considerable amount of computations. In the paper, it is proposed to use the minimax convolution of the partial criteria, according to which the solving of the problem (1) is reduced to the solving of a family of the global optimization problems:

$$min\Big\{F(\lambda, y) = max(\lambda_i f_i(y), 1 \le i \le s) + \rho \sum_{i=1}^{s} (\lambda_i f_i(y))\Big\},$$
$$\lambda \in \Lambda \subset R^s : \sum_{i=1}^{s} \lambda_i = 1, \lambda_i \ge 0, 1 \le i \le s \tag{3}$$

where $\rho > 0$ is a small positive number (the last term in the expression for $F(\lambda, y)$ allows obtaining the Pareto optimal solutions at the appropriate value of the coefficient ρ – see, for example, [22, 23]). It should be noted that the scalar criterion $F(\lambda, y)$ also satisfies the Lipschitz condition

$$|F(\lambda,y') - F(\lambda,y'')| \le L||y' - y''||, y',y'' \in D. \qquad (4)$$

3 Parallel Computations for Solving the Multicriterial Global Optimization Problems

The scalarization of the vector criterion allows reducing the solving of the MCO problem (1) to solving a series of the multiextremal problems (3). And, therefore, the problem of development of the methods for solving the MCO problems is resolved by the possibility of a wide use of the efficient parallel global optimization algorithms.

The proposed approach of parallel computations for solving time-consuming global optimization problems is based on the following main statements:

- The parallelism of the performed computations is provided by means of simultaneous computing the values of the partial criteria $f_i(y)$, $1 \le i \le s$ at several different points of the search domain D. Such an approach provides the parallelization of the most computation costly part of the global optimization and is a general one – it can be applied for many global optimization methods for various global optimization problems.
- In addition, the parallel computations are provided by means of the simultaneous solving of several global optimization problems (3) for various values of the coefficients λ_i, $1 \le i \le s$. For solving the problems of the family (3), a set of computational nodes of the high performance systems with distributed memory can be applied.
- The search information obtained in the course of parallel computations is exchanged between all employed processors because of the information compatibility of the global optimization problems of the family (3).

Below, these statements will be considered in more details.

3.1 Structure of the Global Search Information

For solving the optimization problems, the values of the partial criteria $f^i = f(y^i)$ at the points y^i, $1 \le i \le k$ of the search domain D are computed.

The search information obtained as a result of these computations can be represented in the form of the *Search Information Set* (SIS):

$$\Omega_k = \left\{ \left(y^i, f^i = f(y^i) \right)^T : 1 \le i \le k \right\}. \qquad (5)$$

The availability of SIS allows transforming the results of the previous computations to the values of any next optimization problem (3) being solved without any time-consuming computations of the partial criteria values $f_i(y)$, $1 \leq i \leq s$ from (1) for new values of the convolution coefficients λ'_j, $1 \leq j \leq s$ i.e.

$$z'_i = max\left(\lambda'_j f^j_i, 1 \leq j \leq s\right), 1 \leq i \leq k. \tag{6}$$

As a result, all the search information can be employed for continuing the computations. In general, the reuse of the search information will provide less and less computations for solving every next optimization problem (see Sect. 5).

3.2   General Scheme of Parallel Computations

As it has been mentioned above, when solving a multicriterial optimization problem (1), in order to find several different efficient decisions, solving a series of scalar problems (3) with various values of the coefficients of the minimax convolution of the partial criteria may be required

$$\overrightarrow{\Phi}(y) = \{\varphi_1(y), \dots, \varphi_q(y)\}, \varphi_i(y) = F(\lambda_i, y), 1 \leq l \leq q. \tag{7}$$

The problems of the family $\overrightarrow{\Phi}(y)$ can be solved sequentially, various global optimization methods can be used for solving the problems. On the other hand, these problems can be solved simultaneously with the use of several processors as well. It is important to note that the obtained family of the one-dimensional problems $\overrightarrow{\Phi}(y)$ is an information-linked one – the values of the optimized functions computed for any problem $\varphi_l(y)$, $1 \leq l \leq q$ can be transformed to the values of all the rest problems of the family without the time-consuming recalculations of the partial criteria values $f_i(y)$, $1 \leq i \leq s$ according to (6).

The information compatibility of the problems from the family (7) allows proposing the following scheme of parallel computations. Solving each particular problem can be performed on a separate processor of the computational system; the exchange of the obtained search information between the processors should be performed in the course of computations.

As a result, a unified approach to the parallel computations for the multiprocessor computational systems with the distributed memory can be proposed in the following way.

1. The family of the one-dimensional reduced information-linked problems $\overrightarrow{\Phi}(y)$ from (7) is distributed among the processors of the computational system.
2. For solving the optimization problems from the family (7), the global optimization method applied on each processor should be updated by the following rules:
   (a) Prior to beginning a new global optimization iteration, for any problem $\varphi_l(y)$, $1 \leq l \leq q$ at any point $y'D$, the point y should be transferred to all employed processors in order to exclude the repeated computation of the criteria values $f_i(y)$, $1 \leq i \leq s$ at this point. In order to perform the data transfer, a queue for

receiving the transferred points and the criteria values at these points can be constructed at each computational node.

(b) Upon completing the iteration for any problem $\varphi_l(y)$, $1 \leq l \leq q$ at any point $y'D$, the point y' with the particular criteria values $f_i(y)$, $1 \leq i \leq s$ computed at this point should be transferred to all employed processors.

(c) Prior to beginning the next global search iteration, the method should check the queue of the received messages; if there are any data in this queue, the received information should be included into the search information Ω_k from (5).

The scheme of the parallel computing considered above provides the completeness of the search information Ω_k from (5) for all problems from the family $\overrightarrow{\Phi}(y)$.

The possibility of the asynchronous data transfer is a key feature of such a scheme of the parallel computations. Besides, any single control node is absent in this scheme, and the number of computational nodes can vary in the course of global optimization.

4   Parallel Methods for Solving the Multicriterial Global Optimization Problems

The multiextremal optimization is a research area being developed actively – the current state is presented, for example, in [9–11, 13, 14, 17, 18, 21]. The information-statistical theory of global optimization is one of the promising approaches [10, 11, 15, 27, 31]. The high performance computing systems are used widely for solving the time-consuming global search problems [11, 16, 24–26, 29–31].

4.1   Parallel Algorithm of Multicriterial Global Optimization

The approach is based on the following two statements:

- In order to reduce the complexity of the computational analysis of a multidimensional search information Ω_k from (5), the reduction of the dimensionality of the MCO problems is applied.
- For solving the problems from the family (3), the efficient global search algorithms developed within the framework of the information-statistical theory of the multiextremal optimization [10, 11] are used.

These statements are presented in more details below.

The dimensionality reduction. Within the framework of the proposed approach, the Peano *space-filling curves* or *evolvents* $y(x)$ mapping the interval $[0,1]$ onto the N-dimensional hypercube D unambiguously (see, for example, [10–12]) were used for the dimensionality reduction. As a result of such reduction, the initial multidimensional global optimization problem (3) is reduced to a one-dimensional problem:

$$\varphi(x^*) = \min\{\varphi(x) : x \in [0, 1]\}, \tag{8}$$

where $\varphi(x) = F(\lambda, y(x))$.

The dimensionality reduction scheme reduces the multidimensional problem (3) with the Lipschitzian minimized function to a one-dimensional problem (8), where the corresponding functions satisfy the uniform Hölder condition i.e.

$$\left| F\left(\lambda, y\left(x'\right)\right) - F\left(\lambda, y\left(x''\right)\right) \right| \leq H \left| x' - x'' \right|^{1/N}, x', x'' \in [0, 1], \tag{9}$$

where the Hölder constant H is defined by the relation $H = 4L\sqrt{N}$, L is the Lipschitz constant from (4) and N is the dimensionality of the optimization problem (1).

As a result of the dimensionality reduction, the search information Ω_k from (5) can be transformed into the *Matrix of the Search State* (MSS)

$$A_k = \left\{ (x_i, z_i, l_i)^T : 1 \leq i \leq k \right\}, \tag{10}$$

where x_i, $1 \leq i \leq k$ are the reduced points of the executed global search iterations[1], z_i, $1 \leq i \leq k$ are the values of the scalar criterion of the optimization problem (8) being solved, l_i, $1 \leq i \leq k$ are the indices of the global search iterations, where the points x_i, $1 \leq i \leq k$ were computed.

The matrix of the search state can be used by the optimization algorithms in order to improve the efficiency of the global search – selecting the points for the scheduled iterations can be performed taking into account the results of all computations performed before. Besides, the availability of the MSS allows computing the numerical estimates of the Hölder constant H from (9)

$$m = \begin{cases} rM, & M > 0, \\ 1, & M = 0, \end{cases} M = \max_{1 < i \leq k} \frac{|z_i - z_{i-1}|}{\rho_i} \tag{11}$$

as the maximum values of the relative differences of the minimized function values $z_i = \varphi(x_i)$, $1 \leq i \leq k$ on the set of points x_i, $1 \leq i \leq k$. Hereafter $\rho_i = \sqrt[N]{x_i - x_{i-1}}$, $1 < i \leq k$ and the constant r, $r > 1$ is the *reliability parameter* of the estimate of the constant H.

The parallel algorithm. Within the proposed approach, for solving the reduced one-dimensional multiextremal optimization subproblems (8), it is proposed to use well-known Multidimensional Algorithm of Global Search (MAGS) developed within the framework of the information-statistical theory of the multiextremal optimization [10, 11]. This method has a good theoretical substantiation and has demonstrated a high efficiency as compared to other global search algorithms (see also the results of numerical experiments in Sect. 5).

For the sake of completeness, let us consider briefly the general computational scheme of MAGS, which consists in the following.

Let us suppose that k, $k > 1$ global search iterations have been executed already, the optimization function values have been calculated at the previous iteration points (hereafter these calculations will be called the *trials*), and the obtained search

---

[1] The lower indices denote the increasing order of the coordinate values of the points x_i, $1 \leq i \leq k$.

information has been represented in the form of A_k from (10). The trial point for the next $(k+1)^{th}$ iteration is determined by the following rules.

Rule 1. Compute the *characteristics* $R(i)$ for each interval (x_{i-1}, x_i), $1 < i \leq k$ from A_k

$$R(i) = \rho_i + \frac{(z_i - z_{i-1})^2}{m^2 \rho_i} - \frac{2(z_i + z_{i-1})}{m}, 1 < i \leq k \tag{12}$$

Rule 2. Determine the interval (x_{t-1}, x_t) with the maximum characteristic $R(t)$ i.e.

$$R(t) = \max\{R(i) : 1 < i \leq k\} \tag{13}$$

Rule 3. Compute the trial point of the next global search iteration x^{k+1} within the interval t, $1 < t \leq k$ from (13):

$$x^{k+1} = \frac{x_t + x_{t-1}}{2} - \frac{z_t - z_{t-1}}{2m}.$$

The termination condition is defined by the inequality

$$\rho_t \leq \varepsilon \tag{14}$$

which should be checked for the interval t from (13) and the quantity $\varepsilon > 0$ is the predefined *accuracy* of the problem solution. If the termination condition is not fulfilled, the iteration index k is incremented by 1, and the execution of the algorithm is continued.

As the current estimate of the optimization problem solution, the minimum computed value of the optimization function is accepted i.e.

$$z_k^* = \min\{z_i : 1 \leq i \leq k\}. \tag{15}$$

Additional information on the MAGS algorithm is given in [11]. Here, it should be noted that the characteristics $R(i)$, $1 < t \leq k$ from (12) can be interpreted as some measures of importance of the intervals with respect to the location of the global minimum point in these ones.

Within the framework of the proposed approach, the MAGS algorithm is applied to solving every problem from the family $\overrightarrow{\Phi}(y)$ in combination with the general scheme of the parallel computations presented in Sect. 3. The method obtained as a result of such extension is called hereafter Parallel Multicriterial Global Algorithm (PMGA) for high-performance computing systems with distributed memory.

4.2   Multilevel Parallel Algorithm of Multicriterial Global Optimization

The general scheme of parallel computations considered in Subsect. 4.1 can be extended for the simultaneous computing of several minimized function values for every optimization problem from the family $\overrightarrow{\Phi}(y)$ on a separate multiprocessor multicore node with shared memory. For this purpose, a parallel generalization of the MAGS method can be applied – see, for example, [11, 16, 22]. This generalization consists in the following.

Let p is the number of employed parallel computational units (processors or cores) of a high-performance system node with shared memory. The rules of the parallel algorithm correspond to the computational scheme of the MAGS method (see Subsect. 4.1) except the rules of computation of the next global search iteration points. Below, for the sake of brevity, the modified rules for the parallel algorithm only are given.

Rule 2 (updated). Arrange the characteristics of the intervals obtained in (12) in the decreasing order

$$R(t_1) \geq R(t_2) \geq \ldots \geq R(t_{k-1}) \geq R(t_k) \tag{16}$$

and select p intervals with the indices t_j, $1 \leq j \leq p$ having the maximum values of the characteristics.

Rule 3 (updated). Perform new trials (the computations of the optimization function values (x)) at the points x^{k+j}, $1 \leq j \leq p$ located in the intervals with the maximum characteristics from (16)

$$x^{k+j} = \frac{x_{t_j} + x_{t_j-1}}{2} - sign(z_{t_j} - z_{t_j-1}) \frac{1}{2r} \left[\frac{\left| z_{t_j} - z_{t_j-1} \right|}{m} \right]^N, 1 \leq t_j \leq p.$$

The termination condition (14) in this case should be checked for all intervals from (16), where the scheduled trials are performed i.e.

$$\rho_{t_j} \leq \varepsilon, 1 \leq t_j \leq p.$$

The PMGA algorithm updated by the scheme of parallel computations for the computational nodes with shared memory will be named hereafter Multilevel Parallel Multicriterial Global Algorithm (MPMGA).

5   Results of Numerical Experiments

The numerical experiments have been carried out on the «Lobachevsky» supercomputer at State University of Nizhni Novgorod (the operating system – CentOS 6.4, the supercomputer management system – SLURM). Each supercomputer node had 2 Intel Sandy Bridge E5-2660 2.2 GHz processors, 64 Gb RAM. Each processor had 8 cores (i.e. total 16 CPU cores were available at each node). To generate the executable program code, Intel C++ 14.0.2 compiler was used.

At the very beginning, let us consider the results of comparison of the proposed approach with a number of other multicriterial optimization algorithms presented in [28]. For the comparison, the bi-criterial test problem proposed in [29] was used:

$$f_1(y) = (y_1 - 1)y_2^2 + 1, f_2(y) = y_2, 0 \le y_1, y_2 \le 1. \tag{17}$$

As a solution of a MCO problem, a Pareto domain approximation (PDA) was considered. To evaluate the efficiency of the computed approximations, the completeness and the uniformity of coverage of the Pareto domain were compared using the following two indicators [28, 29]:

- The *hypervolume index* (HV) defined as the volume of the subdomain of the values of the vector criterion $f(y)$ dominated by the points of the Pareto domain approximation. This indicator characterizes the completeness of the Pareto domain approximation (the higher the value, the more complete the coverage of the Pareto domain).
- The *distribution uniformity index* (DU), which characterizes the uniformity of the Pareto domain approximation (the less the value, the more uniform the coverage of the Pareto domain).

Within the framework of the considered experiment, five multicriterial optimization algorithms were compared: the Monte-Carlo (MC) method, the genetic algorithm SEMO from the PISA library [20, 32], the Non-Uniform Coverage (NUC) method [20], the Bi-objective Lipschitz Optimization (BLO) method proposed in [32], and the serial version of the MPMGA algorithm proposed in the present paper. Total 50 problems (3) have been solved by MPMGA with various values of the convolution coefficients λ distributed in Λ from (3) uniformly. The results of experiments from [28] are presented in the complete form in Table 1.

Table 1. Results of numerical experiments from [28] for the test problem (17)

Method	Iterations	PDA points	HV	DU
MC	500	67	0.300	1.277
SEMO	500	104	0.312	1.116
NUC	515	29	0.306	0.210
BLO	498	68	0.308	0.175
MPMGA	370	100	0.316	0.101

The results of the performed experiments have demonstrated that MPMGA has a considerable advantage with respect to the considered multicriterial optimization methods even when solving the relatively simple MCO problems.

The next numerical experiment has been carried out on solving the bi-criterial two-dimensional MCO problems i.e. for $N = 2$, $s = 2$. As the problem criteria, the multiextremal functions defined by the relations [11]:

$$\phi(y_1, y_2) = -(AB + AC)^{\frac{1}{2}},$$

Where

$$AB = \left(\sum_{i=1}^{7}\sum_{j=1}^{7}\left[A_{ij}a_{ij}(y_1,y_2) + B_{ij}b_{ij}(y_1,y_2)\right]\right)^2,$$

$$CD = \left(\sum_{i=1}^{7}\sum_{j=1}^{7}\left[C_{ij}a_{ij}(y_1,y_2) - D_{ij}b_{ij}(y_1,y_2)\right]\right)^2$$

and

$$a_{ij}(y_1,y_2) = \sin(\pi i y_1)\sin(\pi j y_2), \ b_{ij}(y_1,y_2) = \cos(\pi i y_1)\cos(\pi j y_2)$$

were used. These functions are defined in the range $0 \le y_1$, $y_2 \le 1$, and the parameters $-1 \le A_{ij}, B_{ij}, C_{ij}, D_{ij} \le 1$ are the independent random numbers distributed uniformly.

The solving of 100 multicriterial problems has been performed in these experiments. To compute the Pareto domain approximation, each problem has been solved for 50 coefficients λ distributed in Λ from (3) uniformly. The obtained results were averaged over the number of solved MCO problems.

The results of numerical experiments are presented in Table 2. The first two columns in Table 2 denote the numbers of the processors (P) and of the parallel computational cores on each processor (Q) employed. The third column (P*Q) contains the total number of cores employed. In the fourth, fifth, and sixth columns, the numbers of iterations necessary to find the solutions in given groups of problems from the family (3) for the corresponding numbers of the different coefficients λ from (3) are given. The last two columns contain the speedups of the parallel computations obtained with the use of the search information (S_1) and without the one (S_2).

Table 2. Results of a series of experiments for solving the two-dimensional bi-criterial MCO problems

P	Q	P*Q	1–25	26–50	1–50	S_1	S_2
Computations without the reuse of the search information							
1	1	1	8 571,6	8 590,2	17 165,9	–	1
Computations with the reuse of the search information							
1	1	1	1 199,5	573,9	1 773,4	1	9,7
1	25	25	52,1	27,6	79,7	22,2	215,4
25	1	25	135,1	54,8	189,8	9,3	90,4
5	5	25	66,8	37,1	103,9	17,1	165,2
25	25	625	8,6	8,1	16,7	106,3	1 029,1

The obtained results of experiments demonstrate that even simple reuse of the search information allows reducing the total amount of computations 9.7 times without the use of additional computational resources. When using 25 computational cores, one can obtain the speedup from 9.3 up to 22.2 times. If 625 computational cores are used, the speedup with the reuse of the search information reaches 106.3 times. The overall speedup in this case relative to the initial algorithm without the reuse of the search information was more than 1029 times.

6 Conclusion

In the present article, an efficient method for solving the time-consuming multicriterial optimization problems, where the optimality criteria can be multiextremal and computing the criteria values can require a large amount of computations has been proposed. The key aspect of the developed approach consists in the overcoming of the high computational complexity in solving the multicriterial optimization problems. A considerable improvement of the efficiency and a significant reduction of the amount of computations have been provided by means of the intensive use of the search information obtained in the course of computations. Within the framework of the developed approach, the methods for reusing the available search information for the values of current scalar nonlinear programming problem being solved have been proposed. The search information was used by the optimization methods for the adaptive planning of the executed global search iterations.

The results of the numerical experiments have demonstrated such an approach to allow reducing the computation costs of solving the multicriterial optimization problems considerably – tens and hundreds times.

In conclusion, one can note that the developed approach is a promising one and requires further investigations. First of all, it is necessary to continue carrying out the numerical experiments for solving the multicriterial optimization problems with more partial criteria of efficiency and for a greater dimensionality of the optimization problems being solved.

Acknowledgements. This work has been supported by Russian Science Foundation, project No 16-11-10150 "Novel efficient methods and software tools for time-consuming decision making problems using superior-performance supercomputers."

References

1. Mardani, A., Jusoh, A., Nor, K., Khalifah, Z., Zakwan, N., Valipour, A.: Multiple criteria decision-making techniques and their applications – a review of the literature from 2000 to 2014. Econ. Res.-Ekonomska Istraživanja 28(1), 516–571 (2015). doi:10.1080/1331677X.2015.1075139
2. Miettinen K.: Nonlinear Multiobjective Optimization. Springer, New York (1999)
3. Ehrgott, M.: Multicriteria Optimization, 2nd edn. Springer, Heidelberg (2010)
4. Collette, Y., Siarry, P.: Multiobjective Optimization: Principles and Case Studies (Decision Engineering). Springer, Heidelberg (2011)
5. Marler, R.T., Arora, J.S.: Survey of multi-objective optimization methods for engineering. Struct. Multidisciplin. Optim. 26, 369–395 (2004)
6. Figueira,J., Greco, S., Ehrgott, M. (eds.): Multiple Criteria Decision Analysis: State of the art Surveys. Springer, New York (2005)

7. Eichfelder, G.: Scalarizations for adaptively solving multi-objective optimization problems. Comput. Optim. Appl. **44**, 249–273 (2009)
8. Siwale, I.: Practical multi-objective programming. Technical Report RD-14-2013. Apex Research Limited (2014)
9. Pintér, J.D.: Global optimization in action (continuous and Lipschitz optimization: algorithms, implementations and applications). Kluwer Academic Publishers, Dordrecht (1996)
10. Strongin, R.G.: Numerical Methods in Multiextremal Problems: Information-Statistical Algorithms. Nauka, Moscow (1978). (in Russian)
11. Strongin, R., Sergeyev, Ya.: Global Optimization with Non-Convex Constraints. Sequential and Parallel Algorithms. Kluwer Academic Publishers, Dordrecht (2000). 2nd edn. (2013). 3rd edn. (2014)
12. Sergeyev Y.D., Strongin R.G., Lera D.: Introduction to Global Optimization Exploiting Space-Filling Curves. Springer, New York (2013)
13. Floudas, C.A., Pardalos, M.P.: Recent Advances in Global Optimization. Princeton University Press, Princeton (2016)
14. Locatelli, M., Schoen, F.: Global Optimization: Theory, Algorithms, and Applications. SIAM, Philadelphia (2013)
15. Sergeyev, Y.D.: An information global optimization algorithm with local tuning. SIAM J. Optim. **5**(4), 858–870 (1995)
16. Sergeyev, Y.D., Grishagin, V.A.: Parallel asynchronous global search and the nested optimization scheme. J. Comput. Anal. Appl. **3**(2), 123–145 (2001)
17. Törn, A., Žilinskas, A. (eds.): Global Optimization. LNCS, vol. 350. Springer, Heidelberg (1989). doi:10.1007/3-540-50871-6
18. Zhigljavsky, A.A.: Theory of Global Random Search. Kluwer Academic Publishers, Dordrecht (1991)
19. Marler, R.T., Arora, J.S.: Multi-Objective Optimization: Concepts and Methods for Engineering. VDM Verlag, Saarbrucken (2009)
20. Hillermeier, C., Jahn, J.: Multiobjective optimization: survey of methods and industrial applications. Surv. Math. Ind. **11**, 1–42 (2005)
21. Forrester, A.I.J., Keane, A.J.: Recent advances in surrogate-based optimization. Prog. Aerosp. Sci. **45**(1), 50–79 (2009)
22. Krasnoshekov, P.S., Morozov, V.V., Fedorov, V.V.: Decompozition in design problems. Eng. Cybern. **2**, 7–17 (1979). (in Russian)
23. Wierzbicki, A.: The use of reference objectives in multiobjective optimization. In: Fandel, G., Gal, T. (eds.) Multiple Objective Decision Making, Theory and Application, vol. 177, pp. 468–486. Springer, New York (1980)
24. Gergel, V., Sidorov, S.: A two-level parallel global search algorithm for solution of computationally intensive multiextremal optimization problems. In: Malyshkin, V. (ed.) PaCT 2015. LNCS, vol. 9251, pp. 505–515. Springer, Cham (2015). doi:10.1007/978-3-319-21909-7_49
25. Barkalov, K., Gergel, V., Lebedev, I.: Use of Xeon Phi coprocessor for solving global optimization problems. In: Malyshkin, V. (ed.) PaCT 2015. LNCS, vol. 9251, pp. 307–318. Springer, Cham (2015). doi:10.1007/978-3-319-21909-7_31
26. Gergel, V.: An unified approach to use of coprocessors of various types for solving global optimization problems. In: 2nd International Conference on Mathematics and Computers in Sciences and in Industry, pp. 13–18 (2015) doi:10.1109/MCSI.2015.18
27. Gergel, V.P., Grishagin, V.A., Gergel, A.V.: Adaptive nested optimization scheme for multidimensional global search. J. Global Optim. **66**(1), 1–17 (2015)

28. Gergel, V., Kozinov, E.: Accelerating parallel multicriterial optimization methods based on intensive using of search information. Procedia Comput. Sci. **108**, 1463–1472 (2017)
29. Evtushenko, Y.G., Posypkin, M.A.: A deterministic algorithm for global multi-objective optimization. Optim. Methods Softw. **29**(5), 1005–1019 (2014)
30. Gergel, V., Lebedev, I.: Heterogeneous parallel computations for solving global optimization problems. Procedia Comput. Sci. **66**, 53–62 (2015). doi:10.1007/s10898-016-0411-y
31. Strongin, R., Gergel, V., Grishagin, V., Barkalov, K.: Parallel Computations for Global Optimization Problems. Moscow State University, Moscow (2013). (in Russian)
32. Zilinskas, A., Zilinskas, J.: Adaptation of a one-step worst-case optimal univariate algorithm of bi-objective Lipschitz optimization to multidimensional problems. Commun. Nonlinear Sci. Numer. Simul. **21**, 89–98 (2015)

A Functional Approach to Parallelizing Data Mining Algorithms in Java

Ivan Kholod[1(✉)], Andrey Shorov[1], and Sergei Gorlatch[2]

[1] Saint Petersburg Electrotechnical University "LETI", Saint Petersburg, Russia
{iiholod,ashxz}@mail.ru
[2] University of Muenster, Muenster, Germany
gorlatch@uni-muenster.de

Abstract. We describe a new approach to parallelizing data mining algorithms. We use the representation of an algorithm as a sequence of functions and we use higher-order functions to express parallel execution. Our approach generalizes the popular *MapReduce* programming model by enabling not only data-parallel, but also task-parallel implementation and a combination of both. We implement our approach as an extension of the industrial-strength library *Xelopes*, and we illustrate it by developing a multi-threaded Java program for the 1R classification algorithm, with experiments on a multi-core processor.

Keywords: Parallel algorithms · Data mining · Parallel data mining · Multithreads · Multi-core processors · MapReduce, homomorphisms

1 Motivation and Related Work

Data mining algorithms have become especially popular in analyzing Big data, and their parallelization is in increasing demand, because they are often time-intensive.

Recent research in the field of parallel data mining [1] has created several parallelization approaches for particular classes of data mining algorithms, e.g., for search associations [2], clustering [3], building decision tree [4], etc. However, these individual approaches have high complexity and require much development and debugging effort. A popular alternative is the MapReduce programming model [5, 6] that relies on the functions *map* and *reduce* used in functional programming and is especially efficient for data-parallel functions called homomorphisms [7, 8]. MapReduce was used in [9] for a subclass of data mining algorithms that correspond to the Statistical Query Model. The libraries Apache Spark Machine Learning Library [10] and Apache Mahout [11] also contain data mining algorithms based on MapReduce.

The restrictions of MapReduce in the area of data mining are as follows: we need to find customizing functions for the *map* and *reduce* phases; only data parallelism is supported; parallelization is done only within a single loop iteration over the data set, while many data mining algorithms possess a more complex parallel structure.

Our contribution in this paper is a novel approach to parallelizing algorithms of data mining. Like MapReduce, our approach is based on the principles of functional programming, but it facilitates a more flexible parallelization, including task-parallel execution and parallelization across the loop iterations. We implement our approach as

© Springer International Publishing AG 2017
V. Malyshkin (Ed.): PaCT 2017, LNCS 10421, pp. 459–472, 2017.
DOI: 10.1007/978-3-319-62932-2_44

an extension of *Xelopes* – a commercial Java-based library for data mining. The extended library allows the developer to transform a sequential data mining algorithm into several parallel versions, efficiently running on modern multi-core processors.

2   The Formal Functional Approach

2.1   Data Mining Algorithm as a Composition of Functions

A data mining algorithm is represented in our approach as a function that takes a data set $d \in D$ as input and builds a mining model $m \in M$ as output:

$$dma : D \to M \tag{1}$$

where D is a type of data sets, and M is a type of mining models. We use capital letters to denote types and lower case letters for variables of these types and functions.

A data set contains characteristics of objects (e.g., persons, items, or courses) described by attributes (such as age, height, weight, or gender). Thus, a data set is often represented as a 2-dimensional array, e.g., for z objects and p attributes [12, 13]:

$$d = \left(x_{j.k}\right)_{j=1,k=1}^{z,p} \tag{2}$$

where $x_{j.k}$ is the value of kth attribute for j^{th} object. The set of possible values of kth attribute is denoted as Def_k ($x_{j.k} \in Def_k$). A row in matrix (2) is called a vector.

A mining model comprises elements that describe knowledge from a data set, e.g., classification or association rules, cluster centers, decision tree nodes, etc. We represent a mining model $m \in M$ as an array of mining model's elements e_i, $i = 0...w$:

$$m = [e_0, e_1, ..., e_w]. \tag{3}$$

We represent a data mining algorithm as a composition of functions, e.g.:

$$dma = f_n \circ f_{n-1} \circ ... \circ f_r \circ ... \circ f_s ... \circ f_1 \circ f_0, \tag{4}$$

where function $f_0 : D \to M$ takes a data set $d \in D$ as an argument and returns a mining model $m_0 \in M$, adding information about the data set into the mining model, and functions $f_t: M \to M$, $t = 1..n$ take the mining model $m_{t-1} \in M$ created by the previous function f_{t-1} and return the changed mining model $m_t \in M$. We will call that functions as a Functional Mining Block (FMB).

A data mining algorithm as a composition of functions (4) looks like Fig. 1:

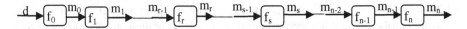

Fig. 1. Data mining algorithm as a composition of functions.

To express loops in our algorithms, we define a loop as a higher-order function that applies a FMB f_t to the mining model's elements starting from index i_s till index $i_{.e.}$:

$$\text{loop} \; : \; I \rightarrow I \rightarrow (M \rightarrow M) \rightarrow M \rightarrow M$$
$$\text{loop } i_s \; i_e \; f_t \; m \; = \; (fti_e \circ fti_{e-1} \circ \ldots \circ fti_s) \; m \tag{5}$$

where I is a set of mining model's arrays indices; fti_h are the FMBs which execute the FMB f_t for the mining model's element with index i_h ($i_s < i_h < i_{.e.}$). The FMB fti_h can be implemented as a composition of f_t and a function selecting a certain mining model's element in the array with index i_h.

2.2 Illustration for the 1R Algorithm

Listing 1 shows a 1R algorithm [14] as a pseudocode of its implementation in the Weka library [15]. Let us illustrate how 1R can be represented in our approach. The 1R algorithm builds mining model's elements as simple classification rules. These rules determine the value $v_{t.p}$ of a dependent attribute a_t ($v_{t.p} \in Def_t$) using values $v_{k.q}$ of independent attributes a_k: $k = 1..p$, where $v_{k.q} \in Def_k$:

$$\text{if } \left(a_k = v_{k.q}\right) \text{ then } \left(a_t = v_{t.p}\right).$$

They are included into the mining model as a tuple $<a_k, v_{k.q}, v_{t.p}>$.

```
1       For k = 1 ...p // loop for each attribute
2           For j = 1 ...z // loop for each vector
            // count frequency of each value of k-th independent attribute x_{j.k}
            // for value dependent attribute x_{j.t}
3               count[k][x_{j.k}][x_{j.t}]++;
4           For q =1 ... |Def_k| // loop for each value of k-th independent attribute
            // find the most frequent value dependent attribute
5               For r =1 ... |Def_t| // loop for each value dependent attribute
6                   if count[k][v_{k.q}][v_{t.r}] > count[k][v_{k.q}][v_{t.mi}]
7                       mi = r; // save index of found value
            // create a candidate rule for indep. attribute v_{k.q} of and dep. attribute v_{k.mi}
8                   CR[k].add({a_k, v_{k.q}, v_{t.mi}, count[k][v_{k.q}][v_{t.mi}]})
            // Calculate number of vectors satisfied by the rule R_k
9                   correctVectors+=count[k][v_{k.q}][v_{t.mi}]
            // Choose the rules with the most number of vectors
10                  if(correctVectors > bestCorrectVectors)
11                      bestCorrectVectors = correctVectors; R = CR[k];
12                  correctVectors = 0;
```

Listing 1. Pseudocode of the 1R algorithm implementation in the Weka library

According to our approach, we represent the 1R algorithm (Listing 1) as a composition (4) of functions f_t, t = 0..n. The array of the mining model's elements for the 1R algorithm (initialized by function f_0) can be split in the following disjoint sets:

$$m_{1R} = m_A \cup m_V \cup m_T \cup m_X \cup m_R \cup m_C \cup m_{CR} \qquad (6)$$

- m_A is the subset of mining model's elements with information about the attributes of the matrix (2), with indices in $[0, p - 1]$ (where p is the number of attributes);
- m_V is the subset of mining model's elements containing information about the independent attribute's values with indices in $[v1, v2]$, where $v1 = p$, $v2 = p + p \cdot |Def_k|$;
- m_T is the subset of mining model's elements, containing information about the dependent attribute's values, with indices in $[t1, t2]$, where $t1 = v1$, $t2 = t1 + |Def_t|$;
- m_X is the subset of mining model's elements with information about the vectors of the matrix (2), with indices in $[x1, x2]$, where $x1 = t2 + 1$, $x2 = x1 + z+1$ (z is the number of vectors);
- m_R is the subset of mining model's elements, containing information about the created rules R (Listing 1), with indices in $[r1, r2]$, where $r1 = x2 + 1$, $r2 = r1 + |R|$;
- m_C is the subset of mining model's elements, containing value of array *count* (Listing 1), with indices in $[c1, c2]$, where $c1 = r2 + 1$, $c2 = c1 + p \cdot |Def_k| \cdot |Def_t|$;
- m_{CR} is subset of mining model's elements, containing information about the candidate-rules CR (Listing 1), with indices in $[cr1, cr2]$, where $cr1 = c2 + 1$, $cr2 = cr1 + |CR|$.

The representation of the 1R algorithm according to the schema (4) comprises the following FMBs:

- f_1 is the loop for the mining model's elements of the subset m_A (line 1 in Listing 1): $f_1 = loop\ 0\ p - 1\ (f_{13}°f_{10}°f_4°f_2)$;
- f_2 is the loop for mining model's elements of the subset m_X (lines 2–3 in Listing 1): $f_2 = loop\ x1\ x2\ f_3$;
- f_3 increments the *count* array element for each independent k-th and dependent t-th attributes and each vector of a data set (line 3);
- f_4 is the loop for mining model's elements of the subset m_V (lines 4–9 in Listing 1): $f_4 = loop\ v1\ v2\ (f_9°f_8°f_5)$;
- f_5 is the loop for mining model's elements of the subset m_T (lines 5–7 in Listing 1): $f_5 = loop\ t1\ t2\ f_6$,
- f_6 searches for the maximum value in matrix *count* (lines 6–7 in Listing 1);
- f_8 adds the rule for the maximum value (line 8 in Listing 1);
- f_9 counts the number of vectors which relevant to rules CR (line 9 in Listing 1);
- f_{10} selects the set of rules (R or CR) with maximum the number of relevant vectors (lines 10–12 in Listing 1);
- f_{11} clears the set of candidate rules (line 13 in Listing 1).

Thus, we can represent the 1R algorithm as a composition of functions as follows:

$$
\begin{aligned}
1R = f_1 \circ f_0 = & \\
& (\text{loop } 0 \text{ p}-1 \text{ } f_{11} \circ f_{10} \circ f_4 \circ f_2) \circ f_0 = \\
& (\text{loop } 0 \text{ p}-1 \text{ } f_{11} \circ f_{10} \circ (\text{loop v1 v2 } f_9 \circ f_8 \circ f_5) \circ (\text{loop x1 x2 } f_3)) \circ f_0 = \\
& (\text{loop } 0 \text{ p}-1 \text{ } f_{11} \circ f_{10} \circ (\text{loop v1 v2 } f_9 \circ f_8 \circ (\text{loop t1 t2 } f_6)) \circ (\text{loop x1 x2 } f_3)) \circ f_0
\end{aligned}
\tag{7}
$$

2.3 Functions for Parallelization

In representation (4), parallel execution of FMBs f_t and f_{t+1} is possible if the data dependency between them allows it. We introduce the following identifications for FMBs: $In(f_t)$ is subset of mining model elements used by FMB f_t; $Out(f_t)$ is subset of mining model elements modified by FMB f_t. According to the Bernstein's conditions [16], two FMBs f_t and f_{t+1} can be executed in parallel in systems with share memory if and only if:

- there is no data anti-dependency: $In(f_t) \cap Out(f_{t+1}) = \varnothing$;
- there is no data flow dependency: $Out(f_t) \cap In(f_{t+1}) = \varnothing$;
- there is no output dependency: $Out(f_t) \cap Out(f_{t+1}) = \varnothing$.

For parallel execution of FMBs, we introduce the higher-order function *fork* that takes a list of FMBs and a mining model, applies each FMB from the list to the mining model and returns the list of the resulting mining models:

$$
\begin{aligned}
\text{fork} &: [M \rightarrow M] \rightarrow M \rightarrow [M] \\
\text{fork } [f_1, \ldots, f_k] \text{ m} &= [f_1 \text{ m}, \ldots, f_k \text{m}] = [m_1, \ldots, m_k]
\end{aligned}
\tag{8}
$$

In this function, all calls to FMBs from the list can be executed in parallel, such that the mining models $(m_1, \ldots, m_k)$ are built in parallel.

We use the function *join* to combine the source mining model (1st argument) and mining models (2th argument) that are built by several FMBs:

$$
\text{join} : M \rightarrow [M] \rightarrow M
\tag{9}
$$

The implementation of the *join* function depends on the structure of the particular mining model.

Using the introduced functions, we define the higher-order function *parallel* for the parallel execution of FMBs:

$$
\begin{aligned}
\text{parallel} &: [(M \rightarrow M)] \rightarrow M \rightarrow M \\
\text{parallel } [f_1, \ldots, f_k] \text{ m} &= \text{join m (fork } [f_1, \ldots, f_k] \text{ m)}
\end{aligned}
\tag{10}
$$

Different FMBs $f_s, \ldots, f_r$ can be computed in parallel using function *parallel* (10) according to the principle of task parallelism.

Data parallelism can be implemented by applying the function *parallel* to the function *loop* (5). If a data mining algorithm is represented by (4), a FMB f_r is a loop $(f_r \equiv loop\ i_s\ i_e\ f_t)$ and the FMB f_t for mining model's elements of array with index i satisfy the Bernstein's conditions, then the FMB f_r can be executed in parallel:

$$loop\ i_s i_e f_t\ =\ parallel[fti_s,\ \ldots,\ fti_{e-1},\ fti_e].$$

Thus parallelizing a loop for vectors is a generalization of MapReduce: FMB f_t is an analog of the *map* function and the function *join* (9) is an analog of the *reduce* function. Additionally, unlike the MapReduce, the *parallel* function can be used multiple times to parallelize different parts of the algorithm.

Thus, a data mining algorithm is parallelized in our approach in three steps:

(1) represent the algorithm as a composition (4) of functions f_t, $t = 0..n$;
(2) verify Bernstein's conditions for FMBs $f_s \ldots f_r$ for sequence of the FMBs;
(3) transform the sequential execution of the FMBs $f_s \ldots f_r$ into the parallel execution by using the *parallel* function:

$$f_n \circ \ldots \circ f_{r+1} \circ f_r \circ \ldots \circ f_s \circ f_{s-1} \circ \ldots \circ f_0 = f_n \circ \ldots \circ f_{r+1} \circ (parallel[f_s, \ldots, f_r]) \circ f_{s-1} \circ \ldots \circ f_1 \circ f_0$$

The parallel execution of a data mining algorithm is shown in the Fig. 2.

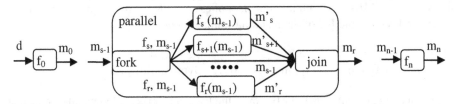

Fig. 2. Parallel execution of the data mining algorithm

2.4   Illustration: The 1R Algorithm

In our approach, the sequential form of the 1R algorithm (7) can be transformed into parallel form as follows. The Bernstein's conditions are verified based on Listing 1. For example, sets of the used and modified mining model's elements for f_8 and f_9 are determined based on line 8 and line 9 of the pseudocode:

$$In(f_8) = \{a_k, v_{k.q}, v_{t.mi}, count[k][v_{k.q}][v_{t.mi}]\},\ Out(f_8) = \{CR[k]\}$$
$$In(f_9) = \{correctVectors, count[k][v_{k.q}][v_{t.mi}]\},\ Out(f_9) = \{correctVectors\}.$$

For these FMBs, the Bernstein's conditions are satisfied: $In(f_8) \cap Out(f_9) = \emptyset$; Out $(f_8) \cap In(f_9) = \emptyset$; $Out(f_8) \cap Out(f_9) = \emptyset$.

Verifying the Bernstein's conditions for all FMBs allows us to obtain the following variants of the 1R algorithm (differences between the variants are indicated by the bold underline font) with parallel execution of:

- the loop for vectors f_2 (variant 1RParVec):

$$1\text{RParVec} = (\text{loop } 0 \text{ p-1 } f_{11}{}^\circ f_{10}$$
$$^\circ(\text{loop v1 v2 } f_9{}^\circ f_8{}^\circ (\text{loop t1 t2 } f_6)) \tag{11}$$
$$^\circ(\textbf{parallel } [\text{loop x1 x2 } f_3]))^\circ f_0$$

- the loop for values of dependent attribute f_5 (variant 1RParVal):

$$1\text{RParVal} = (\text{loop } 0 \text{ p-1 } f_{11}{}^\circ f_{10}$$
$$^\circ (\textbf{parallel } [\text{loop v1 v2 } f_9{}^\circ f_8{}^\circ(\text{loop t1 t2 } f_6)]) \tag{12}$$
$$^\circ (\text{loop x1 x2 } f_3))^\circ f_0$$

- the FMB adding a rule for the number of relevant vectors f_8 and counting the number of relevant vectors for current independent attribute a_k f_9 (variant 1RParFMB):

$$1\text{RParFMB} = (\text{loop } 0 \text{ p-1 } f_{11}{}^\circ f_{10}$$
$$^\circ(\text{loop v1 v2 } (\textbf{parallel } [f_8, f_9])^\circ(\text{loop t1 t2 } f_6)) \tag{13}$$
$$^\circ(\text{loop x1 x2 } f_3))^\circ f_0$$

Note that the variant 1RParVec is the traditional way of parallelizing using the MapReduce, while 1RParVal can be implemented by applying MapReduce to the values of an independent attribute. The variant 1RParFMB realizes task parallelism that cannot be implemented by MapReduce. Additionally, we can combine all there variants (variant 1RParAll) as follows, which is also not possible in MapReduce:

$$1\text{RParAll} = (\text{loop } 0 \text{ p-1 } f_{11}{}^\circ f_{10}$$
$$^\circ (\textbf{parallel } [\text{loop v1 v2 } (\textbf{parallel } [f_8, f_9])^\circ(\text{loop t1 t2 } f_6)]) \tag{14}$$
$$^\circ (\textbf{parallel } [\text{loop x1 x2 } f_3]))^\circ f_0$$

3   Implementation of the Approach

We implement our approach as an extension of the commercial Java-based library Xelopes [17] that comprises a broad variety of data mining algorithms.

3.1   Implementation of Functional Mining Block

Our implementation of the approach is a set of Java classes for parallel data mining algorithms: they hide the details of parallel execution from the developer. Figure 3 shows the class diagram of the basic classes for implementation of the FMBs.

Any FMB in the library is implemented as a subclass of the `MiningBlock` class. This class defines the abstract method `execute()`. In accordance with the definition of FMB, method `execute()` takes one argument - a mining model (implemented as a subclass of the `EMiningModel` class) - and returns the changed mining model.

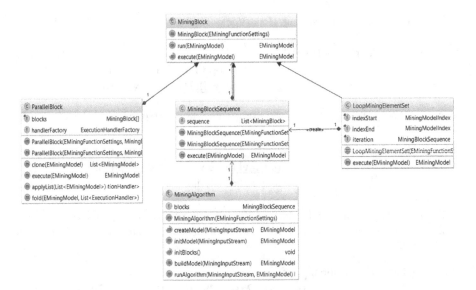

Fig. 3. Class diagram of basic functional mining blocks

To implement a composition of FMBs and our two higher-order functions - *loop* (5) and *parallel* (10) - we added in the library subclasses of the `MiningBlock` class.

The composition of FMBs is implemented by class `MiningBlockSequence`. Its method the `execute()` sequentially calls FMBs from the `sequence` list. The higher-order function *loop* (5) is implemented by the `LoopMiningElementSet` class. The `execute()` method executes calls to the composition – `iteration` for all elements from mining model array from `indexStart` till `indexEnd`.

The higher-order function *parallel* (10) is implemented by the `ParallelBlock` class. The `execute()` method executes the `blocks` composition in parallel. It calls FMBs in parallel with the `fork()` method, and unites the resulting mining models with the `join()` method. The method `fork()` is also implemented in the `ParallelBlock` class. The method `join()` is implemented by the class that describes the mining model for implemented data mining algorithm.

The whole data mining algorithm is implemented by a subclass of the class `MiningAlgorithm`. It contains a sequence of all FMBs of the algorithm – `blocks`. The data mining algorithm structure is formed by creating composition of the FMBs `blocks` in the abstract method `initBlocks()`, which must be implemented in subclass for particular data mining algorithm.

A mining model is built by the `buildModel()` method of the class `MiningAlgorithm`. This method calls the methods `initModel()` and `runAlgorithm()`. The method `initModel()` creates a new mining model and initializes it by arrays of vectors and attributes, i.e., this method implements the f_0 function. The method `runAlgorithm()` executes FMBs from the sequence `blocks`, i.e., it executes the whole data mining algorithm.

3.2   Illustration: The 1R Algorithm

To develop different parallel variants of a data mining algorithm using our approach and the library implementation, the developer performs the following steps:

(1) decompose the data mining algorithm into FMBs;
(2) implement the FMBs as subclasses of the `MiningBlock` class;
(3) implement the sequential data mining algorithm as a subclass of the `MiningAlgorithm` class;
(4) verify the Bernstein's conditions for each pair FMBs;
(5) implement the parallel variants for FMBs which satisfy the Bernstein's conditions using instances of the `ParallelBlock` class.

The step 1 for the 1R algorithm is described in Sect. 2.2; it produces expression (7). In step 2, for example, `IncrementCorrectVectorsCount` class implements f_3.

In step 3, we implement 1R in the method `initBlocks`. To implement 1R according to (7), we create instances of all the FMBs of the algorithm. For example, for the FMB f_2 (the loop for array of vectors $f_2 = \text{loop } x1\ x2\ f_3$) we create the instance of `LoopMiningElementSet` (matches the function *loop*) by a constructor with arguments: x1 and x2 are the start and end indices of the processed mining model's elements; the instance of the `IncrementCorrectVectorsCount` class is the FMB f_3. Therefore, the developer only writes the following Java code:

```
new LoopMiningElementSet(settings, x1, x2
        new IncrementCorrectVectorsCount(settings))
```

The whole algorithm is implemented according to expression (7) in the initBlocks() method of the `OneRAlgorithm` class:

```
protected void initBlocks() throws MiningException
  blocks = new MiningBlockSequence(settings,
    new LoopMiningElementSet(settings, 0, p-1,
      new LoopMiningElementSet(settings, x1, x2,
        new IncrementCorrectVectorsCount(settings)),
      new LoopMiningElementSet(settings, v1, v2
        new LoopMiningElementSet(settings, t1, t2
          new SelectMaxCorrectVectorsCount(settings)),
        new AddBestOneRule(settings),
        new SummCorrectVectorsCount(settings)),
      new SelectBestRuleSet(settings),
      new RemoveAttributeRuleSet(settings)));
}
```

In step 4, the developer verifies the Bernstein's conditions for all FMBs of the algorithm (example is described in Sect. 2.4). The FMBs listed in Sect. 2.4 satisfy them. The parallel variants for these FMBs are given by (11)–(14).

In step 5, the developer transforms the sequential variant of 1R into parallel forms using the instance of the `ParallelBlock` class. For example, to implement variant 1RParVec (13) and parallelize the loop for vectors (*parallel [loop x1 x2 f₃]*), the instance of `LoopMiningElementSet` is passed to the constructor of the `ParallelBlock` class:

```
new ParallelBlock (settings,
        new LoopMiningElementSet(settings, x1, x2
            new IncrementCorrectVectorsCount(settings))).
```

Similarly, other variants (11)–(14) are obtained from the sequential implementation of 1R. For example, variant 1RParAll described by (14) is implemented in the `initBlocks()` of the `OneRAlgorithm` class by changing few lines (these changes for parallel variants are shown by the bold underline font):

```
protected void initBlocks() throws MiningException {
  blocks = new MiningBlockSequence(settings,
    new LoopMiningElementSet(settings, 0, p-1
      new ParallelBlock (settings,
          new LoopMiningElementSet(settings, x1, x2
            new IncrementCorrectVectorsCount(settings))),
      new ParallelBlock (settings,
          new LoopMiningElementSet(settings, v1, v2
            new LoopMiningElementSet(settings, t1, t2
             new SelectMaxCorrectVectorsCount(settings)),
            new ParallelBlock (settings,
               new AddBestOneRule(settings),
               new SummCorrectVectorsCount(settings)))),
      new SelectBestRuleSet(settings),
      new RemoveAttributeRuleSet(settings)));
}
```

Thus, the transformation of a sequential Java implementation to different parallel variants in our library requires only changing few lines of code.

4 Experimental Results

We perform several experiments for the implemented parallel versions of the 1R algorithm. The experiments are conducted on various generated input data sets (Table 1).

The experiments were run on a multi-core computer with the following configuration: CPU Intel Xenon (8 cores), 2.90 GHz, 4 Gb. The parallel algorithms are executed for different numbers of cores: 2, 4 and 8.

Table 1. Parameters of experimental data sets

Name of input data set	W10 m	W50 m	W100 m	C1t	C5t	C10t
Number of vectors	10^7	$5 \cdot 10^7$	10^8	1 000	1 000	1 000
Number of attributes	10	10	10	100	100	100
Avg. number of values of attributes	1 000	1 000	1 000	1 000	5 000	10 000

Figures 4, 5, 6 and 7 show the results of our experiments on the data sets from Table 1 with the following parallel variants of the 1R algorithm: 1R according to the expression (7); 1RParVec according to (11); 1RParVal according to (12); 1RParFMB according to (13); 1RParAll according to (14).

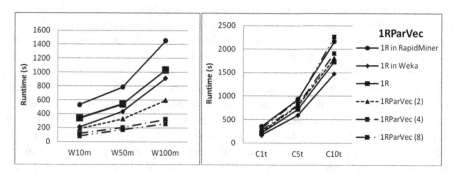

Fig. 4. Runtime of the 1RParVec variant of the 1R algorithm

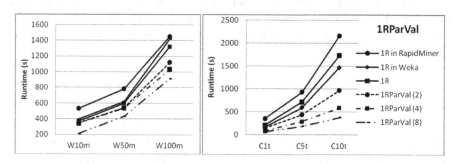

Fig. 5. Runtime of the 1RParVal variant of the 1R algorithm

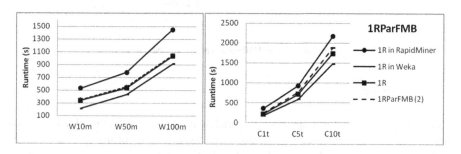

Fig. 6. Runtime of the 1RParFMB variant of the 1R algorithm

We compare to the sequential implementation in Weka [15] and Rapid Miner [18]. The results of the experiments show good scalability: the more threads we employ on the 8-core processor, the better speedup we achieve on up to 8 threads.

Our different variants of parallelism for the 1R data mining algorithm achieve different efficiency (runtime of efficient variants are indicated by the bold font):

- parallelization of the loop for vectors (Fig. 4 left) is more efficient for the data sets with large number of vectors (data sets: W10 m, W50 m, W100 m);
- parallelization of the loop for values of the dependent attribute (Fig. 5 right) is more efficient for the data sets with large number of classes (C1t, C5t, C10t).

These results can be explained by the fact that longest part of the algorithm that comprises the maximum number of iterations is parallelized. The parallel execution compensates the overhead of preparing for parallel execution (creation and running of threads, etc.) and subsequent processing (synchronization of threads, joining of mining model, etc.). Parallelizing loops with a small number of iterations is inefficient: the overhead significantly exceeds the effect of parallelization.

Bigger mining models (w.r.t. the number of vectors or values of attributes) increase the overhead, which is reflected in the loss of performance of the 1RParVec variant for the data set with large number of attribute values (Fig. 4 right) and the 1RParVal variant for the data set with large number of vectors (Fig. 5 left). The 1RParFMB variant (task parallelism) is inefficient for both types of data sets (Fig. 6), because of the comparatively small runtime of parallel FMBs for the 1R algorithm, such that their parallel execution does not compensate the overhead. We note that our experiments with different task parallelism variants for the more time-intensive association algorithm Apriori [2] show much better parallel performance. The 1RParAll variant that combines both data and task parallelism demonstrates good performance for both algorithms (Fig. 7).

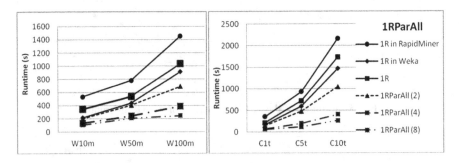

Fig. 7. Runtime of the 1RParAll variant of the 1R algorithm

5   Conclusion

This paper suggests a novel approach to parallelizing data mining algorithms based on their functional representation. We demonstrate the generality of the approach and its applicability for a particular example of the 1R algorithm.

The advantages of our approach are as follows: (1) it is provably correct as based on the formal transformations of functions; (2) it covers both data parallelism and task parallelism and a combination of both; (3) it is implemented as an extension of the commercial library Xelopes and can be effectively used to develop different multi-threaded Java implementations by changing only few lines of program code.

Our library contains parallel implementations of several data mining algorithms (for example association rules algorithms Apriori [2], clustering algorithm k-Means [3], decision tree algorithm C 4.5 [4] and others) based on our approach.

Our experiments with different data sets on an eight-core processor show a good performance of the parallel Java implementations of the 1R data mining algorithm.

Acknowledgments. This work was supported by the Ministry of Education and Science of the Russian Federation in the framework of the state order "Organization of Scientific Research", task #2.6113.2017/BY, and by the German Research Agency (DFG) in the framework of the Cluster of Excellence *Cells-in-Motion* at the University of Muenster.

References

1. Paul, S.: Parallel and distributed data mining. In: Funatsu, K. (ed.) New Fundamental Technologies in Data Mining, Karunya University, Coimbatore, India, pp. 43–54 (2011). ISBN 978-953-307-547-1

2. Zaki, M.: Parallel and distributed association mining : a survey. IEEE Concurrency **7**(4), 14–25 (1999)

3. Kim, W.: Parallel clustering algorithms: survey. In: CSC 8530 Parallel Algorithms. Spring (2009). http://s3-us-west-2.amazonaws.com/mlsurveys/46.pdf

4. Satuluri, V.: A survey of parallel algorithms for classification (2007). http://citeseerx.ist.psu.edu/viewdoc/summary?doi=10.1.1.126.5567

5. Dean, J. Ghemawat, S.: MapReduce: simplified data processing on large clusters. In: Proceedings of Operating Systems Design and Implementation. San Francisco (2004)

6. Lammel, R.: Google's MapReduce programming model—revisited. Sci. Comput. Program. **70**(1), 1–30 (2008)

7. Gorlatch, S.: Extracting and implementing list homomorphism in parallel program development. Sci. Comput. Program. **33**(1), 1–27 (1999)

8. Rasch, A., Gorlatch, S.: Multi-dimensional homomorphisms and their implementation in OpenCL. Int. J. Parallel Prog. **45**, 300–319 (2017)

9. Ng, A.Y., Bradski, G., Chu, C.-T., Olukotun, K., Kim, S.K., Lin, Y.-A., Yu, Y.Y.: Map-Reduce for machine learning on multicore. In: Proceedings of the Twentieth Annual Conference on Neural Information Processing Systems, Vancouver, Canada, pp. 281–288 (2006)

10. Machine learning library (MLlib) guide. http://spark.apache.org/docs/latest/mllib-guide.html

11. Grant ingersoll, introducing apache mahout. http://www.ibm.com/developerworks/java/library/j-mahout/

12. Hastie, T., Tibshirani, R., Friedman, J.: The elements of statistical learning: data mining, inference and prediction, 533 p. Springer, New York (2001)

13. Han, J., Kamber, M.: Data Mining: Concepts and Techniques. Morgan Kaufman, San Francisco (2001)

14. Holte, R.C.: Very simple classification rules perform well on most commonly used datasets. Mach. Learn. **11**, 63–90 (1993)
15. Witten, I.H., Eibe, F., Hall, M.A.: Data Mining Practical Machine Learning Tools and Techniques, 3rd edn., 629 pp. Morgan Kaufmann, San Francisco (2011)
16. Bernstein, A.J.: Program analysis for parallel processing. IEEE Trans. Electron. Comput. **EC-15**, 757–762 (1966)
17. Prudsys Xelopes. https://prudsys.de/en/knowledge/technology/prudsys-xelopes/
18. Rapid Miner. http://rapidminer.com/

Parallel Calculation of Diameter Constrained Network Reliability

Sergei N. Nesterov and Denis A. Migov[✉]

Institute of Computational Mathematics and Mathematical Geophysics SB RAS,
Novosibirsk, Russia
cepera@inbox.ru, mdinka@rav.sscc.ru

Abstract. The problem of network reliability calculation in case of the diameter constraint is studied. The problem of computing this characteristic is known to be NP-hard. We introduce the parallel methods, which are based on the well-known factoring method and on the factoring method modification proposed by H. Cancela and L. Petingi. The analysis of the numerical experiments has allowed us to set some important parameters of the parallel algorithm for speeding up calculations.

Keywords: Network reliability · Parallel algorithm · Random graph · Diameter constraint · Factoring method

1 Introduction

In the present article we consider networks where links are subject to random failures under the assumption that failures are statistically independent. Random graphs are commonly used for modeling of such networks. As a rule, network reliability is defined as some connectivity measure. The most common reliability measure of such networks is the probability that all terminal nodes in the network can keep connected together, given the reliability of each network node and edge. The problem of calculation of network probabilistic connectivity is known to be NP-hard. Nevertheless, it is possible to conduct the exact calculation of reliability for networks with dimension of a practical interest by taking into consideration some special features of real network structures and based on modern high-speed computers [1,2].

Another popular measure of network reliability is the diameter constrained network reliability (Petingi and Cancela, 2001 [3]). Further on we will use the abbreviation DCNR for its notation. DCNR is a probability that every two nodes from a given set of terminals are connected with a path of length less or equal to a given integer. By the length of a path we understand the number of edges in this path. This reliability measure is more applicable in practice, for example, in the case of P2P networks. However, the problems of computing these characteristics

---

Supported by Russian Foundation for Basic Research under grants 16-37-00345, 16-07-00434.

V. Malyshkin (Ed.): PaCT 2017, LNCS 10421, pp. 473–479, 2017.
DOI: 10.1007/978-3-319-62932-2_45

are known to be NP-hard. Moreover, DCNR calculation problem is NP-hard for most combinations of a diameter value and a number of terminals [4]. In our previous studies we have obtained some methods for speeding up DCNR calculations [5].

However, despite the improvements achieved on the efficiency of the computational methods for reliability analysis, they still are ineffective and so their parallel realizations are needed for executing on the modern supercomputers. By now we have in this area only the parallel approach for estimation of network reliability by Monte Carlo technique [6] and the parallel implementation of the well-known factoring method, which was proposed in one of our previous study [7].

In this paper we propose the parallel method for DCNR calculating. The proposed method is based on the well-known sequential factoring method [2]. We have chosen the fastest modification [3] of factoring method for DCNR calculation with the improvements proposed in [5]. For parallel implementation we chosen "Master-Slave" parallel programming model, as we have done for calculation of network probabilistic connectivity [7]. The analysis of the numerical experiments results allows us to optimize some important parameters of the algorithm which further increase its speedup and scalability.

2   The Basic Definitions and Notations

A network with perfectly reliable nodes and unreliable edges is modeled by an undirected random graph $G = (V, E)$ with given presence probabilities $0 \leq r_e \leq 1$ of any edge $e \in E$. There is also a given set of terminals $K \subseteq V$, that is, the nodes of the network G which should be connected via operational edges to the network to operate well.

Assume $Q = (V, E_Q)$ is a subgraph of the graph G where E_Q is defined by existence or absence of each edge $e \in E$. An edge e is called *operational* if it exists in E', otherwise we call it *faulty*. Q is usually called as an *elementary event*. Hence the count of possible elementary events is $2^{|E|}$. The probability of an elementary event Q could be calculated by multiplying the product of probabilities of absence of faulty edges and the product of probabilities of existence of operational edges.

A reliability with diameter constraint d of network G is defined as the sum of probabilities of elementary events in which every pair of terminals $u, v \in K$ can be connected by a path p of length at most d, where the length of the path p is the number of edges belonging to this path. We denote this reliability measure by $R_K^d(G)$.

3   Methods for DCNR Calculation

In practice, it is no use computing of DCNR directly by the definition because this approach should result to an exhaustive search of all graph realizations.

Therefore the other methods are used for calculation of different reliability measures. The most common method among them is the factoring method [2], which can be applied to any network reliability measure, including DCNR. The factoring method divides the probability space into two sets, based on the success or failure of one graph particular unreliable element: a node or an edge. For DCNR we have the following formula:

$$R_K^d(G) = r_e R_K^d(G/e) + (1 - r_e)R_K^d(G\backslash e), \qquad (1)$$

Where $G\backslash e$ is graph G without edge e, G/e is graph G with absolutely reliable edge e. Recursions continue until a graph is obtained, in which at least one pair of terminals cannot be connected by path of limited length (returns 0), or all pairs of terminals are connected by absolutely reliable paths of limited length (returns 1). Further on we refer to this method as SFM (simple factoring method).

A modified factoring method for DCNR calculation was proposed by Cancela and Petingi [3]. Further on we refer to this method as CPFM (Cancela & Petingi factoring method). This method is much faster than basic factoring method in the diameter constrained case (1). The main feature of the modified factoring method is operating with list of paths instead of operating with graphs. In the preliminary step for any pair of terminals s, t the list $P_{st}(d)$ of all paths with limited length between s, t is generated. It automatically removes all edges which do not belong to any such path from consideration. For example, all so called "attached trees" without terminals are no longer considered. By P_d the union of $P_{st}(d)$ for all pairs of terminals is denoted. By $P(e)$ the set of paths of P_d which include link e is denoted. Parameters of the modified factoring procedure are not graphs. Instead of graphs we use 6 parameters, which describe the corresponding graph from the viewpoint of P_d.

One of the main reasons, why the calculation of diameter constrained network reliability much more complicated in comparison with other network reliability measures, is the lack of methods for decreasing of recursions quantity. In our previous studies [5] we have obtained such methods which can make DCNR calculation faster. These methods are the analogue of the well-known series-parallel transformation for CPFM, and the pivot edge selection strategy. Also we have obtained decomposition methods for calculating DCNR in two terminal case. Obtained methods allow to significantly reduce the number of recursive calls in CPFM and complexity of DCNR computation.

4   Parallel Computation of DCNR

In this section we introduce an algorithm with use of MPI for DCNR calculation for supercomputers with distributed memory.

During the factoring procedure two subtasks are created: "contraction" and "removal" of edge e. So one part of this work (for example, "contracting") could be sent to some idle process while another one will be evaluated in current process. If all of the processes are busy now, then both tasks should be performed in current process.

Unfortunately, this approach has a significant disadvantage. Since the algorithm is based on the factorization formula, it is necessary to send the value evaluated in the current process to the parent process to perform other multiplication and summation operations. Therefore this could lead to additional sending operations, and the idle time could increase. For example, "contraction" $R_K(G\backslash e, d)$ of an edge e is performed in the first process and at the same time "removal" $R_K(G/e, d)$ of this edge is performed in the second process. So, if the first process is already finished, it should wait for response from the second process to evaluate $R_K(G, D) = r_e \times R_K(G\backslash e, D) + (1 - r_e) \times R_K(G/e, D)$.

As in our previous work on parallel computing of all-terminal network reliability [7], we suggest sending a subtask to a helping process along with an auxiliary parameter p which is the probability of obtaining this subtask, e.g. the probability of graph realization to be sent. For the initial graph this parameter is equal to 1. For the first "contraction" subtask this parameter equals to r_e and for the first "removal" subtask this parameter equals to $1 - r_e$. Please note that on every step the current task could also have some probability p, so in general the parameters are $p * r_e$ and $p * (1 - r_e)$. This approach allows to avoid back-sending of the probability value calculated, and now it is possible to accumulate on every process its own value. When all processes are finished, all values are summed to get the exact probability value.

To implement the method described above we have chosen "Master-Slave" parallel programming model in which one *master* process controls all the other *guided* processes. This process is also responsible for summation of the involved values. A guided process evaluates tasks received from the master process, sends the results obtained back and then receives a new task. If it is necessary, the guided process can ask the master process for help in the form of an idle process.

Master process

- Controls of all the other processes load.
- Sends the initial graph to one of the guided processes to calculate.
- When it receives a help request, it either declines it or sends the number of an idle process back.
- Sums up all the values obtained on all the guided processes getting the exact reliability value.

Guided process

- Is initialized as an idle at the beginning.
- After receiving a task it performs the factorization procedure (with use of formula (1) or with use of CPFM). One of the subtasks remains on the current process for the further calculations. Thereafter the process sends the help request to the master process. According to the answer it either sends the other subtask to the idle process or calculates this subtask itself.
- When all calculation are finished, it sends the accumulated value to the master process and waits up to get the new task.

The difference between two parallel methods (CPFM and simple factoring) is only the format of a data that should be sent. In the CPFM we work with the list of structures instead of the graph, so the data size is much bigger. Therefore sending of all the required data takes more time than in the simple factoring method. Anyway, the CPFM itself is much faster than the basic factoring method. So in the next chapter we try to figure out which method is better to use in the parallel realization.

5   Case Studies

To compare the scalabilities of two parallel methods (CPFM and SFM) we have choosen a grid 5×5 topology, it contains 25 vertices and 40 edges. In spite of its small dimension, this graph is very hard for DCNR computing because it does not applicable to various accelerating methods. The number of terminals was set to 3 and the diameter was equal to 4 (Fig. 1). The terminals are nodes no. 12, 16, 20 when numbered from the upper left corner to the right. So using the sequential factoring method the reliability value was calculated in 16 min 9.507 s with 249602055 recursions. At the same time CPFM has finished in 0 min 0.061 s with only 74 recursions. It was decided to make a data for CPFM more complex in order to make both algorithms be finished in almost the same time. The following parameters were set for CPFM: the number of terminals: 5, and the diameter: 9. The terminals are nodes no. 1, 12, 16, 20, 23. With the new data the CPFM finished in 16 min 35.16 s with 1154905688 recursions.

The experiments were made on the computing cluster HKC-30T of the Siberian Supercomputer Center. This cluster consists of double-blade servers HP BL2220 G6 with Intel Xeon 5540 2.53 GHz CPUs.

Fig. 1. Tested graph

The scalabilities of the both methods are almost the same: a speedup can be observed till the number of cores is lower than 16. And when it is higher then 16, runtime became slower. Anyway, since the data for the CPFM is much more complicated for the calculations, we can conclude that in parallel realizations CPFM works much faster than the basic factoring method.

In [7] to speedup the calculation time of all-terminal reliability one important parameter N_{Edges} was introduced: the lower limit of a dimension of graph that could be sent to another process. For example, it is useless to send a small dimension graph to another process since it would be faster to calculate it in the one process. The experiments have shown that this parameter is significantly affect the working time of the algorithm. In the diameter constraint case the analogue of N_{Edges} was found: a lower limit of considered edges amount. So when the edges amount in the current task is below this limit, the working process stops sending help requests to the master process and executes all procedures without any help.

Below we try to find optimal value of N_{Edges}, which make the algorithm faster, using the binary search. Figures 2 and 3 show the scalability of the proposed algorithm for different values of N_{Edges}. As we can see the optimal value is between 10 and 20. Continuing the search is worthless because the difference between 10 and 20 is 1 s only.

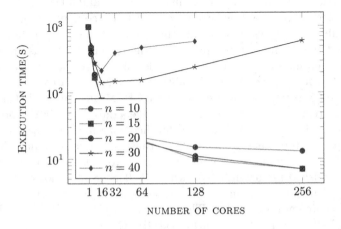

Fig. 2. Scalability of the simple factoring method

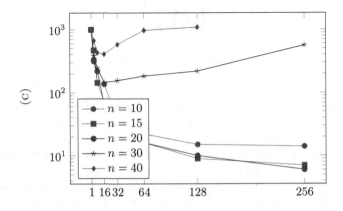

Fig. 3. Scalability of the CPFM

The results show that the CPFM works well in parallel implementation for supercomputers with distributed memory. The algorithm shows a linear speedup for number of cores less than 256.

6   Conclusion

We introduce a parallel implementation of the factoring method for exact calculation of network reliability in case of the diameter constraint. Also we have suggested to set one important parameter of the proposed algorithm which significantly improve its performance. The results of the numerical experiments show that the CPFM works well in parallel implementation for supercomputers with distributed memory. Our next primary goal is to further improve scalability of the proposed algorithm. It seems that there are two ways to do this: to use the several master processes or do not use them at all.

References

1. Won, J.-M., Karray, F.: Cumulative update of all-terminal reliability for faster feasibility decision. IEEE Trans. Reliabil. **59**(3), 551–562 (2010)
2. Page, L.B., Perry, J.E.: A practical implementation of the factoring theorem for network reliability. IEEE Trans. Reliabil. **37**(3), 259–267 (1998)
3. Cancela, H., Petingi, L.: Diameter constrained network reliability: exact evaluation by factorization and bounds. In: International Conference on Industrial Logistics, Okinawa, Japan, pp. 359–356 (2001)
4. Canale, E., Cancela, H., Robledo, F., Romero, P., Sartor, P.: Full complexity analysis of the diameter-constrained reliability. Int. Trans. Oper. Res. **22**(5), 811–821 (2015)
5. Migov, D.A., Nesterov, S.N.: Methods of speeding up of diameter constrained network reliability calculation. In: Gervasi, O., Murgante, B., Misra, S., Gavrilova, M.L., Rocha, A.M.A.C., Torre, C., Taniar, D., Apduhan, B.O. (eds.) ICCSA 2015. LNCS, vol. 9156, pp. 121–133. Springer, Cham (2015). doi:10.1007/978-3-319-21407-8_9
6. Martnez, S.P., Calvino, B.O., Rocco, S.C.M.: All-terminal reliability evaluation through a Monte Carlo simulation based on an MPI Implementation. In: European Safety and Reliability Conference: Advances in Safety, Reliability and Risk Management (PSAM 2011/ESREL 2012), Helsinki, pp. 1–6 (2012)
7. Migov, D.A., Rodionov, A.S.: Parallel Implementation of the factoring method for network reliability calculation. In: Murgante, B., et al. (eds.) ICCSA 2014. LNCS, vol. 8584, pp. 654–664. Springer, Cham (2014). doi:10.1007/978-3-319-09153-2_49

Congestion Game Scheduling Implementation for High-Throughput Virtual Drug Screening Using BOINC-Based Desktop Grid

Natalia Nikitina[1]([✉]), Evgeny Ivashko[1], and Andrei Tchernykh[2]

[1] Institute of Applied Mathematical Research, Karelian Research Center,
Russian Academy of Sciences, Petrozavodsk, Russia
{nikitina,ivashko}@krc.karelia.ru
[2] Computer Science Department, CICESE Research Center,
Ensenada, Baja California, Mexico
chernykh@cicese.mx

Abstract. Virtual drug screening is one of the most common applications of high-throughput computing. As virtual screening is time consuming, a problem of obtaining a diverse set of hits in a short time is very important. We propose a mathematical model based on game theory. Task scheduling for virtual drug screening in high-performance computational systems is considered as a congestion game between computing nodes to find the equilibrium solutions for best balancing between the number of interim hits and their chemical diversity. We present the developed scheduling algorithm implementation for Desktop Grid and Enterprise Desktop Grid, and perform comprehensive computational experiments to evaluate its performance. We compare the algorithm with two known heuristics used in practice and observe that game-based scheduling outperforms them by the hits discovery rate and chemical diversity at earlier steps.

Keywords: Drug discovery · Virtual drug screening · High-performance computing · High-throughput computing · Desktop grid · BOINC · Scheduling · Game theory · Congestion game

1 Introduction

Drug development is a time-consuming process. It takes up to 10–15 years to develop a new market-available drug [1]. One of the first stages of this process is an identification of a set of chemical compounds called *hits* with predicted desired biochemical activity. Hits are identified among a set of ligands, low-molecular compounds able to form a biochemical complex with a protein molecule responsible for disease progression, called a target.

With computer development of highly accurate specific disease models to validate targets and bind ligands to targets, virtual screening [2] (VS) has emerged to aid this stage of research. In the course of VS, one performs computer modeling of the interaction of the candidate ligands with the target and scores the resulting molecular complexes. The ligands with high scores become hits.

© Springer International Publishing AG 2017
V. Malyshkin (Ed.): PaCT 2017, LNCS 10421, pp. 480–491, 2017.
DOI: 10.1007/978-3-319-62932-2_46

The chemical space of all small organic molecules, each of which might become a drug for the studied disease, is estimated to have the order of 10^{60} [3]. The libraries that are used in VS setups typically have size from hundreds of thousands to tens of millions of molecules [4–7]. The largest database GDB-17 contains more than 166 billion molecules [8]. Performing VS over such large databases essentially requires high-performance computational facilities. The exhaustive structure-based VS could take over five months at one of the world's fastest supercomputers Tianhe-2 [9].

With proper VS organization, the interim results can significantly contribute to the research progress boost. Considering the long time interval to screen the whole ligand library, it is important to obtain a diverse set of hits as soon as possible, so that they could proceed to the next stage of research in laboratory without finding the rest.

In this contribution, we propose a method for reducing the explored ligands space on the fly when performing structure-based VS to obtain diverse and useful results in a short time. The method is based on mathematical model of task scheduling. We develop an algorithm which promises the best balancing between the number of interim hits and their diversity. The algorithm does not depend on the docking algorithm implementation, the quality of protein and ligands models used for VS.

The model considers the heterogeneous environment with uncertainty of processing time and workload, and limited knowledge about the input dataset structure. The algorithm can be used to develop software based on a high-performance computing cluster, a Grid system or a Cloud. In this paper, we concentrate on the algorithm implementation for Desktop Grid and Enterprise Desktop Grid.

2 Related Work

In comparison with supercomputers and computational clusters, task scheduling in Desktop Grids is more complicated because of such factors as huge hardware and software heterogeneity, lack of trust, uncertainty, etc. A wide range of algorithms has been proposed in the literature to address these challenges. The most popular optimization parameters for task scheduling in Desktop Grids are the throughput [10, 11], the makespan [12, 13], the probability to obtain a correct result [10, 14], etc.

The task replication in Desktop Grids as a method to achieve optimization aims is a wide area of study [11, 14]. It often comes along with the notion of reliability or credibility of the computing nodes [10, 11, 14]. The availability patterns predictions of the computing nodes are also used to improve task scheduling in Desktop Grids [12].

Game-theoretic methods find their application for task scheduling in Desktop Grids [15, 16]. The thesis [17] presents methods for the fair distribution of resources between heterogeneous computing nodes in order to optimize throughput and average task completion times, basing on optimization methods and game theory.

The problem of drug discovery imposes additional challenges to the task scheduling in Desktop Grids. According to drug development principles, two primary characteristics of the compounds resulting set have to be tested in the laboratory: efficiency of their interaction with the target protein, and the chemical diversity [18, 19]. In general, these two objectives are in conflict: an improvement of one leads to simultaneous deterioration of another. Hence, we have to find solutions that balance them depending on drug developers preferences.

The most conventional method to reduce input dataset for VS down to manageable size is pre-filtering of the chemical space, leaving only representative and chemically diverse compounds that promise to show desired biological activities. In many cases, the resulting library is still large, VS results are redundant and require post-processing. Compound classes that are potentially good for a specific target, but do not comply with general considerations, can be filtered out [19]. Hence, it is important to develop new efficient VS methods for large datasets.

A recent example of genetic algorithms application [20] has been published in 2015. The authors explore the chemical space by creating generations of molecules, filtering them by desirable properties, and selecting maximally diverse sets of results. The algorithm demonstrates high efficiency in selecting diverse subsets of molecules with desired properties from large databases. Apart from genetic algorithms, compound libraries can be prepared using dissimilarity analysis [21], clustering analysis [22], partitioning methods [23] etc.

The mathematical model described in this paper has been elaborated in [24].

3   Congestion Game Model

Due to variations in chemical characteristics, molecules have different chances to show high predicted binding affinity. One can expect that these chances are higher for molecules close in topology to a known ligand [25, 26]. In contrast, molecules with very large number of atoms are less likely to become hits [27]. Thus, non-overlapping subsets of molecules in the library could be ranked by their prospectivity for VS.

The estimated prospectivity can be updated in the course of VS according to interim results. At the same time, results originating from the same subset might be redundant. The idea behind this work is to explore most prospective subsets first while keeping the desired level of diversity by restricting intensity of subsets exploration.

Let us consider a computer system with m computational nodes or players $C = C_1, \ldots C_m$, and a set of computational tasks T (data) for VS processing. Each node C_i is characterized by its computational performance ops_i, which is an average number of operations performed in a time unit.

The data set T is divided into n non-overlapping blocks $T = T_1 \cup \ldots \cup T_n$ of sizes $N_1^{total}, \ldots, N_n^{total}$ such that the estimated portion of VS hits in block T_j will be p_j. We define priority of the block T_j as

$$\sigma_j = \frac{p_j}{p_1 + \ldots + p_n}, 0 \leq \sigma_j \leq 1. \tag{1}$$

The blocks with higher priority have to be chosen first for processing. We assume that all tasks in block T_j have average computational complexity θ_j, i.e., a number of operations to process one task. Each node selects exactly one block.

The nodes make their decisions at time steps $0, \tau, 2\tau, \ldots$. After a node has processed its portion of tasks, it sends the results to the server and is ready for the next portion. Let the utility of node C_i at time step τ express the amount of useful work performed during this step. This amount depends on the number of executed tasks from the chosen block, its computational complexity, priority, and the number of other nodes who have also chosen this block.

The fewer nodes explore block T_j simultaneously, the more valuable their work is. This condition ensures diversification of the interim set of hits. Let n_j be the number of the players who have chosen block T_j at the considered step, and $\delta(n_j)$ be the congestion coefficient for the block:

$$\delta(n_j) = \frac{m + 1 - n_j}{m}. \tag{2}$$

The utility of node C_i that chooses block T_j is

$$U_{ij} = \left(\sigma_j + \delta(n_j)\right)\frac{ops_i}{\theta_j}\tau. \tag{3}$$

Therefore, at each considered time step, we have a singleton congestion game $G = (C, T, U)$, where C is the set of players (computational nodes), T is the set of data blocks of which each node selects exactly one, and U is the set of utility functions. A strategy profile is a schedule $s = (s_1, \ldots s_m)$, where the component $s_i = j$ equals to the block T_j chosen by player C_i.

Such games have been thoroughly studied in literature. The existence of at least one Nash equilibrium in pure strategies has been proven for the case of identical players [28] and identical task blocks [29]. The equilibrium situation means that no node can increase amount of its useful work by unilaterally deviating from the schedule. Moreover, better- and best-response dynamics are guaranteed to converge to equilibrium in polynomial time [29, 30].

Heterogeneity complicates the model: utility of each player depends both on its own performance, and on the task complexity of the chosen block. But due to the form of utility functions, the game G has at least one Nash equilibrium in pure strategies [31]. The Nash equilibrium situation means the best amount of useful work given the current estimates of probabilities and the available set of nodes and molecules. Further in this paper we show that the Nash equilibrium, for the most part, corresponds to the best proportion between the chemical diversity and hits number.

4   Algorithm Implementation

4.1   Desktop Grids

Desktop Grid is a form of distributed high-throughput computing system, which uses idle time of non-dedicated geographically distributed computing nodes connected over low-speed network. The computing nodes are personal desktop computers of volunteers connected over Internet (volunteer computing) or organization desktop computers connected over local area computer network (Enterprise Desktop Grid).

On the high-level, Desktop Grid has the following architecture. The main server holds a large number of tasks that are mutually independent pieces of a computationally heavy problem. Computing nodes are connected to the server. When they are idle, they request a work from the server, receive one or more tasks, and process them independently from each other. When the node finishes the processing, it reports results back to the server. The results are then stored in the database for further usage.

The server does not distribute all available tasks because of internal uncertainties present in Desktop Grid systems. These uncertainties are related to unknown availability periods of the nodes, their speed, node failures, possible computational errors, variation of tasks complexities, etc. With heterogeneous nodes and heterogeneous tasks, the scheduling system has variety of options how to assign different tasks to different nodes or the groups of nodes.

4.2 BOINC Platform

There are a number of middleware systems for Desktop Grid computing. However, the open source BOINC platform [32] is nowadays considered as de facto standard. Since 1990s, BOINC has been a framework for many independent volunteer computing projects. Today, it is the most actively developed Desktop Grid middleware, which supports the widest range of applications.

BOINC is based on server-client architecture, where the workflow proceeds as described in the previous subsection. The client part is able to work at an arbitrary number of computers with various hardware and software characteristics. The server part consists of the Web server that provides functionality to employ volunteer computing power, the database server that monitors the state of tasks and results, stores information about the clients and the whole computational process, and a set of daemon programs that periodically check the database state and implement necessary actions to operate the system and distribute the tasks.

In addition to the standard components of the BOINC server, each computational project might need individually developed utilities. One of them is a *task generator,* which creates computational tasks with specified parameters and attributes. The *client application* is developed individually for the project. This application can be either developed from scratch under the BOINC platform or be adapted using a wrapper program, which allows to run non-native applications under BOINC.

4.3 Implementation

In this section, we describe the Desktop Grid application, which implements our approach to VS in BOINC Desktop Grid middleware based on resources of the Karelian Research Center, RAS. The proposed implementation is intended for the BOINC server version 7.7.0 as well as for earlier versions.

In order to implement task scheduling algorithm proposed in this paper, one needs to implement the task generating program, modify the scheduler and assimilator daemons at the server side, and implement the docking application for the client side.

The BOINC server takes up every other part of the task workflow apart from the modifications proposed in this subsection.

Below, we describe the necessary modifications.

Firstly, the task generator must be able to create computational tasks based on the current knowledge about the input dataset structure, as defined in Sect. 3. According to the mathematical model, the knowledge about input dataset structure is being updated after each step of computations. Therefore, the task generator must consider the Desktop Grid properties as well, in order to supply the necessary amount of tasks for each subsequent step.

The task generator pseudo code is provided in Fig. 1.

```
1:  for Molecule in Database
2:    Properties = get_properties (Molecule)
3:    Block = get_block (Properties)
4:    create_task (Molecule, Block)
```

Fig. 1. The task generator pseudo code.

Secondly, the scheduler must consider the current state of computational process and assign new tasks to the clients who ask for work. The assignment is being performed according to the equilibrium game schedule as described in Sect. 3.

The scheduler pseudo code is provided in Fig. 2.

```
1:  if update signal received
2:    Dataset_structure = get_dataset_structure (Database)
      #s is the initial schedule vector where each client
      #chooses a separate non-empty block
3:    for i = 1 to m   #m is the number of BOINC clients
4:      s[i] = i % n   #n is the number of non-empty blocks
      #The optimal schedule vector is then calculated
      #by finite improvements sequence starting from s
      #according to the formulae (2)-(4)
5:    Schedule = get_equilibrium (s, Dataset_structure)
      #Each component of Schedule is the block number
      #chosen by the corresponding BOINC client
6:  if request signal received
7:    Client = client_id
8:    Block = Schedule[Client]
9:    Timespan = requested_timespan
10: send_work (Client, Block, Timespan)
```

Fig. 2. The scheduler pseudo code.

Thirdly, the assimilator must handle completed tasks. If the result is not erroneous, it must be considered for updating the knowledge of the input dataset structure. The assimilator pseudo code is provided in Fig. 3.

```
1: if Result received
2:   if Result.error_mask != 0
3:     handle_error (Result)
4: else
5:   if Result.value >= Hit_threshold
6:     update Hits table in Database
7:   update Blocks table in Database
8:   send update signal
```

Fig. 3. The assimilator pseudo code.

Finally, the client application must perform the molecular docking. According to the BOINC system workflow for VS, the input files are the target structure, the ligand structure and the docking configuration file. The output value is the predicted binding energy. A variety of docking programs can be used as the client application for VS.

5   Experimental Setup

5.1   Database Preparation

In order to perform computational experiments and evaluate the performance of the developed approach, we divide a molecules database into blocks and simulate VS. The efficiency of the approach can be shown by earlier acquisition and higher diversity of hits at early VS stages comparing with known heuristics used in practice.

We use the database GDB-9 of about 320 thousand enumerated organic molecules with variety of chemical properties. The chosen database is manageable for performing computational experiments and can be unambiguously divided into several non-overlapping blocks. Nevertheless, the set of molecules is rich enough to demonstrate the feasibility and practicability of proposed solutions.

For the experiments, we consider three pre-calculated chemical properties of each molecule: the total number of atoms, the polar surface area PSA, and the partition coefficient $logP$. Basing on these properties, we divide the database into 16 non-overlapping task blocks.

As 10% of molecules in GDB-9 have $logP \geq x = 1.8823$, the value $x = 1.8823$ is taken as a threshold to count a molecule as a hit.

5.2   The Chemical Diversity Measure

We employ the knowledge about input dataset structure to describe and investigate the chemical diversity of interim results. In perfect case, the fraction of hits discovered in each pre-defined block at each computational step should be equal for all the blocks. We define the diversity of a result subset as expression (4) shows. Here, h_i is the number of hits in block T_i, p_i is the estimation of hits fraction in the block, and N_i^{total} is the initial block size.

$$D = \max_{1 \leq i \leq n} \frac{h_i}{p_i N_i^{total}} - \min_{1 \leq i \leq n} \frac{h_i}{p_i N_i^{total}}. \tag{4}$$

In Subsection 5.4 we provide the dynamics of the chemical diversity obtained at each computational step and show that the proposed scheduling algorithm outperforms two scheduling heuristics at early steps.

5.3 Experimental Setup

We performed numerical experiments, simulating homogeneous and heterogeneous Desktop Grid consisting of 64 computing nodes. As a first test case, we consider the case with identical computational nodes and tasks of identical complexities. At the second test case, we consider a Desktop Grid with heterogeneous nodes and heterogeneous tasks. The parameters of the simulations are provided in Table 1.

Table 1. Simulation parameters.

Parameter	Value		Description
	First test case	Second test case	
ops	25	15 (nodes C_1–C_{16})	Performance of a computational
		20 (nodes C_{17}–C_{32})	node (number of conditional
		25 (nodes C_{33}–C_{48})	operations per time unit)
		30 (nodes C_{49}–C_{64})	
θ	100	100 (blocks T_1–T_4)	Complexity of a computational task
		125 (blocks T_5–T_8)	(number of conditional operations)
		125 (blocks T_9–T_{12})	
		150 (blocks T_{13}–T_{16})	

At each VS step, the optimal schedule is computed based on the current knowledge about expected fractions of hits in blocks. After completion of the computations, the expected fractions of hits are updated according to the number of hits discovered in each block. Then a next step is performed, etc.

The performance of the proposed game scheduling, where each node selects a task block defined by the Nash equilibrium, is compared with two simple scheduling strategies: Probabilistic scheduling and Uniform scheduling. The probabilistic scheduling strategy represents the case when the selection of a task block does not depend on the congestion level of the block, but only on the probability to find a hit. Simulation results for the probabilistic scheduling are averaged on 20 runs. On the contrary, the uniform scheduling strategy ensures the least possible level of congestion,

488   N. Nikitina et al.

or the highest diversity, by distributing the nodes across task blocks uniformly, so, as many diverse blocks are explored simultaneously as possible.

5.4 Experimental Analysis

In Figs. 4 and 5, we provide the rate of discovery hits during the first and second test cases, respectively. Each discrete point represents the fraction of the total amount of hits discovered at the corresponding step. The quality measure of such curves is their proximity to the upper left corner of the chart.

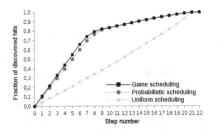

Fig. 4. Fraction of hits discovered at each step of simulations (identical nodes, identical tasks).

Fig. 5. Fraction of hits discovered at each step of simulations (heterogeneous nodes, heterogeneous tasks).

To illustrate the performance in terms of obtained chemical diversity, in Figs. 6 and 7, we provide the normalized chemical diversity obtained during the first and the second test cases, respectively. The chemical diversity is defined by formula (4).

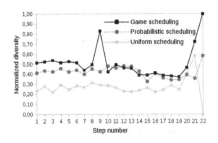
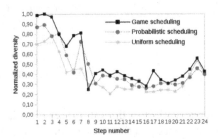

Fig. 6. Normalized diversity obtained at each step of simulations (identical nodes, identical tasks).

Fig. 7. Normalized diversity obtained at each step of simulations (heterogeneous nodes, heterogeneous tasks).

The presented results indicate that the game scheduling algorithm outperforms both heuristics by the fraction of discovered hits at early steps. Performance of the probabilistic algorithm is approximately equal to performance of the game scheduling algorithm in terms of the speed of hits discovery. However, the performances in terms of obtained chemical diversity at early steps differ significantly.

6   Conclusion and Discussion

In this paper, we present an implementation of congestion game-based scheduling algorithm for high-throughput virtual drug screening using BOINC-based Desktop Grid. It is based on the mathematical model of game theory, where task scheduling is considered as a game with computing nodes as players, who choose specific data blocks for processing. We show that the equilibrium solution corresponds to the best balance between the number of interim hits and their chemical diversity. We discuss key points of implementations: the task generating program, scheduler and assimilator daemon at the server side, and present pseudo codes.

We perform computational experiments on the Enterprise Desktop Grid based on resources of the Karelian Research Center, RAS. We compare the algorithm with two known heuristics used in practice and observe that game-based scheduling outperforms them by the hits discovery rate and chemical diversity at earlier steps.

However, further study is required to assess its performance and effectiveness in multi objective domains. This will be the subject of future work. Moreover, game equilibrium solutions stability and selection of the most efficient algorithms among all equilibria are other important issues to be addressed.

Acknowledgments. This work is partially supported by the Russian Fund for Basic Research under grants no. 16-07-00622 and 15-29-07974, and CONACYT (Consejo Nacional de Ciencia y Tecnología, México) under grant no. 178415.

References

1. Pharmaceutical Research and Manufacturers of America (PhRMA). Biopharmaceutical Industry Profile (2016). http://phrma.org/sites/default/files/pdf/biopharmaceutical-industry-profile.pdf accessed 2017/05/14
2. Bielska, E., Lucas, X., Czerwoniec, A., et al.: Virtual screening strategies in drug design — methods and applications. J. Biotechnol. Comput. Biol. Bionanotechnol. **92**(3), 249–264 (2011)
3. Bohacek, R.S., McMartin, C., Guida, W.C.: The art and practice of structure-based drug design: A molecular modeling perspective. Med. Res. Rev. **16**(1), 3–50 (1996)
4. Irwin, J., et al.: ZINC: a free tool to discover chemistry for biology. J. Chem. Inf. Model. **52**, 1757–1768 (2012)
5. Bento, A.P., et al.: The ChEMBL bioactivity database: an update. Nucleic Acids Res. **42**, 1083–1090 (2014)
6. Pence, H.E., Williams, A.: ChemSpider: an online chemical information resource. J. Chem. Educ. **87**(11), 1123–1124 (2010)
7. Bolton, E.E., et al.: Chapter 12 - PubChem: integrated platform of small molecules and biological activities. Annu. Rep. Comput. Chem. **4**, 217–241 (2008). Elsevier

8. Ruddigkeit, L., van Deursen, R., Blum, L.C., Reymond, J.-L.: Enumeration of 166 billion organic small molecules in the chemical universe database GDB-17. J. Chem. Inf. Model. **52**, 2864–2875 (2012)

9. Liu, T., et al.: Applying high performance computing in drug discovery and molecular simulation. Nat. Sci. Rev. **3**(1), 49–63 (2016)

10. Yasuda, S., Nogami, Y., Fukushi, M.: A dynamic job scheduling method for reliable and high-performance volunteer computing. In: 2nd International Conference on Information Science and Security (ICISS 2015), pp. 1–4. IEEE (2015)

11. Sonnek, J., Chandra, A., Weissman, J.: Adaptive reputation-based scheduling on unreliable distributed infrastructures. IEEE Trans. Parallel Distrib. Syst. **18**(11), 1551–1564 (2007)

12. Byun, E., et al.: MJSA: Markov job scheduler based on availability in desktop grid computing environment. Futur. Gener. Comput. Syst. **23**, 616–622 (2007)

13. Gil, J.-M., Kim, S., Lee, J.: Task scheduling scheme based on resource clustering in desktop grids. Int. J. Commun. Syst. **27**(6), 918–930 (2014)

14. Miyakoshi, Y., Watanabe, K., Fukushi, M., Nogami, Y.: A job scheduling method based on expected probability of completion of voting in volunteer computing. In: 2nd International Symposium on Computing and Networking, pp. 399–405. IEEE (2014)

15. Wang, Y., et al.: Toward integrity assurance of outsourced computing — a game theoretic perspective. Futur. Gener. Comput. Syst. **55**, 87–100 (2016)

16. Donassolo, B., et al.: Non-cooperative scheduling considered harmful in collaborative volunteer computing environments. In: Proceedings of 11th IEEE/ACM International Symposium on Cluster, Cloud and Grid Computing (CCGrid), pp. 144–153 (2011)

17. Legrand, A.: Scheduling for large scale distributed computing systems: approaches and performance evaluation issues. Distrib. Parallel, Clust. Comput. [cs.DC], Université Grenoble Alpes, p. 167 (2015)

18. Tanrikulu, Y., Krüger, B., Proschak, E.: The holistic integration of virtual screening in drug discovery. Drug Discov. Today **18**(7/8), 358–364 (2013)

19. Lionta, E., Spyrou, G., Vassilatis, D.K., Cournia, Z.: Structure-based virtual screening for drug discovery: principles, applications and recent advances. Curr. Top. Med. Chem. **14**, 1923–1938 (2014)

20. Rupakheti, C., Virshup, A., Yang, W., Beratan, D.N.: Strategy to discover diverse optimal molecules in the small molecule universe. J. Chem. Inf. Model. **55**, 529–537 (2015)

21. Ashton, M., et al.: Identification of diverse database subsets using property-based and fragment-based molecular descriptions. Quant. Struct. Act. Relationsh. **21**, 598–604 (2002)

22. Downs, G.M., Barnard, J.M.: Clustering methods and their uses in computational chemistry. Rev. Comput. Chem. **18**, 1–40 (2003)

23. Oprea, T.I., Gottfries, J.: Chemography: the art of navigating in chemical space. J. Comb. Chem. **3**, 157–166 (2001)

24. Nikitina, N., Ivashko, E., Tchernykh, A.: Congestion game scheduling for virtual drug screening optimization. J. Comput. Aided Mol. Des. (2017). Manuscript submitted for publication

25. Patterson, D.E., et al.: Neighborhood behavior: a useful concept for validation of "molecular diversity" descriptors. J. Med. Chem. **39**, 3049–3059 (1996)

26. Willet, P., Barnard, J.M., Downs, G.M.: Chemical similarity searching. J. Chem. Inf. Comput. Sci. **38**(6), 983–996 (1998)

27. Hann, M.M., Leach, A.R., Harper, G.: Molecular complexity and its impact on the probability of finding leads for drug discovery. J. Chem. Inf. Comput. Sci. **41**, 856–864 (2001)

28. Rosenthal, R.: A class of games possessing pure-strategy Nash equilibria. Int. J. Game Theor. **2**(1), 65–67 (1973)

29. Milchtaich, I.: Congestion games with player-specific payoff functions. Games Econ. Behav. **13**, 111–124 (1996)

30. Ieong, S. et al.: Fast and compact: a simple class of congestion games. In: Proceedings of AIII, pp. 1–6 (2005)

31. Gairing, M., Klimm, M.: Congestion games with player-specific costs revisited. In: Vöcking, B. (ed.) SAGT 2013. LNCS, vol. 8146, pp. 98–109. Springer, Heidelberg (2013). doi:10. 1007/978-3-642-41392-6_9

32. Anderson, D.P.: BOINC: A system for public-resource computing and storage. In: Proceedings of 5th IEEE/ACM International Workshop on Grid Computing, pp. 4–10 (2004)

Globalizer – A Parallel Software System for Solving Global Optimization Problems

Alexander Sysoyev[✉], Konstantin Barkalov, Vladislav Sovrasov,
Ilya Lebedev, and Victor Gergel

Lobachevsky State University of Nizhni Novgorod, Nizhny Novgorod, Russia
alexander.sysoyev@itmm.unn.ru

Abstract. In this paper, we describe the Globalizer software system for solving global optimization problems. The system implements an approach to solving the global optimization problems using the block multistage scheme of the dimension reduction, which combines the use of Peano curve type evolvents and the multistage reduction scheme. The scheme allows an efficient parallelization of the computations and increasing the number of processors employed in the parallel solving of the global optimization problems many times.

Keywords: Multidimensional multiextremal optimization · Global search algorithms · Parallel computations · Dimension reduction · Block multistage dimension reduction scheme

1 Introduction

The development of optimization methods that use high-performance computing systems to solve time-consuming global optimization problems is an area receiving extensive attention. The theoretical results obtained provide efficient solutions to many applied global optimization problems in various fields of scientific and technological applications. At the same time, the practical software implementation of these algorithms for multiextremal optimization is quite limited. Among the software for the global optimization, one can select the following systems: LGO (Lipschitz Global Optimization) [14], GlobSol [11], LINDO [12], IOSO (Indirect Optimization on the basis of Self-Organization) [3], MATLAB Global Optimization Toolkit [23], TOMLAB system [10], BARON (Branch-And-Reduce Optimization Navigator) [15], GAMS (General Algebraic Modeling System) [2], Global Optimization Library in R [13].

In this paper, a novel Globalizer software system is considered. The development of the system was conducted based on the information-statistical theory of multiextremal optimization aimed at developing efficient parallel algorithms for global search – see, for example, [21, 22]. The Globalizer advantage is that the system is designed to solve time-consuming multiextremal optimization problems. In order to obtain global optimized solutions within a reasonable time and cost, the system efficiently uses modern high-performance computer systems.

© Springer International Publishing AG 2017
V. Malyshkin (Ed.): PaCT 2017, LNCS 10421, pp. 492–499, 2017.
DOI: 10.1007/978-3-319-62932-2_47

2 Statement of Multidimensional Global Optimization Problem

In this paper, the core class of optimization problems which can be solved by using Globalizer is examined. This involves multidimensional global optimization problems without constraints, which can be defined in the following way:

$$\varphi(y) \to \inf, \ y \in D \subset R^N, \tag{1}$$

$$D = \left\{ y \in R^N \colon a_i \leq y_i \leq b_i, 1 \leq i \leq N \right\}. \tag{2}$$

If y^* is an exact solution of problem (1)–(2), the numerical solution of the problem is reduced to building an estimate y^0 of the exact solution matching to some notion of nearness to a point (for example, $\|y^* - y^0\| \leq \varepsilon$ where $\varepsilon > 0$ is a predefined accuracy) based on a finite number k of computations of the optimized function values.

Regarding to the class of problems considered, the fulfillment of the following important conditions is supposed:

1. The optimized function $\varphi(y)$ can be defined by some algorithm for the computation of its values at the points of the domain D.
2. The computation of the function value at every point is a computation-costly operation.
3. Function $\varphi(y)$ satisfy the Lipschitz condition.

3 Globalizer Architecture

The Globalizer expands the family of global optimization software systems successively developed by the authors during the past several years [1, 5].

The Globalizer architecture is presented in Fig. 1. The structural components of the systems are:

- Block 0 is an external block. It consists of the procedures for computing the function values (criteria and constraints) for the optimization problem being solved.
- Blocks 1–4 form the optimization subsystem for solving the global optimization problems (Block 1), nonlinear programming (Block 2), multicriterial optimization (Block 3), and general decision making problems (Block 4).
- Block 5 is a subsystem for accumulating and processing the search information.
- Block 6 contains the dimensional reduction procedures based on the Peano evolvents; this block also provides interaction between the optimization blocks and the initial multidimensional optimization problem.
- Block 7 organizes the choice of parallel computation schemes in the Globalizer system subject to the computing system architecture employed (the numbers of cores in the processors, the availability of shared and distributed memory, the availability of accelerators for computations, etc.) and the global optimization methods applied.

- Block 8 is responsible for managing the parallel processes when performing the global search (determining the optimal configuration of parallel processes, distributing the processes between computing elements, etc.).
- Block 9 is a management subsystem, which controls the computational process when solving global optimization problems.
- Block 10 is responsible for organizing the dialog interaction with users for stating the optimization problem, adjusting system parameters (if necessary), and visualizing and presenting the global search results.
- Block 11 is a set of tools for visualizing and presenting the global search results; the availability of tools for visually presenting the computational results enables the user to provide efficient control over the global optimization process.

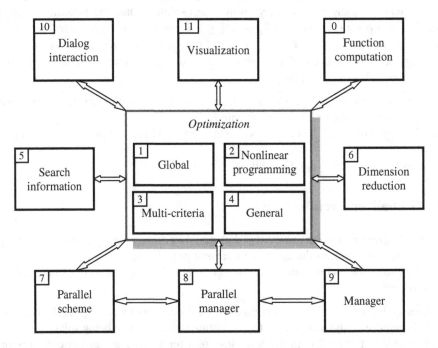

Fig. 1. Program architecture of Globalizer system (Blocks 1–2, 5–7 have been implemented; Blocks 3–4 and 8–11 are under development)

4   Globalizer Approach for Solving the Global Optimization Problems

4.1   Methods of Dimension Reduction

The Globalizer implements a block multistage scheme of dimension reduction [1], which reduces the solving of initial multidimensional optimization problem (1)–(2) to the solving of a sequence of «nested» problems of less dimensionality.

Thus, initial vector y is represented as a vector of the «aggregated» macro-variables

$$y = (y_1, y_2, \ldots, y_N) = (u_1, u_2, \ldots, u_M) \qquad (3)$$

where the i-th macro-variable u_i is a vector of the dimensionality N_i from the components of the vector y taken sequentially i.e.

$$
\begin{aligned}
u_1 &= (y_1, y_2, \ldots, y_{N_1}), \\
u_2 &= (y_{N_1+1}, y_{N_1+2}, \ldots, y_{N_1+N_2}), \ldots \\
u_i &= (y_{p+1}, \ldots, y_{p+N_i}), \ldots
\end{aligned}
\qquad (4)
$$

where $p = \sum_{k=1}^{i-1} N_k$ and $\sum_{k=1}^{M} N_k = N$.

Using the macro-variables, the main relation of the well-known multistage scheme can be rewritten in the form

$$\min_{y \in D} \varphi(y) = \min_{u_1 \in D_1} \min_{u_2 \in D_2} \ldots \min_{u_M \in D_M} \varphi(y), \qquad (5)$$

where the subdomains D_i, $1 \le i \le M$, are the projections of the initial search domain D onto the subspaces corresponding to the macro-variables u_i, $1 \le i \le M$.

It should be pointed that the nested subproblems

$$\varphi_i(u_1, \ldots, u_i) = \min_{u_{i+1} \in D_{i+1}} \varphi_{i+1}(u_1, \ldots, u_i, u_{i+1}), \; 1 \le i \le M, \qquad (6)$$

are the multidimensional ones in the block multistage scheme and this is the key difference from the initial scheme. Thus, this approach can be combined with the reduction of the domain D (for example, with the evolvent based on Peano curve) for the possibility to use the efficient methods of solving the one-dimensional problems of the multiextremal optimization [20].

The Peano curve $y(x)$ lets map the interval of the real axis [0, 1] onto the domain D uniquely:

$$\{y \in D \subset R^N\} = \{y(x) : 0 \le x \le 1\}. \qquad (7)$$

The evolvent is the approximation to the Peano curve with the accuracy of the order 2^{-m} where m is the density of the evolvent.

Application the mappings of this kind allows reducing multidimensional problem (1)–(2) to a one-dimensional one

$$\varphi(y^*) = \varphi(y(x^*)) = min\{\varphi(y(x)) : x \in [0, 1]\}. \qquad (8)$$

4.2   Method for Solving the Reduced Global Optimization Problems

The information-statistical theory of global search formulated in [22] has served as a basis for the development of a large number of efficient multiextremal optimization methods – see, for example, [6–9, 16–19], etc. Within the framework of information-statistical theory, a general approach to parallelization computations when solving global optimization problems has been proposed – the parallelism of computations is

496 A. Sysoyev et al.

provided by means of simultaneously computing the values of the minimized function $\varphi(y)$ at several different points within the search domain.

The general computation scheme of Parallel Multidimensional Algorithm of Global Search (PMAGS) that is implemented in Globalizer is fully described in [1].

4.3 Implementation of Parallel Algorithm of Global Optimization

Let us consider a parallel implementation of the block multistage dimension reduction scheme described in Subsect. 4.1.

For the description of the parallelism in the multistage scheme, let us introduce a vector of parallelization degrees

$$\pi = \left(\pi_1, \pi_2, \dots, \pi_M\right), \tag{9}$$

where π_i, $1 \leq i \leq M$, is the number of the subproblems of the $(i + 1)$-th nesting level being solved in parallel, arising as a result of execution of the parallel iterations at the i-th level. For the macro-variable u_i, the number π_i means the number of parallel trials in the course of minimization of the function

$$\varphi_M\left(u_1, \dots, u_M\right) = \varphi\left(y_1, \dots, y_N\right)$$

with respect to u_i at fixed values of $u_1, u_2, \dots, u_{i-1}$, i.e. the number of the values of the objective function $\varphi(y)$ computed in parallel.

In the general case, the quantities π_i, $1 \leq i \leq M$ can depend on various parameters and can vary in the course of optimization, but we will limit ourselves to the case when all components of the vector π are constant.

At every nested multistage dimension reduction level PMAGS is used. Let us remind that the parallelization is implemented by selection not a single point for the next trial (as in the serial version) but p points, which are placed into p intervals with the highest characteristics. Therefore, if p processors are available, p trials can be executed in these points in parallel. At a result, the solving of the problem at the i-th level generates p subproblems for the $(i + 1)$-th level.

5 Numerical Results

The computational experiments were conducted using the Lobachevsky supercomputer at the State University of Nizhny Novgorod (http://hpc-education.unn.ru/en/resources). The problems generated by the GKLS-generator [4] were selected for the test problems.

The results of the numerical experiments with Globalizer on an Intel Xeon Phi are provided in Table 1. The computations were performed using the Simple and Hard function classes with the dimensions equal to 4 and 5.

Table 1. Average number of iterations

		p	N = 4		N = 5	
			Simple	Hard	Simple	Hard
I	**Serial computations** Average number of iterations	1	11953	25263	15920	>148342(4)
II	**Parallel computations on CPU** Speedup	2	2.51	2.26	1.19	1.36
		4	5.04	4.23	3.06	2.86
		8	8.58	8.79	4.22	6.56
III	**Parallel computations on Xeon Phi** Speedup	60	8.13	7.32	9.87	6.55
		120	16.33	15.82	15.15	17.31
		240	33.07	27.79	38.80	59.31

In the first series of experiments, serial computations using MAGS were executed. The average number of iterations performed by the method for solving a series of problems for each of these classes is shown in row I. The symbol ">" reflects the situation where not all problems of a given class were solved by a given method. It means that the algorithm was stopped once the maximum allowable number of iterations K_{max} was achieved. In this case, the K_{max} value was used for calculating the average number of iterations corresponding to the lower estimate of this average value. The number of unsolved problems is specified in brackets.

In the second series of experiments, parallel computations were executed on a CPU. The relative "speedup" in iterations achieved is shown in row II; the speedup of parallel computations was measured in relation to the serial computations ($p = 1$).

The final series of experiments was executed using a Xeon Phi. The results of these computations are shown in row III; in this case, the speedup factor is calculated in relation to the PMAGS results on a CPU using eight cores ($p = 8$).

6   Conclusion

In this paper, the Globalizer global optimization software system was presented for implementing a general scheme for the parallel solution of globally optimized decision making. The work is devoted to the investigation of the possibility to speedup the process of searching the global optimum when solving the multidimensional multiextremal optimization problems using the approach based on the application of the parallel block multistage scheme of the dimension reduction.

This research was supported by the Russian Science Foundation, project No 16-11-10150 "Novel efficient methods and software tools for the time consuming decision making problems with using supercomputers of superior performance".

References

1. Barkalov, K.A., Gergel, V.P.: Multilevel scheme of dimensionality reduction for parallel global search algorithms. In: Proceedings of the 1st International Conference on Engineering and Applied Sciences Optimization, pp. 2111–2124 (2014)
2. Bussieck, M.R., Meeraus, A.: General algebraic modeling system (GAMS). In: Kallrath, J. (ed.) Modeling Languages in Mathematical Optimization, pp. 137–157. Springer, Boston (2004). doi:10.1007/978-1-4613-0215-5_8
3. Egorov, I.N., Kretinin, G.V., Leshchenko, I.A., Kuptzov, S.V.: IOSO optimization toolkit - novel software to create better design. In: 9th AIAA/ISSMO Symposium on Multidisciplinary Analysis and Optimization, Atlanta, Georgia (2002). http://www.iosotech.com/text/2002_4329.pdf
4. Gaviano, M., Lera, D., Kvasov, D.E., Sergeyev, Y.D.: Software for generation of classes of test functions with known local and global minima for global optimization. ACM Trans. Math. Software 29, 469–480 (2003)
5. Gergel, V.P.: A software system for multiextremal optimization. Eur. J. Oper. Res. 65(3), 305–313 (1993)
6. Gergel, V.P.: A method of using derivatives in the minimization of multiextremum functions. Comput. Math. Math. Phys. 36(6), 729–742 (1996)
7. Gergel, V.P.: A global optimization algorithm for multivariate function with Lipschitzian first derivatives. J. Glob. Optim. 10(3), 257–281 (1997)
8. Gergel, V., Lebedev, I.: Heterogeneous parallel computations for solving global optimization problems. Procedia Comput. Sci. 66, 53–62 (2015)
9. Gergel, V.P., Strongin, R.G.: Parallel computing for globally optimal decision making. In: Malyshkin, V.E. (ed.) PaCT 2003. LNCS, vol. 2763, pp. 76–88. Springer, Heidelberg (2003). doi:10.1007/978-3-540-45145-7_7
10. Holmström, K., Edvall, M.M.: The TOMLAB optimization environment. In: Kallrath, J. (ed.) Modeling Languages in Mathematical Optimization. Applied Optimization, vol. 88, pp. 369–376. Springer, Boston (2004). doi:10.1007/978-1-4613-0215-5_19
11. Kearfott, R.B.: GlobSol user guide. Optim. Methods Softw. 24, 687–708 (2009)
12. Lin, Y., Schrage, L.: The global solver in the LINDO API. Optim. Methods Softw. 24, 657–668 (2009)
13. Mullen, K.M.: Continuous global optimization in R. J. Stat. Softw. 60(6) (2014)
14. Pintér, J.D.: Global Optimization in Action: Continuous and Lipschitz Optimization: Algorithms, Implementations and Applications. Springer, New York (1996). doi:10.1007/978-1-4757-2502-5
15. Sahinidis, N.V.: BARON: a general purpose global optimization software package. J. Glob. Optim. 8(2), 201–205 (1996)
16. Sergeyev, Y.D.: An information global optimization algorithm with local tuning. SIAM J. Optim. 5(4), 858–870 (1995)
17. Sergeyev, Y.D.: Multidimensional global optimization using the first derivatives. Comput. Math. Math. Phys. 39(5), 743–752 (1999)
18. Sergeyev, Y.D., Grishagin, V.A.: A parallel method for finding the global minimum of univariate functions. J. Optim. Theor. Appl. 80(3), 513–536 (1994)
19. Sergeyev, Y.D., Grishagin, V.A.: Parallel asynchronous global search and the nested optimization scheme. J. Comput. Anal. Appl. 3(2), 123–145 (2001)
20. Sergeyev, Y., Strongin, R.G., Lera, D.: Introduction to Global Optimization Exploiting Space-Filling Curves. Springer, New York (2013). doi:10.1007/978-1-4614-8042-6

21. Strongin, R.G., Gergel, V.P., Grishagin, V.A., Barkalov, K.A.: Parallel Compucations for Global Optimization Problems. Moscow State University, Moscow (2013). (In Russian)
22. Strongin, R.G., Sergeyev, Y.D.: Global Optimization with Non-convex Constraints: Sequential and Parallel Algorithms. Springer, New York (2000). doi: 10.1007/978-1-4615-4677-1
23. Venkataraman, P.: Applied Optimization with MATLAB Programming. Wiley, Chichester (2009)

A Novel String Representation and Kernel Function for the Comparison of I/O Access Patterns

Raul Torres$^{(\boxtimes)}$ ⓘ, Julian Kunkel ⓘ, Manuel F. Dolz ⓘ, and Thomas Ludwig

Scientific Computing Research Group, Universität Hamburg, Hamburg, Germany
raul.torres@informatik.uni-hamburg.de

Abstract. Parallel I/O access patterns act as fingerprints of a parallel program. In order to extract meaningful information from these patterns, they have to be represented appropriately. Due to the fact that string objects can be easily compared using Kernel Methods, a conversion to a weighted string representation is proposed in this paper, together with a novel string kernel function called Kast Spectrum Kernel. The similarity matrices, obtained after applying the mentioned kernel over a set of examples from a real application, were analyzed using Kernel Principal Component Analysis (Kernel PCA) and Hierarchical Clustering. The evaluation showed that 2 out of 4 I/O access pattern groups were completely identified, while the other 2 conformed a single cluster due to the intrinsic similarity of their members. The proposed strategy can be promisingly applied to other similarity problems involving tree-like structured data.

Keywords: Kernel functions · Kast spectrum kernel · I/O access pattern comparison · Kernel PCA

1 Introduction

I/O access patterns act as fingerprints of an application. The identification and analysis of these patterns is important in High Performance Computing because it helps, not only to understand the impact factors on the underlying Parallel File System, but also to design better ways of organizing I/O operations. In order to understand the correlation of a collection of patterns, two requirements have to be met: (a) a proper representation able to abstract the relevant features of each pattern and (b) an appropriate strategy to find similarities or dissimilarities between the data in this new representation. To tackle (a) this paper proposes a two-stage string conversion technique for access patterns. The first stage transforms the data and reflects the containment relationships of the pattern in a tree-like data structure. The second stage flattens the resulting tree and simplifies the representation in a weighted string. In order to tackle (b) these weighted strings are compared with a novel string kernel function called Kast Spectrum Kernel.

ⓒ Springer International Publishing AG 2017
V. Malyshkin (Ed.): PaCT 2017, LNCS 10421, pp. 500–512, 2017.
DOI: 10.1007/978-3-319-62932-2_48

2   Background

2.1   Parallel File Systems

Generalities. Parallel File Systems [1] are minded for accessing files in a simultaneous, concurrent and efficient way. The contents of a file are usually scattered among different I/O subsystems in order to take advantage of the highest local performance of each subsystem. These systems should provide, among other capabilities, persistence, consistence, performance, and manageability. Other desired features might include: scalability, fault-tolerance and availability. Different approaches can be used to analyze the performance of a Parallel File System. Checking the patterns of the I/O traces is among the most commonly used ones.

I/O Access Patterns. I/O access patterns depict the behavior of disk access over a period of time. They can be used to determine the overall performance of an I/O system. It is possible to characterize them by the following properties: access granularity, randomness, concurrency, load balance, access type and predictability. Liu et al. [2] mentioned three additional features seen on supercomputing I/O patterns: burstiness, periodicity and repeatability.

2.2   Kernel Methods for Similarity Search

As stated in [3], a typical machine learning systems consists of two subsystems: the feature extraction and clustering/classifier subsystems. On the one hand, the feature extraction subsystem performs the process of conversion of raw data to a meaningful representation. On the other hand, the clustering/classifier subsystem makes reference to the strategy used to distill information from the new representation. There is group of algorithms, among the constellation of machine learning techniques, that have been successfully applied in structured data problems: they are called Kernel Methods. Kernel Methods are well documented in the book of Shawe-Taylor and Cristianini [4]. This group of algorithms are strong enough to detect stable patterns robustly and efficiently from a finite data sample; basically, the idea is to embed the original data into a space where linear relations manifest as patterns. These methods have been successfully applied in problems with structured data types like trees and strings [5]. Kernel methods follow the mentioned two-stages strategy: first, a mapping is made by the Kernel Function, which depends on the specific data type and domain knowledge. Second, a general purpose and robust kernel learning algorithm is applied to find the linear relationships in the induced feature space. The stage of construction of the kernel function can be characterized as follows:

- Original data items are embedded into a vector space called *feature space*.
- The images of data in the feature space have linear relations.
- The learning algorithm does not need to know the coordinates of the feature space data; the pairwise inner products are enough.
- These inner products can be calculated in an efficient way using a kernel function.

The inner products between the training examples conform the *kernel matrix*. The learning algorithms are independent from the kernel function and need only the kernel matrix to extract meaningful information from the data. In this work we used two algorithms: Hierarchical Clustering [6] and Kernel Principal Component Analysis (Kernel PCA) [7].

String Kernels. Usually, data is delivered as a collection of attribute-value tuples; the widely used Polynomial and Gaussian Kernels Functions are ideal for this kind of representation. But for the case of structured data like trees and strings, the design of kernel functions becomes more complex. Despite this complexity, some solutions have been proposed, for example, Convolution Kernels [8–10]. Strings kernels are explained in a comprehensive way in [11]. They basically check for the number of shared substrings among a collection of strings. These substrings must comply with certain weighting factors, which produces different kernel functions; The bag-of-characters kernel only takes into account single-character matching. The bag-of-words kernel searches for shared words among strings. The k-spectrum kernel [12] only counts sub-strings of length k. The k-blended spectrum kernel [4] only counts sub-strings which length are less or equal to a given number k.

3   Methodology

3.1   Creating Strings from I/O Access Patterns

The I/O access pattern files are plain text files where each line corresponds to an operation. Some of these operations are negligible and hence ignored (e.g. **fileno**, **nmap** and **fscanf**). Some other operations keep information of the number of bytes involved on it. The proposed string representation can either use or ignore such byte information (ignoring is made by assuming all byte values are zero), which means that two different type of strings can be generated from a single I/O access pattern. Operations in the I/O access pattern are registered chronologically; with several file handles acting at the same time it is not always possible that all the operations belonging to the same file handle could have been written contiguously. For that reason the patterns are first converted into trees. Trees are ideal data structures for representing containment relationships between objects.

From I/O Access Patterns to Trees. The trees that we use in this paper will have the following levels: The *ROOT* level, the *HANDLE* level, the *BLOCK* level and the operation level (See Fig. 1):

- At the highest level, an imaginary root node groups all the operations of a single I/O access pattern file. Such node is represented as *ROOT*.
- At the second level, imaginary nodes group all the operations belonging to the same file handle. Such nodes are represented as *HANDLE*.

– At the third level, imaginary nodes group all the operations found between an **open** operation and its corresponding **close** operation. Such nodes are represented as *BLOCK*.
– At the deepest level, operations are given nodes, except for **open** and **close**, because the *BLOCK* node already plays the role of a delimiter.

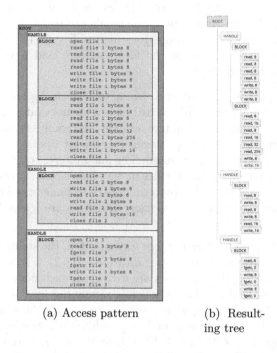

(a) Access pattern   (b) Result-
                              ing tree

Fig. 1. Conversion of a plain text I/O access pattern into a tree

In order to save space, a set of consecutive operation nodes on the same block can be expressed as a single node when they present some simple patterns. A similar approach was applied by Kluge [13]. The resulting node will have an additional field that stores of the number of repetitions. This compression step is based on the following transformations, which are performed in the given order:

– Consecutive operations with the same name and the same number of bytes are simplified to a single operation with the same information. E.g. a **read** operation inside a loop reading a file n bytes per iteration.
– Consecutive operations with the same name but different number of bytes are simplified to a single operation with the same name. The new byte value is a combination of both previous byte numbers. E.g. initializing in a loop an array of C structures compound of a 2-bytes integer and a 4-bytes integer will need a **read** operation extracting two bytes first and another **read** extracting four bytes afterwards.

504   R. Torres et al.

- Consecutive operations with different name but same number of bytes are simplified to a single operations with the same number of bytes. The new operation name is a combination of both previous names. E.g., a series of interlaced **read** and **write** operations with the same number of bytes might indicate a tacit copy operation.
- Consecutive operations with different name and different number of bytes but with one operation having 0 as number of bytes are simplified to a single operation with the non-zero value as the number of bytes. The new operation name is a combination of both previous names. E.g. inside a loop an **lseek** operation moves the pointer in the file descriptor and a **write** operation records the information there.

The previous steps are repeated once again to capture higher level patterns. Some of the operations (e.g. **read**, **write**) have a memory address associated to them. If this values would be taken into account, the compression step would be more precise to capture related operations e.g. a copy operation. However, the degree of compression would be reduced. The main interest of this research is to use patterns for determining in an efficient way how similar a collection of I/O traces are, not to break down the pattern and try to understand the underlying structure of it. For this reason, the memory addresses are ignored completely.

From Trees to Strings. Once the tree is compacted, the string representation can be built. The process is straightforward (See Fig. 2). The tree is traversed in pre-order and each node properties are extracted; each node of the tree corresponds to a token in the string. A token is compound by a literal part and a weight value. For leaf nodes the literal part is formed with the name of the operation and the number of bytes enclosed by *[ ]* while their weight corresponds to the number of repetitions. *ROOT*, *HANDLE* and *BLOCK* nodes are translated as *[ROOT]*, *[HANDLE]* and *[BLOCK]* respectively; their weight is always 1. To preserve information about the tree structure, we introduced a new token that does not correspond to any node but give a notion of distance between nodes. The rational of this design corresponds to the future application of this representation in more complex structures like Abstract Syntax Trees (ASTs). The *[LEVEL_UP]* token represents the change to an upper level when doing the pre-order traversal. Its weight is simply the amount of levels jumped until the next new node is found. Notice that there is no need for a token to indicate a change to a lower level, due to the fact that in the pre-order traversal the number of levels jumped from a parent to a child is always 1, which is implicitly expressed when two tokens are written one after the other.

3.2 Comparing Strings: The Kernel Function

The basic idea is to create a comparison measure for strings conformed by weighted tokens. In theory, the number of different tokens is infinite. In practice, the number of different tokens can be limited to the I/O operations on a program and the number of bytes related to each operation; still, the number is high. In

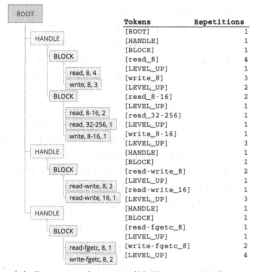

(a) Compacted tree   (b) Extracted tokens

Fig. 2. Creation of a string of tokens from a tree

order to define a proper similarity measure, it is necessary to define first some important concepts:

- A weighted string is a set of consecutive weighted tokens (from here on out referred simply as strings and tokens).
- A substring is a string that is fully contained by another string.
- The weight of a string is the summation of the weights of its tokens.

It is easy to infer here that the number of possible strings is also infinite. In an hypothetical feature space, where every string is characterized by the presence or absence of each possible token with each possible weight, the number of features is still infinite. However, in practice, for a single string, most of the features of this hypothetical space are zero-valued. This is a fact that eases the creation of a feasible kernel function. In this work the *Kast Spectrum Kernel* is proposed. In this kernel, some conditions have to be met to build the new embedding space:

- The algorithm precises a minimum weight value as parameter (from here on out referred simply as cut weight). Strings with a weight value that is smaller than the cut weight are ignored.
- The aim is to find the substrings shared by two strings which weight is greater than or equal to the cut weight.
- The weight of a target substring might be different in each string.
- A target substring might appear more than once in one of the strings.
- A target substring must not be a substring of another matching substring in at least on of the original strings.

For each target substring complying with the previous conditions, a new embedding feature is created. Its value is the summation of the weights of all the substring appearances in a string. This way, a new embedding space with a finite and small number of features can be built. The number of features for both strings is equal to the number of substrings that comply to the above mentioned conditions. It is possible now to calculate a similarity measure using the inner product between the new feature vectors; this is the so-called *kernel value*. The following is an example that illustrates the proposed kernel function: Let A and B be strings as shown in Fig. 3. The function $weight_{w \geq n}(A)$ returns the summation of the weights of all the tokens of A which weight is greater than or equal to n. The function $k_{w \geq n}(A, B)$ returns the evaluation of the Kast Spectrum Kernel between A and B. The function $\bar{k}_{w \geq n}(A, B)$ is the normalized version of the former kernel. For $n = 4$ the respective weights are:

$$weight_{w \geq 4}(A) = 64 \tag{1}$$

$$weight_{w \geq 4}(B) = 52 \tag{2}$$

The target in this example are all substrings with weight greater than or equal to 4 (cut weight). According to the kernel definition, three shared substrings are obtained: S_1, S_2 and S_3 (See Figs. 3, 4 and 5). The respective weights of each feature in A are calculated with:

$$weight_{w \geq 4}(S_1)_A = 19 \tag{3}$$

$$weight_{w \geq 4}(S_2)_A = 7 + 6 = 13 \tag{4}$$

$$weight_{w \geq 4}(S_3)_A = 6 + 9 = 15 \tag{5}$$

The embedding feature vector for A is:

$$f_{w \geq 4}(A) = \{19, 13, 15\} \tag{6}$$

The respective weights of each feature in B are calculated with:

$$weight_{w \geq 4}(S_1)_B = 17 + 18 = 35 \tag{7}$$

$$weight_{w \geq 4}(S_2)_B = 6 + 5 = 11 \tag{8}$$

$$weight_{w \geq 4}(S_3)_B = 8 + 6 = 14 \tag{9}$$

The embedding feature vector for B is:

$$f_{w \geq 4}(B) = \{35, 11, 14\} \tag{10}$$

The inner product of these two vectors gives us the kernel value

$$k_{w \geq 4}(A, B) = < f_{w \geq 4}(A), f_{w \geq 4}(B) > = 1018 \tag{11}$$

A normalization step will use the weights of each string:

$$\bar{k}_{w \geq 4}(A, B) = \frac{k_{w \geq 4}(A, B)}{\sqrt{k_{w \geq 4}(A, A) * k_{w \geq 4}(B, B)}} = \frac{k_{w \geq 4}(A, B)}{weight_{w \geq 4}(A) * weight_{w \geq 4}(B)} \tag{12}$$

$$\bar{k}_{w \geq 4}(A, B) = \frac{1018}{64 * 52} = \frac{1018}{3328} = 0.3059 \tag{13}$$

Fig. 3. S_1 is the largest substring found on both examples

Fig. 4. S_2 appears once as an independent case

Fig. 5. S_3 appears twice as an independent case

4   Evaluation

4.1   Experiment Configuration

The I/O access patterns were taken from two different parallel I/O benchmarks [14,15]. The patterns were generated from 4 different I/O forms of accessing the storage: (A) were those using Flash I/O, (B) were the ones using Random POSIX I/O, (C) were those using Normal I/O and (D) the ones using Random Access I/O. For each pattern 4 additional synthetic copies were created. Such copies introduced small mutations on the pattern; the idea behind these mutations was the need to create access patterns that were, in theory, closer to a determined example than the rest of the category members. So, from 22 examples we ended up with 110, distributed as follows: (A) 50 examples, (B) 20 examples, (C) 20 examples and (D) 20 examples. Each access pattern was converted to the two proposed string representations: the one that took into account the byte information of the operations and the one that totally ignored it. The proposed *Kast Spectrum Kernel* function was applied to them, as well as the *Blended Spectrum Kernel* proposed in the literature. The selected cut weight values were the following: $\{2^1, 2^2, ..., 2^n\} : n = 10$. If the matrices presented negative eigenvalues, they were replaced by zero and the matrices rebuilt. All the similarity matrices were analyzed with both Kernel PCA and Hierarchical Clustering, the latest using the simple linkage method.

4.2   Kast Spectrum Kernel

The application of the proposed kernel function (*Kast Spectrum Kernel*) over strings that preserved the byte information from the I/O operations, achieved the best results when a small cut weight was used. The fact that small cut weights

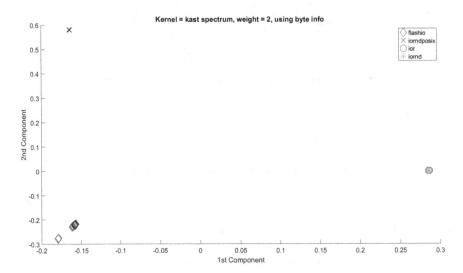

Fig. 6. Kernel PCA for Kast Spectrum Kernel using byte information (cut weight = 2)

were sufficient to achieve a meaningful clustering, eased the parametrization of the comparison process. It was remarkable that both learning algorithms clearly separated the same 3 clusters (See Figs. 6 and 7). While Flash I/O (A) and Random POSIX I/O (B) were separated independently, Normal I/O and Random Access I/O (C-D) were placed on the same group. This corresponded to the structure of each category: (A) examples contained contiguous **write** operations with different byte values that were not present in the other categories. (B) examples contained **lseek** operations not seen elsewhere. (C) and (D) shared roughly the same pattern. Also, it is important to notice that there were not misplaced examples on any of the groups.

In the case of the strings that ignored the byte information, such clear separation of clusters was not so easily achieved. For small cut weights only two clusters were identified: Random POSIX I/O (B) was the only group independently separated, while Flash I/O, Normal I/O and Random Access I/O (A-C-D) conformed a second group. In order to obtain the same three clustering groups identified using the other string category, the weight value had to be increased, which made the parametrization more difficult. Notice that, regardless of the string representation, the smaller the cut weight the most expensive the computation became, because the algorithm always started searching from the substrings with the highest weight. According to the clustering analysis results one can infer that the usage of high cut weights is recommended to focus only on finding general categories and lower cut weights to discriminate better among examples. However, a small cut weight is always preferred, as it eases the parametrization.

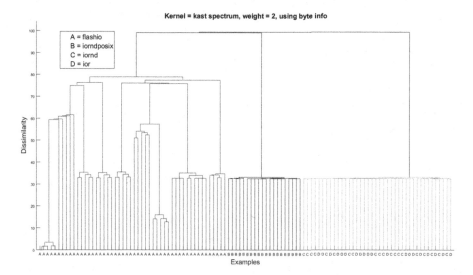

Fig. 7. Hierarchical clustering for Kast Spectrum Kernel using byte information (cut weight = 2)

4.3 Blended Spectrum Kernel

Given the particular form of the string representation we propose, where a group of subsequent tokens can encode more meaningful information than a single one, we discarded the bag-of-characters and the bag-of-words kernels. Experimental evaluation showed also that the k-Spectrum kernel was not successful at finding an acceptable clustering, a task where the Blended Spectrum Kernel had a better performance. However, for strings containing byte information the obtained clusters were not as diverse as those achieved with our solution (See Figs. 8 and 9). In this case only Flash I/O (A) examples were independently separated, while Random POSIX I/O, Normal I/O and Random Access I/O (B-C-D) conformed a single group.

For the case of strings lacking the byte information, both clustering analysis results were not satisfactory.

5 Related Work

Kluge [13] proposed an intermediate representation of I/O events from High Performance Computing (HPC) applications as a Directed Acyclic Graph (DAG). In this DAG vertices are used to represent events while edges are used to depict the chronological order of the events. Kluge also proposed a redundancy elimination step where adjacent synchronization vertices can be merged in a single one. Madhyastha et al. [16] applied two supervised learning algorithms to classify Parallel I/O access patterns: a feed forward neural network and a hidden Markov models based approach. Both strategies require training with previously

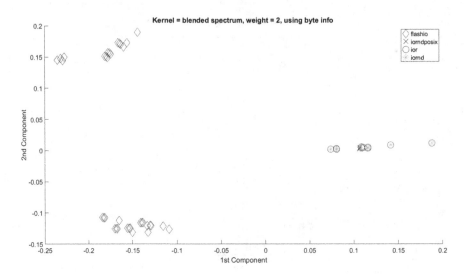

Fig. 8. Kernel PCA for Blended Spectrum Kernel using byte information (cut weight = 2)

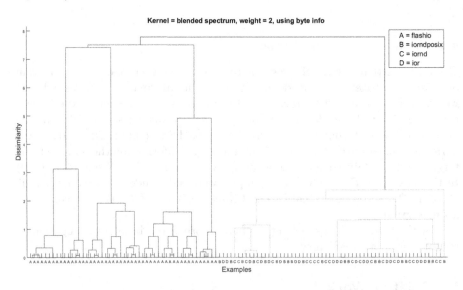

Fig. 9. Hierarchical clustering for Blended Spectrum Kernel using byte information (cut weight = 2)

labeled examples. Behzad et al. [17] proposed and I/O auto tuning framework that extracts the patterns from an application and searches for a match on a database of previously known pattern models. If there is a match, the associated model is adopted on the fly during the execution of the application. A different abstraction approach was made by Liu et al. [2]. They used the I/O bursts registered on noisy server-side logs of an application as a signature to find similarities between I/O samples. The final signature is a 2D grid called CLIQUE [18] that relates a correlation coefficient with time. Because the signature extraction was made over log files there was zero overhead in the application performance. Koller and Rangaswami [19] used disk static similarity and workload static similarity at the block level to analyze the performance of concurrent applications of the same file system. Unfortunately, we couldn't find suitable studies on I/O pattern similarity with kernel methods for comparing our results.

6 Conclusions and Future Work

In this paper we showed how the I/O traces of a parallel program can be used to extract patterns and represent them as a string of tokens. The resulting strings were compared using a novel kernel function proposed by the authors. The Kast Spectrum Kernel emits a similarity matrix between examples that can be later analyzed by a proper algorithm. This kernel was applied to a set of examples taken from a real parallel application, where 4 distinct patterns were present; Kernel PCA and Hierarchical Clustering showed a consistent formation of 3 groups according to the pattern with no misplaced examples. The best results were obtained when the string representation took into account the byte information of the operations and the cut weight was small. It was observed that the cut weight determined the granularity of the search, while the usage of the byte information permitted the separation between examples of the same cluster. These findings clearly show that both the proposed string representation and the comparison method are suitable to compare I/O access patterns of a parallel application. However, due to the fact that the proposed string representation is independent from the domain, it can also be used to compare I/O access of a sequential program. Future efforts of this project will focus on the comparison of the intermediate representation delivered by the LLVM Compiler Infrastructure using the string representation and kernel method here proposed.

Acknowledgements. Raul Torres would like to acknowledge the financial support from the Colombian Administrative Department of Science, Technology and Innovation (Colciencias) as well as the mathematical advisory received from Ruslan Krenzler.

References

1. Kunkel, J.M.: Simulating parallel programs on application and system level. Comput. Sci. Res. Dev. **28**(2), 167–174 (2012)
2. Liu, Y., Gunasekaran, R., Ma, X.S., Vazhkudai, S.S.: Automatic identification of application I/O signatures from noisy server-side traces. In: Proceedings of the 12th USENIX Conference on File and Storage Technologies (FAST 2014), Santa Clara, pp. 213–228 (2014)
3. Kung, S.Y.: Kernel Methods and Machine Learning. Cambridge University Press, Cambridge (2014)

4. Shawe-Taylor, J., Cristianini, N.: Kernel Methods for Pattern Analysis. Cambridge University Press, New York (2004)
5. Bakır, G., Hofmann, T., Schölkopf, B., Smola, A.J., Taskar, B., Vishwanathan, S.V.N.: Predicting Structured Data. The MIT Press, Cambridge (2007)
6. Hastie, T., Tibshirani, R., Friedman, J.: The Elements of Statistical Learning - Data Mining, Inference. Springer Series in Statistics. Springer, New York (2009)
7. Schölkopf, B., Smola, A., Müller, K.-R.: Kernel principal component analysis. In: Gerstner, W., Germond, A., Hasler, M., Nicoud, J.-D. (eds.) ICANN 1997. LNCS, vol. 1327, pp. 583–588. Springer, Heidelberg (1997). doi:10.1007/BFb0020217
8. Gärtner, T., Lloyd, J.W., Flach, P.A.: Kernels for structured data. In: Matwin, S., Sammut, C. (eds.) ILP 2002. LNCS (LNAI), vol. 2583, pp. 66–83. Springer, Heidelberg (2003). doi:10.1007/3-540-36468-4_5
9. Gärtner, T., Lloyd, J.W., Flach, P.A.: Kernels and Distances for Structured Data. Mach. Learn. **57**(3), 205–232 (2004)
10. Haussler, D.: Convolution Kernels on Discrete Structures. Technical Report. University of California at Santa Cruz (1999)
11. Vishwanathan, S.V.N., Smola, A.J.: Fast kernels for string and tree matching. In: Advances in Neural Information Processing Systems 15, pp. 569–576 (2003)
12. Leslie, C., Eskin, E., Noble, W.S.: The spectrum kernel: a string kernel for SVM protein classification. In: Proceedings of the Pacific Symposium on Biocomputing, vol. **7**, pp. 566–575 (2002)
13. Kluge, M.: Comparison and End-to-End Performance Analysis of Parallel Filesystems. Ph.D. Thesis Dissertation. Technische Universität Dresden (2011)
14. Loewe, W., McLarty, T., Morrone, C.: IOR Benchmark (2012)
15. Fryxell, B., Olson, K., Ricker, P., Timmes, F.X., Zingale, M., Lamb, D.Q., MacNeice, P., Rosner, R., Truran, J.W., Tufo, H.: FLASH: an adaptive mesh hydrodynamics code for modeling astrophysical thermonuclear flashes. Astrophys. J. Suppl. Ser. **131**(1), 273 (2000)
16. Madhyastha, T.M., Reed, D.A.: Learning to classify parallel input/output access patterns. IEEE Trans. Parallel Distrib. Syst. **13**(8), 802–813 (2002)
17. Behzad B., Byna S., Prabhat and Snir, M.: Pattern-driven parallel I/O tuning. In: Proceedings of the 10th Parallel Data Storage Workshop, Austin, Texas, pp. 43–48 (2015)
18. Agrawal, R., Gehrke, J., Gunopulos, D., Raghavan, P.: Automatic subspace clustering of high dimensional data for data mining applications. In: Proceedings of the 1998 ACM SIGMOD International Conference on Management of Data, SIGMOD 1998, Seattle, pp. 94–105 (1998)
19. Koller, R., Rangaswami, R.: I/O Deduplication: utilizing content similarity to improve I/O performance. ACM Trans. Storage (TOS) **6**(3), 13:1–13:26 (2010)

Author Index

Cambridge Aerospace Series 14

Editors:
MICHAEL J. RYCROFT AND WEI SHYY

Frontispiece George Hartley Bryan (1864–1928). The originator, with W. E. Williams, of the equations of airplane motion. Bryan's equations are the basis for the analysis of airplane flight dynamics and closed-loop control and for the design of flight simulators. (From *Obit. Notices of Fellows of the Royal Soc.*, 1932–1935)

Airplane Stability and Control, Second Edition

A History of the Technologies That Made Aviation Possible

MALCOLM J. ABZUG

ACA Systems

E. EUGENE LARRABEE

Professor Emeritus, Massachusetts Institute of Technology

CAMBRIDGE
UNIVERSITY PRESS

CAMBRIDGE UNIVERSITY PRESS
Cambridge, New York, Melbourne, Madrid, Cape Town, Singapore, São Paulo

Cambridge University Press
The Edinburgh Building, Cambridge CB2 2RU, UK

Published in the United States of America by Cambridge University Press, New York

www.cambridge.org
Information on this title: www.cambridge.org/9780521809924

© Cambridge University Press 2002

First published 2002
This digitally printed first paperback version 2005

A catalogue record for this publication is available from the British Library

Library of Congress Cataloguing in Publication data

Abzug, Malcolm J.
  Airplane stability and control : a history of the technologies that made aviation possible /
  Malcolm Abzug, E. Eugene Larrabee. – 2nd ed.
    p.   cm. – (Cambridge aerospace series; 14)
  Includes bibliographical references and index.
  ISBN 0-521-80992-4
  1. Stability of airplanes.   2. Airplanes – Design and construction – History.
  3. Airplanes – Control systems – History.   I. Larrabee, E. Eugene.   II. Title.   III. Series.
  TL574.S7 A2   2002
  629.132′36 – dc21   2001052847

ISBN-13 978-0-521-80992-4 hardback
ISBN-10 0-521-80992-4 hardback

ISBN-13 978-0-521-02128-9 paperback
ISBN-10 0-521-02128-6 paperback

"From th'envious world with scorn I spring,
And cut with joy the wond'ring Skies."

From Odes II xx, by Horace, translated by Samuel Johnson in 1726.
St. Martin's Press, New York, 1971.

"When this one feature [balance and steering] has been worked out, the age of flying
machines will have arrived, for all other difficulties are of minor importance."

From *The Papers of Wilbur and Orville Wright,* Vol. 1.
McGraw-Hill Book Co., New York, 1953.

". . . any pilot can successfully fly anything that looks like an airplane."

From "Airplane Stability and Control from a Designer's Point of View."
Otto C. Koppen, *Jour. of the Aero. Sci.*, Feb. 1940.

Contents

Preface

After raising student enthusiasm by a particularly inspiring airplane stability and control lecture, Professor Otto Koppen would restore perspective by saying, "Remember, airplanes are not built to demonstrate stability and control, but to carry things from one place to another." Perhaps Koppen went too far, because history has shown over and over again that neglect of stability and control fundamentals has brought otherwise excellent aircraft projects down, sometimes literally. Every aspiring airplane builder sees the need intuitively for sturdy structures and adequate propulsive power. But badly located centers of gravity and inadequate rudder area for spin recovery, for example, are subtleties that can be missed easily, and have been missed repeatedly.

Before the gas turbine age, much of the art of stability and control design was devoted to making airplanes that flew themselves for minutes at a time in calm air, and responded gracefully to the hands and feet of the pilot when changes in course or altitude were required. These virtues were called flying qualities. They were codified for the first time by the National Advisory Committee for Aeronautics, the NACA, in 1943. Military procurement specifications based on NACA's work followed two years later.

When gas turbine power arrived, considerations of fuel economy drove airplanes into the stratosphere and increased power made transonic flight possible. Satisfactory flying qualities no longer could be achieved by a combination of airplane geometry and restrictions on center-of-gravity location. Artificial stability augmenters such as pitch and yaw dampers were required, together with Mach trim compensators, all-moving tailplanes, and irreversible surface position actuators. At roughly the same time, the Boeing B-47 and the Northrop B-49 and their successful stability augmenters marked the beginning of a new age.

Since then much of the art and science that connected airplane geometry to good low-altitude flying qualities have begun to be lost to a new generation of airplane designers and builders. The time has come to record the lore of earlier airplane designers for the benefit of the kit-built airplane movement, to say nothing of the survivors of the general-aviation industry. Accordingly, this book is an informal, popular survey of the art and science of airplane stability and control. As history, the growth of understanding of the subject is traced from the pre–Wright brothers' days up to the present. But there is also the intention of preserving for future designers the hard-won experience of what works and what doesn't. The purpose is not only to honor the scientists and engineers who invented airplane stability and control, but also to help a few future airplane designers along the path to success.

If this work has any unifying theme, it is the lag of stability and control practice behind currently available theory. Repeatedly, airplanes have been built with undesirable or even fatal stability and control characteristics out of simple ignorance of the possibility of using better designs. In only a few periods, such as the time of the first flights near the speed of sound, theoreticians, researchers, and airplane designers were all in the same boat, all learning together.

The second edition of this book brings the subject up to date by including recent developments. We have also used the opportunity to react to the numerous reviews of the first edition and to the comments of readers. One theme found in many reviews was that the first edition had neglected important airplane stability and control work that took place outside of the United States. That was not intentional, but the second edition has given the authors a new opportunity to correct the problem. In that effort, we were greatly aided by the following correspondents and reviewers in Canada, Europe, and Asia: Michael V. Cook, Dr. Bernard Etkin, Dr. Peter G. Hamel, Dr. John C. Gibson, Bill Gunston, Dr. Norohito Goto, Dr. Gareth D. Padfield, Miss A. Jean Ross, the late Dr. H. H. B. M. Thomas, and Dr. Jean-Claude L. Wanner.

The interesting history of airplane stability and control has not lacked for attention in the past. A number of distinguished authors have presented short airplane stability and control histories, as distinct from histories of general aeronautics. We acknowledge particularly the following accounts:

> Progress in Dynamic Stability and Control Research, by William F. Milliken, Jr., in the September 1947 *Journal of the Aeronautical Sciences*.
>
> Development of Airplane Stability and Control Technology, by Courtland D. Perkins, in the July–August 1970 *Journal of Aircraft*.
>
> Eighty Years of Flight Control: Triumphs and Pitfalls of the Systems Approach, by Duane T. McRuer and F. Dunstan Graham, in the July–August 1981 *Journal of Guidance and Control*.
>
> Twenty-Five Years of Handling Qualities Research, by Irving L. Ashkenas, in the May 1984 *Journal of Aircraft*.
>
> Flying Qualities from Early Airplanes to the Space Shuttle, by William H. Phillips, in the July–August 1989 *Journal of Guidance, Control, and Dynamics*.
>
> Establishment of Design Requirements: Flying Qualities Specifications for American Aircraft, 1918–1943, by Walter C. Vincenti, Chapter 3 of *What Engineers Know and How They Know It*, Johns Hopkins University Press, 1990.
>
> Evolution of Airplane Stability and Control: A Designer's Viewpoint, by Jan Roskam, in the May–June 1991 *Journal of Guidance*.
>
> Recollections of Langley in the Forties, by W. Hewitt Phillips, in the Summer 1992 *Journal of the American Aviation Historical Society*.

Many active and retired contributors to the stability and control field were interviewed for this book; some provided valuable references and even more valuable advice to the authors. The authors wish to acknowledge particularly the generous help of a number of them. Perhaps foremost in this group was the late Charles B. Westbrook, a well-known stability and control figure. Westbrook helped with his broad knowledge of U.S. Air Force–sponsored research and came up with several obscure but useful documents. W. Hewitt Phillips, an important figure in the stability and control field, reviewed in detail several book chapters. His comments are quoted verbatim in a number of places. Phillips is now a Distinguished Research Associate at the NASA Langley Research Center.

We were fortunate to have detailed reviews from two additional experts, William H. Cook, formerly of the Boeing Company, and Duane T. McRuer, chairman of Systems Technology, Inc. Their insights into important issues are used and also quoted verbatim in several places in the book. Drs. John C. Gibson, formerly of English Electric/British Aerospace, and Peter G. Hamel, director of the DVL Institute of Flight Research, Braunschweig, were helpful with historical and recent European developments, as were several other European and Canadian engineers.

Jean Anderson, head librarian of the Guggenheim Aeronautical Laboratory at the California Institute of Technology (GALCIT) guided the authors through GALCIT's impressive aeronautical collections. All National Advisory Committee for Aeronautics (NACA) documents are there, in microfiche. The GALCIT collections are now located at the Institute's Fairchild Library, where the Technical Reference Librarian, Louisa C. Toot, has been most helpful. We were fortunate also to have free access to the extensive stability and control collections at Systems Technology, Inc., of Hawthorne, California. We thank STI's chairman and president, Duane T. McRuer and R. Wade Allen, for this and for very helpful advice.

The engineering libraries of the University of California, Los Angeles, and of the University of Southern California were useful in this project. We acknowledge also the help of George Kirkman, the volunteer curator of the library of the Museum of Flying, in Santa Monica, California, and the NASA Archivist Lee D. Saegesser.

In addition to the European and Asian engineers noted previously, the following people generously answered our questions and in many cases loaned us documents that added materially to this work: Paul H. Anderson, James G. Batterson, James S. Bowman, Jr., Robert W. Bratt, Daniel P. Byrnes, C. Richard Cantrell, William H. Cook, Dr. Eugene E. Covert, Dr. Fred E. C. Culick, Sean G. Day, Orville R. Dunn, Karl S. Forsstrom, Richard G. Fuller, Ervin R. Heald, Robert K. Heffley, Dr. Harry J. Heimer, R. Richard Heppe, Bruce E. Jackson, Henry R. Jex, Juri Kalviste, Charles H. King, Jr., William Koven, David A. Lednicer, Dr. Paul B. MacCready, Robert H. Maskrey, Dr. Charles McCutchen, Duane T. McRuer, Allen Y. Murakoshi, Albert F. Myers, Dr. Gawad Nagati, Stephen Osder, Robert O. Rahn, Dr. William P. Rodden, Dr. Jan Roskam, Edward S. Rutowski, George S. Schairer, Roger D. Schaufele, Arno E. Schelhorn, Lawrence J. Schilling, Dr. Irving C. Statler, and Dr. Terrence A. Weisshaar.

Only a few of these reviewers saw the entire book in draft form, so the authors are responsible for any uncorrected errors and omissions.

This book is arranged only roughly in chronological order. Most of the chapters are thematic, dealing with a single subject over its entire history. References are grouped by chapters at the end of the book. These have been expanded to form an abbreviated or core airplane stability and control bibliography. The rapid progress in computerized bibliographies makes anachronistic a really comprehensive airplane stability and control bibliography.

Malcolm J. Abzug
E. Eugene Larrabee